WITHDRAWN

Methods in Cell Biology
Receptor-Receptor Interactions

Volume 117

Series Editors

Leslie Wilson
Department of Molecular, Cellular and Developmental Biology
University of California
Santa Barbara, California

Paul Matsudaira
Department of Biological Sciences
National University of Singapore
Singapore

Phong Tran
Department of Cell and Developmental Biology
University of Pennsylvania
Philadelphia, Pennsylvania

Methods in Cell Biology
Receptor-Receptor Interactions

Volume 117

Edited by

P. Michael Conn
Senior Vice President for Research
Associate Provost
Professor of Internal Medicine and Cell Biology
Texas Tech University Health Sciences Center
Lubbock, TX, USA

AMSTERDAM • BOSTON • HEIDELBERG • LONDON
NEW YORK • OXFORD • PARIS • SAN DIEGO
SAN FRANCISCO • SINGAPORE • SYDNEY • TOKYO
Academic Press is an imprint of Elsevier

Academic Press is an imprint of Elsevier
525 B Street, Suite 1800, San Diego, CA 92101-4495, USA
225 Wyman Street, Waltham, MA 02451, USA
The Boulevard, Langford Lane, Kidlington, Oxford, OX5 1GB, UK
32 Jamestown Road, London NW1 7BY, UK
Radarweg 29, PO Box 211, 1000 AE Amsterdam, The Netherlands

First edition 2013

Copyright © 2013 Elsevier Inc. All rights reserved

No part of this publication may be reproduced, stored in a retrieval system or transmitted in any form or by any means electronic, mechanical, photocopying, recording or otherwise without the prior written permission of the publisher

Permissions may be sought directly from Elsevier's Science & Technology Rights Department in Oxford, UK: phone (+44) (0) 1865 843830; fax (+44) (0) 1865 853333; email: permissions@elsevier.com. Alternatively you can submit your request online by visiting the Elsevier web site at http://elsevier.com/locate/permissions, and selecting *Obtaining permission to use Elsevier material*

Notice
No responsibility is assumed by the publisher for any injury and/or damage to persons or property as a matter of products liability, negligence or otherwise, or from any use or operation of any methods, products, instructions or ideas contained in the material herein. Because of rapid advances in the medical sciences, in particular, independent verification of diagnoses and drug dosages should be made

ISBN: 978-0-12-408143-7
ISSN: 0091-679X

For information on all Academic Press publications visit our website at store.elsevier.com

Printed and bound in USA
13 14 15 16   11 10 9 8 7 6 5 4 3 2 1

Contents

Contributors xiii
Preface xxi

CHAPTER 1 Spatial Intensity Distribution Analysis (SpIDA): A New Tool for Receptor Tyrosine Kinase Activation and Transactivation Quantification 1
Annie Barbeau, Jody L. Swift, Antoine G. Godin, Yves De Koninck, Paul W. Wiseman, Jean-Martin Beaulieu
Introduction 2
1.1 Theory of Spatial Intensity Distribution Analysis 4
1.2 SpIDA: Examples of Application to RTK 4
1.3 Procedure for SpIDA 6
1.4 Discussion 17
Acknowledgments 18
References 18

CHAPTER 2 Dimerization of Nuclear Receptors 21
Pierre Germain, William Bourguet
Introduction 22
2.1 Methods 24
Acknowledgments 39
References 39

CHAPTER 3 Network Analysis to Uncover the Structural Communication in GPCRs 43
Francesca Fanelli, Angelo Felline, Francesco Raimondi
Introduction 44
3.1 Materials 45
3.2 Methods 45
3.3 Discussion 50
Summary 56
Acknowledgments 59
References 59

CHAPTER 4 Simulating G Protein-Coupled Receptors in Native-Like Membranes: From Monomers to Oligomers 63
Ramon Guixà-González, Juan Manuel Ramírez-Anguita, Agnieszka A. Kaczor, Jana Selent
Introduction 64

	4.1	Membranes 65
	4.2	GPCR Monomers in Membranes 72
	4.3	GPCR Dimers and Oligomers in Membranes 80
		References 83

CHAPTER 5 Structure-Based Molecular Modeling Approaches to GPCR Oligomerization 91

Agnieszka A. Kaczor, Jana Selent, Antti Poso

		Introduction 92
	5.1	Protein–Protein Docking 93
	5.2	MD Simulation 96
	5.3	Normal Mode Analysis 99
	5.4	Electrostatics Studies 100
		Acknowledgments 101
		References 102

CHAPTER 6 Biochemical and Imaging Methods to Study Receptor Membrane Organization and Association with Lipid Rafts 105

Bruno M. Castro, Juan A. Torreno-Piña, Thomas S. van Zanten, Maria F. Gracia-Parajo

		Introduction and Rationale 106
	6.1	Methods to Determine Protein Association with Lipid Rafts 107
	6.2	Optical Techniques to Study Receptor Membrane Organization and Association with Lipid Rafts 110
		Concluding Remarks 118
		Acknowledgments 119
		References 119

CHAPTER 7 Serotonin Type 4 Receptor Dimers 123

Sylvie Claeysen, Romain Donneger, Patrizia Giannoni, Florence Gaven, Lucie P. Pellissier

		Introduction 124
	7.1	Materials 125
	7.2	Methods 127
	7.3	Discussion 136
		Summary 137
		Acknowledgments 137
		References 138

CHAPTER 8 Bioluminescence Resonance Energy Transfer Methods to Study G Protein-Coupled Receptor–Receptor Tyrosine Kinase Heteroreceptor Complexes **141**

Dasiel O. Borroto-Escuela, Marc Flajolet, Luigi F. Agnati, Paul Greengard, Kjell Fuxe

Introduction 142
- 8.1 The Method Principle 143
- 8.2 Setting Up a BRET Assay 146
- 8.3 Methods and Detailed Protocols 148
- 8.4 Discussion and Note on Critical Parameters 159
  Acknowledgments 161
  References 162

CHAPTER 9 A Simple Method to Detect Allostery in GPCR Dimers **165**

Eugénie Goupil, Stéphane A. Laporte, Terence E. Hébert

Introduction and Rationale 166
- 9.1 Materials 169
- 9.2 Methods 170
- 9.3 Discussion 174
  Summary 176
  Acknowledgments 176
  References 176

CHAPTER 10 Fluorescence Correlation Spectroscopy and Photon-Counting Histogram Analysis of Receptor–Receptor Interactions **181**

Katharine Herrick-Davis, Joseph E. Mazurkiewicz

Introduction 182
- 10.1 Materials 184
- 10.2 Methods 185
- 10.3 Discussion 194
  Summary 195
  Acknowledgments 195
  References 195

CHAPTER 11 Monitoring Receptor Oligomerization by Line-Scan Fluorescence Cross-Correlation Spectroscopy **197**

Mark A. Hink, Marten Postma

Introduction 198

11.1	Materials	199
11.2	Methods	202
11.3	Discussion	209
	Acknowledgments	210
	References	210

CHAPTER 12 Biochemical Assay of G Protein-Coupled Receptor Oligomerization: Adenosine A_1 and Thromboxane A_2 Receptors Form the Novel Functional Hetero-oligomer ... 213

Natsumi Mizuno, Tokiko Suzuki, Yu Kishimoto, Noriyasu Hirasawa

	Introduction	214
12.1	Detection of GPCR Oligomerization	216
12.2	Measurement of the Signaling Pathway	220
12.3	Materials and Methods	223
	Conclusion	224
	References	224

CHAPTER 13 Oligomerization of Sweet and Bitter Taste Receptors ... 229

Christina Kuhn, Wolfgang Meyerhof

	Introduction and Rationale	230
13.1	Materials	232
13.2	Methods	233
13.3	Discussion	240
	Summary	241
	Acknowledgments	241
	References	241

CHAPTER 14 Analysis of Receptor–Receptor Interaction by Combined Application of FRET and Microscopy ... 243

Sonal Prasad, Andre Zeug, Evgeni Ponimaskin

	Introduction and Rationale	244
14.1	Theory	245
14.2	Materials / Experimental Setups	249
14.3	Methods (Protocols and Procedure)	253
14.4	Results and Discussion	263
	Acknowledgments	264
	References	264

CHAPTER 15 Site-Specific Labeling of Genetically Encoded Azido Groups for Multicolor, Single-Molecule Fluorescence Imaging of GPCRs 267
He Tian, Thomas P. Sakmar, Thomas Huber
15.1 Purpose 270
15.2 Theory 270
15.3 Protocol 1: Genetically Encoded Azido Groups for Bioorthogonal Conjugation 273
15.4 Protocol 2: Labeling with Fluorescent Probes Using Cyclooctynes 276
15.5 Protocol 3: Preparation of Biotinylated Antibodies 282
15.6 Protocol 4: Preparation of Detergent-Solubilized Lipids 285
15.7 Protocol 5: Single-Molecule Immunoprecipitation on Glass-Bottom Microplates 287
15.8 Protocol 6: Automated Multicolor, Single-Molecule TIRF Microscopy 293
Acknowledgments 302
References 302

CHAPTER 16 Analysis of EGF Receptor Oligomerization by Homo-FRET 305
Cecilia de Heus, Nivard Kagie, Raimond Heukers, Paul M.P. van Bergen en Henegouwen, Hans C. Gerritsen
Introduction 306
16.1 Theory Homo-FRET Quantification 309
16.2 Materials 313
16.3 Methods 316
16.4 Discussion 317
Acknowledgments 319
References 320

CHAPTER 17 Detection of G Protein-Coupled Receptor (GPCR) Dimerization by Coimmunoprecipitation 323
Kamila Skieterska, Jolien Duchou, Béatrice Lintermans, Kathleen Van Craenenbroeck
Introduction 324
17.1 Materials 326
17.2 Methods 327
17.3 Discussion 338
Acknowledgments 339
References 339

CHAPTER 18 Lipid-Dependent GPCR Dimerization 341
Alan D. Goddard, Patricia M. Dijkman, Roslin J. Adamson, Anthony Watts

Introduction 342
18.1 Materials and Method 343
18.2 FRET Measurements and Analysis 351
18.3 Interpreting Results 354
Summary 355
References 355

CHAPTER 19 Monitoring Peripheral Protein Oligomerization on Biological Membranes 359
Robert V. Stahelin

Introduction 360
19.1 Materials and Methods 362
19.2 Considerations 365
Summary and Conclusion 366
Acknowledgments 369
References 369

CHAPTER 20 Single-Molecule Imaging of Receptor–Receptor Interactions 373
Kenichi G.N. Suzuki, Rinshi S. Kasai, Takahiro K. Fujiwara, Akihiro Kusumi

Introduction 374
20.1 Receptor Expression and Fluorescent Labeling 376
20.2 Single-Molecule Imaging 380
20.3 Data Analysis 382
Acknowledgments 387
References 387

CHAPTER 21 Visualization of TCR Nanoclusters via Immunogold Labeling, Freeze-Etching, and Surface Replication 391
Gina J. Fiala, María Teresa Rejas, Wolfgang W. Schamel, Hisse M. van Santen

Introduction 393
21.1 Materials 394
21.2 Equipment 395
21.3 Methods 395
21.4 Analysis 405
21.5 Considerations 405

Acknowledgments 408
References 408

CHAPTER 22 Identification of Multimolecular Complexes and Supercomplexes in Compartment-Selective Membrane Microdomains 411

Panagiotis Mitsopoulos, Joaquín Madrenas

Introduction and Rationale 413
22.1 Detergent-Insoluble Glycosphingolipid (DIG) Microdomain Isolation 415
22.2 Plasma Membrane Isolation 418
22.3 Cardiolipin-Enriched Mitochondrial Membrane Microdomain Isolation 421
22.4 Mitochondrial Supercomplex Identification by Blue-Native Gel Electrophoresis 423
22.5 Discussion 428
Summary 429
Acknowledgments 429
References 430

CHAPTER 23 G Protein-Coupled Receptor Transactivation: From Molecules to Mice 433

Kim C. Jonas, Adolfo Rivero-Müller, Ilpo T. Huhtaniemi, Aylin C. Hanyaloglu

Introduction 434
23.1 Materials 435
23.2 Methods 436
23.3 Discussion 447
Summary 448
Acknowledgments 448
References 448

CHAPTER 24 Crystallization of G Protein-Coupled Receptors 451

David Salom, Pius S. Padayatti, Krzysztof Palczewski

Introduction 452
24.1 Crystallization of Bovine Rhodopsin 453
24.2 Crystallization of β2-AR 460
24.3 Discussion 465
Acknowledgments 465
References 466

Index 469

Contributors

Roslin J. Adamson
Biomembrane Structure Unit, Department of Biochemistry, University of Oxford, Oxford, United Kingdom

Luigi F. Agnati
IRCCS Lido, Venice, Italy

Annie Barbeau
Department of Psychiatry and Neuroscience, Faculty of Medicine, Laval University, and Institut universitaire en santé mentale de Québec, Québec, Canada

Jean-Martin Beaulieu
Department of Psychiatry and Neuroscience, Faculty of Medicine, Laval University, and Institut universitaire en santé mentale de Québec, Québec, Canada

Dasiel O. Borroto-Escuela
Department of Neuroscience, Karolinska Institutet, Stockholm, Sweden

William Bourguet
Inserm U1054, Centre de Biochimie Structurale, and CNRS UMR5048, Universités Montpellier 1 & 2, Montpellier, France

Bruno M. Castro
ICFO—Institut de Ciencies Fotoniques, Mediterranean Technology Park, Castelldefels (Barcelona), Spain

Sylvie Claeysen
CNRS, UMR-5203, Institut de Génomique Fonctionnelle; Inserm, U661, and Universités de Montpellier 1 & 2, UMR-5203, Montpellier, France

Cecilia de Heus
Cell Biology, Department of Biology, Science Faculty, Utrecht University, Utrecht, Netherlands

Yves De Koninck
Department of Psychiatry and Neuroscience, Faculty of Medicine, Laval University, and Institut universitaire en santé mentale de Québec, Québec, Canada

Patricia M. Dijkman
Biomembrane Structure Unit, Department of Biochemistry, University of Oxford, Oxford, United Kingdom

Romain Donneger
CNRS, UMR-5203, Institut de Génomique Fonctionnelle; Inserm, U661, and Universités de Montpellier 1 & 2, UMR-5203, Montpellier, France

Jolien Duchou
Laboratory of Eukaryotic Gene Expression and Signal Transduction (LEGEST), Ghent University-UGent, Ghent, Belgium

Francesca Fanelli
Dulbecco Telethon Institute (DTI), Rome, and Department of Chemistry, University of Modena and Reggio Emilia, Modena, Italy

Angelo Felline
Dulbecco Telethon Institute (DTI), Rome, and Department of Chemistry, University of Modena and Reggio Emilia, Modena, Italy

Gina J. Fiala
Department of Molecular Immunology, Faculty of Biology, BIOSS Centre for Biological Signaling Studies; Centre for Chronic Immunodeficiency (CCI), University Clinics Freiburg and Medical Faculty, and Spemann Graduate School of Biology and Medicine (SGBM), Albert Ludwigs University Freiburg, Freiburg, Germany

Marc Flajolet
Laboratory of Molecular and Cellular Neuroscience, The Rockefeller University, New York, USA

Takahiro K. Fujiwara
Institute for Integrated Cell-Material Sciences (WPI-iCeMS), Kyoto University, Kyoto, Japan

Kjell Fuxe
Department of Neuroscience, Karolinska Institutet, Stockholm, Sweden

Florence Gaven
CNRS, UMR-5203, Institut de Génomique Fonctionnelle; Inserm, U661, and Universités de Montpellier 1 & 2, UMR-5203, Montpellier, France

Pierre Germain
Inserm U1054, Centre de Biochimie Structurale, and CNRS UMR5048, Universités Montpellier 1 & 2, Montpellier, France

Hans C. Gerritsen
Molecular Biophysics, Department of Soft Condensed Matter and Biophysics, Science Faculty, Utrecht University, Utrecht, Netherlands

Patrizia Giannoni
CNRS, UMR-5203, Institut de Génomique Fonctionnelle; Inserm, U661, and Universités de Montpellier 1 & 2, UMR-5203, Montpellier, France

Alan D. Goddard
School of Life Sciences, University of Lincoln, Lincoln, United Kingdom

Antoine G. Godin
University of Bordeaux, and CNRS & Institut d'Optique, LP2N, Talence, France

Eugénie Goupil
Department of Pharmacology, McGill University, Montréal, Québec, Canada

Contributors

Maria F. Gracia-Parajo
ICFO—Institut de Ciencies Fotoniques, Mediterranean Technology Park, Castelldefels (Barcelona), and ICREA—Institució Catalana de Recerca i Estudis Avançats, Barcelona, Spain

Paul Greengard
Laboratory of Molecular and Cellular Neuroscience, The Rockefeller University, New York, USA

Ramon Guixà-González
Research Programme on Biomedical Informatics (GRIB), Department of Experimental and Health Sciences, Universitat Pompeu Fabra/IMIM (Hospital del Mar Medical Research Institute), Dr. Aiguader, Barcelona, Spain

Aylin C. Hanyaloglu
Department of Surgery and Cancer, Institute of Reproductive and Developmental Biology, Imperial College London, London, United Kingdom

Terence E. Hébert
Department of Pharmacology, McGill University, Montréal, Québec, Canada

Katharine Herrick-Davis
Center for Neuropharmacology & Neuroscience, Albany Medical College, Albany, New York, USA

Raimond Heukers
Cell Biology, Department of Biology, Science Faculty, Utrecht University, Utrecht, Netherlands

Mark A. Hink
Section of Molecular Cytology, van Leeuwenhoek Centre for Advanced Microscopy (LCAM), Swammerdam Institute for Life Sciences (SILS), University of Amsterdam, The Netherlands

Noriyasu Hirasawa
Department of Pharmacotherapy of Life-style Related Disease, Graduate School of Pharmaceutical Sciences, Tohoku University, Sendai, Japan

Thomas Huber
Laboratory of Chemical Biology and Signal Transduction, The Rockefeller University, New York, New York, USA

Ilpo T. Huhtaniemi
Department of Surgery and Cancer, Institute of Reproductive and Developmental Biology, Imperial College London, London, United Kingdom, and Department of Physiology, Institute of Biomedicine, University of Turku, Turku, Finland

Kim C. Jonas
Department of Surgery and Cancer, Institute of Reproductive and Developmental Biology, Imperial College London, London, United Kingdom

Agnieszka A. Kaczor
Department of Synthesis and Chemical Technology of Pharmaceutical Substances, Faculty of Pharmacy with Division for Medical Analytics, Lublin, Poland, and School of Pharmacy, University of Eastern Finland, Kuopio, Finland

Nivard Kagie
Molecular Biophysics, Department of Soft Condensed Matter and Biophysics, Science Faculty, Utrecht University, Utrecht, Netherlands

Rinshi S. Kasai
Institute for Integrated Cell-Material Sciences (WPI-iCeMS), and Institute for Frontier Medical Sciences, Kyoto University, Kyoto, Japan

Yu Kishimoto
Department of Pharmacotherapy of Life-style Related Disease, Graduate School of Pharmaceutical Sciences, Tohoku University, Sendai, Japan

Christina Kuhn
Axxam SpA, Bresso (Milan), Italy

Akihiro Kusumi
Institute for Integrated Cell-Material Sciences (WPI-iCeMS), and Institute for Frontier Medical Sciences, Kyoto University, Kyoto, Japan

Stéphane A. Laporte
Department of Pharmacology, and Department of Medicine, McGill University, Montréal, Québec, Canada

Béatrice Lintermans
Laboratory of Eukaryotic Gene Expression and Signal Transduction (LEGEST), Ghent University-UGent, Ghent, Belgium

Joaquín Madrenas
Department of Microbiology and Immunology, McGill University, Montreal, QC, Canada

Joseph E. Mazurkiewicz
Center for Neuropharmacology & Neuroscience, Albany Medical College, Albany, New York, USA

Wolfgang Meyerhof
Department of Molecular Genetics, German Institute of Human Nutrition Potsdam-Rehbruecke, Nuthetal, Germany

Panagiotis Mitsopoulos
Department of Microbiology and Immunology, McGill University, Montreal, QC, Canada

Natsumi Mizuno
Department of Pharmacotherapy of Life-style Related Disease, Graduate School of Pharmaceutical Sciences, Tohoku University, Sendai, Japan

Pius S. Padayatti
Polgenix Inc., Cleveland, Ohio, USA

Krzysztof Palczewski
Polgenix Inc., and Department of Pharmacology, School of Medicine, Case Western Reserve University, Cleveland, Ohio, USA

Lucie P. Pellissier
CNRS, UMR-5203, Institut de Génomique Fonctionnelle; Inserm, U661, and Universités de Montpellier 1 & 2, UMR-5203, Montpellier, France

Evgeni Ponimaskin
Cellular Neurophysiology, Center of Physiology, Hannover Medical School, Hannover, Germany

Antti Poso
Department of Pharmaceutical Chemistry, School of Pharmacy, University of Eastern Finland, Kuopio, Finland

Marten Postma
Section of Molecular Cytology, van Leeuwenhoek Centre for Advanced Microscopy (LCAM), Swammerdam Institute for Life Sciences (SILS), University of Amsterdam, The Netherlands

Sonal Prasad
Cellular Neurophysiology, Center of Physiology, Hannover Medical School, Hannover, Germany

Francesco Raimondi
Dulbecco Telethon Institute (DTI), Rome, and Department of Chemistry, University of Modena and Reggio Emilia, Modena, Italy

Juan Manuel Ramírez-Anguita
Research Programme on Biomedical Informatics (GRIB), Department of Experimental and Health Sciences, Universitat Pompeu Fabra/IMIM (Hospital del Mar Medical Research Institute), Dr. Aiguader, Barcelona, Spain

María Teresa Rejas
Servicio de Microscopía Electrónica, Centro Biología Molecular Severo Ochoa, Consejo Superior de Investigaciones Científicas, Universidad Autónoma de Madrid, Madrid, Spain

Adolfo Rivero-Müller
Department of Physiology, Institute of Biomedicine, University of Turku, Turku, Finland

Thomas P. Sakmar
Laboratory of Chemical Biology and Signal Transduction, The Rockefeller University, New York, New York, USA

David Salom
Polgenix Inc., Cleveland, Ohio, USA

Wolfgang W. Schamel
Department of Molecular Immunology, Faculty of Biology, BIOSS Centre for Biological Signaling Studies, and Centre for Chronic Immunodeficiency (CCI), University Clinics Freiburg and Medical Faculty, Albert Ludwigs University Freiburg, Freiburg, Germany

Jana Selent
Research Programme on Biomedical Informatics (GRIB), Department of Experimental and Health Sciences, Universitat Pompeu Fabra/IMIM (Hospital del Mar Medical Research Institute), Dr. Aiguader, Barcelona, Spain

Kamila Skieterska
Laboratory of Eukaryotic Gene Expression and Signal Transduction (LEGEST), Ghent University-UGent, Ghent, Belgium

Robert V. Stahelin
Department of Biochemistry and Molecular Biology, Indiana University School of Medicine-South Bend, South Bend, and Department of Chemistry and Biochemistry, University of Notre Dame, Notre Dame, Indiana, USA

Kenichi G.N. Suzuki
Institute for Integrated Cell-Material Sciences (WPI-iCeMS), Kyoto University, Kyoto, Japan, and National Centre for Biological Sciences (NCBS)/Institute for Stem Cell Biology and Regenerative Medicine (inStem), Bangalore, India

Tokiko Suzuki
Department of Cellular Signaling, Graduate School of Pharmaceutical Sciences, Tohoku University, Sendai, Japan

Jody L. Swift
Department of Medicine, The University of British Columbia, Vancouver, British Columbia, Canada

He Tian
Laboratory of Chemical Biology and Signal Transduction, The Rockefeller University, New York, New York, USA

Juan A. Torreno-Piña
ICFO—Institut de Ciencies Fotoniques, Mediterranean Technology Park, Castelldefels (Barcelona), Spain

Kathleen Van Craenenbroeck
Laboratory of Eukaryotic Gene Expression and Signal Transduction (LEGEST), Ghent University-UGent, Ghent, Belgium

Thomas S. van Zanten
ICFO—Institut de Ciencies Fotoniques, Mediterranean Technology Park, Castelldefels (Barcelona), Spain

Paul M.P. van Bergen en Henegouwen
Cell Biology, Department of Biology, Science Faculty, Utrecht University, Utrecht, Netherlands

Hisse M. van Santen
Departamento de Biología Celular e Inmunología, Centro Biología Molecular Severo Ochoa, Consejo Superior de Investigaciones Científicas, Universidad Autónoma de Madrid, Madrid, Spain

Anthony Watts
Biomembrane Structure Unit, Department of Biochemistry, University of Oxford, Oxford, United Kingdom

Paul W. Wiseman
Departments of Chemistry and Physics, McGill University, Montreal, Québec, Canada

Andre Zeug
Cellular Neurophysiology, Center of Physiology, Hannover Medical School, Hannover, Germany

Preface

In the early 1980s, our laboratory showed (Conn, Rogers, Stewart, Niedel, & Sheffield, 1982) that bringing receptors of the gonadotropin-releasing hormone (GnRH) receptor (GnRHR) into close proximity (i.e., 100 Å) resulted in their activation. We named this process "microaggregation" to distinguish it from large-scale patching and capping ("macroaggregation"), associated with internalization of GnRH plasma membrane receptors (Hazum, Cuatrecasas, Marian, & Conn, 1980). We proposed that this was a mechanistic step in hormone action. Microaggregation of the GnRHR also increased the potency of GnRH agonists (Conn, Rogers, & McNeil, 1982), supporting that view.

The observation that receptors could come together and take on different properties was not part of the general view of hormone action 35 years ago: a receptor and ligand combine to make a complex that activated a signaling system. "$R + L \rightarrow RL \rightarrow RL^*$" was the dogma of the day. At the time, we did not know that the GnRHR was a G protein-coupled receptor (GPCR). GPCRs were not even a defined class and G proteins were not yet commonly identified.

Times have changed. The ability of receptors to interact is now called "oligomerization" and it is accepted that GPCRs oligomerize both in the anterograde (endoplasmic reticulum to plasma membrane) and retrograde (plasma membrane to intracellular spaces) modes and in the plasma membrane itself. The relevance of this process appears to differ from receptor to receptor and it is clear that GPCRs can exist as monomers, dimers, and oligomers. Hetero-oligomers are also known to exist (Kuner et al., 1999; Rocheville et al., 2000) and may be responsible for cross talk between systems, as suggested by Yogesh Patel (and others), shortly before his death.

Receptor–receptor interactions are not limited to GPCR, of course. Nuclear receptors and kinases also oligomerize within each class.

Watching this area develop and seeing the different ways in which receptor–receptor interactions are utilized by cells is a fascination to me, so I was pleased to be invited to assemble this book. Understanding the relevance by which protein–protein interactions convey information is one of the main ways that information is transferred. It is a timely topic and one that will benefit from the opportunity to compare and contrast different systems.

I was happy that so many leaders in this field were willing to commit time to contribute to this work. I thank the editors at Elsevier and the senior editors of the series for presenting this opportunity.

P. Michael Conn
Portland, Oregon
September 2013

References

Conn, P. M., Rogers, D. C., & McNeil, R. (1982). Potency enhancement of a GnRH agonist: GnRH-receptor microaggregation stimulates gonadotropin release. *Endocrinology, 111,* 335–337.

Conn, P. M., Rogers, D. C., Stewart, J. M., Niedel, J., & Sheffield, T. (1982). Conversion of a gonadotropin-releasing hormone antagonist to an agonist. *Nature, 296,* 653–655.

Hazum, E., Cuatrecasas, P., Marian, J., & Conn, P. M. (1980). Receptor-mediated internalization of fluorescent gonadotropin-releasing hormone by pituitary gonadotropes. *Proceedings of the National Academy of Sciences of the United States of America, 77,* 6692–6695.

Kuner, R., Kohr, G., Grunewald, S., Eisenhardt, G., Bach, A., & Kornau, H. C. (1999). Role of heteromer formation in GABAB receptor function. *Science, 283,* 74–77.

Rocheville, M., Lange, D. C., Kumar, U., Patel, S. C., Patel, R. C., & Patel, Y. C. (2000). Receptors for dopamine and somatostatin: Formation of hetero-oligomers with enhanced functional activity. *Science, 288,* 154–157.

CHAPTER

Spatial Intensity Distribution Analysis (SpIDA): A New Tool for Receptor Tyrosine Kinase Activation and Transactivation Quantification

1

Annie Barbeau[*,†], Jody L. Swift[‡], Antoine G. Godin[§,¶], Yves De Koninck[*,†], Paul W. Wiseman[∥], and Jean-Martin Beaulieu[*,†]

[*]*Department of Psychiatry and Neuroscience, Faculty of Medicine, Laval University, Québec, Canada*
[†]*Institut universitaire en santé mentale de Québec, Québec, Canada*
[‡]*Department of Medicine, The University of British Columbia, Vancouver, British Columbia, Canada*
[§]*University of Bordeaux, LP2N, Talence, France*
[¶]*CNRS & Institut d'Optique, LP2N, Talence, France*
[∥]*Departments of Chemistry and Physics, McGill University, Montreal, Québec, Canada*

CHAPTER OUTLINE

Introduction	2
1.1 Theory of Spatial Intensity Distribution Analysis	4
1.1.1 Theoretical Basis of SpIDA	4
1.2 SpIDA: Examples of Application to RTK	4
1.2.1 Quantification of EGFR–eGFP	4
1.2.2 Application to RTK Transactivation in a Native System: Neurons expressing Dopamine Receptors	6
1.3 Procedure for SpIDA	6
1.3.1 Material and Apparatus	6
1.3.1.1 Sample Preparation	8
1.3.2 Necessary Experimental Controls: Defining the QB for Monomeric moiety	9
1.3.2.1 Determining the Monomeric QB for Receptors Tagged with Fluorescent Proteins	9

1.3.2.2 Determining the Monomeric QB for Antibody Labeling using Detection of a Reference Monomeric Protein 9
1.3.2.3 Determining the Monomeric QB by using Pharmacological Agents that Block Oligomerization and Induce Monomeric Conformation 11
    1.3.3 Image Acquisitions 11
    1.3.4 Analysis with the SpIDA Program Graphical User Interface 12
        1.3.4.1 Description of Histogram Parameters 13
        1.3.4.2 SpIDA GUI Procedures 13
    1.3.5 Determination of Analog Detector Signal Broadening 15
        1.3.5.1 Analog Detector Calibration Procedure 16
    1.3.6 Data Interpretation and Pharmacological Analysis 16
1.4 Discussion 17
Acknowledgments 18
References 18

Abstract

This chapter presents a general approach for the application of spatial intensity distribution analysis (SpIDA) to pharmacodynamic quantification of receptor tyrosine kinase homodimerization in response to direct ligand activation or transactivation by G protein-coupled receptors. A custom graphical user interface developed for MATLAB is used to extract quantal brightness and receptor density information from intensity histograms calculated from single fluorescence microscopy images. This approach allows measurement of monomer/oligomer protein mixtures within subcellular compartments using conventional confocal laser scanning microscopy. Application of quantitative pharmacological analysis to data obtained using SpIDA provides a universal method for comparing studies between cell lines and receptor systems. In addition, because of its compatibility with conventional immunostaining approaches, SpIDA is suitable not only for use in recombinant systems but also for the characterization of mechanisms involving endogenous proteins. Therefore, SpIDA enables these biological processes to be monitored directly in their native cellular environment.

INTRODUCTION

Receptor protein kinases (RTKs) constitute a large superfamily of membrane proteins implicated in several biological processes including cell proliferation, differentiation, motility, and survival (Yarden & Ullrich, 1988). The RTK protein family includes 58 different single-transmembrane-domain glycoproteins that can be subdivided in 20 subfamilies, which include most of the receptors for growth factors (Lemmon & Schlessinger, 2010). The majority of RTK activation depends upon protein–protein interactions and formation of homodimer or higher-order oligomeric states (Lemmon & Schlessinger, 2010).

Ligand binding and dimerization initiate activation of the tyrosine kinase domain, resulting in autophosphorylation of regulatory tyrosine residues. Phosphorylated residues serve as binding sites for several enzymes (e.g., PLCgamma1) or adaptor proteins (e.g., Grb2 or Shc) through Src homology 2 (SH2) domains. Resultant RTK activation proceeds through four main signaling pathways: mitogen-activated protein kinase (MAPK) cascades, p85/p110 phosphatidylinositol-3 kinase (PI3-K) lipid kinase family, signal transducers and activator of transcription (STAT) family, and the phospholipase Cgamma1 (PLCgamma1) pathway.

G protein-coupled receptors (GPCRs) are the most prominent group of cell surface receptors and the principal targets for the majority of therapeutic drugs (Heilker, Wolff, Tautermann, & Bieler, 2009). As indicated by their name, GPCRs exert their cell signaling actions by activating different heterotrimeric G proteins (Gilman, 1987). However, GPCRs are also known to signal via alternative mechanisms, some of which involve the adaptor proteins beta arrestin 1 and 2 (Beaulieu & Caron, 2005; Ferguson, 2001; Luttrell & Gesty-Palmer, 2010). Several lines of evidence have recently suggested the interconnection between RTKs activation and GPCR signaling (Natarajan & Berk, 2006).

RTK transactivation refers to a process by which activation of a GPCR in turn activates an RTK (Schlessinger, 2000). RTK transactivation by GPCRs was first reported in the late 1990s and could be implicated in neurophysiology, cancer, and cardiac function and dysfunction (Dicker & Rozengurt, 1980). To date, the majority of literature has reported qualitative characterization to describe RTK–GPCR heterologous activation. Specifically, the bulk of research directed to study transactivation has been limited to biochemical assays relying on measuring downstream targets, which, in contrast to early events like receptor dimerization, provide limited information on the dynamics and spatial localization of the transactivation process (Eguchi et al., 1998; Rajagopal, Chen, Lee, & Chao, 2004). Currently, there is a great need for more quantitative characterization of this relationship that would allow for understanding mechanisms and relative importance of RTK transactivation leading to biological outcomes.

The investigation of biological processes at the subcellular level is typically done using a myriad of microscopy techniques. Confocal laser scanning microscope (CLSM) has become a centerpiece in cell biology laboratories and remains a popular tool to study many biological questions, albeit with an optical spatial resolution that is limited by the diffraction of light. In this chapter, we present a method to study and quantify RTK cell surface density, dimerization, and internalization that integrates the analysis of standard CLSM images together with the well-established transgene expression of fluorescent protein labels and/or immunofluorescence labeling approaches in native or fixed cells or tissue.

1.1 THEORY OF SPATIAL INTENSITY DISTRIBUTION ANALYSIS

Spatial intensity distribution analysis (SpIDA) is based on fitting super-Poissonian distributions to pixel intensity histograms calculated from CLSM images and returns information on the densities of the underlying fluorescent molecules and their quantal brightness (QB). As described previously (Barbeau et al., 2013; Godin et al., 2011; Swift et al., 2011), SpIDA is inspired by the temporal photon counting histogram (PCH) approach (Chen, Muller, So, & Gratton, 1999) but is applied to the spatial domain rather than the time domain, enabling measurements on subregions within single images collected on conventional fluorescence microscopes equipped with analog photomultiplier tube (PMT) detectors.

1.1.1 Theoretical basis of SpIDA

The intensity histogram of an imaged region of interest (ROI) is simply calculated by counting the number of pixels for each intensity value or intensity bin of values. Each pixel intensity in a CLSM image is the integrated fluorescence intensities collected from detected fluorescence photons originating within the region excited by the laser beam focal volume, the point spread function (PSF), at a given position within the sample.

The theoretical intensity distribution for an infinite region containing N particles on average per PSF, randomly distributed in space, with each particle of QB ε, gives a unique intensity histogram. Recovering the set of parameters (N and ε) that best fit the experimental intensity histogram provides information on the distribution of particles in the image and their QB (which can be used as a measure of oligomerization states).

A more detailed description of the underlying basis of SpIDA is provided in Godin et al. (2011) and Barbeau et al. (2013).

1.2 SpIDA: EXAMPLES OF APPLICATION TO RTK

We provide two examples of the applicability of SpIDA to quantify RTK transactivation. The first example involves the use of stable cell line expressing a green fluorescent protein (eGFP)-tagged epidermal growth factor receptor (EGFR) transactivated by several different GPCRs. In a second example, we demonstrate the transactivation of endogenous neuronal tropomyosin receptor kinase B (TrkB) by dopamine receptors. It should be noted that, in both examples, SpIDA was also used to quantify direct RTK activation by their cognate ligands, validating the use of SpIDA for quantitatively describing dimerization (Sergeev, Swift, Godin, & Wiseman, 2012; Swift et al., 2011).

1.2.1 Quantification of EGFR–eGFP

Using SpIDA, we were able to focus on direct measurement of early transactivation events, specifically EGFR dimerization, following stimulation of a transfected GPCR (Swift et al., 2011). Figure 1.1 illustrates a typical EGFR dimerization curve,

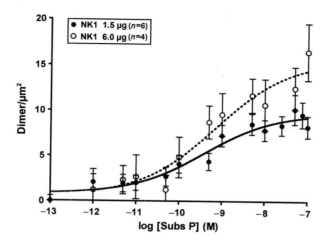

FIGURE 1.1

SpIDA allows for pharmacological characterization of EGFR transactivation by NK1 GPCR. Dose–response curves of EGFR dimer density 1 min after substance P stimulation in CHO-k1 cells. Cells were transfected with either 1.5 μg of GPCR DNA ($n=120$ cells/point from six individual experiments) or 6 μg ($n=60$ cells/point from three individual experiments). Cells were cotransfected with Rab5 S34N to prevent receptor internalization. Data are means ± SEM.

resulting from the stimulation of a cotransfected GPCR, specifically stimulation of neurokinin 1 receptors (NK-1R) by substance P. CHO–EGFR–eGFP cells (Brock, Hamelers, & Jovin, 1999) were transiently transfected with NK-1R DNA, and EGFR–eGFP dimerization was measured by SpIDA following agonist-specific stimulation by increasing doses of substance P. Dimer densities were plotted using the Hill–Langmuir binding isotherm model, and SpIDA results allowed us to extract key parameters including D_{50} (ligand concentration required to obtain 50% maximum dimer density signal) and D_{max} (maximal level of receptor dimer density signal).

The fitted parameters obtained, D_{50} and D_{max}, can be compared to similar pharmacodynamic parameters like Bret50 and Fret50 commonly measured in RET-based pharmacological studies (Ayoub et al., 2007; Masri et al., 2008; Salahpour & Masri, 2007). Therefore, application of SpIDA to CLSM image analysis allows for the direct comparison of transactivation efficiencies across different cell types and tissues. As demonstrated in Fig. 1.1, SpIDA is able to reveal that the abundance of the GPCR may be a limiting factor in RTK transactivation. Indeed, increasing the amount of GPCR cotransfected into cells results in an increased D_{max}, similar to what would be observed in the case of a B_{max} in a binding experiment (Kenakin, 2009; Swift et al., 2011).

The analysis further allows for quantification of receptor internalization, as being evaluated by a reduction in surface receptor density. In the case of RTK transactivation by GPCRs, receptor internalization may be prevented by transfection of a rab5 mutant DNA construct, rab5 S34N (Bucci et al., 1992; Li & Stahl, 1993). Using a rab5 S34N DNA construct, we were able to observe that, for some GPCRs (e.g., D1R, D2R, and NK-1R), internalization was instrumental to obtain a maximum EGFR–GFP dimer density, whereas in other cases (AT1aR and beta2AR), results indicated that internalization was not necessary for observing full transactivation as measured by RTK dimerization.

1.2.2 Application to RTK transactivation in a native system: neurons expressing dopamine receptors

One of the greatest advantages of SpIDA has been demonstrated using endogenous systems (Doyon et al., 2011; Sergeev, Godin, et al., 2012; Swift et al., 2011). For exemple, in one of these systems SpIDA was used to show that the brain-derived neurotrophic factor (BDNF) receptor TrkB is transactivated by dopamine receptors subtypes 1 and 2 (D1R and D2R, respectively). This was acheived using primary culture of striatal neurons (Swift et al., 2011) obtained from transgenic mice expressing fluorescent reporter genes tdTomato, under the control of the D1R gene, and EGFP, under the control of the D2R gene (Shuen, Chen, Gloss, & Calakos, 2008; Zhang, Burke, Calakos, Beaulieu, & Vaucher, 2010). Expression of these reporter genes allowed locating two different populations of neurons expressing predominantly either D1R or D2R.

TrkB dimerization was measured by staining cell surface receptors with a primary antibody on nonpermeabilized neurons and detected using secondary antibodies conjugated to an Alexa Fluor 647 moiety. The SpIDA method allowed for quantification of TrkB oligomeric states in response to either BDNF (direct activation) or the nonselective dopamine receptor agonist apomorphine (transactivation) in specific neurons (D1 in red or D2 in green) (Fig. 1.2).

This result demonstrates the usefulness of this image analysis method to quantify RTK transactivation by GPCR in situ. Given the wide availability of antibodies for many targets of interest and the quality of secondary fluorescent antibody conjugates available commercially, SpIDA can be used to investigate a wide variety of biological systems involving protein homooligomerization in native tissue with endogenous expression.

1.3 PROCEDURE FOR SpIDA

Here, we describe the steps for detector calibration, image acquisition, and SpIDA analysis using CLSM images acquired from fluorescently labeled cell or tissue samples. The overall process is straightforward for any user trained in biological fluorescence imaging (Fig. 1.3).

1.3.1 Material and apparatus

SpIDA was validated using images obtained by CLSM, but, in theory, SpIDA can also be used with total internal reflection fluorescence (TIRF) microscopy. To appropriately perform SpIDA, one needs to determine the suitable parameters to work in

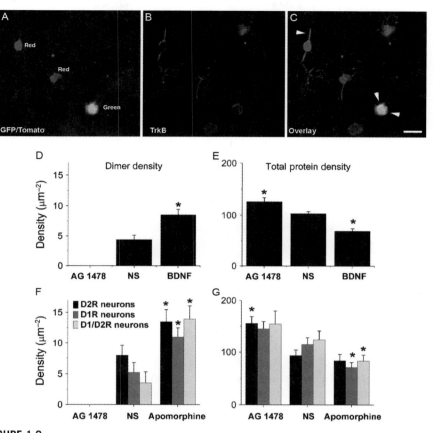

FIGURE 1.2

SpIDA allows the detection and quantification of endogenous TrkB activation by endogenous dopamine receptors in striatal neurons. (A–C) Surface immunodetection of endogenous TrkB in striatal neurons prepared from BAC transgenic mice expressing an EGFP (green) reporter gene in cells having endogenous D2R or a Tomato (red) reporter gene in cells having endogenous D1R. Arrowheads in C indicate surface TrkB labeling on striatal neurons expressing different dopamine receptors. Scale bar: 15 μm. (D–E) SpIDA analysis of TrkB dimer (D) and total surface densities (E) of neurons after incubation with AG1478 (0.2 mM for 30 min, $n=31$ neurons/bar), under nonstimulated condition ($n=93$ neurons/bar), and following direct stimulation with BDNF (50 pM for 3 min, $n=83$ neurons/bar). (F–G) SpIDA analysis of TrkB dimer (F) and total surface densities (G) of neurons after incubation with AG1478 (0.2 mM for 30 min, n(D2R$=30$, D1R$=33$, D1/2R$=16$) neurons/bar), under nonstimulated condition (n(D2R$=19$, D1R$=26$, D1/2R$=14$) neurons/bar), and following stimulation of endogenous dopamine receptors with apomorphine (2 μM for 3 min, n(D2R$=33$, D1R$=38$, D1/2R$=28$) neurons/bar). Data are presented for three subpopulations of striatal neurons expressing D1R, D2R, or both reporter transgenes. Data are means ± SEM.

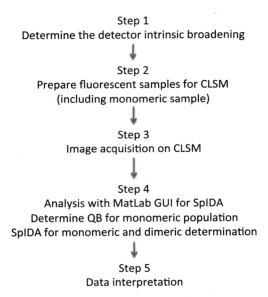

FIGURE 1.3

Overview of SpIDA procedures. (For color version of this figure, the reader is referred to the online version of this chapter.)

the linear regime of the PMT detectors (e.g., avoid saturation and nonlinear PMT response) of the desired system. To date, SpIDA has been successfully applied to quantify mCherry- and GFP-tagged proteins along with proteins visualized by immunofluorescence staining with Alexa Fluor 488, 546, 633, and 647 antibody conjugates. For fluorophore-conjugated antibodies, it is important to use versions that have a high number of fluorescent moieties/antibody molecules. This information is available from data sheet specific for the antibody lot used. Indeed, we have observed that a higher degree of labeling per antibody yields less variation in the establishment of the monomeric QB. This distribution can also be taken into account during the analysis if needed (Godin et al., 2011).

1.3.1.1 Sample preparation

As cell fixation and immunochemistry are techniques routinely used in most cell biology laboratories and are the subject of many protocol papers and manuals, we will not go into details for these steps (Maity, Sheff, & Fisher, 2013). We will focus on image acquisition and SpIDA analysis. Just remember to choose fluorophores accordingly to system performance.

In the case of an upright microscope, cells are grown on glass coverslips and mounted with antifading reagent following fixation. For an inverted microscope, cells may be grown on glass bottom plates (e.g., MatTek culture dishes), chemically fixed (e.g., paraformaldehyde, glutaraldehyde, or methanol) and immersed with phosphate-buffered saline or equivalent buffer for imaging. For live-cell experiments, samples may be imaged through the glass bottom of a culture plate suitable for fluorescence imaging acquisition or an upright microscope with a water immersion objective.

1.3.2 Necessary experimental controls: defining the QB for monomeric moiety

After determination of detector signal broadening due to intrinsic characteristic of the detector (Section 1.3.5), the first step in the analysis is to obtain an estimate of the monomeric QB of the fluorescent label. Remember that the monomeric QB will vary depending on the choice of fluorophore and the microscope image collection settings (PMT voltage, laser power, filters, etc.); thus, it must be characterized for each set of experiments. Different options to determine the monomeric QB are presented in the succeeding text.

1.3.2.1 Determining the monomeric QB for receptors tagged with fluorescent proteins

The main objective when choosing a bright fluorophore suitable for determining QB is to avoid ones that have the tendency to oligomerize, which is a problem for several fluorescent proteins, including early versions of GFP (Zacharias, Violin, Newton, & Tsien, 2002). Several DNA constructs have been developed to express monomeric fluorescent proteins (e.g., monomeric GFP, mGFP). These constructs are either targeted to the membrane (e.g., farnesylated mGFP (Zacharias et al., 2002) or glycosylphosphatidylinositol anchored (Kondoh, 2002)) or expressed in the cytoplasm.

Whether a construct is suitable as a control for the monomeric brightness can be verified. To do this, simply collect an image time series and analyze each frame to verify if the QB stays constant within the detector error. If the fluorophores exist in its oligomerized form, one should observe a time-dependent decrease in the density as reported by SpIDA due to fluorophore photobleaching. In contrast, the QB will remain constant if the construct is truly monomeric. Steps for monomeric QB determination for receptor tagged with fluorescent protein are described in Fig. 1.4. Note that this approach to test the validity of a monomeric sample is not suitable for fluorescently labeled antibodies since most labeled antibodies are labeled by several fluorescent molecules.

1.3.2.2 Determining the monomeric QB for antibody labeling using detection of a reference monomeric protein

Because the fluorescence of labeled antibodies does not react in a dichotomic manner (on/off) and that photobleaching will induce a change in the brightness of single probes, it is important to minimize prealteration of the sample before imaging and also to minimize the photobleaching during imaging. Determining the QB of the probe can be done using a protein known to be monomeric that has been reveled with the same secondary antibody used for the protein of interest.

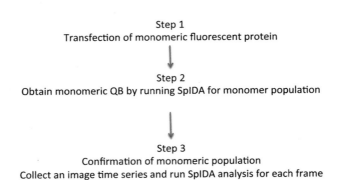

FIGURE 1.4

Monomeric quantal brightness determination for receptor tagged with fluorescent protein. (For color version of this figure, the reader is referred to the online version of this chapter.)

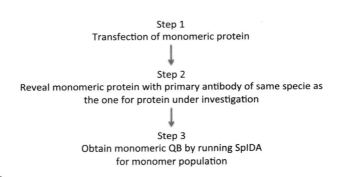

FIGURE 1.5

Monomeric quantal brightness determination for antibody complex labeling using a reference monomeric fluorescent protein. (For color version of this figure, the reader is referred to the online version of this chapter.)

In our experiments, mGPF is often used for assessing monomeric QB. A transfected construct is expressed and detected with a primary antibody from the same species as the protein of interest. For example, if the primary antibody targeting the receptor of interest is raised in the mouse, then a control cell expression sample can be made with farnesylated mGFP, which will then also be labeled using a mouse antibody. The same fluorescently tagged secondary antibody will then be used to visualize the primary antibodies in both samples. Just remember that to apply this method, one has to be selective in the choice of the fluorophore conjugated to the secondary antibodies to ensure clear spectral separation from that of the monomeric fluorescent protein tag (Fig. 1.5).

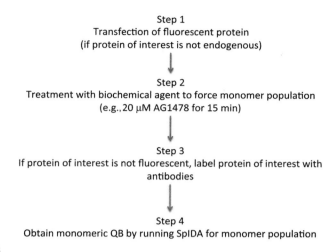

FIGURE 1.6

Monomeric quantal brightness determination using pharmacological agents that block oligomerization and induce monomeric conformation. (For color version of this figure, the reader is referred to the online version of this chapter.)

1.3.2.3 Determining the monomeric QB by using pharmacological agents that block oligomerization and induce monomeric conformation

In our studies with RTKs EGFR and TrkB, the inhibitor AG1478 (Levitzki & Gazit, 1995) was used to block oligomerization and force a monomeric RTK population. In our hands, the use of this inhibitor with CHO–EGFR–GFP cells revealed that following growth factor starvation, most of EGFR–GFP were in monomeric conformation (Fig. 1.6).

In summary, any sample or region of a sample that contains a vast majority of monomers that are labeled with the same fluorescent probe, when imaged with the same microscope using the same experimental settings (e.g., laser intensity, PMT voltage, pixel dwell times, and pixel size), can be used to obtain an estimate of the monomeric QB.

1.3.3 Image acquisitions

Optimal CLSM conditions should be adjusted for each experimental design. CLSM settings should be chosen to achieve a good signal, that is, within the linear dynamic range of the detector and not saturated. Laser intensity and PMT gain are adjusted to minimize pixel saturation and photobleaching. Once found, image collection parameters (settings of laser power, filters, dichroic mirrors, polarization voltage, scan speed and pixel dwell time, PMT gain, laser intensity detector, and pinhole) are kept constant for the image acquisitions of all samples in a given experiment (experiment and monomeric control).

Note that when applying SpIDA to fixed samples for single-image analysis, it is better to increase the laser intensity, always considering the effect of photobleaching, than the PMT gain/voltage because raising the PMT gain will result in increased detector variance.

In all cases, image acquisitions can be single image, time series, or Z-stack. SpIDA may be applied to measure proteins expressed at the plasma membrane or in other cellular organelles, with the caveat that the spatial resolution is set by the three-dimensional (3D) PSF of the CLSM.

1.3.4 Analysis with the SpIDA program graphical user interface

A custom MATLAB graphical user interface (GUI) was designed to perform SpIDA in a user-friendly way (Fig. 1.7). This section provides basic information for operating the GUI.

The GUI can be downloaded free of charge at the website http://www.neurophotonics.ca/en/tools/software. The GUI was tested in the Windows 32- and 64-bit MAC OS X, and UNIX operating systems and was created in MATLAB

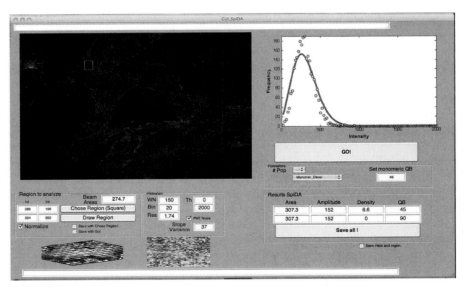

FIGURE 1.7

MATLAB GUI for SpIDA. (For color version of this figure, the reader is referred to the online version of this chapter.)

Version 7.10.0.499 (R2010a) 64-bit (win64) but should work on most of older versions of the MATLAB platform.

1.3.4.1 Description of histogram parameters

- White noise background level: Corresponds to the measured pixel intensity values for an image collected when the laser is off or to the mean pixel intensity calculated for a region of the image where no cells or fluorophores are present (e.g., area between cells). A histogram of an image is produced by discretization of the intensity range into a number of bins.
- Bin: We suggest that the bin size should be chosen in order to be larger than half of the monomeric QB of the control, but not too large, so that the distribution of the curve is not truncated. The smaller the intensity bin size, the longer the computational time for SpIDA analysis.
- Resolution (Res): Is defined in the GUI as being the number of pixels per beam waist radius (i.e., ω_{xy}/pixel size). The resolution and beam waist radius can be estimated by imaging a control sample of subdiffraction-size fluorescent microspheres using the various zoom settings of the CLSM and analyzing intensity lineout plots using ImageJ.
- Threshold (Th): Sets the minimum and maximum values of intensities included in the histogram analyzed.

1.3.4.2 SpIDA GUI procedures

This section is a step-by-step recipe explaining how to use the GUI. Note that all details are also accessible on the website http://www.neurophotonics.ca/en/tools/software. An image showing the GUI is presented in Fig. 1.8. To analyze images using the GUI, follow these steps:

1. Set the Current Folder in MATLAB to the directory containing the GUI files and launch GUI_SpIDA command.
2. An explorer window (select an image) opens and offers you the option to load image files. If the loaded file contains more than one image, a scroll bar will appear to permit navigation through the image stack. To load another image, just click on the "load image" textbox and select the image file.
3. Set the correct parameters for the image file. The pixel size and the beam waist radius in micrometer are needed as inputs. The GUI will attempt to obtain the information from the image file itself. If the program cannot find the spatial information in the header, a dialog box will appear and the correct image information can be manually entered.
4. To select an ROI to analyze, click on the button "Chose Region." Using the mouse, click, drag, and click again to delimitate a rectangular ROI. The two chosen pixel positions are shown left of the two buttons "Chose Region" and "Draw Region." A zoom of the chosen region with enhanced contrast is also presented in the GUI lower left panel. Set the check box "Normalize" to ON

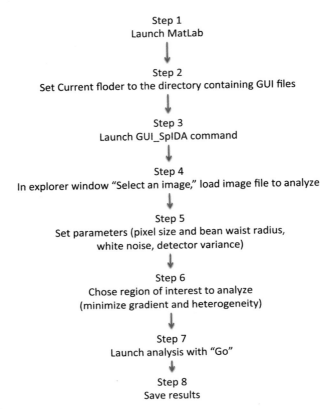

FIGURE 1.8

Procedures for SpIDA GUI. (For color version of this figure, the reader is referred to the online version of this chapter.)

position and adjust the image contrast to make only image pixels that are within the threshold range visible.

Note: Select the ROI in order to minimize gradient and heterogeneity. To do so, change the ROI size in order to minimize any intensity gradients so that they are not visible to the eye when looking at the lower left panel. This is an advantage of SpIDA, as it can be applied robustly to images that contain significant heterogeneity and complex cell morphology or boundaries of fluorescence labeling in ramified cells, which is often the case with immunolabeling in tissue sections. To change the ROI without analyzing it, clicking the button "Draw Region" refreshes the image with the new coordinates and intensity histogram parameters.

5. To run the SpIDA analysis on the selected ROI, click on the "GO!" button.

Note: We strongly suggest running the analysis with noise correction inherent to the PMT implemented. Click ON the box "PMT Noise" and the "Slope Variance"

textbox will appear and set the value for the "Slope Variance." Slope variance determination is described in Section 1.3.5.
6. When a mixture of monomers and dimers is expected in the image, the number of populations "# Pop" can be set to 2. The monomeric QB can also be fixed in the fit, using the previously determined value from control images using a single population fit. A dimer will then be twice as bright as a monomer. If the monomeric QB is set to 0, the QB will also be fit.
7. The results of the fits are shown in the box "Results SpIDA."
   The following results are obtained as output:
   - "Area" is the area of the chosen region in beam areas corresponding to $pi*\omega^2_{xy}$ (beam area is the fundamental unit of sampling where the pixel size is smaller than the beam focus area).
   - "Amplitude" of the fit is the height of the histogram in pixels. The density of each of the population, as units of particles per effective illumination volume, is also given. This effective volume can be a surface if the region chosen is on the cell membrane ($\pi*\omega^2_{xy}$) or a volume if the ROI is completely within a cell (e.g., cytoplasm or nucleus), and then the volume can be approximated as the effective volume of a 3D Gaussian ($\pi^{3/2}*\omega^2_{xy}\omega_z$, where ω_z is the axial radius of the PSF).
   - QB is the average brightness of a fluorescent entity in the effective volume and has units of intensity.
8. Save the results. To save the results of an analysis, just press the button "Save all!" and the GUI will save a ".dat" file with the name of the image with all fit values and set parameters. If "Save with Go!" button is activated, results will be saved automatically in the chosen folder when the analysis is done, whereas when the button "Save with Chosen region!" is set to true, the analysis will be launched and the results will be saved automatically. These options allow for a faster analysis by reducing the number of clicks for iterative analysis.
9. Results are displayed in .txt format and the data may be extracted and opened in other programs such as Excel. We strongly suggest programming a macro command to extract the values from each file.

1.3.5 Determination of analog detector signal broadening

As SpIDA measures the fluorescence intensity fluctuations of the signal in the image to return information on the number of particles and their QB values, it is important to consider only the fluctuations that originate from true signal variations of the fluorescently labeled proteins in the sample and exclude fluctuations inherent to the detector. Analog PMT detector variance is empirically determined as described in the succeeding text. Use either back reflection of the laser from a mirror placed at focus or a bright fluorescent sample (e.g., a solution containing an extremely high concentration of fluorophores or a commercial fluorescent slide). This determination needs to be done with the same settings (PMT voltage, filter sets, and scan speed), which will be used for later image acquisition for samples under investigation.

1.3.5.1 Analog detector calibration procedure

For a given PMT voltage and pixel dwell time, time series are acquired in the spot selection mode.

1. Set pinhole to 1 Airy unit and amplifier offset to 0.
2. Repeated acquisitions are done at increasing laser intensities. The initial acquisition is carried out with a low-laser-intensity setting to avoid detector saturation. Increase the laser intensity until reaching around 90% of detector saturation and then systematically ramp down the laser intensity. A final acquisition is also collected with the laser turned off. Typically, 10–20 acquisitions are performed for each PMT gain and pixel dwell time.
3. Open the different acquisitions in ImageJ and ask software to extract standard deviation of values collected. Variance corresponds to this value squared. Plot variances versus mean intensities. Slopes of the variance versus the mean intensity of curves are calculated only for the linear regime of the PMT. An example is shown in Fig. 1.9. Calculate slope, which corresponds to detector variance number needed to be entered in the GUI.

It is important to note that the slope can depend on many parameters (dwell time, PMT voltage, scan speed, temperature, etc.) and that, for each set of parameters, the detector variance should be determined. Similar calibration can be performed for a CCD camera by generating image time series of constant stable light source and generating a graph of the variance of each pixel as a function of the mean intensity. If the CLSM has photon-counting detectors, then this calibration is not necessary.

1.3.6 Data interpretation and pharmacological analysis

Oligomer density can be expressed as a number of protomers/beam areas. As beam area can be converted to μm^2, protein density can be expressed as protomers/μm^2 (i.e., monomer or dimer/μm^2). In the case of RTKs, when imaging

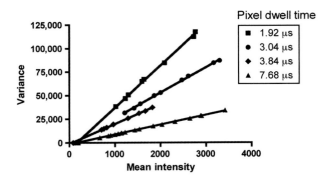

FIGURE 1.9

Example of detector broadening determination for different pixel dwell time.

the plasma membrane, monomer and protomer densities are quantified and the total receptor density can be extrapolated. Receptor internalization can also be observed as a decrease in fluorescently labeled protein density at the plasma membrane for fixed samples. For live-cell studies, photobleaching will also contribute to a decrease in measured density as a function of time. Therefore, SpIDA allows for the simultaneous quantification of oligomerization and internalization. In addition, as shown in Fig. 1.1, data obtained with SpIDA can readily be used for pharmacological analysis with the Hill–Langmuir binding isotherm model.

1.4 DISCUSSION

Overall, the image analysis described in this chapter allows for the quantification of RTK membrane density, oligomeric state, and internalization. Of general importance, SpIDA is versatile and is not restricted to cell surface proteins. In reality, as this is a postacquisition image analysis method, it may be applied to any optical section obtained by CLSM from fixed or live preparations.

Furthermore, it is compatible with the analysis of both fluorescent-tagged proteins and native proteins in primary cultured cells or tissues. However, some limitations exist as with any image quantification method. In its present form, SpIDA can only resolve monochromatic images, which limits its applicability for complex heteromerization schemes. On the other hand, SpIDA can be applied simultaneously to images acquired at different wavelengths (i.e., different fluorophores) from the same preparation and is, therefore, compatible with multiplexing. In theory, higher-order arrangements can be studied with SpIDA (tetramers or pentamers), but the limitation here is the complexity of possible configurations because of either nondetected proteins (for antibody labeling) or nonfluorescent proteins that can be present. Finally, it is important to keep in mind that in contrast to RET, SpIDA is not providing information about direct protein–protein interactions but only about oligomerization within a common complex, which can be dependent upon other intermediates such as adaptor proteins. While this may be a limitation for those seeking to identify direct protein–protein interactions, it offers the advantage of enabling detection of regular, yet indirect, associations between proteins with complex macromolecular arrangements.

A major advantage of SpIDA over RET techniques is that it allows quantification of mixtures of oligomeric arrangements (e.g., provide numbers of both monomer and dimer densities within a mixed distribution). In summary, SpIDA is a low-cost image analysis method enabling quantitative assessment of oligomerization that is compatible with most CLSM systems commonly used in biomedical research laboratories. Its compatibility with CLSM and conventional immunocytochemical methods opens the door to studies of oligomerization in native tissue, which, to date, were largely restricted to expression systems.

Acknowledgments

A. B. is supported by the CIHR *Neurophysics* Training Program. A.G.G. is funded by the *Fondation pour la recherche médicale* (FRM, programme Espoirs de la Recherche). J.-M.B. holds a Canada Research Chair in Molecular Psychiatry. This work was supported by NSERC discovery grants to P.W.W., Y.D.K., and J.-M.B.

References

Ayoub, M. A., Maurel, D., Binet, V., Fink, M., Prezeau, L., Ansanay, H., et al. (2007). Real-time analysis of agonist-induced activation of protease-activated receptor 1/Galphai1 protein complex measured by bioluminescence resonance energy transfer in living cells. *Molecular Pharmacology*, *71*(5), 1329–1340.

Barbeau, A., Godin, A. G., Swift, J. L., De Koninck, Y., Wiseman, P. W., & Beaulieu, J. M. (2013). Quantification of receptor tyrosine kinase activation and transactivation by G-protein-coupled receptors using spatial intensity distribution analysis (SpIDA). *Methods in Enzymology*, *522*, 109–131.

Beaulieu, J. M., & Caron, M. G. (2005). Beta-arrestin goes nuclear. *Cell*, *123*(5), 755–757.

Brock, R., Hamelers, I. H., & Jovin, T. M. (1999). Comparison of fixation protocols for adherent cultured cells applied to a GFP fusion protein of the epidermal growth factor receptor. *Cytometry*, *35*(4), 353–362.

Bucci, C., Parton, R. G., Mather, I. H., Stunnenberg, H., Simons, K., Hoflack, B., et al. (1992). The small GTPase rab5 functions as a regulatory factor in the early endocytic pathway. *Cell*, *70*(5), 715–728.

Chen, Y., Muller, J. D., So, P. T., & Gratton, E. (1999). The photon counting histogram in fluorescence fluctuation spectroscopy. *Biophysical Journal*, *77*(1), 553–567.

Dicker, P., & Rozengurt, E. (1980). Phorbol esters and vasopressin stimulate DNA synthesis by a common mechanism. *Nature*, *287*(5783), 607–612.

Doyon, N., Prescott, S. A., Castonguay, A., Godin, A. G., Kroger, H., & De Koninck, Y. (2011). Efficacy of synaptic inhibition depends on multiple, dynamically interacting mechanisms implicated in chloride homeostasis. *PLoS Computational Biology*, *7*(9), e1002149.

Eguchi, S., Numaguchi, K., Iwasaki, H., Matsumoto, T., Yamakawa, T., Utsunomiya, H., et al. (1998). Calcium-dependent epidermal growth factor receptor transactivation mediates the angiotensin II-induced mitogen-activated protein kinase activation in vascular smooth muscle cells. *Journal of Biological Chemistry*, *273*(15), 8890–8896.

Ferguson, S. S. (2001). Evolving concepts in G protein-coupled receptor endocytosis: The role in receptor desensitization and signaling. *Pharmacological Reviews*, *53*(1), 1–24.

Gilman, A. G. (1987). G proteins: Transducers of receptor-generated signals. *Annual Review of Biochemistry*, *56*, 615–649.

Godin, A. G., Costantino, S., Lorenzo, L. E., Swift, J. L., Sergeev, M., Ribeiro-da-Silva, A., et al. (2011). Revealing protein oligomerization and densities in situ using spatial intensity distribution analysis. *Proceedings of the National Academy of Sciences of the United States of America*, *108*(17), 7010–7015.

Heilker, R., Wolff, M., Tautermann, C. S., & Bieler, M. (2009). G-protein-coupled receptor-focused drug discovery using a target class platform approach. *Drug Discovery Today*, *14*(5–6), 231–240.

Kenakin, T. P. (2009). *A pharmacology primer: Theory, applications, and methods* (3rd ed.). Amsterdam, Boston: Academic Press/Elsevier.

Kondoh, G. (2002). Development of glycosylphosphatidylinositol-anchored enhanced green fluorescent protein. One-step visualization of GPI fate in global tissues and ubiquitous cell surface marking. *Methods in Molecular Biology, 183*, 215–224.

Lemmon, M. A., & Schlessinger, J. (2010). Cell signaling by receptor tyrosine kinases. *Cell, 141*(7), 1117–1134.

Levitzki, A., & Gazit, A. (1995). Tyrosine kinase inhibition: An approach to drug development. *Science, 267*(5205), 1782–1788.

Li, G., & Stahl, P. D. (1993). Structure–function relationship of the small GTPase rab5. *Journal of Biological Chemistry, 268*(32), 24475–24480.

Luttrell, L. M., & Gesty-Palmer, D. (2010). Beyond desensitization: Physiological relevance of arrestin-dependent signaling. *Pharmacological Reviews, 62*(2), 305–330.

Maity, B., Sheff, D., & Fisher, R. A. (2013). Immunostaining: Detection of signaling protein location in tissues, cells and subcellular compartments. *Methods in Cell Biology, 113*, 81–105.

Masri, B., Salahpour, A., Didriksen, M., Ghisi, V., Beaulieu, J. M., Gainetdinov, R. R., et al. (2008). Antagonism of dopamine D2 receptor/beta-arrestin 2 interaction is a common property of clinically effective antipsychotics. *Proceedings of the National Academy of Sciences of the United States of America, 105*(36), 13656–13661.

Natarajan, K., & Berk, B. C. (2006). Crosstalk coregulation mechanisms of G protein-coupled receptors and receptor tyrosine kinases. *Methods in Molecular Biology, 332*, 51–77.

Rajagopal, R., Chen, Z. Y., Lee, F. S., & Chao, M. V. (2004). Transactivation of Trk neurotrophin receptors by G-protein-coupled receptor ligands occurs on intracellular membranes. *Journal of Neuroscience, 24*(30), 6650–6658.

Salahpour, A., & Masri, B. (2007). Experimental challenge to a 'rigorous' BRET analysis of GPCR oligomerization. *Nature Methods, 4*(8), 599–600, author reply 601.

Schlessinger, J. (2000). Cell signaling by receptor tyrosine kinases. *Cell, 103*(2), 211–225.

Sergeev, M., Godin, A. G., Kao, L., Abuladze, N., Wiseman, P. W., & Kurtz, I. (2012). Determination of membrane protein transporter oligomerization in native tissue using spatial fluorescence intensity fluctuation analysis. *PLoS One, 7*(4), e36215.

Sergeev, M., Swift, J. L., Godin, A. G., & Wiseman, P. W. (2012). Ligand-induced clustering of EGF receptors: A quantitative study by fluorescence image moment analysis. *Biophysical Chemistry, 161*, 50–53.

Shuen, J. A., Chen, M., Gloss, B., & Calakos, N. (2008). Drd1a-tdTomato BAC transgenic mice for simultaneous visualization of medium spiny neurons in the direct and indirect pathways of the basal ganglia. *Journal of Neuroscience, 28*(11), 2681–2685.

Swift, J. L., Godin, A. G., Dore, K., Freland, L., Bouchard, N., Nimmo, C., et al. (2011). Quantification of receptor tyrosine kinase transactivation through direct dimerization and surface density measurements in single cells. *Proceedings of the National Academy of Sciences of the United States of America, 108*(17), 7016–7021.

Yarden, Y., & Ullrich, A. (1988). Growth factor receptor tyrosine kinases. *Annual Review of Biochemistry, 57*, 443–478.

Zacharias, D. A., Violin, J. D., Newton, A. C., & Tsien, R. Y. (2002). Partitioning of lipid-modified monomeric GFPs into membrane microdomains of live cells. *Science, 296*(5569), 913–916.

Zhang, Z. W., Burke, M. W., Calakos, N., Beaulieu, J. M., & Vaucher, E. (2010). Confocal analysis of cholinergic and dopaminergic inputs onto pyramidal cells in the prefrontal cortex of rodents. *Frontiers in Neuroanatomy, 4*, 21.

CHAPTER 2

Dimerization of Nuclear Receptors

Pierre Germain[*,†] **and William Bourguet**[*,†]

[*]*Inserm U1054, Centre de Biochimie Structurale, Montpellier, France*
[†]*CNRS UMR5048, Universités Montpellier 1 & 2, Montpellier, France*

CHAPTER OUTLINE

Introduction	22
2.1 Methods	24
2.1.1 Studies of NR–NR Interactions through Protein Crystallization	24
2.1.1.1 Required Materials	26
2.1.1.2 Protocol	26
2.1.2 Expression and Purification of NR–NR Complexes	27
2.1.2.1 Required Materials	28
2.1.2.2 Protocol	29
2.1.3 Monitoring NR–NR Interactions by Noncovalent Electrospray Ionization Mass Spectrometry	29
2.1.3.1 Required Materials	30
2.1.3.2 Protocol	30
2.1.4 Monitoring NR–NR Interactions by Electrophoretic Mobility Shift Assays	31
2.1.4.1 Required Materials	31
2.1.4.2 Protocol	31
2.1.5 Two-hybrid Assays to Define NR–NR Interactions in Living Cells	32
2.1.5.1 Required Materials	34
2.1.5.2 Protocol	34
2.1.6 Fluorescence Cross-Correlation Spectroscopy to Measure the Concentrations and Interactions of NRs in Living Cells	35
2.1.6.1 Required Materials	36
2.1.6.2 Protocol	36
Acknowledgments	39
References	39

Abstract

Multicellular organisms require specific intercellular communication to properly organize the complex body plan during embryogenesis and maintain its properties and functions during the entire life. While growth factors, neurotransmitters, and peptide hormones bind to membrane receptors, thereby inducing the activity of intracellular kinase cascades or the JAK–STAT signaling pathways, other small signaling compounds such as steroid hormones, certain vitamins, and metabolic intermediates enter, or are generated, within the target cells and bind to members of a large family of nuclear receptors (NRs). NRs are ligand-inducible transcription factors that control a plethora of biological phenomena, thus orchestrating complex events like development, organ homeostasis, immune function, and reproduction. NR–NR interactions are of major importance in these regulatory processes, as NRs regulate their target genes by binding to cognate DNA response elements essentially as homo- or heterodimers. A number of structural and functional studies have provided significant insights as to how combinatorial NRs rely on protein–protein contacts that discriminate geometric features of their DNA response elements, thereby allowing both binding site diversity and physiological specificity. Here, we will review our current understanding of NR–NR interactions and provide protocols for a number of experimental approaches that are useful for their study.

INTRODUCTION

Nuclear receptors (NRs) are members of a large superfamily of evolutionarily related transcription factors that regulate genetic programs involved in a broad spectrum of physiological phenomena. The human NR family is classified by sequence into six evolutionary groups of unequal size. The reader is referred to several databases providing comprehensive annotation of the literature on the pharmacology of the 48 human NR gene products, together with relevant information on their structure and function (IUPHAR-DB (http://www.iuphar-db.org/), Nuclear Receptor Signaling Atlas (NURSA, http://www.nursa.org/), and NureXbase (http://nurexbase.prabi.fr)). Among NRs, ligands have been identified for only 24 family members. These receptors are ligand-dependent transcriptional factors that respond directly to a large variety of hormonal and metabolic substances that are hydrophobic, lipid-soluble, and of small size (e.g., retinoic acid or estradiol). The other class of NRs is the group of so-called orphan receptors, for which regulatory ligands are still unknown or may not exist (true orphans) or for which candidates have only recently been identified (adopted orphans). In both groups, NRs can exist as different receptor subtypes that originate from different genes such as the retinoic acid receptors α, β, and γ (RARα (NR1B1), RARβ (NR1B2), and RARγ (NR1B3)), which can exist themselves as several isoforms originating from alternative splicing and differential promoter usage (e.g., RARβ1–RARβ4) (Germain, Staels, Dacquet, Spedding, & Laudet, 2006; Gronemeyer, Gustafsson, & Laudet, 2004).

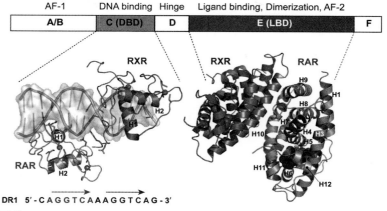

FIGURE 2.1

Structural and functional organization on nuclear receptors. The structural organization of the DBD and LBD is illustrated by the crystal structures of the RXRα–RARα DBD heterodimer bound to a DR1 response element (Protein Data Bank code 1dsz) and the 9-cis-retinoic acid-bound RXRα–RARα LBD heterodimer (Protein Data Bank code 1xdk). Helices are represented as ribbons and labeled from H1 to H12 (LBD) or H1 and H2 (DBD). 9-cis-retinoic acid in RARα and RXRα LBDs is represented by yellow and red spheres. The gray spheres in the DBD indicate Zn^{2+} ions. Residues mediating the weak DBD intersubunit contacts are displayed in yellow. The cysteine residues coordinating the Zn^{2+} ions are shown in violet. (See color plate.)

All NR proteins exhibit a characteristic modular structure that consists of five to six domains of homology (designated A–F, from the N-terminal to the C-terminal) based on regions of conserved sequence and function (Fig. 2.1). The DNA-binding domain (DBD; region C) is the most highly conserved domain and encodes two zinc finger modules. The ligand-binding domain (LBD; region E) is less conserved and mediates ligand binding, dimerization, and a ligand-dependent transactivation function, termed AF-2. The N-terminal A–B region contains a cell- and promoter-specific transactivation function termed AF-1. The region D is considered as a hinge domain. The F region is not present in all receptors and its function is poorly understood. Except for the dosage-sensitive sex reversal, adrenal hypoplasia congenital critical region on the X chromosome gene 1 (DAX-1 (NR0B1)), and the short heterodimer partner (SHP (NR0B2)), which both lack an essential DBD, NRs can bind their cognate sequence-specific promoter elements on target genes either as monomers, homodimers, or heterodimers with retinoid X receptors (RXRα (NR2B1), RXRβ (NR2B2), and RXRγ (NR2B3)). They all recognize the same hexameric DNA core motif, 5′-PuGGTCA (Pu=A or G), but mutation, extension, duplication, and distinct relative orientations of repeats of this motif have generated response elements that are selective for a given class of receptors. For example, the RXR–RAR heterodimers bind to retinoic acid response elements that mainly correspond to direct repeats of the motif 5′-AGGTCA separated by five, two, or one nucleotides (referred to as DR5, DR2, and DR1; Fig. 2.1), whereas steroid receptors such as the androgen receptor (AR, (NR3C4)) bind essentially as homodimers to response elements containing palindromic repeats of the hexamerix half-site sequence 5′-AGAACA-3′ (Khorasanizadeh & Rastinejad, 2001).

Dimerization is a general mechanism to increase binding site affinity, specificity, and diversity due to (i) cooperative DNA binding, (ii) the lower frequency of two hexamer-binding motifs separated by a defined spacer compared to that of single hexamers, and (iii) heterodimers that may have recognition sites distinct from those of homodimers. In this regard, RXRs play a central role in various signal transduction pathways since they can both homodimerize and act as promiscuous heterodimerization partner for almost fifteen NRs. Crystal structures of DBD and LBD homo- and heterodimers have defined the surfaces involved in dimerization. It is important to point out that the response element repertoire for receptor homo- and heterodimers is dictated by the DBD while LBDs stabilize the dimers, but do not contribute to response element selection. Two types of dimerization functions mediate homo- and heterodimerization. One involves several surface residues in the DBD that establish weak response element-specific interfaces with corresponding surfaces in the partner DBD (Fig. 2.1). The second is a strong dimerization function in the LBDs of both partners that differs between homo- and heterodimers and to some extent between the partners of RXR. We present here different procedures, which comprise both cellular and biochemical approaches, to investigate the interactions that occur between the NRs, that is, the formation of homodimers or heterodimers.

2.1 METHODS

2.1.1 Studies of NR–NR interactions through protein crystallization

One of the very effective structural methods to study protein–protein interactions is based on the preparation of crystals of the complexes followed by their analysis using X-ray beams. The determination of the high-resolution three-dimensional (3D) structures that derive from this technique provides atomic-level information on the protein–protein interface. The vast majority of NR–NR interactions are mediated by the dimerization surface located in the LBD of receptors. In this regard, several structures of homo- and heterodimers of NRs have been reported, thereby identifying the structural organization of NR dimers. In particular, crystal structures of several NRs in their homodimeric form (Bourguet, Ruff, Chambon, Gronemeyer, & Moras, 1995; Brzozowski et al., 1997; Greschik et al., 2002; Nolte et al., 1998) or containing RXR LBD in complex with various partner LBDs, including RAR (Bourguet, Vivat, et al., 2000; Pogenberg et al., 2005; Sato et al., 2010), PPAR (Gampe et al., 2000), TR (Putcha, Wright, Brunzelle, & Fernandez, 2012), LXR (Svensson et al., 2003), or CAR (Suino et al., 2004; Xu et al., 2004), have been reported. All these structures demonstrated a topologically conserved dimerization surface with identical structural elements generating the interface in homo- and heterodimers. However, amino acid variations at the surface of the various NRs determine their specific dimerization characteristics. An extensive analysis of the dimerization interfaces of NR LBDs,

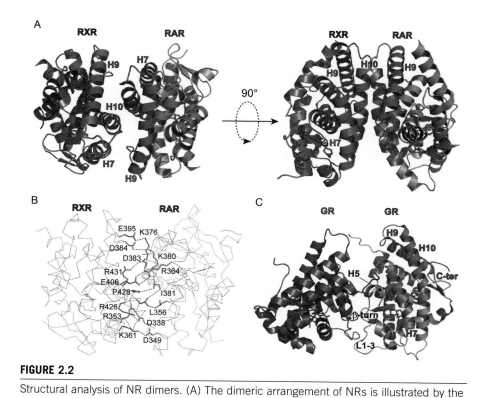

FIGURE 2.2

Structural analysis of NR dimers. (A) The dimeric arrangement of NRs is illustrated by the structure of RARα–RXRα LBD heterodimer (Protein Data Bank code 1dkf) viewed along (left) or perpendicular (right) to the dimer axis. The secondary structural elements involved in the dimer interface are displayed in orange and labeled. The ligands in both subunits are represented as yellow sticks. (B) Some important intersubunit interactions of the RARα–RXRα LBD heterodimer are shown. For clarity, not all the contacts are displayed. (C) The dimeric arrangement of the GR LBD homodimer as seen in the crystal (Protein Data Bank code 1m2z). The secondary structural elements involved in the canonical dimerization surface used by other NRs are labeled together with the C-terminal β-strand that masks a portion of this surface. The structural elements that are involved in the GR–GR interaction are indicated. (See color plate.)

based on crystal structure and sequence alignment, has been reported (Bourguet, Vivat, et al., 2000). The reader is referred to this publication for details and an interpretation of the structural basis that accounts for the homo- and heterodimerization pattern of distinct members of this family. Briefly, the structures show that the dimeric arrangements are closely related, with residues from helices H7, H9, and H10 and loops L8–9 and L9–10 of each protomer forming an interface comprising a network of complementary hydrophobic and charged residues and further stabilized by neutralized basic and acidic surfaces (Fig. 2.2A and 2.2B). Interestingly, the recently reported crystal structures of the entire PPARγ–RXRα heterodimer and HNF-4α homodimer bound to their DNA response elements fully validated the studies on the isolated LBDs (Chandra et al., 2008, 2013). Indeed, both the general dimeric organization and the details of the inter-LBD interactions observed in the LBD dimers are strictly conserved in the context of the full-length receptors. A notable exception to this conserved dimeric arrangement is the case of 3-keto-steroid receptors (namely, the androgene AR, progesterone (PR (NR3C3)), glucocorticoid (GR (NR3C1)), and mineralocorticoid (MR (NR3C2)) receptors) whose dimerization surface is partly masked by a C-terminal extension folding as a β-strand and forming a β-sheet interaction with residues located between helices H8 and H9 (Fig. 2.2C). The 3D structure of the GR LBD homodimer (Bledsoe et al., 2002) suggested an alternative mode of dimerization involving residues from a β-turn located between H5 and H6 and the extended loop between H1 and H3, as well as the last residue of H5 (Fig. 2.2C). However, formation of the GR homodimer buries only 600 $Å^2$ of solvent-accessible surface as compared with the 1000–1700 $Å^2$ of buried surfaces observed in the other NRs so that the biological relevance of this mode of dimerization remains to be confirmed.

Considering that the methods used for the crystallization of protein complexes are diverse and that crystallization conditions are unique to each protein (or protein complex), we provide here a very general procedure that may apply to all kinds of NR LBDs in complex or in isolation. The following protocol is an example given for the crystallization of the RARα–RXRα LBD heterodimers bound to BMS614, a RARα-selective antagonist.

2.1.1.1 Required materials
- Crystallization robot (X8 nanosystem, Cartesian; Freedom EVO, Tecan)
- Crystallization screens (Molecular Dimension, Hampton Research, etc.)
- 96-Well crystallization plates (Greiner Bio-One)
- 24-Well crystallization plates (Greiner Bio-One, Molecular Dimension)
- Siliconized glass coverslips (Greiner Bio-One)
- Stereo microscope (Leica)
- X-ray generator (Rigaku HF 007) with a MAResearch image plate

2.1.1.2 Protocol
Prior to crystallization trials, the purified heterodimer (see the succeeding text for the description of a general purification method of NR heterodimers) is mixed with threefold molar excess of the ligand, concentrated to 5–10 mg/mL, and centrifuged for 30 min at 13,000 rpm to get rid of any aggregated material. An initial search for crystallization conditions is undertaken by mixing the purified protein with sparse matrix screening conditions using crystallization robots and dedicated 96-well crystallization plates. This setup allows for the rapid screening of hundreds of crystallization conditions commercially available in a 96-well format with a reduced amount of protein. The robot automatically performs each crystallization trial by mixing 0.1 μL of protein with 0.1 μL of precipitant condition in a well suspended over a reservoir containing 100 μL of the corresponding condition. Once the 96 conditions have been dispensed, the plates are sealed with a transparent

plastic film and incubated at a fixed temperature, usually around 20 °C. The drops are observed on a daily basis with a stereo microscope to detect any crystal growth. Numerous crystallization kits in a 96-well format are provided by several companies. They correspond to a collection of diverse kinds of buffers, precipitants, and additives designed to cover a wide range of conditions. The number of screens to be used depends on protein availability. The crystals obtained at this step are generally small (<100 μm) or not well shaped and need to be optimized through refinement of the preliminary crystallization conditions. This refinement is performed by generating "home-made" grid screens where all the parameters of the initial hit are varied (pH, salt, concentration of precipitant, etc.). Typically, this optimization step is carried out manually in 24-well crystallization plates by mixing 1 μL of the concentrated protein with 1 μL of each crystallization condition, the reservoir containing 500 μL of the corresponding crystallization condition. The drops are suspended from siliconized glass coverslips, which are used to seal the wells with Vaseline (hanging drop method). This scale-up of the crystallization process generally provides larger crystals suitable for further crystallographic studies. Once they have reached suitable sizes, the crystal are mounted into a cryogenic nylon loop and flash frozen in liquid nitrogen to preserve it before and during the X-ray diffraction experiment. This is achieved by transferring the crystal to a solution containing the mother liquor plus 20–30% cryoprotectant (e.g., glycerol) for a few seconds before scooping up the crystal into the loop again and rapidly freezing by plunging it into liquid nitrogen. The entire process is carried out by looking through a stereo microscope. The frozen crystal is then stored in a dewar maintained at cryogenic temperature until data collection. Regarding the example of the RARα–RXRα LBD heterodimer crystallization, the initial crystals were observed under condition #38 of the Crystal Screen II (Hampton Research) comprising 20% PEG 10,000 and 0.1 M HEPES, pH 7.5. The crystals grew as small hexagonal bipyramids of $50 \times 25 \times 25$ μm^3 within 5 days. Further attempts to optimize the conditions were performed by using PEGs of various molecular weights, additives at different concentrations, and pH levels ranging between 6.0 and 8.5. Optimized crystallization conditions contained 23% PEG 10,000, 0.1 M HEPES, pH 7.25, the reservoir being composed of 20% PEG 10,000, 0.1 M HEPES, pH 7.25. Under such conditions, only a few large crystals ($500 \times 250 \times 250$ μm^3) were obtained. At this step, the structure is solved by conventional crystallographic analysis with the collection of diffraction data using in-house or synchrotron X-ray sources, followed by data processing, determination of the structure, model building, and refinement (see Bourguet, Vivat, et al., 2000 for more details and programs used).

2.1.2 Expression and purification of NR–NR complexes

The availability of homogeneous NR dimers is a prerequisite for biochemical and biophysical studies addressing NR–NR interactions. We and others have developed similar methods of overexpression and purification of high-quality NR–NR complexes (Bourguet, Andry, et al., 2000; Iyer et al., 1999). These protocols apply to both isolated LBDs, which contain the major dimerization function, and full-length receptors. In the succeeding text, we provide the example of a rapid two-step copurification procedure yielding nonaggregated and functionally homogeneous RARα–RXRα LBD heterodimers in large quantities suitable for structural and other *in vitro* studies. The purification strategy is based on the observation that RARα is always isolated as a monomer, whereas RXRα exists as a mixture of monomers, homodimers, and homotetramers (Chen et al., 1998). Therefore, a histidine tag is added to RARα (His-RARα) only but not to RXRα. Upon mixing the extracts of *Escherichia coli* programmed to express His-RARα and RXRα, His-RARα–RXRα heterodimers are formed. In the subsequent immobilized metal affinity chromatography, only His-RARα monomers and His-RARα–RXRα heterodimers are recovered, while excess of RXRα does not bind to the nickel resin. Next, the excess of His-RARα monomers and aggregated proteins is easily removed by a gel-filtration step. The results of a typical purification procedure are shown in Fig. 2.3.

2.1.2.1 Required materials
- pET-15b and pET-3a expression vectors (Novagen)
- Competent *E. coli* BL21(DE3) cells (Novagen)
- Shaker/incubator set at 37 °C
- Sonicator
- Liquid chromatography system (AKTA purifier, GE Healthcare)
- HisTrap chelating columns (GE Healthcare)
- HiLoad 26/60 Superdex 75 column (GE Healthcare)
- UV–Vis spectrophotometer (NanoDrop, Thermo Scientific)

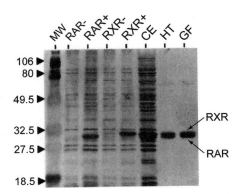

FIGURE 2.3

Generic purification of RARα–RXRα LBD heterodimers. Representative SDS–polyacrylamide gel of noninduced and IPTG-induced bacteria (RARα LBD+/− and RXRα LBD+/−; lanes 2–5), crude extract (CE; lane 6), pool of the heterodimer eluted from the Ni^{2+}-charged HisTrap chelating column (HT; lane 7), and gel-filtration pool (GF; lane 8). MW, molecular weight markers (in kDa).

2.1.2.2 Protocol

The RARα LBD and the RXRα LBD cloned, respectively, into the pET-15b vector (as a histidine-tagged protein) or into the pET-3a vector are expressed in *E. coli* cells. For both transformants, cells are grown on LB plates containing 50 µg ampicillin/mL. A 3 mL LB starter culture containing 200 µg ampicillin/mL is inoculated with RAR LBD/pET-15b- or RXR LBD/pET-3a-transformed BL21(DE3) grown overnight at 37 °C, and 12 mL is used to inoculate 500 mL of LB containing 200 µg ampicillin/mL. Cultures are grown at 37 °C to an OD600 of 0.4–0.5, and expression of T7 RNA polymerase is induced by the addition of IPTG to 1 mM. After an additional incubation for 2.5 h at 25 °C, cells are harvested by centrifugation. The pellets from 3 L of RARα LBD and 2 L of RXRα LBD cultures are suspended together in 62.5 mL of ice-cold buffer A (5 mM imidazole, 500 mM NaCl, 20 mM Tris–HCl, pH 8.0). The suspension is sonicated for 2 min, diluted with 62.5 mL of ice-cold buffer A, and sonicated again. The crude extract is obtained by centrifugation for 30 min at 40,000 rpm. A 5 mL HisTrap chelating column is equilibrated with 10 volumes of sterile deionized water, 50 mM NiSO4, sterile deionized water, and finally 10 volumes of buffer A. The crude extract is passed over the column at a flow rate of 5 mL/min, followed by washing with 20 volumes of buffer A and 20 volumes of 50 mM imidazole in buffer A. The heterodimer is eluted with 20 volumes of buffer B (150 mM imidazole, 500 mM NaCl, 20 mM Tris–HCl, pH 8.0). Fractions are then analyzed by SDS–polyacrylamide gel electrophoresis, pooled, dialyzed against 5 mM dithiothreitol, 150 mM NaCl, and 10 mM Tris–HCl, pH 7.5, and concentrated. The heterodimer is further purified by gel filtration using a HiLoad 26/60 Superdex 75 column preequilibrated with the same buffer. Fractions of 2 mL are analyzed by SDS–polyacrylamide gel electrophoresis, pooled, and concentrated to 5 mg/mL (Centriprep 30) for subsequent *in vitro* analysis. At this step, the structural homogeneity of the preparations is evaluated by electrophoretic analysis using nondenaturating conditions (native gels), dynamic light scattering (DLS), or circular dichroism (CD).

2.1.3 Monitoring NR–NR interactions by noncovalent electrospray ionization mass spectrometry

Since the 1990s, electrospray ionization mass spectrometry (ESI-MS) has been regularly used to study noncovalent complexes and offers new possibilities for the study of such complexes, providing direct evidence for their formation and an accurate determination of their binding stoichiometry. For instance, ESI-MS was previously used to detect the ERα LBD in its dimeric noncovalently bound form (Witkowska et al., 1997; Witkowska, Green, Carlquist, & Shackleton, 1996) or RXRα–RARα heterodimers (Sanglier et al., 2004). Hence, supramolecular mass spectrometry is a powerful tool to rapidly and unambiguously determine if two NRs can interact. The following protocol is an example given for the use of ESI-MS under nondenaturating conditions to monitor the interaction between the RXRα and RARα LBDs (Fig. 2.4).

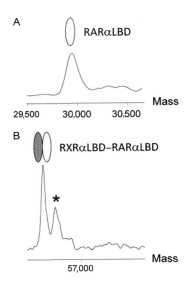

FIGURE 2.4

(A) Positive ESI mass spectra of RARα LBD. (B) Positive ESI mass spectra of RXRα LBD–RARα LBD. The mass spectra are acquired at $V_c = 50$ V and at a pressure in the interface of 2.5 mbar. The following molecular weights are measured: $29,930 \pm 1.8$ Da for RARα LBD and $56,664 \pm 0.3$ Da for RXRα LBD–RARα LBD (with deletion of the N-terminal methionine in RXRα LBD; peak labeled with an asterisk corresponds to species with an additional N-terminal methionine in RXRα). These results are in agreement with the molecular weights calculated from the known amino acid sequences.

2.1.3.1 Required materials
- Electrospray time-of-flight mass spectrometer (LCT, Waters)
- Buffer A: 50 mM ammonium acetate pH 6.5
- Purified RXRα–RARα LBD heterodimer at 5–10 mg/mL (see the preceding text for the description of a purification method)

2.1.3.2 Protocol
The instrument is calibrated using the multiply charged ions produced by an injection of horse heart myoglobin diluted to 2 pmol/mL in a water/acetonitrile mixture (1:1, v/v) acidified with 1% (v/v) formic acid. Prior to ESI-MS analysis, samples are desalted on Centricon PM30 microconcentrators (Amicon, Millipore) in buffer A. Purity and homogeneity of the samples in denaturing conditions are verified by diluting the complex solution to 5 pmol/mL in a water/acetonitrile mixture (1:1, v/v) acidified with 1% (v/v) formic acid. Spectra are recorded in the positive ion mode on the mass range 500–2500 m/z. Verify that the measured molecular masses are in agreement with those calculated from the amino acid sequences. Samples are diluted to 10 pmol/mL in buffer A and are continuously infused into the ESI ion source at a flow rate of 6 mL/min through a Harvard syringe pump.

The accelerating voltage (V_c) must be set to 50 V in order to preserve ternary complex formation and good mass accuracy. ESI-MS data are acquired in the positive ion mode on the mass range 1000–5000 m/z. The relative abundance of the different species present on ESI mass spectra from their respective peak intensities is measured, assuming that the relative intensities displayed by the different species reflect the actual distribution of these species in solution.

2.1.4 Monitoring NR–NR interactions by electrophoretic mobility shift assays

The ability of NRs to bind to DNA as homodimers and heterodimers can be assayed by electrophoretic mobility shift assays (EMSAs). Solutions of protein and nucleic acid are combined and the resulting mixtures are subjected to electrophoresis under native conditions through polyacrylamide gel. After electrophoresis, the distribution of species containing nucleic acid is determined, usually by autoradiography of ^{32}P-labeled nucleic acid. In general, protein–nucleic acid complexes migrate more slowly than the corresponding free nucleic acid. A major advantage of studying protein–DNA interactions by an electrophoretic assay is the ability to resolve complexes of different stoichiometry or conformation. It is also useful for studying higher-order complexes containing several proteins, observed as a "supershift" assay. Another advantage is that target proteins may be obtained from a crude nuclear or whole cell extract, *in vitro* transcription product, or a purified preparation from *E. coli* or mammalian expression systems. On the other hand, identification of the protein bound to the probe can be accomplished by adding specific antibody in the binding reaction. If the protein of interest binds to the target DNA, the antibody will form an antibody–protein–DNA complex, further reducing its mobility relative to unbound DNA. In some cases, the antibody may disrupt the protein–DNA interaction, resulting in loss of the characteristic shift but not causing a "supershift." Here, a protocol is given as example of EMSA to determine the effect of a single mutation of tyrosine 402 to alanine in helix H9 of RXRα on the RXRα dimerization function. This mutation of RXRα substantially weakens RARα heterodimerization while concomitantly increasing homodimerization (Vivat-Hannah, Bourguet, Gottardis, & Gronemeyer, 2003; Fig. 2.5).

2.1.4.1 Required materials
- TNT rabbit reticulocyte lysate system (Promega)
- ^{32}P-DR5 (5′-TCGAGGGTAG<u>GGGTCA</u>CCGAA<u>AGGTCA</u>CTCG-3′; direct repeat underlined) oligonucleotide
- Binding buffer: (10 mM Tris–HCl (pH 8.0), 0.1 mM EDTA, 0.4 mM DTT, 5% glycerol)

2.1.4.2 Protocol
2.1.4.2.1 *In vitro* transcription–translation
DNA-binding proteins generated by cell-free expression systems offer an excellent format for performing EMSA assays. Recombinant proteins (mouse RXRα, mutant mouse RXRαY402A, and human RARα) are prepared by *in vitro* transcription–translation using the TNT rabbit reticulocyte lysate system programmed with

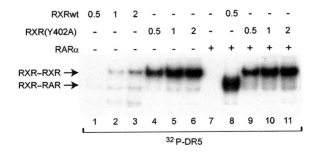

FIGURE 2.5

Increase-gain of binding affinity of RXRY402A homodimers on DR5 response element and parallel loss-impairment of heterodimerization of RXRY402A with RARα as shown by EMSA. RXRY402A homodimers bind to DR5 elements more efficiently than do RXR (compare lanes 1–3 with lanes 4–6), while RXR–RAR heterodimerization is strongly disfavored when RXR is replaced by RXRY402A (compare lane 8 with lanes 9–11). The indicated volume (microliters) of RXR proteins is incubated with or without 0.2 µL of *in vitro*-translated mRARα for 15 min at 4 °C before ^{32}P-DR5 oligonucleotide is added. Upper bands, homodimers; lower bands, heterodimers.

pSG5-based expression vectors according to the instructions provided by the manufacturer.

2.1.4.2.2 Electrophoretic mobility shift assay

Proteins (volumes of TNT reaction products containing the receptor are indicated in the legend of the Fig. 2.5) and 10 fmol preannealed DR5 that is labeled with ^{32}P (125,000 c.p.m.) are incubated on ice for 15 min in a final volume of 20 µL binding buffer, containing 2 µg poly(dI-dC)(dI-dC), and 150 mM KCl and 20 mg/mL BSA. The protein–DNA complexes are resolved through 6% polyacrylamide gels (0.5 × TBE buffer; prerun for 2 h) at 10 V/cm and 4 °C. The gel is dried for at least 45 min and exposed to X-ray film for desired period of time.

2.1.5 Two-hybrid assays to define NR–NR interactions in living cells

The mammalian two-hybrid system is a very powerful tool to investigate protein–protein interactions in terms of functional domains and identify partners of a protein in a cellular environment. In these assays, a chimeric luciferase-based reporter gene ((17-mer)$_{5x}$-βGlobin-Luc) is transfected together with two expression vectors. For instance, in the case of RXRα–RARα heterodimer, one expresses a fusion protein that binds through the Gal DNA-binding domain (Gal) to the pentamer of the "17-m" DNA recognition site in the reporter gene and contains the LBD of RXRα, which can interact with RARα. The other vector expresses a second fusion protein composed of the VP16 acidic transcription activation domain (VP16), a domain that confers constitutive transcription activation if it is brought close to a promoter, and the LBD of RARα. If RARα interacts with RXRα, this results in (indirect) recruitment of the VP16 activation domain to the promoter of the luciferase reporter gene, thereby inducing transcription and luciferase protein production that is quantitated in a luminometer. The following protocols are examples given to determine by a mammalian two-hybrid system the effect of the single mutation introduced into the dimerization interface of RXRα (tyrosine 402 to alanine in helix H9) on the dimerization function of this receptor, namely, the formation of heterodimers with RARα and TRα (NR1A1). The study led to the conclusion that the RXRαY402A mutant exhibits loss of heterodimerization abilities (Vivat-Hannah, Bourguet, Gottardis, & Gronemeyer, 2003; Fig. 2.6).

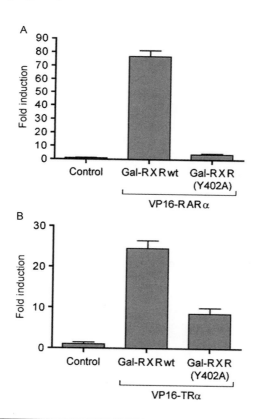

FIGURE 2.6

Example of mammalian two-hybrid analysis that reveals impaired heterodimerization of mRXRαY402A with RARα and TRα in live cells. While the heterodimerization of RXRα–RARα (A) and RXRα–TRα (B) is obvious from the strong two-hybrid signals (wt bars), this signal is ~95% decreased for RARα when Gal-RXRαY402A is used (Y402A bars). TRα shows a significant, albeit 65% reduced, interaction with the Y402 mutant. The control corresponds to the reporter gene activity transfected alone. The transcriptional activity exerted by each heterodimer is expressed as fold induction of reporter gene activity. The results shown are the averages ± standard errors of the means of three independent experiments.

2.1.5.1 Required materials
- HeLa cells; cells have to be maintained in DMEM containing 5% charcoal-stripped fetal calf serum (FCS)
    - Luciferase-based reporter gene ((17-mer)$_{5x}$-βGlobin-Luc)
    - Gal-mRXRα, VP16-mRARα, and VP16-cTRα receptor chimera expression vectors
- Lysis buffer: 25 mM Tris–phosphate (pH 7.8), 2 mM EDTA, 1 mM DTT, 10% glycerol, and 1% Triton X-100
- Luciferin buffer: 20 mM Tris–phosphate (pH 7.8), 1.07 mM $MgCl_2$, 2.67 mM $MgSO_4$, 0.1 mM EDTA, 33.3 mM DTT
- Opaque white Optiplate-96-well plates (Perkin Elmer)
- MicroLumat LB96P luminometer (Berthold)

2.1.5.2 Protocol
2.1.5.2.1 Transient transfection of HeLa cells
10^5 cells are seeded per P24 well 2 h before the transfection with jetPEI reagent (Ozyme, Saint-Quentin-en-Yvelines, France). Per well, add the plasmid mix diluted in 50 μL of 150 mM NaCl buffer. The mix contains 50 ng of Gal4 chimera (e.g., Gal-mRXRα wild-type (wt) or mutant expression vectors Gal-mRXRαY402A), 40 ng of VP16 chimera (e.g., VP16-hRARα or VP16-cTRα expression vectors), 200 ng of (17-mer)$_{5x}$-βglob-luc, and 50 ng of cytomegalovirus–β-galactosidase (CMVβgal; used as an internal control to normalize for variations in the transfection efficiency). The total quantity of DNA was adjusted at 1 μg with pBluescript plasmid. Vortex gently and spin down briefly. Per well, 2 μL of jetPEI is diluted in 150 mM NaCl buffer to a final volume of 50 μL (N/P=5). Vortex gently and spin down briefly. The 50 μL jetPEI solution is added to the 50 μL DNA solution all at once (do not mix the solution in the reverse order). Vortex-mix the solution immediately and spin down briefly. Incubate for 15–20 min at room temperature. Add the 100 μL jetPEI/DNA mixture dropwise onto the cells in 1 mL of medium in each well and homogenize by gently swirling the plate. The cells are incubated at 37 °C in a 5% CO_2 incubator for 24 h. Culture medium is removed from the wells and replaced with fresh medium. The cells are incubated at 37 °C in a 5% CO_2 incubator again for 24 h. Three wells per assay are transfected to have triplicate measurements.

2.1.5.2.2 Cell lysis
The cell lysis is performed with the 5× passive lysis buffer from Promega. A sufficient quantity of 1× working concentration is prepared by adding 1 volume of 5× passive lysis buffer to 4 volumes of distilled water. The growth medium is removed from the cells and a sufficient volume of phosphate saline buffer (PBS) is gently applied to wash the surface of the well. Swirl briefly and completely remove the PBS from the well. Dispense 100 μL of 1× passive lysis buffer into each well. The culture plate is placed on an orbital shaker with gentle rocking at room temperature for 20 min. Then, the lysate is transferred to a 96-well plate.

2.1.5.2.3 Preparation of luciferase assay

50 μL per sample + 3 mL of luciferase assay reagent are prepared just before use. For 1 mL, the luciferase assay reagent contains 500 μL 2× luciferase buffer (40 mM Tris–phosphate, 2.14 mM $MgCl_2$, 5.4 $MgSO_4$, 0.2 mM EDTA, 66.6 mM DTT, pH 7.8; stable at $-20\,°C$), 47 mL luciferin 10^{-2} M, 53 mL ATP 10^{-2} M, 27 mL coenzyme A (lithium salt) 10^{-2} M, and 430 mL distilled water. This buffer is light-sensitive.

2.1.5.2.4 Luciferase measurement

10 μL of lysate is distributed per well of a 96-well white plate. Wash the injector and prime with the luciferase assay reagent. The luminometer is programmed to perform 2 s premeasurement delay followed by a 10 s measurement period with 50 μL luciferase assay reagent. The measurement is performed and the luciferase activity measurement is recorded.

2.1.5.2.5 Measurement of βgal activity

Per sample, the preparation of βgal assay buffer requires the mix of 150 μL βgal buffer (Na_2HPO_4 ($12H_2O$) 60 mM, NaH_2PO_4 40 mM, KCl 10 mM, $MgCl_2$ ($6H_2O$) 1 mM, and β-mercaptoethanol 50 mM) and 30 μL ONPG (4 mg/mL). 10 mL lysate is distributed in a 96-well plate and 180 μL of βgal assay buffer is added. The plate is incubated at 37 °C and the time until the yellow color has developed is measured. The reaction is stopped by adding 75 μL of 1 M Na_2CO_3 (bubbles have to be avoided). The absorbance is immediately read at 420 nm in a plate reader (βgal unit = 100 × total volume × absorbance/(assay volume × time (in hours))).

2.1.5.2.6 Normalization

The luciferase values are normalized for transfection efficiency with the βgal values.

2.1.6 Fluorescence cross-correlation spectroscopy to measure the concentrations and interactions of NRs in living cells

Novel fluorescence microscopy techniques have advanced the frontiers of quantitative measurements in live cells. Among these approaches, fluorescence fluctuation and correlation spectroscopy is especially appealing for situations in which one wishes to measure the degree of interaction between two protein partners. This approach involves simultaneous excitation, using a variety of strategies, among them a femtosecond-pulsed infrared (IR) laser for simultaneous two-photon excitation of two biomolecules bearing two fluorescent dyes (Bacia, Kim, & Schwille, 2006). The time correlation of their intensity in two different detection channels, along with the autocorrelation of their individual fluorescence in single channels, allows calculation of the total concentration of each species, free and bound, and, hence, the degree of complex formation between two proteins. These measurements can also allow for the determination of complex stoichiometries. Two-photon, two-color, fluorescence cross-correlation spectroscopy (FCCS) in transiently transfected cells

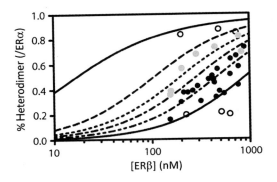

FIGURE 2.7

Degree of heterodimer formation by FCCS with respect to the concentration of total ERα (expressed in terms of monomer) as a function of the concentration of mCherry-ERβ (also expressed in terms of monomer). The concentrations of the homo- and heterodimers are calculated by solving the set of equations for the Go values of the auto- and cross-correlation functions as described in the Protocol section. Lines correspond to simulations assuming homo- and heterodimerization constants of 7 and 2.8 nM and 10, 100, 200, 300, 500, and 1000 nM ERα in descending order. Black circles correspond to experimental results for which the ERα concentration falls between the closest corresponding simulation lines. The empty circles correspond to points for which the ERβ concentration falls well beyond the corresponding simulation lines. The gray circles correspond to points for which the ERβ concentrations are about twofold higher than those predicted by the simulations.

by cerulean and mCherry fusions of ERα (NR3A1) and ERβ (NR3A2) can be used to measure the heterodimerization between the two ER subtypes. Using this approach in a systematic fashion, it can be observed that coexpression of the two ERs revealed substantial receptor heterodimer formation (Fig. 2.7). The results allow for the estimation of the affinities between these protein complexes for the full-length proteins in a live-cell environment. Estimated homo- and heterodimerization constants are found to be similar and in the low nanomolar range.

2.1.6.1 Required materials
- COS-7 cells are maintained in Dulbecco's modified Eagle's–F12 medium (DMEM:F12) supplemented with 10% FCS (Invitrogen, Carlsbad, CA) and an antibiotic solution (Invitrogen)
- Lab-Tek chambered coverglass two-well dishes (Nunc Thermo Fisher Scientific, Roskilde, Denmark)

2.1.6.2 Protocol
2.1.6.2.1 Cell culture and transfection
For imaging, COS-7 cells are deprived of steroids for 5 days in phenol red-free medium supplemented with 3% dextran-coated charcoal-stripped serum (DCC–DMEM:F12), plated in Lab-Tek chambered coverglass two-well dishes at

3×10^5 cells/well, and transfected after 1 day using jetPEI according to the instructions provided by the manufacturer (Ozyme, Saint-Quentin-en-Yvelines, France). After 24 h and at least 2 h before imaging, the medium is replaced two times by PBS and then by DCC–DMEM:F12. pH is stabilized by 10 mM HEPES buffer (Gibco, Paisley, Scotland). Cells are then kept at 37 °C before imaging.

2.1.6.2.2 Fluorescence cross-correlation microscopy

FCCS measurements are conducted using a setup based on a dual-channel ISS Alba fluorescence correlation detector (ISS, Champaign IL) with avalanche photodiodes and a Zeiss Axiovert 200 microscope (Zeiss, Jena, Germany). The sample is excited by means of a Mai Tai HP fs IR tunable laser (Spectra-Physics, Newport, Mountain View, CA) tuned to 935 or 950 nm. The excitation light is focused into the sample by a Zeiss Apochromat 63X oil immersion objective (numerical aperture 1.4), and an E700 SP 2P dichroic filter (Chroma Technology Corporation, Rockingham, VT) is used to eliminate the contribution of the IR excitation light in the detected signal. A 505 nm dichroic mirror is used to split the detected light onto two channels, and additional 653 ± 50 and 455 ± 50 nm band pass filters are set before channels 1 and 2, respectively (Chroma Technology Corporation, Rockingham, VT). The excitation power is set at <10 mW at the scope entrance in all FCCS measurements. This power is determined to be a threshold, one at which the autocorrelation traces became excitation power-independent, in order to avoid excitation saturation effects (Nagy, Wu, & Berland, 2005a,2005b) and photobleaching (Petrasek & Schwille, 2008). The microscope is equipped with a Piezoelectric stage (Mad City, Madison, WI) for 3D imaging.

The two ERs are expressed in COS-7 cells as N-terminal fusions with the blue FP, cerulean (cer-ERα and mCherry-ERβ). More importantly, previous results (Savatier, Jalaguier, Ferguson, Cavailles, & Royer, 2010) for expression of cer-ERα alone showed no bleed-through from the blue detection channel into the red channel. This is important, since such bleed-through is 100% cross-correlated and, hence, would introduce a substantial artifact into the FCCS measurements. Reasonable detection levels for both cerulean and mCherry are achieved with two-photon excitation at 950 nm, using <10 mW at the microscope entrance to limit photobleaching. Under these excitation conditions, cer-ERα exhibits a molecular brightness of 3000 cpspm (dimer); that of cer-ERβ is slightly higher, 4500 cpspm (dimer).

2.1.6.2.3 Data analysis

Three curves are generated by time correlation of the fluorescence intensity fluctuations detected, two autocorrelation functions arising from the blue and red fluorophores $G_B(\tau)$ and $G_R(\tau)$, and one cross-correlation function, which accounts for the molecules in which the two fluorophores diffuse together $G_x(\tau)$. The autocorrelation and cross-correlation functions can be written as follows, in the absence of cross talk:

$$G_i(\tau) = \frac{\langle \delta F(t) \delta F(t+\tau) \rangle}{\langle F(t) \rangle^2} \quad (2.1)$$

$$G_x(\tau) = \frac{\langle \delta F_B(t) \delta F_R(t+\tau) \rangle}{\langle F_B(t) \rangle \langle F_R(t) \rangle} \quad (2.2)$$

where τ is the lag time. In the particular case of a system of freely diffusing species and assuming a 3D Gaussian excitation profile, closed expressions are derived from Eqs. (2.1) and (2.2):

$$G_i(\tau) = G_i(0) \left(1 + \frac{\tau}{\tau_{Di}}\right)^{-1} \left(1 + \frac{\omega_0^2 \tau}{z_0^2 \tau_{Di}}\right)^{-1/2} \quad (2.3)$$

$$G_x(\tau) = G_x(0) \left(1 + \frac{\tau}{\tau_{Dx}}\right)^{-1} \left(1 + \frac{\omega_0^2 \tau}{z_0^2 \tau_{Dx}}\right)^{-1/2} \quad (2.4)$$

where $i = $ B or R (blue or red). ω_0 and z_0 are the waist and length, respectively, of the 3D Gaussian excitation volume at which the intensity drops to $1/e^2$. Here, τ_{Di} and τ_{Dx} are the translational diffusion time of species i and of the complex, respectively, and $G_i(0)$ and $G_x(0)$ are the amplitudes of the auto- and cross-correlation functions, respectively.

If no changes in the brightness of the fluorophores occur upon complexation, the amplitudes of the auto- and cross-correlation curves can be expressed as follows:

$$G_i(0) = \frac{1}{V_{eff} C_{i,t}} \quad (2.5)$$

$$G_x(0) = \frac{C_x}{V_{eff} C_{G,t} C_{R,t}} \quad (2.6)$$

where $i = $ B or R, $V_{eff} = (\pi/2)^{3/2} \omega_0^2 z_0$, and $C_{i,t}$ and C_x are the total concentration of species i and of the complex, respectively.

In general, the amplitudes of the auto- and cross-correlation curves for similarly diffusing species of different brightness can be expressed as

$$G_B(0) = \frac{\sum_i \eta_{i,B}^2 C_i}{V_{eff} \left(\sum_i \eta_{i,B} C_i\right)^2} \quad (2.7)$$

$$G_R(0) = \frac{\sum_i \eta_{i,R}^2 C_i}{V_{eff} \left(\sum_i \eta_{i,R} C_i\right)^2} \quad (2.8)$$

$$G_x(0) = \frac{\sum_i \eta_{i,B} \eta_{i,R} C_i}{V_{eff} \left(\sum_i \eta_{i,B} C_i\right) \left(\sum_i \eta_{i,R} C_i\right)} \quad (2.9)$$

In the case of cer-ERα and mCherry-ERβ, it is necessary to take the brightness into account in the determination of concentrations from the correlation function amplitudes. We assume three diffusing species. The cer-ERα–mCherry-ERβ heterodimer has concentration $C_{\alpha\beta}$ and brightness $\eta_{B,\alpha\beta}$ and $\eta_{R,\alpha\beta}$. The cer-ERα homodimer has concentration $C_{2\alpha}$ and brightness $\eta_{B,2\alpha}$. The mCherry-ERβ homodimer has concentration $C_{2\beta}$ and brightness $\eta_{R,2\beta}$. Since total concentrations measured are above the expected nanomolar binding affinity for both hetero and homodimers, cer-ERα and mCherry-ERβ monomers are not taken into account. The brightness of the homodimer is taken to be twice that of the heterodimer for each detector. With this assumption, the equations reduce to

$$G_B(0) = \frac{(C_{\alpha\beta} + 4C_{2\alpha})}{V_{eff}(C_{\alpha\beta} + 2C_{2\alpha})^2} \quad (2.10)$$

$$G_R(0) = \frac{(C_{\alpha\beta} + 4C_{2\beta})}{V_{eff}(C_{\alpha\beta} + 2C_{2\beta})^2} \quad (2.11)$$

$$G_x(0) = \frac{C_{\alpha\beta}}{V_{eff}(C_{\alpha\beta} + 2C_{2\alpha})(C_{\alpha\beta} + 2C_{2\beta})} \quad (2.12)$$

Equations (2.10)–(2.12) are solved using the function SOLVE in Maxima Version 5.18.1 to determine the concentrations of each species. Heterodimer and homodimer dissociation constants are estimated using the BIOEQS software by global analysis (randomly by groups of 8 data sets) (Royer, Smith, & Beechem, 1990).

Acknowledgments

We thank Sarah Cianferani-Sanglier for the ESI-MS protocol and Catherine Royer for the FCCS protocol.

References

Bacia, K., Kim, S. A., & Schwille, P. (2006). Fluorescence cross-correlation spectroscopy in living cells. *Nature Methods*, *3*(2), 83–89.

Bledsoe, R. K., Montana, V. G., Stanley, T. B., Delves, C. J., Apolito, C. J., McKee, D. D., et al. (2002). Crystal structure of the glucocorticoid receptor ligand binding domain reveals a novel mode of receptor dimerization and coactivator recognition. *Cell*, *110*(1), 93–105.

Bourguet, W., Andry, V., Iltis, C., Klaholz, B., Potier, N., Van Dorsselaer, A., et al. (2000). Heterodimeric complex of RAR and RXR nuclear receptor ligand-binding domains: Purification, crystallization, and preliminary X-ray diffraction analysis. *Protein Expression and Purification*, *19*(2), 284–288.

Bourguet, W., Ruff, M., Chambon, P., Gronemeyer, H., & Moras, D. (1995). Crystal structure of the ligand-binding domain of the human nuclear receptor RXR-alpha. *Nature*, *375*(6530), 377–382.

Bourguet, W., Vivat, V., Wurtz, J. M., Chambon, P., Gronemeyer, H., & Moras, D. (2000). Crystal structure of a heterodimeric complex of RAR and RXR ligand-binding domains. *Molecular Cell, 5*(2), 289–298.

Brzozowski, A. M., Pike, A. C., Dauter, Z., Hubbard, R. E., Bonn, T., Engstrom, O., et al. (1997). Molecular basis of agonism and antagonism in the oestrogen receptor. *Nature, 389*(6652), 753–758.

Chandra, V., Huang, P., Hamuro, Y., Raghuram, S., Wang, Y., Burris, T. P., et al. (2008). Structure of the intact PPAR-gamma-RXR-nuclear receptor complex on DNA. *Nature, 456*(7220), 350–356.

Chandra, V., Huang, P., Potluri, N., Wu, D., Kim, Y., & Rastinejad, F. (2013). Multidomain integration in the structure of the HNF-4alpha nuclear receptor complex. *Nature, 495*(7441), 394–398.

Chen, Z. P., Iyer, J., Bourguet, W., Held, P., Mioskowski, C., Lebeau, L., et al. (1998). Ligand- and DNA-induced dissociation of RXR tetramers. *Journal of Molecular Biology, 275*(1), 55–65.

Gampe, R. T., Jr., Montana, V. G., Lambert, M. H., Miller, A. B., Bledsoe, R. K., Milburn, M. V., et al. (2000). Asymmetry in the PPARgamma/RXRalpha crystal structure reveals the molecular basis of heterodimerization among nuclear receptors. *Molecular Cell, 5*(3), 545–555.

Germain, P., Staels, B., Dacquet, C., Spedding, M., & Laudet, V. (2006). Overview of nomenclature of nuclear receptors. *Pharmacological Reviews, 58*(4), 685–704.

Greschik, H., Wurtz, J. M., Sanglier, S., Bourguet, W., van Dorsselaer, A., Moras, D., et al. (2002). Structural and functional evidence for ligand-independent transcriptional activation by the estrogen-related receptor 3. *Molecular Cell, 9*(2), 303–313.

Gronemeyer, H., Gustafsson, J. A., & Laudet, V. (2004). Principles for modulation of the nuclear receptor superfamily. *Nature Reviews Drug Discovery, 3*(11), 950–964.

Iyer, J., Bonnier, D., Granger, F., Iltis, C., Andry, V., Schultz, P., et al. (1999). Versatile copurification procedure for rapid isolation of homogeneous RAR–RXR heterodimers. *Protein Expression and Purification, 16*(2), 308–314.

Khorasanizadeh, S., & Rastinejad, F. (2001). Nuclear–receptor interactions on DNA-response elements. *Trends in Biochemical Sciences, 26*(6), 384–390.

Nagy, A., Wu, J., & Berland, K. M. (2005a). Characterizing observation volumes and the role of excitation saturation in one-photon fluorescence fluctuation spectroscopy. *Journal of Biomedical Optics, 10*(4), 44015.

Nagy, A., Wu, J., & Berland, K. M. (2005b). Observation volumes and {gamma}-factors in two-photon fluorescence fluctuation spectroscopy. *Biophysical Journal, 89*(3), 2077–2090.

Nolte, R. T., Wisely, G. B., Westin, S., Cobb, J. E., Lambert, M. H., Kurokawa, R., et al. (1998). Ligand binding and co-activator assembly of the peroxisome proliferator-activated receptor-gamma. *Nature, 395*(6698), 137–143.

Petrasek, Z., & Schwille, P. (2008). Precise measurement of diffusion coefficients using scanning fluorescence correlation spectroscopy. *Biophysical Journal, 94*(4), 1437–1448.

Pogenberg, V., Guichou, J. F., Vivat-Hannah, V., Kammerer, S., Perez, E., Germain, P., et al. (2005). Characterization of the interaction between retinoic acid receptor/retinoid X receptor (RAR/RXR) heterodimers and transcriptional coactivators through structural and fluorescence anisotropy studies. *Journal of Biological Chemistry, 280*(2), 1625–1633.

Putcha, B. D., Wright, E., Brunzelle, J. S., & Fernandez, E. J. (2012). Structural basis for negative cooperativity within agonist-bound TR:RXR heterodimers. *Proceedings of the National Academy of Sciences of the United States of America, 109*(16), 6084–6087.

Royer, C. A., Smith, W. R., & Beechem, J. M. (1990). Analysis of binding in macromolecular complexes: A generalized numerical approach. *Analytical Biochemistry*, *191*(2), 287–294.

Sanglier, S., Bourguet, W., Germain, P., Chavant, V., Moras, D., Gronemeyer, H., et al. (2004). Monitoring ligand-mediated nuclear receptor–coregulator interactions by noncovalent mass spectrometry. *European Journal of Biochemistry*, *271*(23–24), 4958–4967.

Sato, Y., Ramalanjaona, N., Huet, T., Potier, N., Osz, J., Antony, P., et al. (2010). The "Phantom Effect" of the Rexinoid LG100754: Structural and functional insights. *PLoS One*, *5*(11), e15119.

Savatier, J., Jalaguier, S., Ferguson, M. L., Cavailles, V., & Royer, C. A. (2010). Estrogen receptor interactions and dynamics monitored in live cells by fluorescence cross-correlation spectroscopy. *Biochemistry*, *49*(4), 772–781.

Suino, K., Peng, L., Reynolds, R., Li, Y., Cha, J. Y., Repa, J. J., et al. (2004). The nuclear xenobiotic receptor CAR: Structural determinants of constitutive activation and heterodimerization. *Molecular Cell*, *16*(6), 893–905.

Svensson, S., Ostberg, T., Jacobsson, M., Norstrom, C., Stefansson, K., Hallen, D., et al. (2003). Crystal structure of the heterodimeric complex of LXRalpha and RXRbeta ligand-binding domains in a fully agonistic conformation. *EMBO Journal*, *22*(18), 4625–4633.

Vivat-Hannah, V., Bourguet, W., Gottardis, M., & Gronemeyer, H. (2003). Separation of retinoid X receptor homo- and heterodimerization functions. *Molecular Cell. Biology*, *23*(21), 7678–7688.

Witkowska, H. E., Carlquist, M., Engstrom, O., Carlsson, B., Bonn, T., Gustafsson, J. A., et al. (1997). Characterization of bacterially expressed rat estrogen receptor beta ligand binding domain by mass spectrometry: Structural comparison with estrogen receptor alpha. *Steroids*, *62*(8–9), 621–631.

Witkowska, H. E., Green, B. N., Carlquist, M., & Shackleton, C. H. (1996). Intact noncovalent dimer of estrogen receptor ligand-binding domain can be detected by electrospray ionization mass spectrometry. *Steroids*, *61*(7), 433–438.

Xu, R. X., Lambert, M. H., Wisely, B. B., Warren, E. N., Weinert, E. E., Waitt, G. M., et al. (2004). A structural basis for constitutive activity in the human CAR/RXRalpha heterodimer. *Molecular Cell*, *16*(6), 919–928.

CHAPTER 3

Network Analysis to Uncover the Structural Communication in GPCRs

Francesca Fanelli*,†, Angelo Felline*,†, and Francesco Raimondi*,†

Dulbecco Telethon Institute (DTI), Rome, Italy
†*Department of Chemistry, University of Modena and Reggio Emilia, Modena, Italy*

CHAPTER OUTLINE

Introduction 44
3.1 Materials 45
3.2 Methods 45
    3.2.1 Workflow of the PSN–MD and PSN–ENM Approaches 45
    3.2.2 Building the PSG 46
    3.2.3 Search for the Shortest Communication Paths 48
3.3 Discussion 50
Summary 56
Acknowledgments 59
References 59

Abstract

Protein structure network (PSN) analysis is one of the graph theory-based approaches currently used to investigate the structural communication in biomolecular systems. Information on system dynamics can be provided by atomistic molecular dynamics simulations or coarse-grained Elastic Network Models paired with Normal Mode Analysis (ENM–NMA).

This chapter describes the application of PSN analysis to uncover the structural communication in G protein-coupled receptors (GPCRs). Strategies to highlight changes in structural communication upon misfolding mutations, dimerization, and activation are described.

Focus is put on the ENM–NMA-based strategy applied to the crystallographic structures of rhodopsin in its inactive (dark) and signaling active (meta II (MII)) states, highlighting clear changes in the PSN and the centrality of the retinal chromophore in differentiating the inactive and active states of the receptor.

INTRODUCTION

G protein-coupled receptors (GPCRs) function rests on intramolecular and intermolecular structural communication (reviewed in Fanelli & De Benedetti, 2005). Indeed, GPCRs regulate most aspects of cell activity by transmitting extracellular signals inside the cell (reviewed in Lefkowitz, 2000; Pierce, Premont, & Lefkowitz, 2002). GPCRs share an up-down bundle of seven transmembrane helices (H1–H7) connected by three intracellular (I1, I2, and I3) and three extracellular loops (E1, E2, and E3), an extracellular N-term, and an intracellular C-term. Upon activation by extracellular signals, the receptors activate the α-subunit in heterotrimeric guanine nucleotide-binding proteins (G proteins) by catalyzing the exchange of bound GDP for GTP, that is, they act as guanine nucleotide exchange factors (GEFs). Thus, GPCRs are allosteric proteins that transform extracellular signals into promotion of nucleotide exchange in intracellular G proteins. Regulated protein–protein interactions are key features of many aspects of GPCR function, and there is increasing evidence that these receptors act as part of multicomponent units comprising a variety of signaling and scaffolding molecules (Brady & Limbird, 2002; Pierce et al., 2002).

The representation of biomolecular structures as networks of interacting amino acids/nucleotides is ever increasingly employed to investigate and elucidate complex phenomena such as protein folding and unfolding, protein stability, the role of structurally and functionally important residues, protein–protein and protein–DNA interactions as well as intraprotein and interprotein communication, and allosterism (Amitai et al., 2004; Angelova et al., 2011; Bode et al., 2007; Brinda & Vishveshwara, 2005; Chennubhotla & Bahar, 2007; Chennubhotla, Yang, & Bahar, 2008; del Sol, Fujihashi, Amoros, & Nussinov, 2006; Fanelli & Felline, 2011; Fanelli & Seeber, 2010; Pandini, Fornili, Fraternali, & Kleinjung, 2012; Papaleo, Lindorff-Larsen, & De Gioia, 2012; Raimondi, Felline, Portella, Orozco, & Fanelli, 2013; Raimondi, Felline, Seeber, Mariani, & Fanelli, 2013; Tang et al., 2007; Vendruscolo, Paci, Dobson, & Karplus, 2001; Vishveshwara, Ghosh, & Hansia, 2009).

These studies rely on methods that differ in the set of graph construction rules. The graph-based approach proposed by Vishveshwara et al. (2009) and defined as protein structure network (PSN) is the one that we have recently implemented in the Wordom software (Seeber et al., 2011). It basically computes network features (e.g., nodes, hubs (i.e., hyperconnected nodes), and links) and shortest communication pathways from molecular dynamics (MD) trajectories (herein indicated as PSN–MD). The employment of MD trajectories instead of a single structure serves to provide a dynamic description of the network as links break and form with atomic fluctuations. We have recently developed a strategy to infer a dynamic structure network even when dealing with a single structure rather than a trajectory. In this case, system dynamics is inferred from the coarse-grained Elastic Network Model paired with Normal Mode Analysis (ENM–NMA) (Raimondi, Felline, Seeber, et al., 2013). The approach is herein defined as PSN–ENM.

This chapter describes the application of PSN analysis to uncover the structural communication in GPCRs. Strategies to highlight changes in structural communication upon misfolding mutations, dimerization, and activation are described.

Focus is put on the PSN–ENM strategy applied to the crystallographic structures of rhodopsin in its inactive (dark) and signaling active (meta II (MII)) states, highlighting clear changes in the PSN and the centrality of the chromophore in differentiating the inactive and active states of the receptor.

3.1 MATERIALS

1. The freely available Wordom software (Seeber et al., 2011) is employed for performing both PSN–MD and PSN–ENM analyses. Postprocessing of the Wordom output relies on a number of in-house made programs and scripts. The latter provide numerical output and scripts for 3D output visualization by the PyMOL software (www.pymol.org).
2. ENM–NMA is carried out by Wordom (Seeber et al., 2011) as well. Three different ENM approaches have been implemented, the rotation translation block being the best performer one for PSN–ENM (Raimondi, Felline, Seeber, et al., 2013).
3. The PSN–ENM method is being also implemented in a web server with a friendly graphical user interface (manuscript in preparation). The server allows the user to easily set up the calculation, perform postprocessing analyses, and both visualize and download numerical and 3D representations of the output.
4. The GBSW implicit water/membrane model implemented in the CHARMM molecular simulation software (Brooks et al., 1983) was used for GPCR simulations.
5. The GROMACS4 simulation package (Hess, Kutzner, Van Der Spoel, & Lindahl, 2008) with the AMBER03 all-atoms force field (Case et al., 2005; Sorin & Pande, 2005) was used for equilibrium MD simulations in explicit water.

3.2 METHODS

3.2.1 Workflow of the PSN–MD and PSN–ENM approaches

The first step in PSN analysis consists in computing the protein structure graph (PSG), that is, an ensemble of nodes and links (Fig. 3.1). PSG can be computed either on an MD trajectory (PSN–MD approach) or on a single high-resolution structural model (PSN–ENM approach). This step provides the basis to search for the shortest paths between pairs of nodes, that is, linked nodes connecting two extremities. In this framework, once the two extremities of interest have been specified, the algorithm first defines all possible shortest communication paths between such extremities and then it filters the results according to the cross correlation of atomic motions derived

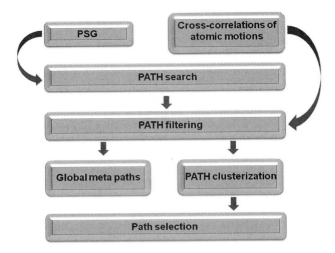

FIGURE 3.1

Flowchart of the PSN analysis approach.

from MD trajectories or ENM–NMA. Outcome of this stage is the total pool of paths for the system under study. Meta paths made of the most recurrent nodes and links in the path pool (i.e., global meta paths) are worth computing to infer a coarse/global picture of the structural communication in the considered system. More resolved information on the most likely communication pathways can be additionally inferred from cluster analysis of the path pool. Path clusters can be analyzed through cluster meta paths, cluster centers, and computational descriptors of path features. All parameters defined for the PSN and PSN–PATH analyses in the case study shown herein are summarized in Table 3.1.

3.2.2 Building the PSG

Building of the PSG is carried out by means of the PSN module implemented in the Wordom software (Seeber et al., 2011). PSN analysis is a product of graph theory applied to protein structures (Vishveshwara, Brinda, & Kannan, 2002). A graph is defined by a set of vertices (nodes) and connections (edges) between them. In a PSG, each amino acid residue is represented as a node and these nodes are connected by edges based on the strength of noncovalent interactions between residues (Vishveshwara et al., 2009). The strength of interaction between residues i and j (I_{ij}) is evaluated as a percentage given by the following equation:

$$I_{ij} = \frac{n_{ij}}{\sqrt{N_i N_j}} \times 100$$

where I_{ij} is the percentage interaction between residues i and j; n_{ij} is the number of atom–atom pairs between the side chains of residues i and j within a distance cutoff

Table 3.1 Network Parameters Concerning the Dark and MII Rhodopsin States

	Dark	MII
Imin[a]	4.33	3.98
Nodes[b]	230	226
CommNodes[b]	84.35	85.84
Links[c]	250	235
CommLinks[c]	53.60	57.02
Hubs[d]	37	35
CommHubs[d]	45.95	48.57
Hlinks[e]	131	121
CommHLinks[e]	39.69	42.98
Paths[f]	13,828	9476
RET[g]	10,006	5638
RET%[g]	72.36	59.50
MaxLen[h]	24 (4%)	31 (2%)
AvgLen[i]	10.73	11.77
MaxScr[j]	1.00 (30%)	1.00 (25%)
AvgScr[k]	0.31	0.28
Hubs50[l]	7084 (51.23%)	2927 (30.89%)
AvgMSDF[m]	0.0004	0.0007

[a] Imin cutoff.
[b] Total number of nodes and fraction (%) of nodes shared with the other functional state.
[c] Total number of links and fraction (%) of links shared with the other functional state.
[d] Total number of hubs and fraction (%) of hubs shared with the other functional state.
[e] Number of hub-mediated links and fraction (%) of hub-mediated links shared with the other functional state.
[f] Total number of shortest pathways.
[g] Number of pathways passing through the retinal and fraction (%) of such retinal-mediated paths over the total number of paths.
[h] Maximal path length (i.e., excluding the two extremities); the number in parenthesis indicates the fraction of paths over the total holding such length.
[i] Average path length.
[j] Maximal correlation score (i.e., number of internal nodes correlated with at least one extremity node, divided by path length); the number in parenthesis indicates the fraction of paths over the total holding such correlation score value.
[k] Average correlation score.
[l] Number of paths, in which $\geq 50\%$ of nodes are hubs: the number in parenthesis indicates the fraction of paths over the total holding such node composition.
[m] Average MSDF index.

(4.5 Å); N_i and N_j are normalization factors for residue types i and j, which account for the differences in size of the amino acid side chains and their propensity to make the maximum number of contacts with other amino acids in protein structures. The normalization factors for the 20 amino acids were derived from the work by Kannan and Vishveshwara (1999). The normalization index for retinal (i.e., 170.13) was computed as the average number of contacts done by the molecule in a dataset of 83 crystallographic structures concerning the different photointermediate states of bacteriorhodopsin, bovine rhodopsin, sensory rhodopsin, and squid rhodopsin. Finally, the normalization factor for water (i.e., 27) was computed on the crystal structures of rhodopsin and four Gα proteins (PDB codes: 1GZM, 1CIP, 1CUL, 1TAG, and 1TND).

Thus, I_{ij} are generally calculated for all node pairs, excluding those made of adjacent nodes. An interaction strength cutoff I_{min} is then chosen and any residue pair ij for which $I_{ij} \geq I_{min}$ is considered to be interacting and hence is connected in the PSG. Therefore, it is possible to obtain different PSGs for the same protein structure depending on the selected I_{min}. Consequently, I_{min} can be varied to obtain graphs with strong or weak interactions forming the edges between the residues. The residues making zero edges are termed as orphans and those that make at least four edges are referred to as hubs at that particular I_{min}.

As previously demonstrated (Vishveshwara et al., 2009), the optimal I_{min} is the one at which the size of the largest cluster of nodes at I_{min} 0% halves. Incidentally, a node cluster is a set of connected nodes in a graph. We approximate the I_{min} value to the second decimal place. In the case study shown herein, the I_{min} cutoffs are 4.33% for dark rhodopsin and 3.98% for MII.

With the PSN–ENM method, all edges at the selected I_{min} are considered in the PSG, whereas with the PSN–MD method, only edges occurring in a given fraction of the trajectory frames, that is, link frequency, enter in the PSG (Angelova et al., 2011; Fanelli & Felline, 2011; Fanelli & Seeber, 2010; Mariani, dell'Orco, Felline, Raimondi, & Fanelli, 2013; Raimondi, Felline, Portella, et al., 2013; Raimondi, Felline, et al., 2013).

Different states of a molecular system, for example, free or bound, wild type or mutated, inactive or active, and monomeric or oligomeric, can be compared in terms of PSGs; PSG differences can be either plotted in histograms or mapped onto the 3D structure (Table 3.1 and Figs. 3.3 and 3.4).

3.2.3 Search for the shortest communication paths

The search for the shortest path(s) between pairs of nodes as implemented in the PSN–PATH module of Wordom relies on the Dijkstra's algorithm (Dijkstra, 1959). Paths are searched by combining PSN data with cross correlation of atomic motions calculated by using the LMI method, for PSN–MD, or by the covariance matrix inferred from ENM–NMA, for the PSN–ENM method (Raimondi, Felline, Seeber, et al., 2013). When dealing with GPCRs, pathways are worth searching between all possible residues in the intracellular and extracellular portions (Angelova et al., 2011; Fanelli & Felline, 2011) or between all residue pairs in the protein except those at sequence distance ±5. The latter setup has been employed in the case study shown herein leading to 52,003 investigated pairs (i.e., the first and last amino acids in the path).

Following calculation of the PSG and of correlated motions, the procedure to search for the shortest path(s) between each residue pair consists of (a) searching for the shortest path(s) between each selected amino acid pair based upon the

PSN connectivities and (b) selecting the shortest path(s) that contains at least one residue correlated (e.g., with a correlation coefficient ≥ 0.8 in the case study shown herein) with either one of the two extremities. With the PSN–MD approach, all the shortest paths that pass the filter of correlation of motions are subjected to a further filter based upon path frequency, that is, number of frames containing the selected path divided by the total number of frames in the trajectory.

Collectively, the main differences between PSN–MD and PSN–ENM in terms of path search include the way in which the cross correlations of atomic motions for path filtering are computed and the application of a frequency-based extra filter. In detail, whereas with PSN–MD the cross correlations of atomic motions are computed on the trajectory frames by means of the LMI method, with PSN–ENM, they are extracted from the covariance matrix of the deformation modes computed by ENM–NMA. As for the path filtering issue, PSN–MD refilters those paths that pass the motion correlation filter by finally keeping only those that exceed a recurrence cutoff (i.e., presence in a given number of trajectory frames); in contrast, PSN–ENM applies only the motion correlation filter. Thus, whereas with PSN–MD, recurrence of network parameters (i.e., links and paths) in the trajectory frames dictates the composition of both PSG and path pool, recurrence-based filtering does not apply to PSN–ENM. In spite of these significant differences, the two different approaches tend to produce overlapping outcomes concerning those nodes and links that recur the most in the predicted pathways (Raimondi, Felline, Seeber, et al., 2013). The PSN–ENM approach tends to predict a more extended communication, likely due to the less heavy filtering applied to the communication pathways.

Thus, the paths that pass the filtering stage(s) constitute the pool of paths of a system at given I_{min} and correlation coefficient cutoffs. The statistical analysis of such pool of paths can lead to the building of global meta paths constituted by the most recurrent nodes and links in the pool. In this case study, the communication pathways characterizing the dark and MII states of rhodopsin are, indeed, represented as meta paths made of nodes that recur in $\geq 20\%$ of the retrieved paths (i.e., "frequent nodes") and of links satisfying conditions both of being present in $\geq 20\%$ of the paths and of connecting "frequent nodes."

Cluster analysis may provide finer information on the predicted pathways. We implemented two path clusterization methods differing both in the clusterization algorithms and in the score employed to evaluate the similarity between path pairs (i.e., similarity score) (Raimondi, Felline, Portella, et al., 2013; Raimondi, Felline, Seeber, et al., 2013). Irrespective of the clusterization method, for each cluster, following a pairwise comparison of all cluster members, the center is computed, which is the path with the highest average similarity among all the paths in the cluster. The center can be employed as a representative of a given cluster (Raimondi, Felline, Seeber, et al., 2013). Computational indices describing path features can be used as well to choose representative paths. These indices include the mean square distance fluctuation (MSDF) either computed between the extreme nodes (this case study) or averaged over all node pairs in a path (Raimondi, Felline, Seeber, et al., 2013).

3.3 DISCUSSION

GPCRs are allosteric proteins whose functioning fundamentals is the communication between the two poles of the helix bundle, that is, the extracellular side receives and transfers extracellular signals to the intracellular side deputed to recognize and activate intracellular proteins, primarily the G protein transducers to which these receptors owe their name. Activating or misfolding mutations, functionally different ligands (e.g., agonists, biased agonists, inverse agonists, antagonists, and allosteric modulators), and different homo/heterooligomeric states are likely to exert differential impacts on such communication.

In the last two decades or so, we paid a lot of effort in setting computational strategies to infer the mechanisms of intramolecular and intermolecular communication in a number of GPCRs of the rhodopsin family and in members of the Ras GTPase superfamily in functionally different states (reviewed in Fanelli & De Benedetti, 2011; Fanelli, De Benedetti, Raimondi, & Seeber, 2009; Fanelli & Raimondi, 2013). As for GPCRs, the extensive investigations done so far led us to infer that the main effects of activating signals, from either mutations or activating ligands, include perturbations in the interaction pattern of the arginine in the conserved E/DRY motif and increases in solvent exposure of selected amino acids in the neighborhoods of such motif (Fanelli & De Benedetti, 2011). The latter effect marks the opening of a solvent-accessible crevice between H3 and H6. These effects were suggested to require the integrity of conserved amino acids in H2 and H7 (e.g., D2.50 as well as N7.49 and Y7.53 of the NPxxY motif; here, the numbering scheme by Ballesteros & Weinstein (1995) is used). We also predicted that such cytosolic crevice would form a receptor docking site for the C-term of the G protein α-subunit (Fanelli & De Benedetti, 2011). All together, these predictions found validation in recent advances from structure determinations, showing that the arginine of the E/DRY motif can act both as a structural hallmark of receptor functionality and as a recognition point for the G protein C-term (Park, Scheerer, Hofmann, Choe, & Ernst, 2008; Rasmussen et al., 2011; Scheerer et al., 2008). MD simulations on a ternary complex between agonist-bound thromboxane A2 receptor (TP) and GDP-bound heterotrimeric Gq were the first attempt in the literature to investigate at the atomic detail GPCR's impact on the G protein dynamics (Raimondi, Seeber, De Benedetti, & Fanelli, 2008). The study suggested that the formation of a composite receptor–G protein interface, dominated by receptor contacts with the C-term of the α-subunit, favors concerted motions of selected G protein loops in the nucleotide-binding site and of the α-helical domain with respect to the Ras-like domain, features intrinsic to the G protein structure, but amplified by receptor binding. Such interdomain uncoupling was related to increases in solvent exposure of GDP. In spite of the extremely short length of simulations, the study highlighted for the first time the displacement of the α-helical domain with respect to the Ras-like domain as one of the early events in receptor-catalyzed GDP release. These inferences find now support in the crystallographic complex between the β2AR and nucleotide-free Gs heterotrimer (Rasmussen et al., 2011) and in the spectroscopic measurements

3.3 Discussion

on Gi activation by rhodopsin (Van Eps et al., 2011). We also speculated that the establishment of receptor–G protein contacts is also instrumental in preventing the establishment of intrasubunit or intersubunit interactions that would occur in receptor-free heterotrimeric G protein. Preventing the formation of such interactions upon receptor binding would contribute to weaken the control exerted by the β-subunit on certain intradomain and interdomain motions of the α-subunit.

The representation of GPCR structures as networks of interacting amino acids can be a meaningful way to decipher the impact of mutation, ligand binding, and/or formation of multiprotein complex on the structural communication of the protein. Indeed, we applied the two different variants of the PSN analysis, PSN–MD and PSN–ENM, to investigate different aspects of GPCR function. In deep detail, the PSN–MD method served to infer (a) the structural bases of rhodopsin mutations associated with autosomal dominant retinitis pigmentosa (ADRP) (Fanelli & Seeber, 2010), (b) the effect of highly conserved amino acids in the structural communication of the luteinizing hormone receptor (LHR) both in its inactive and mutation-induced active states (Angelova et al., 2011), and (c) the effect of ligand binding and dimerization on the structural communication of the A_{2A} adenosine receptor ($A_{2A}R$) (Fanelli & Felline, 2011). As for ADRP-linked rhodopsin mutations, steered MD simulations were instrumental in simulating the unfolding process, whereas PSN was used to infer the effects of mutations on the native fingerprint of the hyperlinked and most stable amino acids in the structure network, that is, native stable hubs, that oppose resistance to connectivity loss in response to an external force (Fanelli & Seeber, 2010). Thus, the analysis focused on mutational effects on the native stable hub frequency. The study showed that native stable hubs essentially group in the two poles of the helix bundle and along the main axes of H3 and H6, thus suggesting that they play a role both in protein stability and in signal transfer between extracellular and intracellular sides (Fanelli & Seeber, 2010). The high concentration of hubs in the retinal-binding site is consistent with computational studies highlighting this receptor portion as a part of the stability core and a hinge site in the dynamics of the protein (Fanelli & Seeber, 2010; Isin, Rader, Dhiman, Klein-Seetharaman, & Bahar, 2006; Tastan et al., 2007). Irrespective of their location, misfolding mutations tend to impair selected native stable hubs in the retinal-binding site. The extent of this structural effect was found related with the extent of the biochemical defect associated with the mutation (Fanelli & Seeber, 2010). As for LHR, PSN analyses allowed the identification of key amino acids that are part of the regulatory network responsible for propagating communication between the extracellular and intracellular poles of the receptor. We found that the number of hubs and link-involving hubs in the wild type is higher compared to the two constitutively active mutants, consistent with the demonstrated lower stability of the active GPCR states compared to the inactive ones (Angelova et al., 2011). The study emphasized the role of highly conserved amino acids both in protein stability and in intraprotein allosteric communication. Indeed, such amino acids behaved as stable hubs in both the inactive and active states. Moreover, they participate as the most frequent nodes in the communication paths between the extracellular and intracellular sides in both functional states, with emphasis on the active one (Angelova et al., 2011). As for the $A_{2A}R$, the dynamic network of intramolecular interactions characterizing the complex with the ZM241385 (i.e., ZMA) antagonist in its monomeric state was compared with that of the same complex in three different dimeric forms, as well as with that of the apo-protomer. The results of the study emphasized the roles of H1 in $A_{2A}R$ homodimerization and of highly conserved amino acids in H1, H2, H6, and H7 in maintaining the structure network of the $A_{2A}R$. $A_{2A}R$ dimerization resulted to affect the communication networks intrinsic to the receptor fold in a way dependent on the dimer architecture. Certain architectures retained the most recurrent communication paths with respect to the monomeric antagonist-bound form but enhancing path numbers and frequencies, whereas some others impaired ligand-mediated communication. Ligand binding turned out to affect the network as well. Collectively, the study suggested that the communication network that pertains to the functional dynamics of a GPCR is influenced by ligand functionality, oligomeric order, and architecture of the supramolecular assembly.

The results discussed above were achieved by the PSN–MD approach. We have recently undertaken a comparative PSN analysis on all the GPCR crystallographic structures released so far by the PSN–ENM approach (manuscript in preparation). The goal of the study is to infer functional state-dependent and state-independent commonalties and differences in the structural communication features at the family and subfamily levels. Herein, the case study of dark rhodopsin and MII has been extracted from that analysis to show what PSN analysis, carried out on a single structure, can tell us in terms of network differences between inactive and active states.

The structure networks of dark and MII states of rhodopsin are quite similar in number of nodes, hubs, links, and hub-mediated links (Table 3.1); the only difference is that the number of links is slightly higher in the inactive than the active state, consistent with the higher stability of the former (Khan, Bole, Hargrave, Santoro, & McDowell, 1991). In spite of numerical similarities in network parameters, up to 57% of links and less than 50% of hubs and link-mediated hubs are shared in common by the two forms (Table 3.1). Major differences in the PSG concern the retinal chromophore, which acts as a hub in both forms. However, whereas 11-cis-retinal is part of a network community, all-trans-retinal is not (Fig. 3.2). Incidentally, communities are sets of highly interconnected vertices such that nodes belonging to the same community are densely linked to each other and poorly connected to nodes outside the community. Thus, photoactivated retinal is involved in a less dense network compared to the 11-cis form. Remarkably, in the dark state, retinal interacts with nodes in three different communities. This is suggestive of retinal and the extracellular regions making a dense network likely involved in the stability of the protein. Such network does not occur in the MII state (Figs. 3.2 and 3.3). Collectively, the receptor portions that hold major specificity in links between inactive and active states include H2, H3, E2, H6, and H7 (Fig. 3.3). Structural water molecules participate in such differences as well.

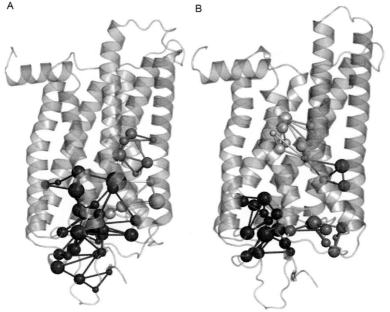

FIGURE 3.2

3D representation of the communities holding ≥4 nodes concerning dark (left) and MII (right) states. Nodes are represented as spheres centered on the Cα-atoms of the protein, on the C8 atom of retinal, and the O atom of the structural water molecules. Sphere diameter is proportional to the number of links made by the node. Red, orange, yellow, and green refer to the first, second, third, and fourth largest communities, respectively. (See color plate.)

As for hubs, as already found by the PSN–MD approach coupled with mechanical unfolding simulations (Fanelli & Seeber, 2010), rhodopsin hubs crowd in the retinal-binding site and involve both highly conserved amino acids and ADRP mutation sites, the latter being localized essentially in the extracellular half of the protein and in the N-term (Fig. 3.4). Remarkably, the number of misfolding mutation sites behaving as hubs is higher in the inactive state than the active one. This is likely linked to 11-cis-retinal making an extended community of nodes including ADRP mutation sites (Figs. 3.2–3.4). Another remarkable difference between dark and MII states concerns the hub behavior of highly conserved amino acids in the cytosolic regions. In particular, whereas the arginine of the E/DRY motif and Y306 of the NPxxY motif are hubs in the dark state and not in MII, the contrary happens for Y223 and F312, likely linked to the structural rearrangements in the cytosolic regions following photoisomerization of 11-cis-retinal. In general, in the dark state, hub specificity essentially locates in the N-term, E2, and H7, whereas in MII, it essentially resides in H2, E2, and H5 (Fig. 3.4).

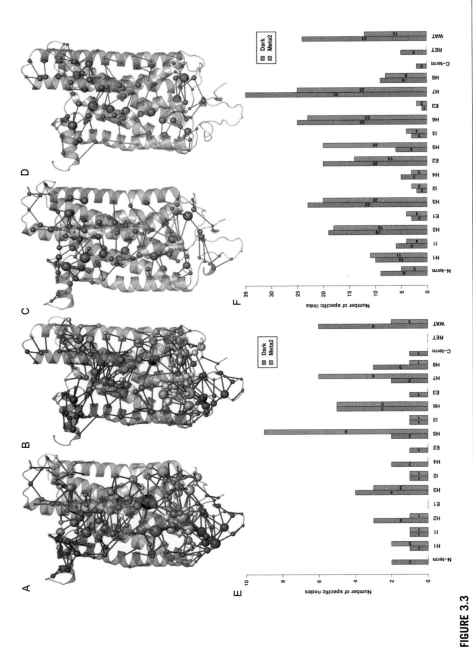

FIGURE 3.3

3D PSGs and network parameter distribution in dark and MII rhodopsin. The 3D PSGs concerning dark (A) and MII (B) states are shown. Nodes are represented as spheres centered on the Cα-atoms of the protein, on the C8 atom of retinal, and the O atom of the structural water molecules. Nodes are colored according to their location (i.e., according to the different receptor regions). In this respect, helices 1, 2, 3, 4, 5, 6, 7, and 8 are, respectively, blue, orange, green, pink, yellow, cyan, violet, and red. I1 and E1 are lime, I2 and E2 are gray, I3 and E3 are magenta, and N-term and C-term, including H8, are red. The diameter of the circle is proportional to the number of links made by the considered node, with the lowest value corresponding to one link. Links are red. In panels (C) and (D), only links that are specific to the dark (orange) and MII (violet) states are, respectively, shown. Specific nodes hold the same link color, whereas common nodes are green. The contribution of each receptor portion to those nodes and links that are specific to the dark (orange) and MII (violet) states is plotted as well in panels (E) and (F), respectively. (See color plate.)

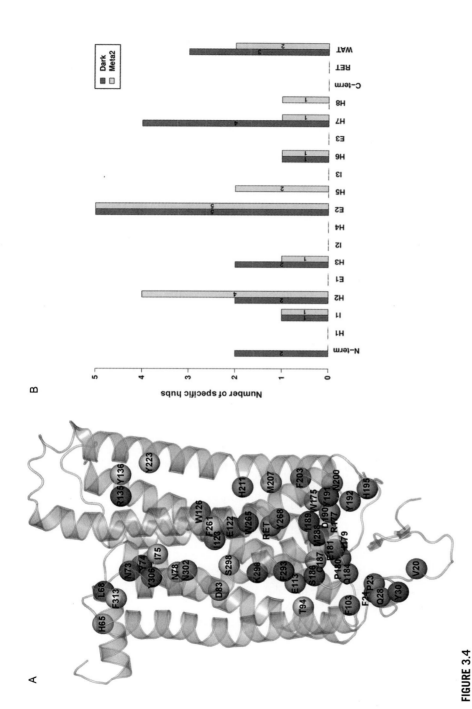

FIGURE 3.4

Comparative hub distribution. In the left panel, hubs are represented as spheres centered on the Cα-atoms of the dark rhodopsin structure, on the C8 atom of retinal, and the O atom of the structural water molecules. Hubs specific to the dark and MII states are, respectively, orange and violet, whereas those shared by the two forms are green. The contribution of each receptor portion to those hubs that are specific to the dark (orange) and MII (violet) states is plotted as well in the right panel. (For interpretation of the references to color in this figure legend, the reader is referred to the online version of this chapter.)

As for the shortest communication pathways, rhodopsin's fold is such that, in both functional states, the most likely pathways express a vertical communication between the two poles of the transmembrane helix bundle, in line with the fact that nature made GPCRs competent to transfer signals from the outside to the inside of the cell (Fig. 3.5). Consistent with the more dense retinal-involving network in the dark state, the number of pathways that pass through retinal is higher in the dark state (i.e., 73%, Table 3.1) compared to MII (i.e., 59%, Table 3.1). On average, paths characterizing the inactive state are shorter and more stiff and tend to involve a major number of hubs than those characterizing the MII state (Table 3.1 and Fig. 3.6). Striking different communication modes are expressed by the meta paths of the two functionally different states (Fig. 3.5). In the dark state, the gross of the communication involves E2, retinal, H6, and H7. The most recurrent nodes in the helices and their respective links in the meta path of such state sketch the following pathway: RET-W6.48-F6.44-WAT-N7.49-M6.40-M6.46-Y7.53 (Fig. 3.5). In contrast, the meta path characterizing the MII state describes a vertical communication essentially involving retinal, H6, H5, and the C-terms of H3 and H8 (Fig. 3.5). In this case, the most recurrent nodes in the helices and their respective links sketch the following pathway: F5.43-RET-W6.48-F6.44-WAT-Y7.53-M6.40-Y5.58-R3.50. Remarkably, in both states, the W6.48-F6.44 aromatic pair in H6 mediates the communication between chromophore-binding site and cytosolic regions. Meta path comparisons highlight a shift of the communication towards the intracellular regions and a reduction in the contribution by the extracellular portions on going from the dark to the MII state. In the latter, the cytosolic end of the pathways involves Y5.58 and the E/DRY arginine. These data suggest that whereas in the inactive states the structural communication serves essentially to ensure structural stability, in the active form, it serves to maintain a binding site for the G protein in the cytosolic regions of the receptor.

Collectively, PSN–ENM applied to the dark and MII states of rhodopsin reveals the structure networks as expression of enhanced stability of the inactive state compared to the active one, highlighting also the receptor portions responsible for such stability differences, which strongly the retinal-binding site. Finally, the structural communication appears to hold different functional competences in the inactive and active states.

SUMMARY

PSN analysis is a powerful tool to unveil the structure communication in biomolecular systems and hence suitable for GPCRs that are designed to receive and transfers inputs from the outside to the inside of the cell. Applications of two different variants of the PSN approach, that is, PSN–MD and PSN–ENM, served to investigate the roles of misfolding or activating mutations, ligand binding, and receptor dimerization on the structural communication features of a number of GPCRs in different functional forms. Comparison of network features is a powerful tool to decipher the structural signatures of functionally different states in different sets of homologous proteins at varying degree of similarity.

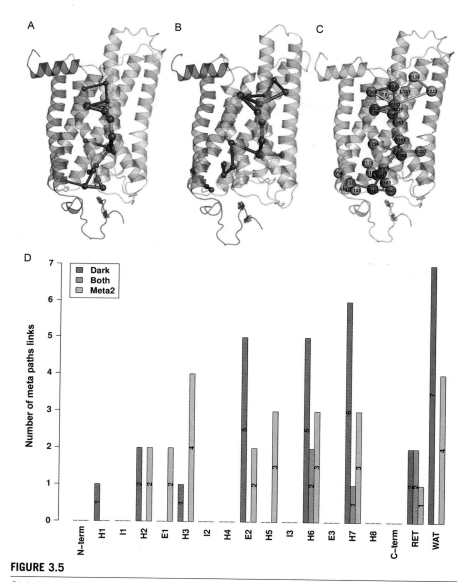

FIGURE 3.5

Global meta paths. 3D representations of the global meta paths concerning dark and MII states are shown in panels (A) and (B), respectively. The receptors are represented as cartoons colored according to the receptor portion (see the legend to Fig. 3.3 for the coloring scheme). Meta paths are black. Nodes are represented as spheres centered on the Cα-atoms of the protein, on the C8 atom of retinal, and the O atom of the structural water molecules. Node diameter and link thickness are proportional to node and link recurrence in the global pool of paths. A recurrence cutoff $\geq 20\%$ is employed for both nodes and links. In panel (C), the difference meta path is shown onto the dark rhodopsin structure. Meta path nodes and links specific to the dark and MII states are orange and green, respectively, whereas those shared by the two different states are green. The contribution of each receptor portion to those meta path links that are specific to dark (orange) and MII (violet) states, as well as those meta path links that are shared by the two forms (green) is plotted as well in panel (D). (See color plate.)

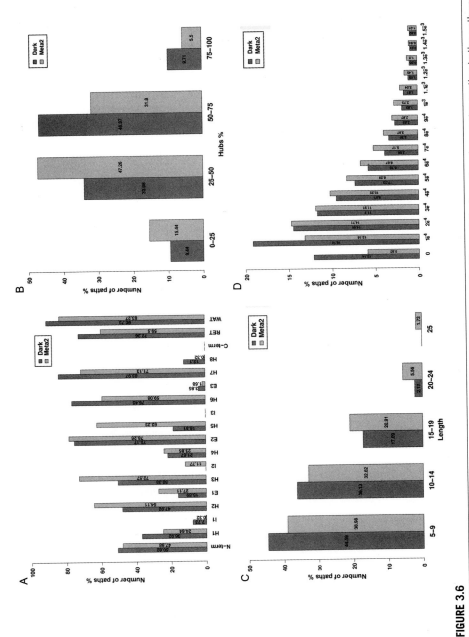

FIGURE 3.6

Selected features of the communication pathways in dark and MII rhodopsin. The contribution of each receptor portion to the pathways characterizing the dark (orange) and MII (violet) states is plotted in panel (A). The distributions of hub content, length, and MSDF concerning the shortest pathways of dark (orange) and MII (violet) states are shown in panels (B), (C), and (D), respectively. (For interpretation of the references to color in this figure legend, the reader is referred to the online version of this chapter.)

Acknowledgments

This study was supported by an Airc-Italy grant [IG10740] and a Telethon-Italy grant [S00068TELC] to FF.

References

Amitai, G., Shemesh, A., Sitbon, E., Shklar, M., Netanely, D., Venger, I., et al. (2004). Network analysis of protein structures identifies functional residues. *Journal of Molecular Biology, 344*, 1135–1146.

Angelova, K., Felline, A., Lee, M., Patel, M., Puett, D., & Fanelli, F. (2011). Conserved amino acids participate in the structure networks deputed to intramolecular communication in the lutropin receptor. *Cellular and Molecular Life Sciences, 68*, 1227–1239.

Ballesteros, J. A., & Weinstein, H. (1995). Integrated methods for the construction of three-dimensional models and computational probing of structure–function relations in G protein-coupled receptors. *Methods in Neurosciences, 25*, 366–428.

Bode, C., Kovacs, I. A., Szalay, M. S., Palotai, R., Korcsmaros, T., & Csermely, P. (2007). Network analysis of protein dynamics. *FEBS Letters, 581*, 2776–2782.

Brady, A. E., & Limbird, L. E. (2002). G protein-coupled receptor interacting proteins: Emerging roles in localization and signal transduction. *Cellular Signalling, 14*, 297–309.

Brinda, K. V., & Vishveshwara, S. (2005). A network representation of protein structures: Implications for protein stability. *Biophysical Journal, 89*, 4159–4170.

Brooks, B. R., Bruccoleri, R. E., Olafson, B. D., States, D. J., Swaminathan, S., & Karplus, M. (1983). Charmm: A program for macromolecular energy, minimization and dynamics calculations. *Journal of Computational Chemistry, 4*, 187–217.

Case, D. A., Cheatham, T. E., Darden, T., Gohlke, H., Luo, R., Merz, K. M., et al. (2005). The Amber biomolecular simulation programs. *Journal of Computational Chemistry, 26*, 1668–1688.

Chennubhotla, C., & Bahar, I. (2007). Signal propagation in proteins and relation to equilibrium fluctuations. *PLoS Computational Biology, 3*, 1716–1726.

Chennubhotla, C., Yang, Z., & Bahar, I. (2008). Coupling between global dynamics and signal transduction pathways: A mechanism of allostery for chaperonin GroEL. *Molecular BioSystems, 4*, 287–292.

del Sol, A., Fujihashi, H., Amoros, D., & Nussinov, R. (2006). Residues crucial for maintaining short paths in network communication mediate signaling in proteins. *Molecular Systems Biology, 2*(2006), 0019.

Dijkstra, E. W. (1959). A note on two problems in connexion with graphs. *Numerische Mathematik, 1*, 269–271.

Fanelli, F., & De Benedetti, P. G. (2005). Computational modeling approaches to structure–function analysis of G protein-coupled receptors. *Chemical Reviews, 105*, 3297–3351.

Fanelli, F., & De Benedetti, P. G. (2011). Update 1 of: Computational modeling approaches to structure–function analysis of g protein-coupled receptors. *Chemical Reviews, 111*, PR438–PR535.

Fanelli, F., De Benedetti, P. G., Raimondi, F., & Seeber, M. (2009). Computational modeling of intramolecular and intermolecular communication in GPCRs. *Current Protein & Peptide Science, 10*, 173–185.

Fanelli, F., & Felline, A. (2011). Dimerization and ligand binding affect the structure network of A(2A) adenosine receptor. *Biochimica et Biophysica Acta*, *1808*, 1256–1266.

Fanelli, F., & Raimondi, F. (2013). Nucleotide binding affects intrinsic dynamics and structural communication in Ras GTPases. *Current Pharmaceutical Design*, *19*, 4214–4225.

Fanelli, F., & Seeber, M. (2010). Structural insights into retinitis pigmentosa from unfolding simulations of rhodopsin mutants. *The FASEB Journal*, *24*, 3196–3209.

Hess, B., Kutzner, C., Van Der Spoel, D., & Lindahl, E. (2008). GROMACS 4: Algorithms for highly efficient, load-balanced, and scalable molecular simulation. *Journal of Chemical Theory and Computation*, *4*, 435–447.

Isin, B., Rader, A. J., Dhiman, H. K., Klein-Seetharaman, J., & Bahar, I. (2006). Predisposition of the dark state of rhodopsin to functional changes in structure. *Proteins*, *65*, 970–983.

Kannan, N., & Vishveshwara, S. (1999). Identification of side-chain clusters in protein structures by a graph spectral method. *Journal of Molecular Biology*, *292*, 441–464.

Khan, S. M. A., Bole, W., Hargrave, P. A., Santoro, M. M., & McDowell, J. H. (1991). Differential scanning calorimetry of bovine rhodopsin in rod-outer-segment disk membranes. *European Journal of Biochemistry*, *200*, 53–59.

Lefkowitz, R. J. (2000). The superfamily of heptahelical receptors. *Nature Cell Biology*, *2*, E133–E136.

Mariani, S., dell'Orco, D., Felline, A., Raimondi, F., & Fanelli, F. (2013). Multi-scale simulations to unveil the structural determinants of mutations linked to retinal diseases. *PLOS Computational Biology*, *9*, e1003207.

Pandini, A., Fornili, A., Fraternali, F., & Kleinjung, J. (2012). Detection of allosteric signal transmission by information-theoretic analysis of protein dynamics. *The FASEB Journal*, *26*, 868–881.

Papaleo, E., Lindorff-Larsen, K., & De Gioia, L. (2012). Paths of long-range communication in the E2 enzymes of family 3: A molecular dynamics investigation. *Physical Chemistry Chemical Physics*, *14*, 12515–12525.

Park, J. H., Scheerer, P., Hofmann, K. P., Choe, H. W., & Ernst, O. P. (2008). Crystal structure of the ligand-free G-protein-coupled receptor opsin. *Nature*, *454*, 183–187.

Pierce, K. L., Premont, R. T., & Lefkowitz, R. J. (2002). Seven-transmembrane receptors. *Nature Reviews Molecular Cell Biology*, *3*, 639–650.

Raimondi, F., Felline, A., Portella, G., Orozco, M., & Fanelli, F. (2013). Light on the structural communication in Ras GTPases. *Journal of Biomolecular Structure & Dynamics*, *31*, 142–157.

Raimondi, F., Felline, A., Seeber, M., Mariani, S., & Fanelli, F. (2013). A mixed protein structure network and elastic network model approach to predict the structural communication in biomolecular systems: The PDZ2 domain from tyrosine phosphatase 1E as a case study. *Journal of Chemical Theory and Computation*, *9*, 2504–2518.

Raimondi, F., Seeber, M., De Benedetti, P. G., & Fanelli, F. (2008). Mechanisms of inter- and intra-molecular communication in GPCRs and G proteins. *Journal of the American Chemical Society*, *130*, 4310–4325.

Rasmussen, S. G., Devree, B. T., Zou, Y., Kruse, A. C., Chung, K. Y., Kobilka, T. S., et al. (2011). Crystal structure of the beta(2) adrenergic receptor-Gs protein complex. *Nature*, *469*, 175–181.

Scheerer, P., Park, J. H., Hildebrand, P. W., Kim, Y. J., Krauss, N., Choe, H. W., et al. (2008). Crystal structure of opsin in its G-protein-interacting conformation. *Nature*, *455*, 497–502.

Seeber, M., Felline, A., Raimondi, F., Muff, S., Friedman, R., Rao, F., et al. (2011). Wordom: A user-friendly program for the analysis of molecular structures, trajectories, and free energy surfaces. *Journal of Computational Chemistry, 32*, 1183–1194.

Sorin, E. J., & Pande, V. S. (2005). Exploring the helix-coil transition via all-atom equilibrium ensemble simulations. *Biophysical Journal, 88*, 2472–2493.

Tang, S., Liao, J. C., Dunn, A. R., Altman, R. B., Spudich, J. A., & Schmidt, J. P. (2007). Predicting allosteric communication in myosin via a pathway of conserved residues. *Journal of Molecular Biology, 373*, 1361–1373.

Tastan, O., Yu, E., Ganapathiraju, M., Aref, A., Rader, A. J., & Klein-Seetharaman, J. (2007). Comparison of stability predictions and simulated unfolding of rhodopsin structures. *Photochemistry and Photobiology, 83*, 351–362.

Van Eps, N., Preininger, A. M., Alexander, N., Kaya, A. I., Meier, S., Meiler, J., et al. (2011). Interaction of a G protein with an activated receptor opens the interdomain interface in the alpha subunit. *Proceedings of the National Academy of Sciences of the United States of America, 108*, 9420–9424.

Vendruscolo, M., Paci, E., Dobson, C. M., & Karplus, M. (2001). Three key residues form a critical contact network in a protein folding transition state. *Nature, 409*, 641–645.

Vishveshwara, S., Brinda, K. V., & Kannan, N. (2002). Protein structure: Insights from graph theory. *Journal of Theoretical and Computational Chemistry, 1*, 187–211.

Vishveshwara, S., Ghosh, A., & Hansia, P. (2009). Intra and inter-molecular communications through protein structure network. *Current Protein & Peptide Science, 10*, 146–160.

CHAPTER 4

Simulating G Protein-Coupled Receptors in Native-Like Membranes: From Monomers to Oligomers

Ramon Guixà-González*, Juan Manuel Ramírez-Anguita*, Agnieszka A. Kaczor[†,‡], and Jana Selent*

Research Programme on Biomedical Informatics (GRIB), Department of Experimental and Health Sciences, Universitat Pompeu Fabra/IMIM (Hospital del Mar Medical Research Institute), Dr. Aiguader, Barcelona, Spain
[†]*Department of Synthesis and Chemical Technology of Pharmaceutical Substances, Faculty of Pharmacy with Division for Medical Analytics, Lublin, Poland*
[‡]*School of Pharmacy, University of Eastern Finland, Kuopio, Finland*

CHAPTER OUTLINE

Introduction	64
4.1 Membranes	65
4.1.1 The Importance of Modeling Physiological Membranes	65
4.1.2 Building and Simulating Membranes	66
4.1.3 Analysis and Interpretation of Membrane Simulations	68
4.2 GPCR Monomers in Membranes	72
4.2.1 Modeling GPCRs	73
4.2.2 Building and Simulating GPCR–Membrane Complexes	74
4.2.3 Studying GPCR–lipid Interactions from Simulations: Quantification of Membrane Remodeling and Residual Mismatch	75
4.3 GPCR Dimers and Oligomers in Membranes	80
4.3.1 Modeling GPCR–GPCR Interactions	80
4.3.2 Simulating GPCR Heteromers: All-Atom Versus CG Simulations	81
References	83

Abstract

G protein-coupled receptors (GPCRs) are one of the most relevant superfamilies of transmembrane proteins as they participate in an important variety of biological events. Recently, the scientific community is witnessing an advent of a GPCR crystallization age along with impressive improvements achieved in the field of computer simulations during the last two decades. Computer simulation techniques such as molecular dynamics (MD) simulations are now frequent tools to study the dynamic behavior of GPCRs and, more importantly, to model the complex membrane environment where these proteins spend their lifetime. Thanks to these tools, GPCRs can be simulated not only longer but also in a more "physiological" fashion. In this scenario, scientists are taking advantage of such advances to approach certain phenomena such as GPCR oligomerization occurring only at timescales not reachable until now. Thus, despite current MD simulations having important limitations today, they have become an essential tool to study key biophysical properties of GPCRs and GPCR oligomers.

INTRODUCTION

G protein-coupled receptors (GPCRs) are transmembrane (TM) proteins involved in a wide range of biological events such as transducing chemical signals, brain signaling, modulating heart rate, or even mediating the immune response (Katritch, Cherezov, & Stevens, 2013). They are the main protein superfamily in human and mammalian genomes and all share a common structural feature: a seven-TM topology (Katritch et al., 2013). The GPCR superfamily is nowadays one of the major drug targets used in research and therapeutic activities (Katritch et al., 2013). The boom of high-performance molecular dynamics (MD) simulations and the deposit of the first GPCR crystal structures have finally enabled computational scientists to simulate GPCRs for longer timescales in an attempt to understand the signaling mode of these proteins (Dror et al., 2011). Likewise, the improved MD techniques make it possible to model and simulate more complex membrane environments in microsecond-regime simulations. Nowadays, one of the most challenging aspects of simulating GPCRs is the study of GPCR–GPCR interactions. Whereas the dimerization and oligomerization of GPCRs have been widely accepted by the scientific community, the dynamics of such interaction remains largely elusive. Despite the study of GPCR dimers and oligomers being still quite a challenging task in computational biosciences, MD simulations are arising as a helpful tool for this purpose.

Among different methods used in the simulation of biomolecular systems, MD is the most commonly used technique. There are three main pillars of an MD simulation, namely, the initial structure of the system, the force field, and the MD simulation software. The initial structure is represented by the Cartesian coordinates of the initial position of the whole system, for example, a Protein Data Bank (PDB) file or a homology model, along with the network of connections that defines such system. The force field is used to describe the type of interaction existing between each of the atoms or a group of atoms connected in the initial structure. While in all-atom simulations all single atoms in the systems are represented, coarse-grained (CG) structures and force fields use a reduced level of description by mapping specific sets of atoms in one group. CHARMM (MacKerell et al., 1998), AMBER (Cornell et al., 1995), GROMOS (Christen et al., 2005), and OPLS (Jorgensen, Maxwell, & Tirado-Rives, 1996) are force fields commonly used in all-atom MD simulations, whereas the MARTINI force field (Marrink, Risselada, Yefimov, Tieleman, & de Vries, 2007) is the most widely used force field in CG simulations.

Although most of the published work on MD simulations of GPCRs typically uses GROMACS (Hess, Kutzner, van der Spoel, & Lindahl, 2008), NAMD (Phillips et al., 2005), ACEMD (Harvey, Giupponi, & Fabritiis, 2009), or CHARMM (Brooks et al., 1983), currently, there exist a vast number of MD engines developed for this purpose. In essence, MD engines use the aforementioned force field to numerically solve the classical equations of motion through a sampling algorithm. The forces needed to solve these equations are obtained from the potential energy function, which depends on the position of the atoms of the system. Despite there are different force fields, the form of the potential energy function is similar in most of the ordinary ones and consists of a sum of both bonded (e.g., bond, angles, and rotation) and nonbonded terms (e.g., van der Waals and electrostatic interactions).

The intention of this chapter is providing an overview of the methodology and the analysis approaches currently used for MD simulations of membranes and GPCR monomers or heteromers in their native-like environment. Thus, firstly, we focus on the simulation of biological membranes; secondly, we describe the process of modeling and simulating GPCR monomers embedded in membranes; and, lastly, we discuss current approaches to study GPCR oligomerization by means of MD simulations.

4.1 MEMBRANES
4.1.1 The importance of modeling physiological membranes

As all TM proteins, GPCRs live permanently surrounded by membrane lipids. They look just like plain membrane solvent or a physical barrier for the whole set of proteins therein confined. Indeed, the amphipathic character of membrane lipids enables cells not only to segregate internal cell organelles but also to separate the whole cell from the interior milieu. However, it is becoming more and more evident that considering the main role of lipids as being only structural is a misconception of the true global mission of these molecules. Cells synthesize a colorful assortment of lipid species across the body (Van Meer, Voelker, & Feigenson, 2008), generating cells and tissues with specific lipid profiles. As a result of such segregation, cell membranes have diverse biophysical properties (Feigenson, 2007) maintained through a careful exchange of lipids between cell organelles and the plasma membrane. Importantly, such lipid regulation also affects the composition of each leaflet of the membrane bilayer, and, thus, the lipid composition of cell membranes is also asymmetric within the membrane itself. Furthermore, certain pathologies such as Alzheimer's (Martín et al., 2010) or Parkinson's disease (Fabelo et al., 2011) can trigger a change in the membrane lipid composition. This finding could be behind the particular behavior or certain GPCRs in such relevant diseases (Guixà-González, Bruno, Marti-Solano, & Selent, 2012).

In this context, membrane simulations need to inevitably make room for an adequate representation of such heterogeneity. The first step towards a native-like representation of cell membranes is to incorporate cholesterol in membrane models. Apart from being a key element of biological membranes, cholesterol concentration is typically different across cells and tissues. This molecule is created in the endoplasmic reticulum but it is largely found within the plasma membrane (Van Meer et al., 2008). Membranes enriched in cholesterol display different biophysical properties primarily in terms of membrane fluidity and condensation (Hung, Lee, Chen, & Huang, 2007). This effect has an impact not only in lipid properties (De Meyer & Smit, 2009; Lindblom & Orädd, 2009) but also in the dynamics of other membrane components such as membrane proteins. In addition, the lateral segregation of cholesterol and specific membrane components into lipid microdomains, known as lipid rafts, is today an exciting line of research in cell biology (Lingwood & Simons, 2010). But along with cholesterol concentration, cells delicately regulate important chemical aspects of other membrane lipids such as the type of polar head or the carbon length and degree of unsaturation of lipid tails (Van Meer et al., 2008). Although this might represent just subtle changes in the lipid structure, these aspects highly influence the biophysical properties of cell membranes (Niemelä, Hyvönen, & Vattulainen, 2009).

Therefore, at least from the theoretical point of view, membrane simulations should have a propensity to use native-like membranes, where reality is surely better represented. On the other hand, the rapid advances made in computer sciences during the last years have greatly increased the timescale reachable by computer simulations and have helped studying more complex lipid mixtures (Bennett & Tieleman, 2013). Nevertheless, the increasing complexity of these realistic environments makes it still an extremely challenging task where one has to be very cautious upon interpretation of results.

4.1.2 Building and simulating membranes

Most simulations are done at constant temperature and pressure (Tieleman, 2010). While constant volume may lead to artificial system dimensions and surely restrict thickness fluctuations, constant pressure is frequently applied when simulating membrane systems. The last update of the CHARMM force field (Klauda et al., 2010) is an important step forward in membrane simulations. In the former study, the validation of this force field in the tensionless ensemble (NPT, number of particles (N), pressure (P), and temperature (T) remain constant) yielded real experimental values. Simulating a membrane system at constant temperature and pressure is a must if one aims to study changes in structural parameters such as area per lipid or membrane thickness (see the succeeding text). In contrast to pressure coupling, the approach used for controlling the temperature is frequently more standard. Nevertheless, the choice of the temperature will obviously impact the timescale of the simulation needed to reach convergence. Since both lipid composition and temperature will determine the phase behavior of our systems, we need to know in advance what lipid is holding the highest chain-melting transition temperature (T_m) in the mixture we are simulating (Koynova & Caffrey, 1998; Maulik & Shipley, 1996).

Although frequently overlooked, the hydration level of lipid bilayers is another important parameter, particularly if one attempts to study structural parameters such as order or hydrogen-bonding networks (Ho, Slater, & Stubbs, 1995). Water ordering at membrane interfaces plays an important role in different biological events (Kasson, Lindahl, & Pande, 2011; Robinson, Besley, O'Shea, & Hirst, 2011), so using adequate hydration levels seems a more physiological approach than just coarsely hydrating. Again, thanks to the last CHARMM force field update (Klauda et al., 2010), the hydration of lipids can be realistically modeled. Experimental hydration levels are available for different bilayers, although these values frequently correspond to phosphatidylcholine bilayers (Nagle & Tristram-Nagle, 2000; Tristram-Nagle, Petrache, & Nagle, 1998). Yet, the number of water molecules per lipid (nw), a common experimental value that represents hydration, is known to range approximately between 20 and 32.5 (Kasson et al., 2011; Robinson et al., 2011; Steinbauer, Mehnert, & Beyer, 2003). Often, 30 water molecules per phospholipid ($nw=30$) is taken as an adequate hydration level for membrane systems.

Building and simulating pure membranes (i.e., one-component mixtures) is always a much simpler task when compared to membrane-protein systems. Despite building a starting structure of a pure membrane has not always been easy, today, we can even do this with a molecular visualization software such as VMD (visual molecular dynamics) (Humphrey, Dalke, & Schulten, 1996). The last releases of this tool offer a plug-in to build pure membranes already solvated. In contrast to earlier years where scientist could only simulate pure membranes, today, we have access to different automated tools that can build for us starting structures of multicomponent membranes. Out of these set of tools, one web-based graphical interface, the CHARMM-GUI (http://www.charmm-gui.org/) (Jo, Kim, Iyer, & Im, 2008), is becoming very popular in both pure membrane and membrane-protein simulations using the CHARMM force field. In particular, this tool offers a specific module (http://www.charmm-gui.org/?doc=input/membrane) intended to help building pure and mixed membranes (Jo, Lim, Klauda, & Im, 2009) or membrane-protein systems (Jo, Kim, & Im, 2007) (see Section 4.2.2). This module is very useful to generate input structures ready for MD simulations and provide a flexible and straightforward set of options for this purpose.

The large number of lipid types available in the CHARMM-GUI membrane builder, 77 to date, allows building membrane systems as heterogeneous as needed. Four straightforward steps guide the user during the process. Firstly, the type and size of the system is determined based on the lipid composition of the membrane, the initial size, and the thickness of the water layer required by the user. Interestingly, the user can also adjust in this step the area per lipid of the different lipids starting from the suggested values based, when available, on experimental values. Lipid asymmetry is also allowed between leaflets, giving the user the opportunity to account for such biologically relevant matter, as discussed earlier. Secondly, the user can choose one out of the two most common methods used to build membrane components (i.e., replacement or insertion) and also the type of ionization needed. A huge structural library of each lipid molecule generated from membrane simulations is used in this step to randomly place lipids and generate the initial structure. Lastly, the membrane is first assembled and then slightly preequilibrated to yield a structure ready for an adequate equilibration.

Finally, it is very important to point out that despite heterogeneous membrane simulation being richer in information than classical pure membrane environments, mixing such complex mixtures requires really high simulation time lengths. Certain simulation techniques, such as the so-called simulated annealing, can slightly favor the mixing lipid components, mainly in those systems containing high-T_m lipids (Metcalf & Pandit, 2012). This technique starts from high temperatures to a gradual cooldown of the system until reaching the final temperature of the production runs. The purpose is favoring a little the mixing and ensuring an adequate thermalization of hydrocarbon tails prior to the equilibration phase. However, on the one hand, the efficacy of fast heating and cooling cycles when sampling the phase space on membrane simulations is still not validated. On the other hand, this protocol can compromise the structure of lipid molecules, needs to be thoroughly validated prior to its use in production runs, and, above all, does not dramatically increase lateral diffusion. All in all, it is necessary to bear in mind that the more complex our membrane is, the longer it will take to undergo an adequate mixing.

4.1.3 Analysis and interpretation of membrane simulations

Studying the biophysical properties of membrane lipids is the first step towards a complete understanding of GPCR–membrane systems. A wide range of biophysical parameters can be obtained from molecular simulations to characterize membrane structure and dynamics. One of these parameters is the thickness of the bilayer, normally measured as the so-called "phosphate-to-phosphate" distance (Fig. 4.1). Membrane thickness provides important information on what is the potential extent of the interaction between the GPCR and the membrane, that is, the regions potentially exposed to the membrane. Electron density profiles of membranes are very useful to obtain information on the general structure of the bilayer and spot changes in membrane thickness between bilayers (Fig. 4.1). Thus, certain changes in lipid composition can impact membrane thickness, for example, high-cholesterol membranes are normally thicker (Hung et al., 2007). In addition, these profiles are able to describe other events such as the interaction between upper and lower leaflets also known as "membrane interdigitation" (Fig. 4.1). Longer lipids display higher levels of membrane interdigitation

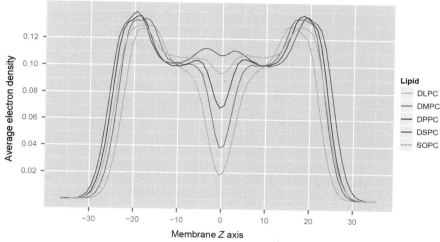

FIGURE 4.1

Electron density profiles of different membranes. This figure, reprinted with permission from Sadiq et al. (2013), shows the average electron density profile of five different membrane systems simulated over 200 ns across the lipid bilayer (x-axis in Å). All systems contain the cholesterol (1:3), POPC (1:3), plus the phospholipid under study, namely, DLPC, DMPC, DPPC, DSPC, or SOPC (1:3). The center of the bilayer is located at 0 Å. Membrane thickness is normally measured as the "phosphate-to-phosphate" distance, that is, the distance in Å between maximum peak values (normally at ±15–25 Å from the bilayer center). In contrast, membrane interdigitation is assessed around the center of the bilayer. Both membrane thickness and membrane interdigitation increase for longer chains; however, a subtle interplay between chain length and degree of chain saturation exists. In this sense, despite having the same carbon length, SOPC and DSPC chains display a different profile of membrane interdigitation. DLPC, 1,2-dilauroyl phosphatidylcholine; DMPC, 1,2-dimyristoyl phosphatidylcholine; DPPC, 1,2-dipalmitoyl phosphatidylcholine; DSPC, 1,2-distearoyl phosphatidylcholine; POPC, 1-palmitoyl-2-oleoyl phosphatidylcholine; SOPC, 1-stearoyl-2-oleoyl phosphatidylcholine. (See color plate.)

as they have to inevitably accommodate the bottom part of their tails (Fig. 4.1). However, when dealing with protein–membrane systems, the local deformation of the membrane around the protein needs to be taken into account when describing membrane thickness. In this line, Mondal, Khelashvili, Shan, Andersen, and Weinstein (2011) recently developed a tool to compute, among other things, the local deformation profile of GPCR–membrane simulations (see Section 4.2.3).

Yet, the structure of the membrane is ultimately determined by the conformation of the individual lipid species within the bilayer. Therefore, the area per lipid is an important parameter that holds information on the condensation degree of our system and that can serve as a monitor of the equilibration state. This parameter is easily calculated as the average of the area of the simulation box, XY, divided by the number of lipids present in the system, N:

$$A = \left\langle \frac{XY}{N} \right\rangle$$

However, the presence of other molecules (e.g., proteins) or the heterogeneity of complex lipid mixtures hampers such calculation. For heterogeneous mixtures, certain approaches such as employing Voronoi tessellations can help to obtain approximate values of area per lipid for individual species (Mori, Ogushi, & Sugita, 2012). This approach uses certain atoms of each lipid molecule and projects them into a plane so that we obtain their x- and y-coordinates. Such layout containing all points is then tessellated in order to fill the plane with polygons of a certain number of vertices, V, where the area of each polygon corresponds to the area of each lipid molecule:

$$A_{\text{polygon}} = \left\langle \frac{1}{2} \sum_{i=0}^{V-1} (x_i + y_{i+1} - y_i + x_{i+1}) \right\rangle$$

Various studies (Pandit et al., 2004; Shushkov, Tzvetanov, Velinova, Ivanova, & Tadjer, 2010) have used this approach in an attempt to obtain values of areas per lipid in ternary or more complex lipid mixtures. Values of areas per lipid obtained thereby need to be used with caution, as, unfortunately, we still cannot compare these values with experiments of complex mixtures.

In addition to condensation and thickness, membrane fluidity is one of the key structural properties one needs to analyze to understand the behavior of lipid mixtures. As discussed earlier, the lipid composition has a big impact on membrane fluidity and indirectly on the dynamics of membrane proteins. A finer analysis of the structure of the lipid bilayer requires studying the chain structure of the hydrophobic tails of membrane phospholipids. Assessing the order of these tails is a way to characterize lipid chain structure as in NMR experiments, where this is habitually measured through the S_{CD} or deuterium order parameter. This parameter quantifies the disorder of phospholipid hydrocarbon tails by averaging the angle of each C—H bond, θ, present in each tail with respect to the bilayer normal (Vermeer, de Groot, Réat, Milon, & Czaplicki, 2007).

$$S_{\text{CD}} = \left| \left\langle \frac{3\cos^2\theta - 1}{2} \right\rangle \right|$$

Highly ordered lipid bilayers show higher values of S_{CD} when compare to more fluid membranes. This parameter, which is normally in agreement with experiments, is very useful to understand the structure of our membrane and to study the impact of different lipid composition on membrane order. In this way, theoretical S_{CD} values are gradually higher when increasing amounts of cholesterol are added to the membrane system, a rigidifying effect already shown in membrane experiments using this sterol (Bartels, Lankalapalli, Bittman, Beyer, & Brown, 2008). Higher chain order values are normally linked to lower areas per lipid and, generally, to higher membrane packing. Such packing should in turn affect the average tilt angle of cholesterol molecules with respect to the bilayer normal. For this reason, the distribution of cholesterol tilt angles across the simulation is frequently used to assess membrane condensation and phase behavior (De Joannis et al., 2011; Zhao et al., 2011).

However, it should be emphasized that the structural behavior of any system is ultimately described by the behavior of its particles. In statistical mechanics, one parameter called radial distribution function (RDF) is used to characterize the density of molecules as a function of distance from a reference molecule. Thus, the RDF can provide information on the probability of finding certain lipid species around a reference lipid drawing the radial symmetry of our system during the simulation. This radial information can help in understanding the probability density of the different coordination shells built around a certain molecule type. However, the standard RDF parameter only gives density values in one dimension, namely, the distance from the reference molecule (Fig. 4.2). By considering two or three variables to represent lipid probability densities, one can visualize a more complete view of the symmetry or solvation profile across the simulation. For example, the planar steroid ring of cholesterol is defined by a smooth face (α-face) versus a rough face (β-face) that contains off-plane methyl groups. Phospholipids do not interact indistinctly with both cholesterol faces and we need more than positional information to characterize this

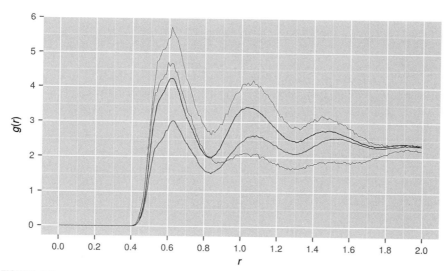

FIGURE 4.2

Radial distribution function, $g(r)$, as a function of the distance to a reference molecule. The plot shows the density of phospholipid tails around cholesterol molecules averaged over 500 ns of simulation for four different membranes. The x-axis represents the distance in nm to the center of mass of each cholesterol molecule of the simulation. First and second coordination shells show up around 0.6 and 0.85 nm, respectively. (See color plate.)

interaction. Thus, bivariate correlation functions based on both orientation and position of lipid tails around cholesterol molecules have been used in the recent years to define the amount and the preferential interaction sites of phospholipids around cholesterol in 2 (Martinez-Seara, Róg, Karttunen, Vattulainen, & Reigada, 2010) or even 3 dimensions (Pitman, Suits, Mackerell, & Feller, 2004).

Apart from membrane structure, biophysical parameters describing the lateral motion of lipids and proteins are needed for a more complete characterization of the membrane system under study. Values of root mean square displacement and diffusion coefficients serve as a reference of how changes in membrane composition affect membrane dynamics. We do not have, however, the computational power needed to generate reliable data on protein or lipid diffusion. On the one hand, all current models still lack the adequate level of representation or the necessary timescale for an accurate interpretation (Javanainen et al., 2013). On the other hand, various studies point towards an existing concerted diffusion in membrane (Apajalahti et al., 2010) and membrane-protein systems (Niemelä et al., 2010). As a result, the interpretation of diffusion data on lipid or protein diffusion needs to be used with extreme caution.

4.2 GPCR MONOMERS IN MEMBRANES

The relationship between lipids and GPCRs is today a well-known fact. The GPCR surface is the most common and intuitive place of interaction between the lipids and GPCRs. Although such interaction has been always thought to be unspecific, various studies describe certain specificity between cholesterol and GPCRs (Thompson et al., 2011). On the one hand, cholesterol seems to affect the GPCR architecture and various specific interaction sites have been postulated (Cang et al., 2013; Sengupta & Chattopadhyay, 2012). Hanson et al. (2008) have recently described for the first time a specific cholesterol binding site at the crystal structure of the β_2-adrenergic receptor. The authors even define what they call a cholesterol consensus motif present in most of the class A GPCRs. Membrane phospholipids also seem to interact to a certain extent with the surface of GPCRs. Both computational (Pitman, Grossfield, Suits, & Feller, 2005) and experimental (Gawrisch, Soubias, & Mihailescu, 2008) works show how preferential interactions between GPCRs and membrane phospholipids are linked to the degree of chain unsaturation of the membrane. A well-known interaction between lipids and GPCRs, although only occurring within the Golgi apparatus, is the so-called palmitoylation. GPCR palmitoylation is a posttranslational modification that many GPCRs undergo during their process of maturation. A palmitic acid is covalently linked to one or more cysteines of the GPCR, increasing its hydrophobicity and driving it to membrane and subcellular localization (Goddard & Watts, 2012). Without a doubt, the most intimate relationship between lipids and GPCRs ever crystallized can be observed, however, at the recent structure of the sphingosine-1-phosphate receptor (Parrill, Lima, & Spiegel, 2012). This receptor can accommodate a physiological phosphosphingolipid in its binding pocket, in contrast to the surface-based interactions described earlier. Therefore, adequately representing the membrane environment in MD simulations of GPCRs is a key factor to successfully characterize the communication between GPCRs and membrane lipids.

In the following section, we give an overview of GPCR modeling, the process of embedding GPCRs into membranes, facts about simulating receptor–membrane complexes, and a detailed description of assessing relevant lipid–GPCR interactions.

4.2.1 Modeling GPCRs

All GPCRs share a general architecture in which seven TM helices are connected by intra- and extracellular loops. However, upon the lack of structural information on a particular target receptor, a first milestone is the construction of a reliable three-dimensional (3D) structure by homology modeling. Homology modeling involves the generation of an all-atom model of the target structure based on experimentally derived high-resolution structures of a closely related (homologous) protein (template). Continuous advances in solving high-resolution GPCR structures keep on providing excellent templates for the prediction of 3D structures. At the time of writing this chapter, 44 GPCR structures of 16 different receptor subtypes were deposited at the PDB (Berman et al., 2000). According to a recent assessment of GPCR modeling (GPCR DOCK2010), the selection of a correct template is extremely important for an accurate prediction of the target structure and should have at least 35–40% sequence identity with the target protein (Kufareva, Rueda, Katritch, Stevens, & Abagyan, 2011). It should be stressed, however, that there is a lack of reliable templates for certain GPCRs, even between some class A GPCRs such as the cannabinoid receptor.

The modeling procedure generally starts with a sequence alignment, where the sequence of the target GPCR, typically retrieved from the UniProt database (http://www.uniprot.org), is aligned to the sequence of the selected template structure using a sequence alignment tool (e.g., ClustalW—http://www.ebi.ac.uk/Tools/msa/clustalw2/). Usually, the resulting alignment needs to be manually refined to guarantee a perfect alignment of the highly conserved residues of the GPCR superfamily. In a next step, the obtained alignment along with the 3D structure of a template serves as an input for specific modeling software, for example, MODELLER software (Eswar et al., 2007), which is able to yield a pool of initial 3D structural models of the target GPCR. Disulfide bonds, such as the highly conserved one established between cysteines C3.25 (Ballesteros–Weinstein nomenclature (Ballesteros & Weinstein, 1995)) at the beginning of the TM helix 3 and the one located in the middle of the extracellular loop 2, have to be assigned and maintained as constraints for geometric optimization. Best models can be selected by both using the MODELLER objective function and visual inspection. An issue of high relevance that could affect the dynamic properties or the stability GPCR target structures in later simulation experiments is the protonation state of titratable groups. A useful method that can help in assessing this issue is based on the PROPKA algorithm (Li, Robertson, & Jensen, 2005), which is also available as a web server application (http://propka.ki.ku.dk/).

After a correction of the protonation state, resulting structures are minimized using a force field suitable for proteins (e.g., CHARMM; Mackerell, Feig, & Brooks (2004)) and any of the available simulation software (e.g., NAMD and ACEMD). Finally, a decent stereochemical quality of the minimized structures in terms of a good distribution of Ψ and Φ angles can be confirmed and visualized in the form of Ramachandran plots by using a software such as PROCHECK (Laskowski, MacArthur, Moss, & Thornton, 1993).

The final structures are now ready to be embedded into a lipid bilayer with a view to simulate GPCR–membrane systems, a process described in the succeeding text.

4.2.2 Building and simulating GPCR–membrane complexes

As described in Section 4.1.2 earlier, the structure of shortly equilibrated phospholipid bilayers of customizable composition can be easily obtained using the membrane-builder module of the CHARMM-GUI (http://www.charmm-gui.org/) web interface. This module automatizes the process by generating the lipid bilayer, embedding the protein into it, solvating and ionizing the membrane-protein complex, and performing a short equilibration of the whole system. As for other membrane-protein simulations, the size of the box is an important parameter since GPCRs need sufficient space to evolve under periodic boundary conditions (i.e., a membrane patch of about 94×94 Å size is a typical choice for GPCR–membrane simulations). Despite the CHARMM-GUI membrane builder can take care of the protein insertion for us, which is useful to quickly build systems out of different initial starting structures, the receptor insertion can also be handled manually by using a molecular visualization software such as VMD. The latter approach gives more control over the construction of the complete receptor–membrane system and enables to avoid artifacts such as certain undesirable interactions between molecules (i.e., lipid tails stuck into the cholesterol ring).

Thus, in the first step, the modeled target GPCR is placed into the center of the membrane patch. In the second step, those lipids that overlap with the receptor or that are in close contact with the receptor (e.g., below 1 Å distance from any protein atoms) are deleted to avoid clashes during the simulation. This procedure generates a hole that perfectly accommodates the target receptor. The receptor–membrane system thereby obtained can then be solvated and neutralized by applying the solvate and autoionize VMD plug-ins, which create a customized water box and assign the type of ions needed to neutralize the system. As mentioned in Section 4.1.2, the hydration level of the system is an important parameter that is frequently overlooked. Similarly, the number and type of ions need to be carefully chosen at this step in order to avoid unexpected results due to nonphysiological ion concentrations.

Once the GPCR is solvated by lipids, water, and ions, the system can be subjected to a relaxation phase before starting the production phase. A typical relaxation phase starts with a minimization procedure of 1000 steps before entering an equilibration phase using an NPT ensemble with a target pressure equal to 1.01325 bar, a temperature of 300 K, and a time step of 2 fs. In this relaxation phase, it is very important to first equilibrate the membrane and solvent around the protein prior to the equilibration of the whole system. The standard procedure is based on the application of positional restraints of 10 kcal mol^{-1} Å^{-2} to all heavy atoms of the receptor to progressively reduce such restraints to zero until the complete release of the protein. Such simulation conditions allow the closure of small holes at the lipid–protein interface, frequently generated at this step while the system is adopting an adequate box size. Nowadays, the available computational resources enable to equilibrate GPCR–membrane systems during 20–100 ns, mostly depending on the complexity of the membrane environment. The use of coupled pressure baths in NPT simulations puts a limit to the speed of the simulation as the integration time step cannot be higher than 2 fs in such simulations. Nevertheless, once the system is equilibrated, we can release the pressure coupling and switch to the canonical ensemble (NVT, number of particles (N), volume (V), and pressure (T) remain constant) where larger integration time steps (i.e., 4 fs) can be used and, as a result, the system can be simulated for longer times.

Studying the dynamic behavior of GPCRs in such native-like membrane systems can give valuable and meaningful insights into GPCR functioning. However, as for any other protein–membrane system, the statistical significance of a particular result or event obtained through MD simulations of GPCRs is based on the number of times one is able to reproduce it. In this line, production runs need to be replicated as many times as possible to support the relevance of the observed dynamic events. Since most of the systems are coupled to temperature bath (i.e., NVT and NPT simulations), we can take advantage of the randomness introduced by the "friction" term of certain thermostats and simply restart the production to yield different replicates. Ideally, any property studied in the reference system (i.e., the first system simulated) should be calculated for all the replicated production runs.

4.2.3 Studying GPCR–lipid interactions from simulations: quantification of membrane remodeling and residual mismatch

The lipid environment plays an important role in the membrane GPCR interactions. The membrane composition modulates both function and conformation of membrane proteins (Engelman, 2005; Phillips, Ursell, Wiggins, & Sens, 2009). Despite the relevance of membrane-mediated GPCR modulation, many aspects remain still unclear at a molecular level. Specific lipid–receptor interactions and unspecific interactions that are mediated via bulk properties of the membrane (e.g., membrane thickness and elasticity) have been reported to modulate receptor functioning. One particular phenomenon of an unspecific membrane effect is the hydrophobic mismatch, where the shape of the membrane changes by getting deformed according to, among other factors, the type and conformation of the protein embedded in this membrane (Huang, 1986; Nielsen, Goulian, & Andersen, 1998). Such adjustment between membrane and proteins can be explained in terms of the energy cost due to the aforementioned hydrophobic mismatch (Lundbaek, Collingwood, Ingólfsson, Kapoor, & Andersen, 2010; Mondal et al., 2011). Thus, two different contributions affect this mismatch: on the one hand, the remodeling of the membrane upon interacting with proteins and, on the other hand, the residual mismatch due to any incomplete adaptation to the membrane in order to avoid highly energetic or unfavorable contacts, that is, hydrophobic–hydrophilic contacts. The purpose of this section is to illustrate how to assess such hydrophobic mismatch, in particular for the simulation of GPCR–membrane complexes.

The consideration of the membrane as an elastic continuum medium has been widely used (Choe, Hecht, & Grabe, 2008; Huang, 1986; Mondal et al., 2011; Nielsen et al., 1998) for the quantification of membrane remodeling due to the assembly of TM proteins. This continuum medium deforms according to several forces specified by means of empirical moduli that depend on the composition of the membrane. The free energy cost of such deformation is described to second order as a surface integral,

$$\Delta G_{def} = \int_\Omega \frac{1}{2} \left\{ \frac{K_a}{d_0^2} 4u^2 + K_c \left(\frac{\partial^2 u}{\partial x^2} + \frac{\partial^2 u}{\partial y^2} - C_0 \right) + \alpha \left(\left(\frac{\partial u}{\partial x} \right)^2 + \left(\frac{\partial u}{\partial y} \right)^2 \right) \right\} d\Omega$$

where K_a and K_c stand for the compression–extension and the splay distortion moduli, respectively; C_0 stands for the monolayer spontaneous curvature and α for the surface tension; Ω represents the membrane area; and u stands for the deformation function per bilayer leaflet, which is defined locally as

$$u(x,y) = \frac{1}{2}(d(x,y) - d_0)$$

without assuming a radial symmetry and where $d(x,y)$ and d_0 are the local thickness and the bulk bilayer thickness, respectively.

The conventional elastic models, which have been devoted to the study of membrane–single TM helix systems, assume a radial symmetry of the system (Huang, 1986; Nielsen & Andersen, 2000; Nielsen et al., 1998; Owicki & McConnell, 1979; Owicki, Springgate, & McConnell, 1978). The usual procedure to assess these systems starts with a self-consistent minimization of the ΔG_{def} with regard to the deformation function u and based on several initial boundary conditions (i.e., u at the membrane-protein boundaries, $u \to 0$ and $\nabla u \to 0$ at the unperturbed regions of the membrane, etc.). Therefore, only a few pieces of information are needed to complete the assessment. In this line, a recent implementation enables the inclusion of asymmetries in the description of a membrane-protein boundary of singly charged TM segments (Choe et al., 2008). However, an adequate application of the elastic continuum theory to membrane–GPCR systems requires a more detailed description of that boundary due to the anisotropic character of the membrane deformation at the membrane-protein contour. Nowadays, this issue has been solved since the deformation function u and the necessary boundary conditions can be obtained from MD trajectories. However, the deformation function u obtained from MD simulations has been claimed to be too noisy for an accurate evaluation of ΔG_{def} (Mondal et al., 2011). Thereby, the combination of MD with continuum elastic models has made possible the accurate study of the remodeling of the membrane due to the GPCR insertion. This philosophy is conceived in a recent model known as three-dimensional continuum MD (3D-CTMD) (Mondal et al., 2011). The 3D-CTMD model is a generalization of conventional continuum models (Huang, 1986; Nielsen et al., 1998; Owicki & McConnell, 1979; Owicki et al., 1978) that allows for the consideration of radial asymmetries, needed for systems with several TM segments such as GPCRs. In addition, the 3D-CTMD model is able to go beyond the assumption of strong hydrophobic coupling (Huang, 1986; Nielsen & Andersen, 2000; Nielsen et al., 1998; Owicki & McConnell, 1979; Owicki et al., 1978), which assumes a total alleviation of the residual mismatch.

The 3D-CTMD model has been implemented in the recently developed CTMDapp software that provides a user-friendly interface (http://memprotein.org). The input required by the CTMDapp is an equilibrated MD trajectory file containing those atoms used for measuring membrane thickness (i.e., P or C2 atoms). As for most membrane-protein simulation analyses, the membrane normal needs to be previously aligned to the z-axis, whereas reference coordinates, or zero point, must be set to the center of mass of the protein. The initial step of the procedure consists in the calculation of the MD-averaged deformation profile u out of the time-averaged leaflets fitted in a grid (Fig. 4.3). Thereafter, the CTMDapp asks the user for a PDB file containing the coordinates of the protein at any point of the simulation in order to accurately define membrane-protein boundaries. At this point, the boundary conditions needed for the optimization procedure are already created. Once the user sets the aforementioned empirical moduli according to the composition of the membrane, the setup is finished and the optimization procedure starts. To check the convergence of the calculation, several extraiterations can be requested after the optimization procedure is finished. Then, the self-consistently optimized deformation profile u and ΔG_{def} are finally obtained (Fig. 4.3). The preparation of all inputs needed in this pipeline can be straightforwardly performed by a visualization software such as VMD.

The complete adaptation of the membrane in order to avoid unfavorable lipid–protein contacts is not always possible. In particular, this adaptation is nearly unreachable for complex TM proteins such as GPCR–membrane systems. GPCRs are proteins with a minimum of seven adjacent TM segments of different lengths. The complete alleviation of hydrophobic–hydrophilic contacts for all segments would likely need a huge amount of energy devoted to severe local deformations of the membrane. Thereby, the so-called residual mismatch is the remaining set of unfavorable interactions occurring due to hydrophobic–hydrophilic contacts in the equilibrated protein–membrane system.

According to the aforementioned continuum formulation, the penalty energy due to the membrane-protein remodeling and the residual mismatch should be optimal in the equilibrium state of the system. Only recently, the quantification of the residual mismatch in GPCR–membrane systems has been performed ("Interactions of the cell membrane with integral proteins," n.d.; Mondal et al., 2011; Mondal, Khelashvili, Shi, & Weinstein, 2013) by determining the accessible surface area involved in unfavorable hydrophilic–hydrophobic (i.e., exposed surface area) contacts between membrane and protein. Here, the authors assess the residual mismatch energy by AG_{res}:

$$AG_{res} = \sum_{i=1}^{N_{TM}} AG_{res,i} \sim \sum_{i=1}^{N_{TM}} \sigma_{res} SA_{res,i}$$

where N_{TM} is the number of TM segments in the protein, $SA_{res,i}$ is the exposed surface area in each TM segment, and σ_{res} is a constant value, namely, 0.028 kcal mol^{-1} Å^{-2} (Ben-Tal, Ben-Shaul, Nicholls, & Honig, 1996; Choe et al., 2008), describing the surface transfer energy of peptide residues between hydrophobic and polar environments.

Frequently, the assessment of the $SA_{res,i}$ involved in unfavorable hydrophobic–hydrophilic contacts is performed by means of solvent-accessible surface area (SASA) calculations. These calculations can be carried out by a variety of available software such as the NACCESS software (http://www.bioinf.manchester.ac.uk/naccess/) used by Mondal et al. (2011) in their recent studies of GPCRs. Herein, two types of unfavorable contacts are considered as contributors for the residual energy penalty, namely, hydrophilic residues facing the hydrophobic part of the bilayer and hydrophobic residues facing the hydrophilic part of the bilayer or water. To identify these contacts, Mondal et al. compare the hydrophobic length of each helical segment along the z-axis with that of the surrounding membrane. It is important to note that $SA_{res,i}$ is calculated from the SASA value of the whole residue. However, such residue-wise consideration is prone to bias the SASA calculation by including favorable contacts such as nonpolar atoms of polar residues in contact with the hydrophobic part of the bilayer. Thus, an atom-wise consideration of residues could generate a more sensitive filtering of the unfavorable hydrophobic–hydrophilic contacts. In spite of these pitfalls, the herein described assessment of the $SA_{res,i}$ is still a useful method to qualitatively study residual exposure in GPCR–membrane systems.

Due to the relatively recent release of the 3D-CTMD model, this powerful tool has not been widely used. To our knowledge, only a few published works perform this type of analysis (Khelashvili et al., 2012; Mondal et al., 2011, 2013). The original work where Mondal et al. describe this methodology (Mondal et al., 2011) is a nice example showing its capabilities. In this study, the CTMD model is used to describe how the lipid composition affects membrane remodeling in GPCR–membrane systems. Thus, they studied the GPCR rhodopsin embedded in membranes of different lipid composition. One of the conclusions of this study is that thicker bilayers (namely, pure diC$_{20:1}$PC and 7:7:6 C$_{18:0}$–C$_{22:6}$PC/C$_{16:0}$–C$_{18:1}$PC/cholesterol membranes) become thinner at the membrane-protein boundary, whereas thinner bilayers (namely, pure diC$_{14:1}$PC and diC$_{16:1}$PC) become thicker. After quantification of the residual mismatch energy for these systems, the authors show how the TM segment 1 (TM1) in rhodopsin embedded in pure diC$_{14:1}$PC is most likely involved in a potential mismatch-driven oligomerization interface. In addition, the authors describe a higher predominance of the TM4 segment when rhodopsin is embedded in thicker membranes, suggesting this segment could also be involved in a potential oligomerization interface (i.e., 7:7:6 C$_{18:0}$–C$_{22:6}$PC/

4.2 GPCR Monomers in Membranes

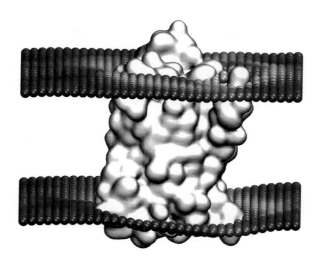

FIGURE 4.3

Membrane-deformation profile $u(x,y)$ of the simulation of a serotonin 5-HT$_{2A}$ receptor embedded in a pure $C_{18:0}$–$C_{22:6}$PC bilayer. The figure depicts a membrane-deformation profile calculated out of the last 200 ns of this trajectory. Beads represent the averaged membrane leaflets based on the positions of lipid P atoms. A color gradient scale illustrates for this particular simulation the minimum (deep blue) and maximum (deep red) values of membrane deformation, $u(x,y)$, being 3.85 and −0.62 Å, respectively. Interestingly, the membrane-to-protein adaptation is higher for the intracellular leaflet when compared to the extracellular one. It is important to note that the formulation of the membrane-deformation variable $u(x,y)$ herein calculated does not account for local adaptations of each leaflet but for the overall adaptation of the bilayer. (For interpretation of the references to color in this figure legend, the reader is referred to the online version of this chapter.)

$C_{16:0}$–$C_{18:1}$PC/cholesterol) (Fig. 4.3). Interestingly, these predictions are in good agreement with experimental results (Palczewski et al., 2000) and other recent CG MD simulations of this protein (Periole, Huber, Marrink, & Sakmar, 2007).

Recently, the same group (Mondal et al.) has studied the effect of ligands with different pharmacological properties on the remodeling of membrane in GPCR–membrane simulations (Mondal et al., 2011; Shan, Khelashvili, Mondal, Mehler, & Weinstein, 2012). In these studies, they analyzed a set of MD trajectories they had previously published (Shan, Khelashvili, Mondal, & Weinstein, 2011). Therein, they performed MD simulations of the serotonin (5-hydroxytryptamine, 5-HT) 5-HT$_{2A}$ receptor in complex with either the full agonist 5-HT, the partial agonist lysergic acid diethylamide (LSD), or the inverse agonist ketanserin (KET) and inserted into a multicomponent membrane (7:7:6 $C_{18:0}$–$C_{22:6}$PC/$C_{16:0}$–$C_{18:1}$PC/cholesterol). The MD trajectories showed conformational changes in response to the different ligand binding at TM1, TM4, and TM6 helices. The 3D-CTMD analysis of those trajectories revealed different patterns of bilayer deformations around the membrane-protein boundary based on different protein conformations induced by ligand. In addition, the authors describe how each system displays particular residual mismatch landscapes as a result of these different deformation patterns. It was concluded that the segments of the 5-HT$_{2A}$ receptor with the largest drive for mismatch-driven oligomerization were TM5 in the 5HT-bound configuration, TM4 and TM5 in the LSD-bound configuration, and TM1 in the KET-bound configuration. In addition, the authors postulate potential dimerization interfaces for this receptor that go along with scientific data and previously reported data for other GPCR systems (Fotiadis et al., 2006, 2003; Guo, Shi, Filizola, Weinstein, & Javitch, 2005; Mancia, Assur, Herman, Siegel, & Hendrickson, 2008).

Therefore, the hydrophobic mismatch between receptor and membrane is being suggested to be a regulatory factor used by nature to modulate GPCR functioning. It seems that one way for the GPCR–membrane system to minimize this energetic penalty in nature is forming higher-order GPCR complexes that allow matching the unfavorable exposed receptor region and burying them in the newly formed receptor–receptor interface. Hence, the residual hydrophobic mismatch might be an important driving factor to facilitate receptor oligomerization, an issue of high relevance for fine-tuning GPCR signaling. Computational approaches to study GPCR oligomerization are described in the next section.

4.3 GPCR DIMERS AND OLIGOMERS IN MEMBRANES
4.3.1 Modeling GPCR–GPCR interactions

The interaction between GPCRs has been largely controversial mainly because the GPCR architecture allows certain monomers to signal individually (Whorton et al., 2007). Today, we know that an altered communication between GPCR monomers can influence the regulation of important diseases such as schizophrenia or Parkinson's disease (Guixà-González et al., 2012). Thus, while the functional communication between GPCRs is a fact, only during the last decade have scientists searched for a potential physical interaction between GPCR monomers that confirms the existence of GPCR heteromers. Various biochemical and biophysical techniques have greatly enriched the information we currently hold on GPCR–GPCR interactions and have helped in defining specific GPCR dimers and oligomers (Kaczor & Selent, 2011). GPCRs, as many other membrane receptors, seem to form dimers and higher-order oligomers in cells (Han, Moreira, Urizar, Weinstein, & Javitch, 2009; Milligan, 2009). The first crystal of a mammalian GPCR could not come to light until 2000, when Palczewski et al. (2000) determined the structure of bovine rhodopsin. Thereafter, an interesting series of class A GPCR crystals have been solved, including a β_2-adrenergic receptor bound to a G protein heterotrimer (Rasmussen et al., 2011). However, only a few of these crystals help in understanding the nature of GPCR oligomers (Sadiq et al., 2013); among them, the first crystal structure of an adrenoceptor homomer has been very recently solved (Huang,

Chen, Zhang, & Huang, 2013), showing dimerization interfaces previously described in computational studies. Therefore, despite all the challenges, assessing GPCR oligomerization by MD simulations allows studying certain aspects of this phenomenon that today we cannot evaluate experimentally.

Nowadays, two main concerns govern the study of GPCR oligomerization by all-atom standard MD simulations. On the one hand, a large amount of simulation data is needed to assess the rotational and diffusional energy profile of the two GPCR monomers. Therefore, describing the energetic landscape of the GPCR dimerization process requires very long timescales, namely, in the order of milliseconds (Sadiq et al., 2013). Such timescales are not reachable by the current technological resources and, particularly, when using highly descriptive simulations such as all-atom standard MD. However, certain attempts have been made to overcome the timescale problem such as reducing the level of description of the simulation (e.g., CG simulations; Periole et al. (2007)) or just performing a reduced sampling by biasing such simulation (e.g., umbrella sampling (Provasi, Johnston, & Filizola, 2010) and/or metadynamics (Johnston et al., 2011)). On the other hand, a very limited number of GPCR dimers have been solved and deposited in the PDB. These crystal structures contain relevant data on potentially physiological dimerization interfaces between GPCRs that could help in reducing the timescale needed to characterize the energy of the different binding modes. The lack of such information hampers the possibility of adequately biasing the simulation towards low-energy dimers (i.e., physiological interfaces).

4.3.2 Simulating GPCR heteromers: all-atom versus CG simulations

Both factors, namely, high timescales and lack of dimer structures, have tipped the scale in favor of CG simulations when dealing with GPCR dimerization. In CG force fields, a certain number of heavy atoms (including water molecules) are represented as a single interaction center, for example, the four-to-one mapping used by the MARTINI force field (Periole & Marrink, 2013). Such underrepresentation allows CG simulations to reach much longer timescales when compared to all-atom simulations, although a prudent interpretation of the results is needed. The main drawback of CG simulations applied to receptor–receptor interactions lies on the set of restraints that one needs to use to preserve the secondary structure of proteins (Periole, Cavalli, Marrink, & Ceruso, 2009). Since the secondary structure needs to be restrained, CG simulations of this nature do not account for any protein-folding event or any change in the predefined secondary structure. In contrast, CG simulations are very useful to measure protein–protein aggregation (Javanainen et al., 2013) or even reveal preferred dimerization interfaces, as recently shown for rhodopsin oligomers in an inspiring computational study made by Periole, Knepp, Sakmar, Marrink, and Huber (2012).

The high performance of CG simulations gives way to a better sampling of the energetic landscape, and, therefore, this method can simulate much more heterogeneous systems when compared to all-atom simulations. Thus, the MARTINI force field has already been used to simulate important challenges in membrane simulations such as estimating timescales for concerted lipid motions (Apajalahti et al., 2010), characterizing lipid rafts from a molecular perspective (Risselada & Marrink, 2008) or studying preferential interaction between lipid species (Marrink, de Vries, Harroun, Katsaras, & Wassall, 2008). In addition, along with pure membrane simulations, various studies have used the MARTINI force field to characterize the tendency of certain TM proteins to partition into lipid domains of different composition (Schäfer et al., 2011). Interestingly, Domański, Marrink, and Schäfer (2012) have recently simulated the lateral heterogeneity of biological membranes by using the MARTINI force field and showed how the protein bacteriorhodopsin partitions into liquid-disordered lipid domains. This study is a perfect example of how CG simulations can be used to simulate GPCRs in its native-like environment prior to study receptor–receptor interactions.

CG simulations enable to simulate a relatively large set of GPCRs embedded in a membrane patch (Fig. 4.4). In these simulations, GPCRs can freely diffuse across the membrane and transiently or stably interact with other GPCRs, that is, we can simulate the self-aggregation of GPCRs (Periole et al., 2007) and, what is more, we can use more native-like membranes when modeling such self-aggregation. As for all-atom simulations, CG simulations of GPCRs benefit from the accuracy of both an adequate protein input structure and a sufficiently equilibrated lipid environment. Thus, an ideal protein input structure should be based on the information contained in any of the currently solved GPCR crystal structures. One of the interesting tools made available by the MARTINI force field team (http://md.chem.rug.nl/cgmartini/)

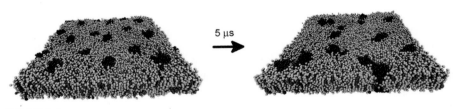

FIGURE 4.4

Coarse-grained (CG) simulations of GPCRs embedded in membranes. This figure represents two snapshots of a CG simulation of 18 GPCR monomers embedded in a membrane. This system was created by multiplying a unit cell consisting of two GPCR monomers embedded in a smaller membrane patch. Membrane lipids were first equilibrated around proteins by restraining all GPCR monomers and running the system during 100 ns. The final snapshot of this equilibration run (left picture) is the starting configuration of the production run. GPCR monomers can now freely diffuse and rotate to find each other and ultimately self-assemble (right picture). Simulation runs longer than 5 μs (i.e., above 10 μs) are normally needed to consistently study GPCR oligomerization. Water and ions are not displayed for clarity; GPCR monomers are depicted in black, whereas phospholipids and cholesterol are depicted in yellow and red, respectively. (See color plate.)

is an automated script for generating a CG representation of a protein from an atomistic structure (i.e., a PDB structure). The generated CG model can be then embedded into a membrane bilayer following similar methods as the one explained earlier. Alternatively, certain molecular engines such as GROMACS have already created specific tools for this purpose, namely, the "g_membed" command, incorporated within the last version of this software. Similarly to all-atom simulations, once the GPCR monomer is embedded in the membrane, the system is solvated and ionized. However, at this point, the unit cell created is multiplied with a view to create a larger system (Fig. 4.4) containing as many GPCRs as allowed by the computational resources available. For example, in their recent study Periole et al. (2012) were able to simulate ten independent systems containing 16 rhodopsin monomers each and, in addition, they created a larger system containing 64 rhodopsin monomers. While all 10 systems were simulated up to 10 μs* (please note an asterisk is used to indicate an effective time that accounts for the speedup factor of four typically used MARTINI CG simulations; Ramadurai et al., 2010), they simulated the large system up to 100 μs*, confirming the valuable performance of CG simulations when dealing with large systems.

All in all, CG simulations remain the preferred simulation technique for generating long trajectories of systems that are otherwise outsized for all-atom simulations, as in the case of GPCR oligomerization studies. Nevertheless, the combination of both CG and all-atom simulations can enrich this approach even further. In this line, various studies describe the use of biased and nonbiased CG and all-atom MD to assess to study the dimerization interface of certain GPCRs such as the δ-opioid receptor (Johnston et al., 2011; Provasi et al., 2010). The aim of these studies is to either structurally describe the most energetically favorable interface (Johnston et al., 2011) or assess the stability of such interface in an attempt to characterize the association lifetime of GPCR dimers (Provasi et al., 2010). Yet, studying the role that membrane lipids have on GPCR dimerization should be the one of the focuses of attention, not only because other studies are pointing towards their potential involvement, as we discuss herein, but also because, today, we do have some tools available for this aim such as CG simulations.

References

Apajalahti, T., Niemelä, P., Govindan, P. N., Miettinen, M. S., Salonen, E., Marrink, S.-J., et al. (2010). Concerted diffusion of lipids in raft-like membranes. *Faraday Discussions*, *144*, 411–430, discussion 445–481.

Ballesteros, J., & Weinstein, H. (1995). Integrated methods for the construction of three dimensional models and computational probing of structure function relations in G-protein coupled receptor. *Methods in Neurosciences*, *25*, 366–428.

Bartels, T., Lankalapalli, R. S., Bittman, R., Beyer, K., & Brown, M. F. (2008). Raftlike mixtures of sphingomyelin and cholesterol investigated by solid-state 2H NMR spectroscopy. *Journal of the American Chemical Society*, *130*(44), 14521–14532. http://dx.doi.org/10.1021/ja801789t.

Bennett, W. F. D., & Tieleman, D. P. (2013). Computer simulations of lipid membrane domains. *Biochimica et Biophysica Acta*, *1828*(8), 1765–1776. http://dx.doi.org/10.1016/j.bbamem.2013.03.004.

Ben-Tal, N., Ben-Shaul, A., Nicholls, A., & Honig, B. (1996). Free-energy determinants of alpha-helix insertion into lipid bilayers. *Biophysical Journal*, *70*(4), 1803–1812. http://dx.doi.org/10.1016/S0006-3495(96)79744-8.

Berman, H. M., Westbrook, J., Feng, Z., Gilliland, G., Bhat, T. N., Weissig, H., et al. (2000). The Protein Data Bank. *Nucleic Acids Research*, *28*(1), 235–242.

Brooks, B. R., Bruccoleri, R. E., Olafson, B. D., States, D. J., Swaminathan, S., & Karplus, M. (1983). CHARMM: A program for macromolecular energy, minimization, and dynamics calculations. *Journal of Computational Chemistry*, *4*(2), 187–217. http://dx.doi.org/10.1002/jcc.540040211.

Cang, X., Du, Y., Mao, Y., Wang, Y., Yang, H., & Jiang, H. (2013). Mapping the functional binding sites of cholesterol in β2-adrenergic receptor by long-time molecular dynamics simulations. *The Journal of Physical Chemistry. B*, *117*(4), 1085–1094. http://dx.doi.org/10.1021/jp3118192.

Choe, S., Hecht, K. A., & Grabe, M. (2008). A continuum method for determining membrane protein insertion energies and the problem of charged residues. *The Journal of General Physiology*, *131*(6), 563–573. http://dx.doi.org/10.1085/jgp.200809959.

Christen, M., Hünenberger, P. H., Bakowies, D., Baron, R., Bürgi, R., Geerke, D. P., et al. (2005). The GROMOS software for biomolecular simulation: GROMOS05. *Journal of Computational Chemistry*, *26*(16), 1719–1751. http://dx.doi.org/10.1002/jcc.20303.

Cornell, W. D., Cieplak, P., Bayly, C. I., Gould, I. R., Merz, K. M., Ferguson, D. M., et al. (1995). A second generation force field for the simulation of proteins, nucleic acids, and organic molecules. *Journal of the American Chemical Society*, *117*(19), 5179–5197. http://dx.doi.org/10.1021/ja00124a002.

De Joannis, J., Coppock, P. S., Yin, F., Mori, M., Zamorano, A., & Kindt, J. T. (2011). Atomistic simulation of cholesterol effects on miscibility of saturated and unsaturated phospholipids: Implications for liquid-ordered/liquid-disordered phase coexistence. *Journal of the American Chemical Society*, *133*(10), 3625–3634. http://dx.doi.org/10.1021/ja110425s.

De Meyer, F., & Smit, B. (2009). Effect of cholesterol on the structure of a phospholipid bilayer. *Proceedings of the National Academy of Sciences of the United States of America*, *106*(10), 3654–3658. http://dx.doi.org/10.1073/pnas.0809959106.

Domański, J., Marrink, S. J., & Schäfer, L. V. (2012). Transmembrane helices can induce domain formation in crowded model membranes. *Biochimica et Biophysica Acta*, *1818*(4), 984–994. http://dx.doi.org/10.1016/j.bbamem.2011.08.021.

Dror, R. O., Arlow, D. H., Maragakis, P., Mildorf, T. J., Pan, A. C., Xu, H., et al. (2011). Activation mechanism of the β2-adrenergic receptor. *Proceedings of the National Academy of Sciences of the United States of America*, *108*(46), 18684–18689.

Engelman, D. M. (2005). Membranes are more mosaic than fluid. *Nature*, *438*(7068), 578–580. http://dx.doi.org/10.1038/nature04394.

Eswar, N., Webb, B., Marti-Renom, M. A., Madhusudhan, M. S., Eramian, D., Shen, M.-Y., et al. (2007). Comparative protein structure modeling using MODELLER. *Current Protocols in Protein Science/Editorial Board, John E. Coligan ... [et al.]*, http://dx.doi.org/10.1002/0471140864.ps0209s50, Chapter 2, Unit 2.9.

Fabelo, N., Martín, V., Santpere, G., Marín, R., Torrent, L., Ferrer, I., et al. (2011). Severe alterations in lipid composition of frontal cortex lipid rafts from Parkinson's disease and incidental Parkinson's disease. *Molecular Medicine (Cambridge, Mass.)*, *17*(9–10), 1107–1118. http://dx.doi.org/10.2119/molmed.2011.00119.

Feigenson, G. W. (2007). Phase boundaries and biological membranes. *Annual Review of Biophysics and Biomolecular Structure*, *36*, 63–77. http://dx.doi.org/10.1146/annurev.biophys.36.040306.132721.

Fotiadis, D., Jastrzebska, B., Philippsen, A., Muller, D. J., Palczewski, K., & Engel, A. (2006). Structure of the rhodopsin dimer: A working model for G-protein-coupled receptors. *Current Opinion in Structural Biology*, *16*(2), 252–259. http://dx.doi.org/10.1016/j.sbi.2006.03.013.

Fotiadis, D., Liang, Y., Filipek, S., Saperstein, D. A., Engel, A., & Palczewski, K. (2003). Atomic-force microscopy: Rhodopsin dimers in native disc membranes. *Nature*, *421*(6919), 127–128. http://dx.doi.org/10.1038/421127a.

Gawrisch, K., Soubias, O., & Mihailescu, M. (2008). Insights from biophysical studies on the role of polyunsaturated fatty acids for function of G-protein coupled membrane receptors. *Prostaglandins, Leukotrienes, and Essential Fatty Acids*, *79*(3–5), 131–134. http://dx.doi.org/10.1016/j.plefa.2008.09.002.

Goddard, A. D., & Watts, A. (2012). Regulation of G protein-coupled receptors by palmitoylation and cholesterol. *BMC Biology*, *10*, 27. http://dx.doi.org/10.1186/1741-7007-10-27.

Guixà-González, R., Bruno, A., Marti-Solano, M., & Selent, J. (2012). Crosstalk within GPCR heteromers in schizophrenia and Parkinson's disease: Physical or just functional? *Current Medicinal Chemistry*, *19*(8), 1119–1134.

Guo, W., Shi, L., Filizola, M., Weinstein, H., & Javitch, J. A. (2005). Crosstalk in G protein-coupled receptors: Changes at the transmembrane homodimer interface determine activation. *Proceedings of the National Academy of Sciences of the United States of America*, *102*(48), 17495–17500. http://dx.doi.org/10.1073/pnas.0508950102.

Han, Y., Moreira, I. S., Urizar, E., Weinstein, H., & Javitch, J. A. (2009). Allosteric communication between protomers of dopamine class A GPCR dimers modulates activation. *Nature Chemical Biology*, *5*(9), 688–695. http://dx.doi.org/10.1038/nchembio.199.

Hanson, M. A., Cherezov, V., Griffith, M. T., Roth, C. B., Jaakola, V.-P., Chien, E. Y. T., et al. (2008). A specific cholesterol binding site is established by the 2.8 A structure of the human beta2-adrenergic receptor. *Structure (London, England: 1993)*, *16*(6), 897–905. http://dx.doi.org/10.1016/j.str.2008.05.001.

Harvey, M. J., Giupponi, G., & Fabritiis, G. D. (2009). ACEMD: Accelerating biomolecular dynamics in the microsecond time scale. *Journal of Chemical Theory and Computation*, *5*(6), 1632–1639. http://dx.doi.org/10.1021/ct9000685.

Hess, B., Kutzner, C., van der Spoel, D., & Lindahl, E. (2008). GROMACS 4: Algorithms for highly efficient, load-balanced, and scalable molecular simulation. *Journal of Chemical Theory and Computation*, *4*(3), 435–447. http://dx.doi.org/10.1021/ct700301q.

Ho, C., Slater, S. J., & Stubbs, C. D. (1995). Hydration and order in lipid bilayers. *Biochemistry*, *34*(18), 6188–6195.

Huang, H. W. (1986). Deformation free energy of bilayer membrane and its effect on gramicidin channel lifetime. *Biophysical Journal*, *50*(6), 1061–1070.

Huang, J., Chen, S., Zhang, J. J., & Huang, X.-Y. (2013). Crystal structure of oligomeric β1-adrenergic G protein-coupled receptors in ligand-free basal state. *Nature Structural & Molecular Biology*, *20*(4), 419–425. http://dx.doi.org/10.1038/nsmb.2504.

Humphrey, W., Dalke, A., & Schulten, K. (1996). VMD: Visual molecular dynamics. *Journal of Molecular Graphics*, *14*(1), 33–38, 27–28.

Hung, W.-C., Lee, M.-T., Chen, F.-Y., & Huang, H. W. (2007). The condensing effect of cholesterol in lipid bilayers. *Biophysical Journal*, *92*(11), 3960–3967. http://dx.doi.org/10.1529/biophysj.106.099234.

Javanainen, M., Hammaren, H., Monticelli, L., Jeon, J.-H., Miettinen, M. S., Martinez-Seara, H., et al. (2013). Anomalous and normal diffusion of proteins and lipids in crowded lipid membranes. *Faraday Discussions*, *161*, 397–417. http://dx.doi.org/10.1039/C2FD20085F.

Jo, S., Kim, T., & Im, W. (2007). Automated builder and database of protein/membrane complexes for molecular dynamics simulations. *PloS One*, *2*(9), e880. http://dx.doi.org/10.1371/journal.pone.0000880.

Jo, S., Kim, T., Iyer, V. G., & Im, W. (2008). CHARMM-GUI: A web-based graphical user interface for CHARMM. *Journal of Computational Chemistry*, *29*(11), 1859–1865. http://dx.doi.org/10.1002/jcc.20945.

Jo, S., Lim, J. B., Klauda, J. B., & Im, W. (2009). CHARMM-GUI Membrane Builder for mixed bilayers and its application to yeast membranes. *Biophysical Journal*, *97*(1), 50–58. http://dx.doi.org/10.1016/j.bpj.2009.04.013.

Johnston, J. M., Aburi, M., Provasi, D., Bortolato, A., Urizar, E., Lambert, N. A., et al. (2011). Making structural sense of dimerization interfaces of delta opioid receptor homodimers. *Biochemistry*, *50*(10), 1682–1690. http://dx.doi.org/10.1021/bi101474v.

Jorgensen, W. L., Maxwell, D. S., & Tirado-Rives, J. (1996). Development and testing of the OPLS all-atom force field on conformational energetics and properties of organic liquids. *Journal of the American Chemical Society*, *118*(45), 11225–11236. http://dx.doi.org/10.1021/ja9621760.

Kaczor, A. A., & Selent, J. (2011). Oligomerization of G protein-coupled receptors: Biochemical and biophysical methods. *Current Medicinal Chemistry*, *18*(30), 4606–4634.

Kasson, P. M., Lindahl, E., & Pande, V. S. (2011). Water ordering at membrane interfaces controls fusion dynamics. *Journal of the American Chemical Society*, *133*(11), 3812–3815. http://dx.doi.org/10.1021/ja200310d.

Katritch, V., Cherezov, V., & Stevens, R. C. (2013). Structure-function of the G protein-coupled receptor superfamily. *Annual Review of Pharmacology and Toxicology*, *53*, 531–556. http://dx.doi.org/10.1146/annurev-pharmtox-032112-135923.

Khelashvili, G., Albornoz, P. B. C., Johner, N., Mondal, S., Caffrey, M., & Weinstein, H. (2012). Why GPCRs behave differently in cubic and lamellar lipidic mesophases. *Journal of the American Chemical Society*, *134*(38), 15858–15868. http://dx.doi.org/10.1021/ja3056485.

Klauda, J. B., Venable, R. M., Freites, J. A., O'Connor, J. W., Tobias, D. J., Mondragon-Ramirez, C., et al. (2010). Update of the CHARMM all-atom additive force field for lipids: Validation on six lipid types. *The Journal of Physical Chemistry. B*, *114*(23), 7830–7843. http://dx.doi.org/10.1021/jp101759q.

Koynova, R., & Caffrey, M. (1998). Phases and phase transitions of the phosphatidylcholines. *Biochimica et Biophysica Acta*, *1376*(1), 91–145.

Kufareva, I., Rueda, M., Katritch, V., Stevens, R. C., & Abagyan, R. (2011). Status of GPCR modeling and docking as reflected by community-wide GPCR Dock 2010 assessment. *Structure (London, England: 1993)*, *19*(8), 1108–1126. http://dx.doi.org/10.1016/j.str.2011.05.012.

Laskowski, R. A., MacArthur, M. W., Moss, D. S., & Thornton, J. M. (1993). PROCHECK: A program to check the stereochemical quality of protein structures. *Journal of Applied Crystallography, 26*, 283–291.

Li, H., Robertson, A. D., & Jensen, J. H. (2005). Very fast empirical prediction and rationalization of protein pKa values. *Proteins, 61*(4), 704–721. http://dx.doi.org/10.1002/prot.20660.

Lindblom, G., & Orädd, G. (2009). Lipid lateral diffusion and membrane heterogeneity. *Biochimica et Biophysica Acta, 1788*(1), 234–244. http://dx.doi.org/10.1016/j.bbamem.2008.08.016.

Lingwood, D., & Simons, K. (2010). Lipid rafts as a membrane-organizing principle. *Science (New York, N.Y.), 327*(5961), 46–50. http://dx.doi.org/10.1126/science.1174621.

Lundbaek, J. A., Collingwood, S. A., Ingólfsson, H. I., Kapoor, R., & Andersen, O. S. (2010). Lipid bilayer regulation of membrane protein function: Gramicidin channels as molecular force probes. *Journal of the Royal Society, Interface/The Royal Society, 7*(44), 373–395. http://dx.doi.org/10.1098/rsif.2009.0443.

MacKerell, A. D., Bashford, D., Bellott, M., Dunbrack, R. L., Evanseck, J. D., Field, M. J., et al. (1998). All-atom empirical potential for molecular modeling and dynamics studies of proteins. *The Journal of Physical Chemistry. B, 102*(18), 3586–3616. http://dx.doi.org/10.1021/jp973084f.

Mackerell, A. D., Jr., Feig, M., & Brooks, C. L., 3rd. (2004). Extending the treatment of backbone energetics in protein force fields: Limitations of gas-phase quantum mechanics in reproducing protein conformational distributions in molecular dynamics simulations. *Journal of Computational Chemistry, 25*(11), 1400–1415. http://dx.doi.org/10.1002/jcc.20065.

Mancia, F., Assur, Z., Herman, A. G., Siegel, R., & Hendrickson, W. A. (2008). Ligand sensitivity in dimeric associations of the serotonin 5HT2c receptor. *EMBO Reports, 9*(4), 363–369. http://dx.doi.org/10.1038/embor.2008.27.

Marrink, S. J., de Vries, A. H., Harroun, T. A., Katsaras, J., & Wassall, S. R. (2008). Cholesterol shows preference for the interior of polyunsaturated lipid membranes. *Journal of the American Chemical Society, 130*(1), 10–11. http://dx.doi.org/10.1021/ja076641c.

Marrink, S. J., Risselada, H. J., Yefimov, S., Tieleman, D. P., & de Vries, A. H. (2007). The MARTINI force field: Coarse grained model for biomolecular simulations. *The Journal of Physical Chemistry. B, 111*(27), 7812–7824. http://dx.doi.org/10.1021/jp071097f.

Martín, V., Fabelo, N., Santpere, G., Puig, B., Marín, R., Ferrer, I., et al. (2010). Lipid alterations in lipid rafts from Alzheimer's disease human brain cortex. *Journal of Alzheimer's Disease, 19*(2), 489–502. http://dx.doi.org/10.3233/JAD-2010-1242.

Martinez-Seara, H., Róg, T., Karttunen, M., Vattulainen, I., & Reigada, R. (2010). Cholesterol induces specific spatial and orientational order in cholesterol/phospholipid, membranes. *PLoS ONE, 5*(6), e11162. http://dx.doi.org/10.1371/journal.pone.0011162.

Maulik, P. R., & Shipley, G. G. (1996). Interactions of N-stearoyl sphingomyelin with cholesterol and dipalmitoylphosphatidylcholine in bilayer membranes. *Biophysical Journal, 70*(5), 2256–2265. http://dx.doi.org/10.1016/S0006-3495(96)79791-6.

Metcalf, R., & Pandit, S. A. (2012). Mixing properties of sphingomyelin ceramide bilayers: A simulation study. *The Journal of Physical Chemistry. B, 116*(15), 4500–4509. http://dx.doi.org/10.1021/jp212325e.

Milligan, G. (2009). G protein-coupled receptor hetero-dimerization: Contribution to pharmacology and function. *British Journal of Pharmacology*, *158*(1), 5–14. http://dx.doi.org/10.1111/j.1476-5381.2009.00169.x.

Mondal, S., Khelashvili, G., Shan, J., Andersen, O. S., & Weinstein, H. (2011). Quantitative modeling of membrane deformations by multihelical membrane proteins: Application to G-protein coupled receptors. *Biophysical Journal*, *101*(9), 2092–2101. http://dx.doi.org/10.1016/j.bpj.2011.09.037.

Mondal, S., Weinstein, H., & Khelashvili, G. (2012). Interactions of the cell membrane with integral proteins. In Egelman E. H. (Ed.), *Comprehensive Biophysics, Vol 9, simulation and modeling, Harel Weinstein* (pp. 229–242). Oxford: Academic Press. ISBN: 978-0-12-374920-8. Copyright 2012 Elsevier B.V. Academic Press.

Mondal, S., Khelashvili, G., Shi, L., & Weinstein, H. (2013). The cost of living in the membrane: A case study of hydrophobic mismatch for the multi-segment protein LeuT. *Chemistry and Physics of Lipids*, *169*, 27–38. http://dx.doi.org/10.1016/j.chemphyslip.2013.01.006.

Mori, T., Ogushi, F., & Sugita, Y. (2012). Analysis of lipid surface area in protein-membrane systems combining Voronoi tessellation and Monte Carlo integration methods. *Journal of Computational Chemistry*, *33*(3), 286–293. http://dx.doi.org/10.1002/jcc.21973.

Nagle, J. F., & Tristram-Nagle, S. (2000). Structure of lipid bilayers. *Biochimica et Biophysica Acta*, *1469*(3), 159–195.

Nielsen, C., & Andersen, O. S. (2000). Inclusion-induced bilayer deformations: Effects of monolayer equilibrium curvature. *Biophysical Journal*, *79*(5), 2583–2604.

Nielsen, C., Goulian, M., & Andersen, O. S. (1998). Energetics of inclusion-induced bilayer deformations. *Biophysical Journal*, *74*(4), 1966–1983.

Niemelä, P. S., Hyvönen, M. T., & Vattulainen, I. (2009). Atom-scale molecular interactions in lipid raft mixtures. *Biochimica et Biophysica Acta*, *1788*(1), 122–135. http://dx.doi.org/10.1016/j.bbamem.2008.08.018.

Niemelä, P. S., Miettinen, M. S., Monticelli, L., Hammaren, H., Bjelkmar, P., Murtola, T., et al. (2010). Membrane proteins diffuse as dynamic complexes with lipids. *Journal of the American Chemical Society*, *132*(22), 7574–7575. http://dx.doi.org/10.1021/ja101481b.

Owicki, J. C., & McConnell, H. M. (1979). Theory of protein–lipid and protein–protein interactions in bilayer membranes. *Proceedings of the National Academy of Sciences of the United States of America*, *76*(10), 4750–4754.

Owicki, J. C., Springgate, M. W., & McConnell, H. M. (1978). Theoretical study of protein–lipid interactions in bilayer membranes. *Proceedings of the National Academy of Sciences of the United States of America*, *75*(4), 1616–1619.

Palczewski, K., Kumasaka, T., Hori, T., Behnke, C. A., Motoshima, H., Fox, B. A., et al. (2000). Crystal structure of rhodopsin: A G protein-coupled receptor. *Science (New York, N.Y.)*, *289*(5480), 739–745.

Pandit, S. A., Vasudevan, S., Chiu, S. W., Mashl, R. J., Jakobsson, E., & Scott, H. L. (2004). Sphingomyelin-cholesterol domains in phospholipid membranes: Atomistic simulation. *Biophysical Journal*, *87*(2), 1092–1100. http://dx.doi.org/10.1529/biophysj.104.041939.

Parrill, A. L., Lima, S., & Spiegel, S. (2012). Structure of the first sphingosine 1-phosphate receptor. *Science Signaling*, *5*(225), pe23. http://dx.doi.org/10.1126/scisignal.2003160.

Periole, X., Cavalli, M., Marrink, S.-J., & Ceruso, M. A. (2009). Combining an elastic network with a coarse-grained molecular force field: Structure, dynamics, and intermolecular recognition. *Journal of Chemical Theory and Computation*, *5*(9), 2531–2543. http://dx.doi.org/10.1021/ct9002114.

Periole, X., Huber, T., Marrink, S.-J., & Sakmar, T. P. (2007). G protein-coupled receptors self-assemble in dynamics simulations of model bilayers. *Journal of the American Chemical Society, 129*(33), 10126–10132. http://dx.doi.org/10.1021/ja0706246.

Periole, X., Knepp, A. M., Sakmar, T. P., Marrink, S. J., & Huber, T. (2012). Structural determinants of the supramolecular organization of G protein-coupled receptors in bilayers. *Journal of the American Chemical Society, 134*(26), 10959–10965. http://dx.doi.org/10.1021/ja303286e.

Periole, X., & Marrink, S.-J. (2013). The Martini coarse-grained force field. *Methods in Molecular Biology (Clifton, N.J.), 924*, 533–565. http://dx.doi.org/10.1007/978-1-62703-017-5_20.

Phillips, J. C., Braun, R., Wang, W., Gumbart, J., Tajkhorshid, E., Villa, E., et al. (2005). Scalable molecular dynamics with NAMD. *Journal of Computational Chemistry, 26*(16), 1781–1802. http://dx.doi.org/10.1002/jcc.20289.

Phillips, R., Ursell, T., Wiggins, P., & Sens, P. (2009). Emerging roles for lipids in shaping membrane-protein function. *Nature, 459*(7245), 379–385. http://dx.doi.org/10.1038/nature08147.

Pitman, M. C., Grossfield, A., Suits, F., & Feller, S. E. (2005). Role of cholesterol and polyunsaturated chains in lipid-protein interactions: Molecular dynamics simulation of rhodopsin in a realistic membrane environment. *Journal of the American Chemical Society, 127*(13), 4576–4577. http://dx.doi.org/10.1021/ja042715y.

Pitman, M. C., Suits, F., Mackerell, A. D., Jr., & Feller, S. E. (2004). Molecular-level organization of saturated and polyunsaturated fatty acids in a phosphatidylcholine bilayer containing cholesterol. *Biochemistry, 43*(49), 15318–15328. http://dx.doi.org/10.1021/bi048231w.

Provasi, D., Johnston, J. M., & Filizola, M. (2010). Lessons from free energy simulations of delta-opioid receptor homodimers involving the fourth transmembrane helix. *Biochemistry, 49*(31), 6771–6776. http://dx.doi.org/10.1021/bi100686t.

Ramadurai, S., Holt, A., Schäfer, L. V., Krasnikov, V. V., Rijkers, D. T. S., Marrink, S. J., et al. (2010). Influence of hydrophobic mismatch and amino acid composition on the lateral diffusion of transmembrane peptides. *Biophysical Journal, 99*(5), 1447–1454. http://dx.doi.org/10.1016/j.bpj.2010.05.042.

Rasmussen, S. G. F., DeVree, B. T., Zou, Y., Kruse, A. C., Chung, K. Y., & Kobilka, T. S. (2011). Crystal structure of the β_2 adrenergic receptor-Gs protein complex. *Nature, 477*(7366), 549–555. http://dx.doi.org/10.1038/nature10361.

Risselada, H. J., & Marrink, S. J. (2008). The molecular face of lipid rafts in model membranes. *Proceedings of the National Academy of Sciences of the United States of America, 105*(45), 17367–17372. http://dx.doi.org/10.1073/pnas.0807527105.

Robinson, D., Besley, N. A., O'Shea, P., & Hirst, J. D. (2011). Water order profiles on phospholipid/cholesterol membrane bilayer surfaces. *Journal of Computational Chemistry, 32*(12), 2613–2618. http://dx.doi.org/10.1002/jcc.21840.

Sadiq, S. K., Guixa-Gonzalez, R., Dainese, E., Pastor, M., De Fabritiis, G., & Selent, J. (2013). Molecular modeling and simulation of membrane lipid-mediated effects on GPCRs. *Current Medicinal Chemistry, 20*(1), 22–38.

Schäfer, L. V., de Jong, D. H., Holt, A., Rzepiela, A. J., de Vries, A. H., Poolman, B., et al. (2011). Lipid packing drives the segregation of transmembrane helices into disordered lipid domains in model membranes. *Proceedings of the National Academy of Sciences of the United States of America, 108*(4), 1343–1348. http://dx.doi.org/10.1073/pnas.1009362108.

Sengupta, D., & Chattopadhyay, A. (2012). Identification of cholesterol binding sites in the serotonin1A receptor. *The Journal of Physical Chemistry. B, 116*(43), 12991–12996. http://dx.doi.org/10.1021/jp309888u.

Shan, J., Khelashvili, G., Mondal, S., Mehler, E. L., & Weinstein, H. (2012). Ligand-dependent conformations and dynamics of the serotonin 5-HT(2A) receptor determine its activation and membrane-driven oligomerization properties. *PLoS Computational Biology, 8*(4), e1002473. http://dx.doi.org/10.1371/journal.pcbi.1002473.

Shan, J., Khelashvili, G., Mondal, S., & Weinstein, H. (2011). Pharmacologically distinct ligands induce different states of 5-HT2AR and trigger different membrane remodeling: Implications for GPCR oligomerization. *Biophysical Journal, 100*(3, Suppl. 1), 254a. http://dx.doi.org/10.1016/j.bpj.2010.12.1605.

Shushkov, P., Tzvetanov, S., Velinova, M., Ivanova, A., & Tadjer, A. (2010). Structural aspects of lipid monolayers: Computer simulation analyses. *Langmuir: The ACS Journal of Surfaces and Colloids, 26*(11), 8081–8092. http://dx.doi.org/10.1021/la904734b.

Steinbauer, B., Mehnert, T., & Beyer, K. (2003). Hydration and lateral organization in phospholipid bilayers containing sphingomyelin: A 2H-NMR study. *Biophysical Journal, 85*(2), 1013–1024. http://dx.doi.org/10.1016/S0006-3495(03)74540-8.

Thompson, A. A., Liu, J. J., Chun, E., Wacker, D., Wu, H., Cherezov, V., et al. (2011). GPCR stabilization using the bicelle-like architecture of mixed sterol-detergent micelles. *Methods (San Diego, Calif.), 55*(4), 310–317. http://dx.doi.org/10.1016/j.ymeth.2011.10.011.

Tieleman, D. P. (2010). Molecular simulations and biomembranes: From biophysics to function. In Sansom, M. S. P. & Biggin, P. C. (Eds.). *RSC biomolecular sciences* (Vol. 20, Chapter 1). Royal Society of Chemistry

Tristram-Nagle, S., Petrache, H. I., & Nagle, J. F. (1998). Structure and interactions of fully hydrated dioleoylphosphatidylcholine bilayers. *Biophysical Journal, 75*(2), 917–925. http://dx.doi.org/10.1016/S0006-3495(98)77580-0.

Van Meer, G., Voelker, D. R., & Feigenson, G. W. (2008). Membrane lipids: Where they are and how they behave. *Nature Reviews. Molecular Cell Biology, 9*(2), 112–124. http://dx.doi.org/10.1038/nrm2330.

Vermeer, L. S., de Groot, B. L., Réat, V., Milon, A., & Czaplicki, J. (2007). Acyl chain order parameter profiles in phospholipid bilayers: Computation from molecular dynamics simulations and comparison with 2H NMR experiments. *European Biophysics Journal, 36*(8), 919–931. http://dx.doi.org/10.1007/s00249-007-0192-9.

Whorton, M. R., Bokoch, M. P., Rasmussen, S. G. F., Huang, B., Zare, R. N., Kobilka, B., et al. (2007). A monomeric G protein-coupled receptor isolated in a high-density lipoprotein particle efficiently activates its G protein. *Proceedings of the National Academy of Sciences of the United States of America, 104*(18), 7682–7687. http://dx.doi.org/10.1073/pnas.0611448104.

Zhao, G., Subbaiah, P. V., Mintzer, E., Chiu, S.-W., Jakobsson, E., & Scott, H. L. (2011). Molecular dynamic simulation study of cholesterol and conjugated double bonds in lipid bilayers. *Chemistry and Physics of Lipids, 164*(8), 811–818. http://dx.doi.org/10.1016/j.chemphyslip.2011.09.008.

Structure-Based Molecular Modeling Approaches to GPCR Oligomerization

Agnieszka A. Kaczor*,†, Jana Selent‡, and Antti Poso*

*Department of Pharmaceutical Chemistry, School of Pharmacy, University of Eastern Finland, Kuopio, Finland

†Department of Chemical Technology of Pharmaceutical Substances with Computer Modeling Lab, Faculty of Pharmacy with Division of Medical Analytics, Medical University of Lublin, Lublin, Poland

‡Research Unit on Biomedical Informatics (GRIB), PRBB, Barcelona, Spain

CHAPTER OUTLINE

Introduction 92
5.1 Protein–Protein Docking 93
    5.1.1 Available Docking Programs 94
    5.1.2 Practical Aspects 95
    5.1.3 Reliability 96
5.2 MD Simulation 96
    5.2.1 Recent Milestones in Simulating GPCRs 97
    5.2.2 Technical Aspects 98
5.3 Normal Mode Analysis 99
    5.3.1 Background and Examples of Applications 99
    5.3.2 Software 100
5.4 Electrostatics Studies 100
    5.4.1 Background and Examples of Application 100
    5.4.2 Software 101
Acknowledgments 101
References 102

Abstract

Classical structure-based drug design techniques using G-protein-coupled receptors (GPCRs) as targets focus nearly exclusively on binding at the orthosteric site of a single receptor. Dimerization and oligomerization of GPCRs, proposed almost

30 years ago, have, however, crucial relevance for drug design. Targeting these complexes selectively or designing small molecules that affect receptor–receptor interactions might provide new opportunities for novel drug discovery. In order to study the mechanisms and dynamics that rule GPCRs oligomerization, it is essential to understand the dynamic process of receptor–receptor association and to identify regions that are suitable for selective drug binding, which may be determined with experimental methods such as Förster resonance energy transfer (FRET) or Bioluminescence resonance energy transfer (BRET) and computational sequence- and structure-based approaches. The aim of this chapter is to provide a comprehensive description of the structure-based molecular modeling methods for studying GPCR dimerization, that is, protein–protein docking, molecular dynamics, normal mode analysis, and electrostatics studies.

INTRODUCTION

Human membrane proteins, mainly metabotropic and ionotropic transmembrane receptors, constitute currently the largest group of targets for drugs on the market. This includes G-protein-coupled receptors (GPCRs), which are drug targets for about 50% of all marketed drugs. GPCRs are complex signaling molecules occurring at the cell surface. They mediate signal transduction and convert extracellular stimuli, such as hormones and neurotransmitters, into intracellular responses through the activation of heterotrimeric G proteins (Ding, Zhao, & Watts, 2013). Some GPCRs are also involved in the processes of vision, olfaction, and taste. The molecular architecture of GPCRs is constituted by seven relatively conserved membrane-spanning α-helical segments connected with alternating intracellular and extracellular loop regions. GPCRs are grouped into five families on the basis of their sequence and structural similarity: rhodopsin (family A), secretin (family B), glutamate (family C), adhesion, and frizzled/taste. The rhodopsin family is the largest and most diverse of all families.

X-ray structures of some GPCRs have become available recently, including the structures in the active state. Knowledge of a GPCR structure enables us to gain a mechanistic insight into its function and dynamics and further aid rational drug design. The new structures make it also possible to design studies aimed to investigate receptor dimerization and ligand-biased signaling (Audet & Bouvier, 2012). It may lead to more selective and safer drugs as despite the advances in drug design; however, promiscuity between drug molecules and targets including GPCRs often leads to undesired signaling effects, which result in unintended side effects (McNeely, Naranjo, & Robinson, 2012).

Dimerization and oligomerization of GPCRs, proposed almost 30 years ago, have been already accepted by the scientific community and have crucial relevance for drug design. It is, however, still controversial how many receptors exist as monomers, dimers, and oligomers in the cell membrane and if they are transient or rather stable complexes. It should be emphasized that formation of GPCR oligomers may influence the range, diversity, and performance by which extracellular signals are transferred to G proteins in the process of receptor transduction. As a consequence, the control of oligomer formation and signaling will be a powerful pharmacological tool. In order to design substances that modify GPCR dimer functioning, a few conditions must be fulfilled. Firstly, it is needed to determine that oligomerization takes place between particular receptors; secondly, that the oligomer has pharmacological importance; and thirdly, that the availability of the oligomer three-dimensional (3D) structure is required (Kaczor & Selent, 2011). Dimerization of GPCRs is studied with experimental approaches, such as FRET and BRET (Kaczor & Selent, 2011), and computational methods (Selent & Kaczor, 2011). Molecular modeling approaches can be classified into sequence- and structure-based methods. The first group is based on the GPCR sequence analysis performed in order to detect evolutionary changes of the GPCR interfaces. Structure-based methodologies are a potent complement to sequence-based approaches and involve protein–protein docking and molecular dynamics (MD) techniques as well as electrostatics analysis with adaptive Poisson–Boltzmann solver (APBS) and normal mode analysis (NMA).

The aim of this chapter is to present a background, examples of applications, and some technical aspects of structure-based molecular modeling approaches to study GPCR oligomerization.

5.1 PROTEIN–PROTEIN DOCKING

While small-molecular docking is a well-established molecular modeling method (the first truly successful docking method DOCK was published in 1982 by Kuntz, Blaney, Oatley, Langridge, & Ferrin, 1982), protein–protein docking is quite often an ignored method. Somehow surprisingly, protein–protein docking is as old method (or even older) as traditional small-molecular docking, as the first true example of protein–protein docking originates to as early as 1970s (Wodak & Janin, 1978). Currently, protein–protein docking is a well-established method, albeit not as widely used as small-molecular docking. Although one could easily believe that the performance of protein–protein docking is far below the small-molecular docking, an excellent review by Janin (2010) clearly does not support this assumption. In the past nine years, 42 protein–protein complexes have been modeled under the CAPRI (critical assessment of predicted interactions) experiment and of these complexes only 6 have not been successfully modeled.

In protein–protein docking, a common assumption is that protein–protein interaction can be modeled based on shape complementarity and using simplified amino acid models (in a similar fashion as typically used in early day molecular mechanics united-atom force fields). While originally most of the protein–protein docking methods were based on rigid-docking approach, currently most of the methods are able to accommodate at least protein side-chain flexibility. Flexibility is clearly a critical point to increase the quality of protein–protein docking, as in the CAPRI experiment most of the failures are connected with protein conformational changes upon protein–protein interaction.

Another typical approach in protein–protein docking is that protein complexation will occur in biological solvent, not inside the cell membrane as in the case of GPCRs. As many of our potential protein–protein complexes are located in membrane environment, this will surely affect the performance of protein–protein docking methods. While earlier-mentioned CAPRI experiment is typically targeting nonmembrane protein–protein complexes, in a recent study (Kaczor, Selent, Sanz, & Pastor, 2013), it was shown that membrane environment needed for GPCR docking is indeed a major problem requiring future work.

Like in the case of small-molecular docking, scoring functions are of utmost importance for the quality of the results. Commonly, scoring functions used in protein–protein docking are based on weighted sums of different parameters including typically chemical and physical properties important for protein–protein interactions. In addition, different geometric properties and atom–atom or residue–residue potential are evaluated. Desolvation effects are usually taken into account indirectly, but it is also possible to calculate this term using explicitly approach (although this latter is quite slow approach).

Docking experiments to create GPRC dimers (both homo- and heterodimers) has not been reported often in recent years, as the phenomenon of GPRC dimerization has gained attention only during the last few years. One of the first reviews devoted just to GPCR dimerization is just 15 years old (Hébert & Bouvier, 1998), and even at the moment (May 2013), PubMed search with keywords "GPRC dimerization docking" yields only eight publications. In spite of this (or actually due to this), it is important to look at those topics, which are important for usage of protein–protein docking programs if and when those are utilized to study the dimerization of GPCRs. In this part, we aim to give some general guidelines regarding available programs, preparation of protein structures (docking partners) prior to the docking exercise and analysis of the docking results. This text should not be taken as a cooking book recipe for protein–protein docking but more as a practical checklist to support someone who is willing to try first time protein–protein dockings within the realm of GPCRs.

5.1.1 Available docking programs

Most (but surely not all) important protein–protein docking programs and their performance to recreate protein–protein complexes found in transmembrane environment have been analyzed by Kaczor et al. (2013). Of the tested methods, GRAMM-X (Tovchigrechko & Vakser, 2006) method gave the best overall performance with RMSD lower than 0.7 Å in 8 out of 12 cases. The problem was that with four cases (pdb-ids: 2I37, 3CAP, 3OE9, and 4DJH), which all where seven-transmembrane receptors, RMSD was at least 9 Å, which basically equals to total failure to detect the true protein–protein interaction mode. The even worse topic is the fact that all the tested methods were no better (with the only exception HADDOCK (De Vries, van Dijk, & Bonvin, 2010) in the case of chemokine receptor

CXCR4, pdb-id 3OE9). It seems that the main problem here is not the sampling of possible docking poses but more on the scoring function.

5.1.2 Practical aspects

Most of the molecular modeling methods can be run using free web servers and this is also true for protein–protein docking methods. Unfortunately, not all the methods are available via web and some other methods are not fully implemented in web servers (e.g., Rosetta server (Schueler-Furman, Wang, & Baker, 2005) is only carrying out local docking when using free web server). Due to this, it is usually a good idea to install the software on a local computer. The best possible environment in the sense of software availability is a linux-based system, as almost all systems can be used under linux. The problem is that several noncomputer-oriented scientists may find linux a little bit challenging environment and also in many organizations linux is typically not supported. If for some reasons, linux is not an option, one can, in most cases, use Mac OS X without any major problems (most of the software packages are also available for Mac) and also Windows-based systems are usually reasonable well supported, although clearly with smaller number of available software packages. Based on our own experience, none of these environments is perfect and thus in our laboratories all operating systems are in use.

Usually most of the software packages are available from the developer web page as ready compiled files, and these are basically always the best option (e.g., GRAMM-X (Tovchigrechko & Vakser, 2006) is available as linux and Windows-ready packages; see http://vakser.bioinformatics.ku.edu/main/resources_gramm1.03.php). Only if truly needed (like in the case of RosettaCommons (Leaver-Fay et al., 2011), see www.rosettacommons.org/home), one needs to compile the software. In those cases, it is a best practice to first read the Readme.txt file and follow the instructions as accurately as possible.

Once the software is properly installed, protein docking experiments can be carried out. For all cases, it is of utmost importance to first prepare the files needed for protein–protein docking. Typically pdb-file format is used as input files for software. Unfortunately, there are several versions of pdb-file and most programs require that the input file is processed properly. The best approach is to manually edit the pdb file and to remove all unnecessary parts, basically leaving only protein residue and atom coordinates plus possible heteroatom data. One should search the current tutorials for given software packages to be sure how the file preparation should be carried out.

Another topic to be checked is the effect of possible ligands/cofactors and ionization of charged residues for docking. Again, this is quite software-dependent topic and it must be taken into account. In general, ligands that are needed for protein–protein docking should be kept within the input. Also, most of the docking methods do not require explicit information for ionization of protein residues, although this is not the case with ligands/cofactors where this data usually must be explicitly defined.

A specific question with GPCR–GPCR docking is the nature of possible complex. It is clear that the interaction will happen within the cell membrane and thus there is no need to run full rotation/translation search where one protein is rotated/translated around the second protein structure. In most of the cases, there is no easy way to define a proper boundary rotation and translation conditions; so the only way is to carry out full search. However, in some cases, the user can define (usually with complicated scripting) the instructions so that only rotations and translations within membrane are possible. Naturally, one can also use the preexisting knowledge and give to software predocked complex and ask the algorithm to carry out only local docking and refinement. This option is typically easily available and also it will speed up the computations remarkably.

Once all the files have been created, the actual docking run will be carried out. The CPU time required for docking can vary a much, but at least 1–2 h is usually needed.

5.1.3 Reliability

As stated earlier, scoring is the main problem when GPRC docking is carried out. To overcome this, one should use all the possible empirical data. We have also developed an iterative approach to tackle the problem. In our approach, we apply protein–protein docking with Rosetta software to obtain populations of dimers as present in membranes with all possible interfaces (28 interfaces for a homodimer). As the number of possible interfaces is reasonably small, it is possible to carry out detailed local docking to optimize each possible docking pose. At the next stage, consensus scoring procedure according to (i) Rosetta score, (ii) surface of the dimer interface, (iii) polar contribution to the dimer interface, (iv) fractal dimension of the dimer interface (Kaczor et al., 2012), (v) evolutionary conservation score (Ashkenazy, Erez, Martz, Pupko, & Ben-Tal, 2010), (vi) shape complementarity, (vii) electrostatic complementarity, (viii) potential energy, and (ix) free energy of binding is applied. The best models are minimized, and the whole cycle is iteratively repeated until the results converge to a consistent dimer formation.

We do not believe that any of the current protein–protein docking scoring functions is capable of reliable predictions. This does not mean that protein–protein docking methods cannot be used, but one cannot claim that results can be used as such. The situation is somehow analog to small-molecular docking where the so-called consensus scoring has been used over a long period of time.

5.2 MD SIMULATION

Nowadays, MD is an important computational tool to study GPCRs in complex systems under native-like conditions at a molecular level. The part A aims to give a general overview about the capability of MD simulation in GPCR research by highlighting recent milestones, while part B provides insights into some technical aspects. Part B is kept in a short format as a detailed description is provided in the chapter 4 within this edition.

5.2.1 Recent milestones in simulating GPCRs

The "insuperable" barrier that impeded the obtention of further structural insights into GPCRs after the release of the first X-ray crystal structure of bovine rhodopsin in 2000 is linked to the inherent flexibility of GPCRs. A tremendous work in protein engineering was undertaken to stabilize these flexible receptors preluding eventually a new era of novel high-resolution structures for different receptor types. These new receptor structures give an extraordinary glimpse into one possible conformational state in the life cycle of GPCRs. However, the latest paradigm assumes the existence of a vast amount of conformational receptor states (Park, 2012), which underlies their ability to respond to diverse extracellular signals in a distinct manner. In this scenario, MD simulations have been exploited to elucidate the dynamic nature of GPCRs taking into account interaction with chemically diverse ligands, solvent effects, allosteric modulation of ions, membrane effects, and last but not least formation of higher-order complexes.

One of the most important question in GPCR drug discovery is how do drugs bind to GPCRs: from initial association, followed by drug entry and adoption of the final binding pose. Recently, this pharmaceutically critical process has been captured using the first unbiased MD simulations for several beta-blockers for the β_1 and β_2 in atomic detail (Dror, Pan, et al., 2011). A surprising finding is that the first drug association with the receptor at the extracellular part (15 Å distant from the binding pocket) constitutes the largest energetic barrier to ligand binding. This high barrier seems to be a cause of drug dehydration that happens during the course of first drug association to the extracellular receptor part. In a subsequent step, drug entrance requires a receptor deformation and squeezing of the ligand through a narrow opening. The final binding pose occurs after microseconds—in case of alprenolol after 3.5 μs—which is still, using modern simulation techniques, a challenging timescale for a complex GPCR-membrane system. This first atomic description of the pathway and mechanism of drug binding to GPCRs provides suggestions for optimization of drug binding as well as ideas for allosteric receptor modulation (Dror, Pan, et al., 2011). In this context, also the allosteric binding of cations at the dopaminergic D_2 receptor was computationally observed via microsecond simulation (Selent, Sanz, Pastor, & De Fabritiis, 2010). Similar to the aforementioned beta-blockers, sodium ions enter the receptor from the extracellular receptor side. Once inside the receptor, sodium ions bind first to the orthosteric site D3.32 before penetrating even deeper into the receptor occupying the well-known allosteric site D2.50. Remarkably, the final observed cation binding pose at D2.50 was later confirmed by an experimentally obtained X-ray structure (Liu et al., 2012), stressing the predictive value of their structural model.

Another milestone in simulating GPCRs is a first molecular insight into GPCR activation. Soon after the release of the activated β_2-adrenergic receptor in complex with a G-protein-mimetic nanobody, Dror, Arlow, et al. tried to elucidated the dynamic steps that lead to receptor activation (Dror, Arlow, et al., 2011). For this purpose, they started from the activated conformation observing its inactivation by monitoring the closure of the ionic lock close to the intracellular region within several microseconds. Their computational study points to the intriguing finding that first activation steps take part in the intracellular receptor region, which is implicated in G-protein binding. In addition, they observe that signal propagation between ligand binding site and the intracellular site of G-protein binding involves a rather smooth coupling between microdomains. This indicates that an agonist may not lock the receptor in an entire active conformation (i.e., active binding pocket state and inactive closed ionic lock). In fact, this is supported by experimental data that captured such partially activated GPCR in complex with an agonist (Ghanouni et al., 2001).

An important issue that is often neglected when simulating GPCRs is the complex membrane environment in which GPCRs are embedded in living cells. Plenty of data indicate that the type and composition of the environment is crucial for the way GPCRs signal (Bruno, Costantino, de Fabritiis, Pastor, & Selent, 2012; Sadiq et al., 2013). This stresses the urgent need to adequately represent the membrane environment in MD simulations for successfully characterizing a physiologically relevant dynamic behavior of GPCRs. First computational attempts have been reported elucidating specific as well as unspecific membrane effects (Sadiq et al., 2013). One particular phenomenon of an unspecific membrane effect is the hydrophobic mismatch, which occurs when the length of the hydrophobic part of transmembrane protein does not match the hydrophobic bilayer thickness (Periole, Huber, Marrink, & Sakmar, 2007). Computationally, this has been addressed for GPCRs by Mondal et al., which build a framework to quantify this membrane effect (Mondal, Khelashvili, Shan, Andersen, & Weinstein, 2011). Noteworthy, a residual mismatch could be an important regulatory key factor for GPCRs functioning in terms of forming of higher-order GPCR complexes. Higher-order GPCR complexes and inherent receptor cross talk have recently attracted high interest as potential drug targets for diseases such as schizophrenia or Parkinson's disease (Guixa-Gonzalez, Bruno, Marti-Solano, & Selent, 2012). In this respect, an important landmark in computationally simulating GPCR complex formation has been set by Periole, Knepp, Sakmar, Marrink, & Huber (2012) who studied the self-assembly of rhodopsin into supramolecular structures using coarse-grained molecular dynamics (CGMD). The predicted dimer orientation is in line with earlier electron microscopy and stresses the potential of this approach for studying complex GPCR-membrane systems.

5.2.2 Technical aspects

Simulating successfully GPCR in a native environment at an all-atom scale involves several critical steps. In case no structural information is available, the first step is dedicated to the construction of a 3D structure of the target GPCR using homology modeling. At this stage, an initial model structure of the target structure is generated based on a closely related protein (template) for which high-resolution data are available in the Protein Data Bank (Berman et al., 2000). Importantly, the template and target structure should have a sequence identity of at least a 35–40% according to a recent assessment (Kufareva, Rueda, Katritch, Stevens, & Abagyan, 2011). In a second step, the model structure is inserted into a membrane system, solvated, and ionized (neutralized) up to a physiological concentration. Nowadays, tremendous efforts have been undertaken to model native-like cell membranes instead of just monocomponent membranes. A detailed description of building, simulating, and analyzing membranes (e.g., hydrophobic mismatch) is given in the chapter 4.

Unfortunately, all-atom simulations are linked to certain limitations in system size and timescale and make the study of physiological relevant invents such as receptor oligomerization not feasible. An alternative approach is provided by applying CGMD as recently shown in an elegant study by Periole et al. (2007, 2012). Thereby, the all-atom system is simplified by combining groups of atoms into one single beat, which allows simulating much longer timescales compared to all-atom systems.

All in all, using modern computing infrastructure (parallel computing), we can simulate the evolution of GPCR target structures in a native-like environment (all-atom or coarse-grained) up to microseconds. Importantly, replicates from individual starting structures are inevitable to proof the relevance of an observed dynamic event.

Readers who would like to learn more about the technical details should refer to the chapter 4 within this edition.

5.3 NORMAL MODE ANALYSIS
5.3.1 Background and examples of applications

As discussed earlier, molecular modeling techniques supply effective approaches for investigation of dynamics of proteins. Besides MD simulations, NMA can be used for this purpose, in particular when the protein is relatively big (several thousand amino acids) and the timescale of the dynamical events of interest is longer than what MD simulations can reach, typically a few nanoseconds (Hollup, Salensminde, & Reuter, 2005). NMA approach assumes that the vibrational normal modes exhibiting the lowest frequencies (also named soft modes) describe the largest movements in a protein and are the ones functionally relevant (Hollup et al., 2005). NMA calculations are based on the diagonalization of the matrix of the second derivatives of the energy with respect to the displacements of the atoms, in mass-weighted coordinates (Hessian matrix) (Skjaerven, Jonassen, & Reuter, 2007). The eigenvectors of the Hessian matrix are the normal modes, and the associated eigenvalues are the squares of the associated frequencies (Skjaerven et al., 2007).

There are not many examples of application of NMA to GPCRs. These include a study of conformational changes occurring in the rhodopsin monomer upon activation (Isin, Rader, Dhiman, Klein-Seetharaman, & Bahar, 2006; Niv, Skrabanek, Filizola, & Weinstein, 2006), an analysis of the low-frequency modes of cone opsins (Thirumuruganandham & Urbassek, 2009), studies of the activation of the ghrelin receptor (Floquet et al., 2010), and an analysis of rhodopsin dimerization (Niv & Filizola, 2008), which is shortly reviewed in the succeeding text.

To help understand the effect of oligomerization on the dynamics of GPCRs, Niv and Filizola (Niv & Filizola, 2008) compared the motion of monomeric, dimeric, and tetrameric arrangements of the prototypic GPCR rhodopsin. They used approximate, however powerful, NMA technique, that is, elastic network model (ENM), to differentiate between putative dynamic mechanisms, which may be responsible for the recently reported conformational rearrangement of the TM4,5–TM4,5 dimerization interface of GPCRs, which takes place during the process of receptor activation. They also found the considerable perturbation of the normal modes of the rhodopsin monomer during oligomerization, which was visible at the studied interface. Furthermore, they noticed increased positive correlation among the transmembrane domains and between the extracellular loop and transmembrane regions of the rhodopsin protomer and highest interresidue positive correlation at the interfaces between protomers (Niv & Filizola, 2008).

5.3.2 Software

NMA can be performed with MD software such as GROMACS (Hess, Kutzner, van der Spoel, & Lindahl, 2008). Also, there are web-based tools available that may be used by nonexperts. These include ElNémo server (Suhre & Sanejouand, 2004) and NOMAD-Ref server (Lindahl, Azuara, Koehl, & Delarue, 2006).

ElNémo server is a web interface to the ENM that is designed as a fast and simple tool to compute, visualize, and analyze low-frequency normal modes of large macromolecules (Suhre & Sanejouand, 2004). Furthermore, there is no upper limit to the size of the proteins that can be calculated. ElNémo server may be used for computation of 100 lowest-frequency modes for a protein and results in descriptive parameters and visualizations, including degree of collectivity of movement, residue mean square displacements, distance fluctuation maps, and the correlation between observed and normal mode-derived atomic displacement parameters (β-factors) (Suhre & Sanejouand, 2004).

NOMAD-Ref server enables online calculation of the normal modes of large molecules (up to 100,000 atoms), enabling a full all-atom representation of their structures, parallel with an access to several programs that utilize these collective motions for deformation and refinement of biomolecular structures (Lindahl et al., 2006).

5.4 ELECTROSTATICS STUDIES
5.4.1 Background and examples of application

Electrostatic forces are one of the main determinants of molecular interactions. They help guiding the folding of proteins, increase the binding of one protein to another, and facilitate protein–DNA and protein–ligand binding (Callenberg et al., 2010). It has been also shown that electrostatics influences various aspects of nearly all biochemical reactions (Baker, Sept, Joseph, Holst, & McCammon, 2001). The relationship between the electric field and the charge in a system is described by Maxwell's equations (Callenberg et al., 2010). One of the most representative models for the evaluation of electrostatic properties is the Poisson–Boltzmann equation. This is a second-order nonlinear elliptic partial differential equation that links the electrostatic potential to the dielectric properties of the solute and solvent, the ionic strength of the solution and the accessibility of ions to the solute interior, and the distribution of solute atomic partial charges (Baker et al., 2001).

Similarly as for NMA, there are not many examples of applications of electrostatics studies for GPCRs and their dimerization. As little is known about the mechanism and stability of the class A GPCR dimerization, we used APBS for modeling biomolecular solvation through resolution of the Poisson–Boltzmann equation for describing electrostatic interactions between molecular solutes in a given medium to study the various dimerization properties of these receptors (Kaczor, Lopez, Pastor, & Selent, 2010). Recent studies suggest that stable complexes are formed by the β_2-adrenergic receptor, while the β_1-adrenergic, the dopaminergic D_2, and the muscarinic M_1 receptor form rather transient complexes. Going into details, the low dielectric membrane-like model was incorporated with the APBS membrane tool in order to crudely represent the nonpolar membrane environment; dielectric constants of 80, 10, and 2 for water, protein, and membrane interiors were introduced, respectively. Thereby, computation of the electrostatic potential for different GPCRs suggests that the stability of the receptor dimer corresponds to the distribution pattern of the electrostatic potential. The computed electrostatic potential for various monomers of class A GPCRs suggests that receptors, forming transient dimer complexes (β_1, D_2, and M_1 receptor), exhibit a surface dominated by a positive (blue) electrostatic potential, while the surface of the β_2-adrenergic receptor, forming a stable dimer complex, exhibits a more balanced set of positive and negative electrostatic patches.

5.4.2 Software

One of the most popular software for electrostatics calculations is APBS, which is used for modeling biomolecular solvation through solution of the Poisson–Boltzmann equation (Baker et al., 2001). Another freely available program is called APBSmem, and it is designed for carrying out the electrostatics calculations in the presence of a membrane (Callenberg et al., 2010).

Acknowledgments

This chapter was partially prepared during the postdoctoral fellowship of Agnieszka A. Kaczor, under Marie Curie IEF fellowship. Jana Selent acknowledges support from the Instituto de Salud Carlos III FEDER (CP12/03139) and the La MARATÓ de TV3 Foundation, Grant number 091010. Part of the calculations was performed under a computational grant by

Interdisciplinary Center for Mathematical and Computational Modelling (ICM), Warsaw, Poland, grant number G30-18 and under resources of CSC, Finland.

References

Ashkenazy, H., Erez, E., Martz, E., Pupko, T., & Ben-Tal, N. (2010). ConSurf 2010: Calculating evolutionary conservation in sequence and structure of proteins and nucleic acids. *Nucleic Acids Research, 38*(Web Server), W529–W533. http://dx.doi.org/10.1093/nar/gkq399.

Audet, M., & Bouvier, M. (2012). Restructuring G-protein-coupled receptor activation. *Cell, 151*(1), 14–23. http://dx.doi.org/10.1016/j.cell.2012.09.003.

Baker, N. A., Sept, D., Joseph, S., Holst, M. J., & McCammon, J. A. (2001). Electrostatics of nanosystems: Application to microtubules and the ribosome. *Proceedings of the National Academy of Sciences of the United States of America, 98*(18), 10037–10041. http://dx.doi.org/10.1073/pnas.181342398.

Berman, H. M., Westbrook, J., Feng, Z., Gilliland, G., Bhat, T. N., Weissig, H., et al. (2000). The Protein Data Bank. *Nucleic Acids Research, 28*(1), 235–242.

Bruno, A., Costantino, G., de Fabritiis, G., Pastor, M., & Selent, J. (2012). Membrane-sensitive conformational states of helix 8 in the metabotropic Glu2 receptor, a class C GPCR. *PLoS One, 7*(8), e42023. http://dx.doi.org/10.1371/journal.pone.0042023.

Callenberg, K. M., Choudhary, O. P., de Forest, G. L., Gohara, D. W., Baker, N. A., & Grabe, M. (2010). APBSmem: A graphical interface for electrostatic calculations at the membrane. *PLoS One, 5*(9). http://dx.doi.org/10.1371/journal.pone.0012722.

De Vries, S. J., van Dijk, M., & Bonvin, A M J J (2010). The HADDOCK web server for data-driven biomolecular docking. *Nature Protocols, 5*(5), 883–897. http://dx.doi.org/10.1038/nprot.2010.32.

Ding, X., Zhao, X., & Watts, A. (2013). G-protein-coupled receptor structure, ligand binding and activation as studied by solid-state NMR spectroscopy. *The Biochemical Journal, 450*(3), 443–457. http://dx.doi.org/10.1042/BJ20121644.

Dror, R. O., Arlow, D. H., Maragakis, P., Mildorf, T. J., Pan, A. C., Xu, H., et al. (2011). Activation mechanism of the β_2-adrenergic receptor. *Proceedings of the National Academy of Sciences of the United States of America, 108*(46), 18684–18689. http://dx.doi.org/10.1073/pnas.1110499108.

Dror, R., Pan, A., Arlow, D. H., Borhani, D. W., Maragakis, P., Shan, Y., et al. (2011). Pathway and mechanism of drug binding to G-protein-coupled receptors. *Proceedings of the National Academy of Sciences of the United States of America, 108*(32), 13118–13123. http://dx.doi.org/10.1073/pnas.1104614108/-/DCSupplemental.www.pnas.org/cgi/doi/10.1073/pnas.1104614108.

Floquet, N., M'Kadmi, C., Perahia, D., Gagne, D., Bergé, G., Marie, J., et al. (2010). Activation of the ghrelin receptor is described by a privileged collective motion: A model for constitutive and agonist-induced activation of a sub-class A G-protein coupled receptor (GPCR). *Journal of Molecular Biology, 395*(4), 769–784. http://dx.doi.org/10.1016/j.jmb.2009.09.051.

Ghanouni, P., Gryczynski, Z., Steenhuis, J. J., Lee, T. W., Farrens, D. L., Lakowicz, J. R., et al. (2001). Functionally different agonists induce distinct conformations in the G protein coupling domain of the beta 2 adrenergic receptor. *Journal of Biological Chemistry, 276*(27), 24433–24436. http://dx.doi.org/10.1074/jbc.C100162200.

Guixa-Gonzalez, R., Bruno, A., Marti-Solano, M., & Selent, J. (2012). Crosstalk within GPCR heteromers in schizophrenia and Parkinson's disease: Physical or just functional? *Current Medicinal Chemistry*, *19*(8), 1119–1134.

Hébert, T. E., & Bouvier, M. (1998). Structural and functional aspects of G protein-coupled receptor oligomerization. *Biochemistry and Cell Biology/Biochimie et biologie cellulaire*, *76*(1), 1–11.

Hess, B., Kutzner, C., van der Spoel, D., & Lindahl, E. (2008). GROMACS 4: Algorithms for highly efficient, load-balanced, and scalable molecular simulation. *Journal of Chemical Theory and Computation*, *4*(3), 435–447. http://dx.doi.org/10.1021/ct700301q.

Hollup, S. M., Salensminde, G., & Reuter, N. (2005). WEBnm@: A web application for normal mode analyses of proteins. *BMC Bioinformatics*, *6*, 52. http://dx.doi.org/10.1186/1471-2105-6-52.

Isin, B., Rader, A. J., Dhiman, H. K., Klein-Seetharaman, J., & Bahar, I. (2006). Predisposition of the dark state of rhodopsin to functional changes in structure. *Proteins*, *65*(4), 970–983. http://dx.doi.org/10.1002/prot.21158.

Janin, J. (2010). Protein-protein docking tested in blind predictions: The CAPRI experiment. *Molecular BioSystems*, *6*(12), 2351–2362. http://dx.doi.org/10.1039/c005060c.

Kaczor, A. A., Guixà-González, R., Carrió, P., Obiol-Pardo, C., Pastor, M., & Selent, J. (2012). Fractal dimension as a measure of surface roughness of G protein-coupled receptors: Implications for structure and function. *Journal of Molecular Modeling*, *18*(9), 4465–4475. http://dx.doi.org/10.1007/s00894-012-1431-2.

Kaczor, A. A., Lopez, L., Pastor, M., & Selent, J. (2010). Exploring G-protein coupled receptors oligomerization interface with adaptive Poisson–Boltzmann solver (APBS). *Drugs Future*, *35*(Suppl. A), 122–123.

Kaczor, A. A., & Selent, J. (2011). Oligomerization of G protein-coupled receptors: Biochemical and biophysical methods. *Current Medicinal Chemistry*, *18*(30), 4606–4634.

Kaczor, A. A., Selent, J., Sanz, F., & Pastor, M. (2013). Modeling complexes of transmembrane proteins: Systematic analysis of protein–protein docking tools. *Molecular Informatics*, *32*, 717–733.

Kufareva, I., Rueda, M., Katritch, V., Stevens, R. C., & Abagyan, R. (2011). Status of GPCR modeling and docking as reflected by community-wide GPCR Dock 2010 assessment. *Structure (London, England: 1993)*, *19*(8), 1108–1126. http://dx.doi.org/10.1016/j.str.2011.05.012.

Kuntz, I. D., Blaney, J. M., Oatley, S. J., Langridge, R., & Ferrin, T. E. (1982). A geometric approach to macromolecule–ligand interactions. *Journal of Molecular Biology*, *161*(2), 269–288.

Leaver-Fay, A., Tyka, M., Lewis, S. M., Lange, O. F., Thompson, J., Jacak, R., et al. (2011). ROSETTA3: An object-oriented software suite for the simulation and design of macromolecules. *Methods in Enzymology*, *487*, 545–574. http://dx.doi.org/10.1016/B978-0-12-381270-4.00019-6.

Lindahl, E., Azuara, C., Koehl, P., & Delarue, M. (2006). NOMAD-Ref: Visualization, deformation and refinement of macromolecular structures based on all-atom normal mode analysis. *Nucleic Acids Research*, *34*(Web Server), W52–W56. http://dx.doi.org/10.1093/nar/gkl082.

Liu, W., Chun, E., Thompson, A. A., Chubukov, P., Xu, F., Katritch, V., et al. (2012). Structural basis for allosteric regulation of GPCRs by sodium ions. *Science*, *337*(6091), 232–236. http://dx.doi.org/10.1126/science.1219218.

McNeely, P. M., Naranjo, A. N., & Robinson, A. S. (2012). Structure-function studies with G protein-coupled receptors as a paradigm for improving drug discovery and development of therapeutics. *Biotechnology Journal*, *7*(12), 1451–1461. http://dx.doi.org/10.1002/biot.201200076.

Mondal, S., Khelashvili, G., Shan, J., Andersen, O. S., & Weinstein, H. (2011). Quantitative modeling of membrane deformations by multihelical membrane proteins: Application to G-protein coupled receptors. *Biophysical Journal*, *101*(9), 2092–2101. http://dx.doi.org/10.1016/j.bpj.2011.09.037.

Niv, M. Y., & Filizola, M. (2008). Influence of oligomerization on the dynamics of G-protein coupled receptors as assessed by normal mode analysis. *Proteins*, *71*(2), 575–586. http://dx.doi.org/10.1002/prot.21787.

Niv, M. Y., Skrabanek, L., Filizola, M., & Weinstein, H. (2006). Modeling activated states of GPCRs: The rhodopsin template. *Journal of Computer-Aided Molecular Design*, *20*(7–8), 437–448. http://dx.doi.org/10.1007/s10822-006-9061-3.

Park, P. S. (2012). Ensemble of G protein-coupled receptor active states. *Current Medicinal Chemistry*, *19*(8), 1146–1154.

Periole, X., Huber, T., Marrink, S.-J., & Sakmar, T. P. (2007). G protein-coupled receptors self-assemble in dynamics simulations of model bilayers. *Journal of the American Chemical Society*, *129*(33), 10126–10132. http://dx.doi.org/10.1021/ja0706246.

Periole, X., Knepp, A. M., Sakmar, T. P., Marrink, S. J., & Huber, T. (2012). Structural determinants of the supramolecular organization of G protein-coupled receptors in bilayers. *Journal of the American Chemical Society*, *134*(26), 10959–10965. http://dx.doi.org/10.1021/ja303286e.

Sadiq, S. K., Guixa-Gonzalez, R., Dainese, E., Pastor, M., De Fabritiis, G., & Selent, J. (2013). Molecular modeling and simulation of membrane lipid-mediated effects on GPCRs. *Current Medicinal Chemistry*, *20*(1), 22–38.

Schueler-Furman, O., Wang, C., & Baker, D. (2005). Progress in protein–protein docking: Atomic resolution predictions in the CAPRI experiment using RosettaDock with an improved treatment of side-chain flexibility. *Proteins*, *60*(2), 187–194. http://dx.doi.org/10.1002/prot.20556.

Selent, J., & Kaczor, A. A. (2011). Oligomerization of G protein-coupled receptors: Computational methods. *Current Medicinal Chemistry*, *18*(30), 4588–4605.

Selent, J., Sanz, F., Pastor, M., & De Fabritiis, G. (2010). Induced effects of sodium ions on dopaminergic G-protein coupled receptors. *PLoS Computational Biology*, *6*(8), pii: e1000884. http://dx.doi.org/10.1371/journal.pcbi.1000884.

Skjaerven, L., Jonassen, I., & Reuter, N. (2007). TMM@: A web application for the analysis of transmembrane helix mobility. *BMC Bioinformatics*, *8*, 232. http://dx.doi.org/10.1186/1471-2105-8-232.

Suhre, K., & Sanejouand, Y.-H. (2004). ElNemo: A normal mode web server for protein movement analysis and the generation of templates for molecular replacement. *Nucleic Acids Research*, *32*(Web Server), W610–W614. http://dx.doi.org/10.1093/nar/gkh368.

Thirumuruganandham, S. P., & Urbassek, H. M. (2009). Low-frequency vibrational modes and infrared absorbance of red, blue and green opsin. *Journal of Molecular Modeling*, *15*(8), 959–969. http://dx.doi.org/10.1007/s00894-008-0446-1.

Tovchigrechko, A., & Vakser, I. A. (2006). GRAMM-X public web server for protein–protein docking. *Nucleic Acids Research*, *34*(Suppl. 2), W310–W314. http://dx.doi.org/10.1093/nar/gkl206.

Wodak, S. J., & Janin, J. (1978). Computer analysis of protein–protein interaction. *Journal of Molecular Biology*, *124*(2), 323–342.

CHAPTER 6

Biochemical and Imaging Methods to Study Receptor Membrane Organization and Association with Lipid Rafts

Bruno M. Castro*, Juan A. Torreno-Piña*, Thomas S. van Zanten*, and Maria F. Gracia-Parajo*,[†]

*ICFO—Institut de Ciencies Fotoniques, Mediterranean Technology Park, Castelldefels (Barcelona), Spain
[†]ICREA—Institució Catalana de Recerca i Estudis Avançats, Barcelona, Spain

CHAPTER OUTLINE

Introduction and Rationale 106
6.1 Methods to Determine Protein Association with Lipid Rafts 107
    6.1.1 Lipid Raft Disruption by Cholesterol-Depleting Agents 107
        6.1.1.1 Methyl-β-Cyclodextrin 108
        6.1.1.2 Cholesterol Oxidase 108
    6.1.2 Colocalization with Putative Lipid Raft Markers 109
6.2 Optical Techniques to Study Receptor Membrane Organization and
    Association with Lipid Rafts 110
    6.2.1 Confocal Fluorescence Microscopy 110
    6.2.2 Homo-Förster Resonance Energy Transfer Microscopy 110
    6.2.3 Superresolution Techniques to Map the Nanolandscape of the
          Cell Membrane 114
        6.2.3.1 Far-Field Localization Techniques (STORM/PALM) 114
        6.2.3.2 Far-Field Patterned Illumination Techniques (STED) 115
        6.2.3.3 Near-Field Techniques (NSOM) 117
Concluding Remarks 118
Acknowledgments 119
References 119

Abstract

Lipid rafts, cell membrane domains with unique composition and properties, modulate the membrane distribution of receptors and signaling molecules facilitating the assembly of active signaling platforms. However, the underlying mechanisms that link signal transduction and lipid rafts are not fully understood, mainly because of the transient nature of these membrane assemblies. Several methods have been used to study the association of membrane receptors with lipid rafts. In the first part of this chapter, a description of how biochemical methods such as raft disruption by cholesterol depletion agents are useful in qualitatively establishing protein association with lipid rafts is presented. The second part of this chapter is dedicated to imaging techniques used to study membrane receptor organization and lipid rafts. We cover conventional approaches such as confocal microscopy to advanced imaging techniques such as homo-FRET microscopy and superresolution methods. For each technique described, their advantages and drawbacks are discussed.

INTRODUCTION AND RATIONALE

Cell membranes are presently described to be dynamically compartmentalized into physically, compositionally, and functionally distinct regions, which, according to their lipid and protein content, fulfill specific cellular roles (Simons & Gerl, 2010). These regions, called membrane domains, are formed due to cooperative behavior between certain membrane components. A special type of membrane domains, which has been associated with a multitude of cellular processes such as endocytosis, intracellular trafficking, lipid and protein sorting, and cell signaling, is lipid rafts (Simons & Gerl, 2010). These structures are defined as heterogeneous and highly dynamic nanoscale assemblies (5–200 nm) enriched in sphingolipids and certain sterols such as cholesterol or ergosterol, which present average lifetimes in subsecond timescales (Pike, 2006). These small and transient domains can be stabilized by interactions with particular lipids, proteins, and actin cytoskeleton, resulting in long-lived (ms timescales) micrometer-sized raft regions (Pike, 2006; Simons & Gerl, 2010). Due to their unique content in high temperature melting sphingolipids and cholesterol, lipid rafts are more ordered than the bulk membrane (Owen, Williamson, Magenau, & Gaus, 2012), imposing the localization or exclusion of specific membrane proteins to these domains. Moreover, protein diffusion is reduced in these regions (Lenne et al., 2006), promoting and stabilizing protein–protein interactions. Adding up, these features make lipid rafts particularly suitable for biochemical processes requiring a specific compartmentalization of macromolecular complexes, such as those occurring in cell signaling. In this context, it has been postulated that rafts spatially organize and concentrate signaling molecules at particular membrane regions while excluding inhibitor molecules. In this way, the favorable interactions between signaling molecules necessary for signal transduction are promoted, increasing signaling efficiency (Simons & Gerl, 2010). Indeed, several receptors such as Fas/CD95, T-cell receptor, epidermal growth factor receptor (EGFR), or G protein-coupled receptors translocate and oligomerize in lipid rafts (Allen, Halverson-Tamboli, & Rasenick, 2007; Chini & Parenti, 2004; Mollinedo & Gajate, 2006; Suzuki, 2012), a process that is vital for efficient signal transduction. However, the molecular mechanisms that govern signal transduction in these domains are still not fully understood. This is due to the highly dynamic and heterogeneous nature of these signaling assemblies, which make their study quite challenging using conventional optical techniques (Munro, 2003). Moreover, there is an intimate link between cell cytoskeleton and the formation/inhibition of receptor oligomers in lipid rafts (Kusumi et al., 2012; Simons & Gerl, 2010), which further contributes to its complexity. Several techniques have been used to investigate lipid rafts and their participation in signal transduction. However, due to their intrinsic limitations, an implicit general rule is to use complementary methods to assess membrane protein association/interactions with lipid rafts. Traditional approaches such as cholesterol depletion and colocalization with raft markers have provided important qualitative data regarding this relationship, whereas significant quantitative insights were obtained from advanced optical imaging techniques, namely, photoactivated localization microscopy (PALM), stimulated emission depletion (STED), and near-field scanning optical microscopy (NSOM). A description of these methods, covering their advantages and handicaps, is given in the succeeding text.

6.1 METHODS TO DETERMINE PROTEIN ASSOCIATION WITH LIPID RAFTS

Historically, protein association with lipid rafts was evaluated by searching for target protein presence in the insoluble membrane fraction obtained after gradient centrifugation of cell membranes lysed with cold (4 °C) nonionic detergents (Brown, 2006). However, the different composition of the insoluble fractions and variable yields of protein resistance to solubilization obtained by this method have led researchers to progress to more robust experimental approaches. In the succeeding text a description of the most commonly used techniques to study membrane protein association with lipid rafts is presented.

6.1.1 Lipid raft disruption by cholesterol-depleting agents

A typical method to determine protein association with lipid rafts is to assess if target protein membrane localization/organization is affected by lipid raft disruption induced by cholesterol depletion. This can be achieved by different approaches such as (i) depletion and removal of cell cholesterol by cyclodextrins (Zidovetzki & Levitan, 2007); (ii) use of microbial cholesterol oxidases that catalyze cholesterol to 4-cholesten-3-one; (iii) cholesterol sequestration by polyene macrolide antibiotics, namely, filipin, nystatin, or amphotericin (Gimpl & Gehrig-Burger, 2007);

and (iv) treatment with statins, which inhibit HMG-CoA reductase activity, thus blocking cholesterol biosynthesis (Allen et al., 2007). These techniques are usually combined with protein extraction or cellular imaging studies, providing relevant information about protein interactions with lipid rafts. Due to their relatively easy implementation and short incubation times, cyclodextrin and cholesterol oxidase treatment are the most widely used methods to disrupt rafts.

6.1.1.1 Methyl-β-cyclodextrin

Cyclodextrins are water-soluble oligosides composed of six to eight β(1–4)-glucopyranose units arranged in a closed ring structure. Their external surface is hydrophilic, while their inner cavity is hydrophobic. These molecules enhance the solubility of hydrophobic compounds (e.g., cholesterol) by encapsulating them in their inner cavity (Gimpl & Gehrig-Burger, 2007; Zidovetzki & Levitan, 2007). Methyl-β-cyclodextrin (MβC), which contains seven methylated oligoside units, has the highest affinity for cholesterol, being also the most selective and efficient in extracting it from cells. Depending on concentration, incubation time, temperature, and cell type, the amount of cholesterol extracted from the cells by these molecules is variable. For lipid raft disruption studies cells are typically incubated with 5–10 mM of MβC for 0.5–2 h at 37 °C. This results in cholesterol extraction percentages between 30 and 60% of the cell's total free cholesterol (Zidovetzki & Levitan, 2007). However, these levels of extraction submit cells to a stress situation, which is further aggravated by the higher permeability of their cholesterol-poor membranes (Gimpl & Gehrig-Burger, 2007; Zidovetzki & Levitan, 2007). Moreover, other membrane components might be also extracted from the cells by cyclodextrin. Therefore, a careful selection of the experimental conditions and rigorous control experiments should be done. A common approach to assess the existence of artifacts in these types of experiments is to add cholesterol back to drug-treated cells. If the effects of the drug treatment are reversed, then it is likely that the experimental observations are not being affected by cholesterol extraction pleiotropic effects. This can be done by loading cyclodextrins with cholesterol *in vitro* and presenting them to drug-treated cells to replenish their cholesterol levels (Allen et al., 2007).

6.1.1.2 Cholesterol oxidase

Cholesterol oxidase (EC 1.1.3.6) is a microbial flavin adenine dinucleotide-containing enzyme (flavoenzyme), belonging to the oxidoreductase family (Gimpl & Gehrig-Burger, 2007). It catalyzes cholesterol oxidation to cholest-5-en-3-one and subsequent isomerization to cholest-4-en-3-one, which does not have the same membrane-ordering properties as cholesterol (Castro, Silva, Fedorov, de Almeida, & Prieto, 2009). Since its first application to track cholesterol cellular localization and to probe cell membrane heterogeneities (Lange, 1992), this enzyme has become a general tool in cell membrane organization studies. Depending on the cell type, raft disruption is normally achieved by incubating cells with 0.5–2 U/mL of cholesterol oxidase in a serum-free medium for 0.5–2 h at 37 °C (Cahuzac et al.,

2006; Le Lay et al., 2009). Compared to cyclodextrins, this treatment is less harmful, since no cholesterol is extracted from the cells, only converted. Nonetheless, cholesterol present on the membranes of living cells is usually a poor subtract for cholesterol oxidase and thus this treatment is not as efficient in disrupting rafts as that with cyclodextrin.

6.1.2 Colocalization with putative lipid raft markers

A direct detection of target protein interactions with lipid rafts on intact cell membranes can be done by studying its membrane localization and comparing it with that of putative raft markers. This colocalization approach overcomes the intrinsic limitations of the invasive methods described earlier. The identification of the protein of interest and putative raft markers is done by specifically labeling with antibodies, chemicals, or fusion proteins, which are chosen according to the experimental technique applied. Putative raft markers are selected among the class of molecules found in high percentages in the low-density fractions resulting from membrane solubilization studies (Simons & Gerl, 2010). As a control, target protein membrane localization is usually also compared to that of a nonraft marker. In Table 6.1 is a list of putative raft and nonraft markers commonly used in this type of studies. Recently, important aspects of the molecular mechanisms governing lipid raft formation and targeting membrane protein to these domains have been unraveled by combining this approach with fluorescence imaging of cell-derived giant plasma membrane vesicles (Levental, Lingwood, Grzybek, Coskun, & Simons, 2010; Lingwood, Ries, Schwille, & Simons, 2008). These structures maintain the compositional complexity and protein content of the cell membrane, not exhibiting the complexity introduced by the cell's cytoskeleton and intracellular proteins, which are not present in these vesicles.

Although valuable information about protein localization with lipid rafts can be gathered by this approach, especially when combined with advanced imaging

Table 6.1 Lipid Raft or Nonlipid Raft Markers Commonly Used in Colocalization Studies

Marker	Lipid Raft Associated
Ganglioside (G_{M1}) (labeled with cholera toxin B)	Yes
GPI-anchored proteins (PLAP, CD59, DAF, etc.)	Yes
Caveolin-1	Yes
Flotillin	Yes
EGFR	Yes
Transferrin receptor (TfR)	No
Calnexin	No
Unsaturated glycerophospholipids (DOPC, DOPE, etc.)	No

techniques, this method is not without pitfalls. As in any technique requiring labeling, protein or raft/nonraft marker membrane localization/organization might be influenced by their labels. Therefore, any result obtained through this approach should always be confirmed by at least one of the techniques described earlier.

6.2 OPTICAL TECHNIQUES TO STUDY RECEPTOR MEMBRANE ORGANIZATION AND ASSOCIATION WITH LIPID RAFTS

6.2.1 Confocal fluorescence microscopy

Confocal fluorescence microscopy has been widely used to image the colocalization between membrane receptors and lipid rafts on cell membranes (Bournazos, Hart, Chamberlain, Glennie, & Dransfield, 2009; Im et al., 2009). In this case, fluorescent molecules are used to label the target receptors and lipid rafts. Normally, cells are fixed and stained with fluorescently labeled antibodies that bind directly to the protein of interest or to primary antibodies that recognize the molecules of interest, whereas lipid rafts are labeled by targeting putative raft markers (Table 6.1). In most occasions, the ganglioside G_{M1} is marked using fluorescently labeled pentameric Cholera toxin B subunit (CTxB). After imaging, standard colocalization analysis methods such as the Pearson or Manders coefficients (Dunn, Kamocka, & McDonald, 2011) are normally performed to quantify the degree of spatial colocalization of the signals corresponding to the target protein and lipid rafts. During raft-mediated signaling, membrane receptors oligomerize in lipid rafts, forming large assemblies that are easily identified by confocal microscopy. Nonetheless, due to the dynamic nature of lipid raft–receptor assemblies, this relationship is sometimes difficult to assess. To overcome this, the molecules of interest are commonly copatched with putative raft markers prior to cell fixation by secondary antibodies, which enhances their interactions, thus facilitating their detection (Hao, Mukherjee, & Maxfield, 2001). However, this labeling scheme should be used with caution, since antibody copatching may lead to a false association of the target protein with lipid rafts.

The main limitation of confocal microscopy in the study of lipid raft-associated processes is its low optical resolution set by the diffraction of light (≈ 200 nm in the lateral axis). Resolving entities of a few nanometers and separated by interparticle distances below the resolution limit is simply impossible using this technique. Nevertheless, due to its standardization, simplicity, versatility, and minimal invasiveness, confocal microscopy continues to be broadly used in this type of studies, providing relevant quantitative data about protein interactions with lipid rafts.

6.2.2 Homo-Förster resonance energy transfer microscopy

Many aspects of membrane receptor spatiotemporal organization, including insights about the link between lipid rafts and receptor oligomerization (Fig. 6.1), have been obtained using homo-FRET (Förster resonance energy transfer) microscopy

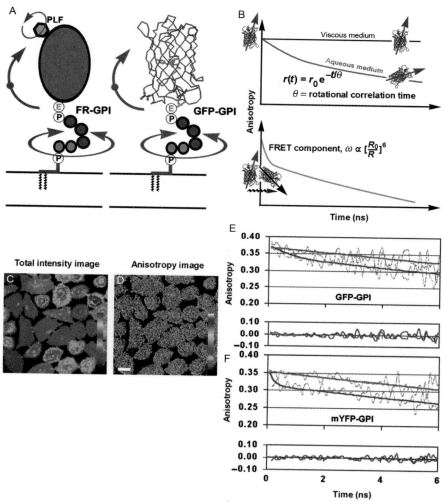

FIGURE 6.1

Time-resolved anisotropy imaging reveals GPI-anchored protein nanoclusters on the cell membranes. (A) Rotational motions affecting the fluorescent folic acid analog (Pteroyl-lysyl-folate, PLF) and GFP that label GPI-anchored proteins. (B) Cartoon depicting how the rotational motions in (A) and homo-FRET can be followed using time-resolved fluorescence anisotropy. (C–D) Mean fluorescence intensity and anisotropy images of GPI-anchored proteins on the cell membrane. (E–F) GPI-anchored protein time-resolved fluorescence anisotropy decays in cholesterol-depleted cells (upper decay lines) and nontreated cells (lower decay lines). In the presence of cholesterol, GPI-anchored proteins undergo efficient homo-FRET, resulting in a rapid component in the anisotropy decay (bottom decay lines), which is absence in cholesterol-depleted cells. This is consistent with a loss in GPI-anchored protein nanoclustering upon disruption of lipid rafts. (For color version of this figure, the reader is referred to the online version of this chapter.)

Adapted from Sharma et al. (2004).

(Goswami et al., 2008; Sharma et al., 2004). This technique takes advantage of the nonradiative energy transfer that occurs between an excited donor fluorophore and an acceptor molecule separated at distances of 1–10 nm (i.e., FRET) to gather information on how these molecules interact at the nanoscale (Jares-Erijman & Jovin, 2003). The efficiency of the energy transfer process is highly dependent on the distance between the donor and the acceptor molecules and other physical parameters, such as the donor quantum yield and spectral overlap between the emission spectrum of the donor and the absorption spectrum of the acceptor, which are included in the donor–acceptor Förster radius (R_0) (Rao & Mayor, 2005). Most applications of FRET in biological sciences monitor the energy transfer process between two different donor and acceptor molecules. This process, usually termed hetero-FRET, is experimentally quantified by determining its efficiency from the ratio of the relative fluorescence emission of the donor in the presence (I_{AD}) and in the absence (I_D) of the acceptor (Rao & Mayor, 2005). However, energy transfer between identical fluorophores, that is homo-FRET, can also occur. In this case, the ratiometric approach used to study hetero-FRET is not applicable because both the acceptor and the donor molecules have identical spectroscopic properties. Instead, homo-FRET is experimentally measured by quantifying how the anisotropy of the fluorescence emission changes after fluorophores excitation. This is achieved by polarizing the excitation light and by separating the fluorescence emission based on two 90°-shifted polarization components with respect to the polarization of the excitation light. The fluorescence anisotropy is then calculated according to the following expression (Bader et al., 2011):

$$r = \frac{I_{par} - I_{per}}{I_{par} + 2I_{per}} \quad (6.1)$$

where I_{par} and I_{per} are, respectively, the intensities of the fluorescence emission parallel and perpendicular to the polarization of the excitation light. By exciting with polarized light, sample fluorophores that have their absorption transition moments parallel to the orientation of the excitation light polarization are photoselected. If the fluorophores do not rotate during their excited-state lifetime and their absorption and emission transition moments are parallel, emission polarization has the same direction to that of the excitation light and the anisotropy reaches its maximum value. However, by a homo-FRET process, excited donors may transfer their excess energy to nearby acceptors. As a result, fluorescence emission originates from molecules that have not been photoselected by the polarized excitation, that is, distinctly oriented from the donors. Both donor and acceptor molecules contribute for the emission polarization, lowering the fluorescence anisotropy value. However, the rotational diffusion of the fluorophores can also critically reduce the resulting anisotropy. For this reason, slowly rotating dyes such as fluorescent proteins are ideally suited for this type of studies.

The combination of homo-FRET with fluorescence microscopy is an elegant approach to gather information of cellular processes that occur at distances smaller than the diffraction limit. This technique is particularly suited to quantify the number of molecules within clusters, having been successfully used to estimate the number of protein molecules in cell membrane oligomers (Bader, Hofman, Voortman, van Bergen en Henegouwen, & Gerritsen, 2009). From steady-state fluorescence anisotropy images, the number of fluorophores in a cluster is given by Bader et al. (2009) and Runnels and Scarlata (1995):

$$r = r_{mono}\frac{1+\omega\tau}{1+N\omega\tau} + r_{et}\frac{(N-1)\omega\tau}{1+N\omega\tau} \qquad (6.2)$$

where r_{mono} and r_{et} are the anisotropy values of a single directly excited molecule and of an excited molecule by a homo-FRET process, $\omega\tau$ accounts for the efficiency of the energy transfer and can be defined as $\omega\tau = E/E-1$, and N is the number of fluorophores. r_{et} and $\omega\tau$ have to be determined in order to extract a quantitative value of the steady-state anisotropy measurement. Nonetheless, the calculation of the number of protein molecules per cluster can be simplified if time-resolved anisotropy images are performed. In this case, multiple homo-FRET processes occur within the protein cluster in the few nanoseconds that follow excitation. As a result, the anisotropy is leveled at its limiting value r_{inf} and every fluorophore has the same probability to emit a photon. The number of molecules in the cluster can thus be determined directly from r_{inf}, according to Bader et al. (2009):

$$r_{inf} = r_{mono}\frac{1}{N} + r_{et}\frac{N-1}{N} \qquad (6.3)$$

The size of the clusters and their spatial distribution can be estimated using the loss of anisotropy due to controlled photobleaching of the fluorophores. Sequential photobleaching of the fluorophores leads to a gradual loss of the fluorescence anisotropy since the probability of a homo-FRET process diminishes with increasing photobleaching time. Proper modeling of the resulting anisotropy curve can be used to estimate the size of the cluster (Sharma et al., 2004). Moreover, hetero-FRET can be used at varying concentrations of donor/acceptor molecules as a complementary method to quantify cluster size.

Although time-resolved homo-FRET microscopy is a powerful tool to study the cell membrane organization and protein clustering, some cares should be taken when using it. Due to the length scale of the interactions involved ($\approx 1-10$ nm), homo-FRET can also occur in samples presenting a high density of molecules displaying a purely Brownian distribution. This, however, should not be interpreted as molecular clustering. A study of how the anisotropy changes when reducing the concentration of molecules might allow distinguishing between actual clustering of molecules and a high density of molecules with a Brownian distribution and interparticle distances close to their characteristic R_0. True clustering of molecules can be identified if the scatterplot of fluorescence intensity versus anisotropy displays a constant anisotropy value independent on the molecule density of the sample (Varma & Mayor, 1998). In contrast, if the anisotropy value anticorrelates with the fluorescence intensity of the sample, homo-FRET can be explained by a high density of dye molecules. This technique might also underestimate clustering when interparticle distances of the fluorophores within the clusters are slightly larger than the R_0, resulting in poor homo-FRET efficiencies.

6.2.3 Superresolution techniques to map the nanolandscape of the cell membrane

The diffraction limit in conventional microscopy is determined by the minimum spot size to which a light beam can be focused with normal lens elements. In practice, the diffraction limit implies that the minimum distance Δx required to independently resolve two distinct objects is dependent on the wavelength of the light used to observe the specimen λ and the overall objective lens system, through the numerical aperture (NA) of the objective, as $\Delta x \sim \lambda/2 \times NA$. With modern objectives having an NA as high as ~ 1.4, the resolution of conventional microscopy becomes 250–300 nm in the case of visible light.

The landscape of the cell membrane is composed out of a vast selection of different proteins and lipids that organize at length scales well below the diffraction limit of light. To inquire on the nanoscale positions of these protein and lipids or directly visualize their organization on the cell membrane, one should employ superresolution techniques. Nowadays, a palette of superresolution approaches are available to the bioscience community. The various techniques such as NSOM, STED, (F-)PALM, and stochastic optical reconstruction microscopy (STORM) are complementary to each other in terms of specific advantages and limitations.

6.2.3.1 Far-field localization techniques (STORM/PALM)

Although the diffraction limit of light poses a restriction on the immediate number of molecules one can see per unit area, the position of single-molecule spots separated at distances larger than the diffraction limit of light can actually be established to a precision of tens of nanometers. The accuracy of determining its center-of-mass (r_{com}) depends essentially on the number of photons emitted through $r_{com} \sim r_d \times N^{-1/2}$, where r_d is the diffraction limited resolution and N the number of photons (Thompson, Larson, & Webb, 2002). This means that in order to reconstruct a superresolution image of a densely populated cell membrane, the challenge is to have at each given time only a subset of molecules in the "on"-state and determine r_{com} for each molecule. This process is repeated many times such that all the calculated r_{com} are used to reconstruct a "superresolution" image. This indeed is the concept of the superresolution technique called stochastic optical reconstruction microscopy (Rust, Bates, & Zhuang, 2006), which is essentially based on carbocyanine dyes that reversibly switch between on and off states. An analogous method that also uses image reconstruction but is based on fluorescent proteins, and therefore directly applicable in live systems, is called (fluorescent) photoactivatable localization microscopy (FPALM/PALM) (Betzig et al., 2006; Hess, Girirajan, & Mason, 2006). In here, the fluorescent proteins are engineered such that they can be switched on by illumination with a 405 nm light source. After the few photoactivated fluorescent spot photobleach, a subsequent flash of the 405 nm light activates another set of random fluorophores such that the final superresolution image is built up similarly as described earlier for STORM.

Over the last few years, invaluable information about the nature of cell membrane nano- and microdomains has been obtained using localization-based superresolution techniques. For instance, the organization of the Lck protein into 120 nm membrane clusters (Fig. 6.2) was shown to be dependent on lipid raft integrity and protein conformational state, both affecting cell signaling (Owen et al., 2010; Rossy, Owen, Williamson, Yang, & Gaus, 2013). To obtain more quantitative data on cluster or domain formation, the dataset of single-molecule localizations can be analyzed using Ripley´s function (Ripley, 1977) or pair correlation function (Sengupta et al., 2011). From these algorithms, relevant data such as cluster sizes and the receptor densities inside those clusters can be obtained.

To reconstruct high-quality superresolution images, a single-molecule-sensitive camera should detect all emitters labeling a specific structure with sufficient number of photons to get position accuracies down to tens of nanometers. Moreover, the imaging conditions and the reconstruction algorithms are critical to obtain biologically meaningful data and thus optimized imaging protocols for each subject under investigation are usually required. A precise quantitative assessment of protein clustering is still challenging due to the possible multiple localization assignments of a single molecule that is photoblinking on long timescales, which will therefore erroneously appear as a cluster. Furthermore, since no true optical imaging is being generated, all the optical contrast mechanisms associated with light, that is, intensity, polarization, wavelength, and lifetime, are lost on the reconstructed image.

6.2.3.2 Far-field patterned illumination techniques (STED)

An alternative superresolution method that also relies on the photophysical properties of fluorophores is STED microscopy (Klar & Hell, 1999). Conceptually, the method proposes to reduce the spatial extent of the excitation light by stimulating the depletion of the emission on the outer regions of the diffraction-limited excitation profile. The principle is based on the fact that an excited fluorophore can be stimulated back to the ground state without photon emission. This can be achieved by illuminating the fluorophore with red-shifted light after excitation. With enough power for the depletion beam, it is possible to reduce the probability of fluorescence significantly. Then, by using a conventional diffraction-limited spot to excite the fluorophores and overlay this with a donut-shaped deexcitation (or depletion) profile, one can reshape the region where emission is allowed down to a few nanometers (Rittweger, Han, Irvine, Eggeling, & Hell, 2009). To form a true nanoscopic image, this effective nanosized spot is raster-scanned through the sample. For STED microscopy, the resolution, r_{STED}, can be tuned by the intensity of the STED beam, I, through $r_{STED} \sim r_d*(1 + I/I_s)^{-1/2}$, where r_d is the size of the diffraction-limited spot and I_s the intensity of the STED beam needed to deplete the fluorescence to $1/e$. However, because of its mere principle, STED requires accurate control of the position, phase, and amplitude of two laser beams (for single color fluorescence), and

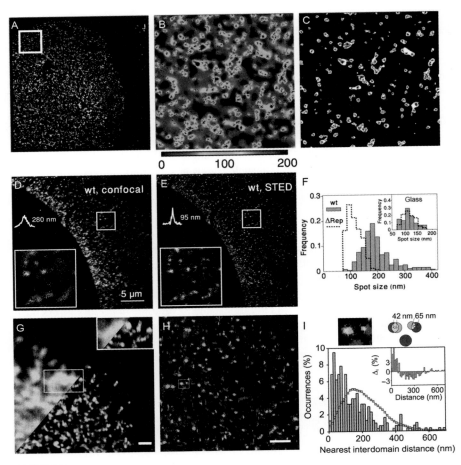

FIGURE 6.2

Superresolution techniques map protein distribution and clustering on the cell membrane. (A) Single-molecule PALM images of the raftophilic protein Lck. (B) Protein cluster maps generated from local point-pattern analysis of outline region. (C) Maps of Lck clusters and molecules inside clusters. Adapted from Rossy et al. (2013). (D–E) Confocal and STED images of the pathogen recognition receptor Dendritic Cell-Specific Intercellular adhesion molecule-3-Grabbing Non-integrin (DC-SIGN). (F) Normalized intensity distribution of fluorescent spots obtained by STED for DC-SIGN and a mutant (ΔRep) compared to the intensity of spots found on glass. Adapted from Manzo et al. (2012). (G) Confocal versus NSOM image of cholera toxin-G_{M1} clusters at the cell membrane. (H) Dual-color NSOM of CD55 (green) and CTxB-GM1 nanodomains (red). (I) Nearest-neighbor distributions analysis of CD55 to its closest CTxB-GM1 nanodomain (bars) compared to simulations of random spatial distribution of proteins and CTxB-GM1 clusters. (See color plate.)

Adapted from van Zanten et al. (2010).

its best resolution is restricted to certain dyes able to withstand repeated cycles of excitation and depletion at extremely high intensities.

Using properly chosen fluorophores, this technique has been used to estimate protein distribution and cluster sizes on the cell membrane (Fig. 6.2; Manzo et al., 2012; Sieber et al., 2007). To obtain information regarding the nanoscale dynamics in living cells, STED has been combined with fluorescence correlation spectroscopy (FCS) (Eggeling et al., 2009). Initially, it was shown that unlike randomly diffusing nonraft lipids, glycosyl-phosphatidylinositol-anchored proteins (GPI-APs) and sphingolipids are transiently trapped for about 10–20 ms in <20 nm-sized cholesterol-dependent molecular complexes. Recently, these observations have been extended to more active involvement of cytoskeleton-mediated interactions (Mueller et al., 2011).

STED can take advantage of standard confocal geometries and therefore allows optical sectioning at the nanoscale and even provides video-rate imaging (Westphal et al., 2008). However, STED requires fluorophores that can cycle many times between dark and bright states. In addition, since the two beams that reshape the excitation field should be aligned properly in time and space, the optical design is rather complex and multicolor imaging challenging.

6.2.3.3 Near-field techniques (NSOM)

A different concept that breaks the diffraction limit of light providing optical superresolution at the nanometer scale is NSOM. In NSOM, a sharp probe physically scans the sample surface generating simultaneous optical and topographic imaging of the sample under study. The most generally applied near-field optical probe consists of a small aperture, typically 20–120 nm in diameter (i.e., much smaller than the wavelength of the excitation light), at the end of a metal-coated tapered optical fiber. The probe funnels the incident light to dimensions that are substantially below the diffraction limit. This results in a light source that has the size of the aperture. However, in contrast to common light sources such as lightbulbs and lasers, the light emitted by the probe is predominantly composed of evanescent rather than propagating waves. The intensity of the evanescent light decays exponentially to insignificant levels at ~100 nm away from the aperture. Effectively, the probe can excite fluorophores only within a layer of <100 nm from the probe, that is, in the "near-field" region. The sample fluorescence can subsequently be collected by conventional optics and transformed into an optical image of the sample surface in which the resolution is now primarily dictated by the aperture dimensions rather than by the wavelength of the light. Noteworthy, NSOM imaging can be obtained with any fluorophore and the illumination geometry allows straightforward implementation of multicolor excitation. However, like other raster-scanning techniques, NSOM imaging is inherently slow.

NSOM has been used to gather information on membrane protein organization, protein clusters sizes, and also on membrane proteins colocalization (de Bakker et al., 2007, 2008). Recently, single-molecule near-field nanoscopy has been used to visualize the nanolandscape of G_{M1} after binding by its ligand cholera toxin (CTxB) on intact monocyte membranes (van Zanten et al., 2010). Pentavalent binding of CTxB to G_{M1} was sufficient to initiate a minimal raft coalescence unit, resulting in the formation of cholesterol-dependent G_{M1} nanodomains <120 nm in size (Fig. 6.2). Simultaneous dual-color high-resolution images revealed that the canonical raft component CD55 (a GPI-anchored protein) was recruited to regions proximal (<150 nm) to CTxB-G_{M1} nanodomains without physical intermixing, but not the nonraft protein CD71. These results demonstrated the existence of raft-based compositional connectivity at the nanoscale crucially mediated by cholesterol. An earlier NSOM study had shown that a similar nanoscale organization of GPI-anchored protein and adhesion receptors was essential for successful cell adhesion, showing the functional relevance of preformed nanoscale platforms (van Zanten et al., 2009). More conventional techniques such as FRET are unable to report on such a spatial proximity at distances >10 nm. On the other extreme, diffraction-limited techniques such as confocal microscopy would misleadingly show colocalization between different components located at distances <300 nm. This shows that superresolution techniques are in fact bridging the gap between 10 and 300 nm and provide exquisite information at these important spatial scales.

Nanoscale dynamics on living cell membranes can be measured by combining NSOM with FCS (Manzo, van Zanten, & Garcia-Parajo, 2011). Additionally, probe design based on optical antennas further improves lateral resolution below 30 nm (e.g., Mivelle, van Zanten, Neumann, van Hulst, & Garcia-Parajo, 2012). For these reasons, NSOM constitutes a valuable tool to characterize spatiotemporal details of many biological processes occurring on the cell membrane.

CONCLUDING REMARKS

Over the last decade, the experimental approaches used in the study of receptor membrane distribution, arrangement in clusters, association with lipid rafts, and the implications on this organization to cell signaling have changed substantially. The development of advanced imaging techniques greatly contributed to this progress. Important features of the molecular mechanism governing cell membrane (nano-) domain formation and their functions have been obtained using these imaging methods. Nonetheless, this would not have been possible if biochemical approaches to study the organization of intact cell membranes have not been available. This was especially important in studies of protein association with lipid rafts, since for long years, this relationship was almost exclusively assayed by testing membrane protein resistance to solubilization by nonionic detergents. Nowadays, protein association with lipid rafts is mainly investigated using nondestructive or at least not fully destructive cell methods, for example, using cholesterol depletion agents or colocalization with raft markers, allowing the study of these interactions on the compositional complex environment of the cell membrane. However, there are no perfect methods to study cell membrane protein organization and its relation with lipid rafts. As described above, current techniques used to investigate this problematic have their intrinsic limitations, which, associated with transient nature of cell membrane assemblies, make their study difficult. Improvements in labeling methods' and imaging techniques' spatial and temporal resolution will certainly overcome some of those drawbacks, making the study of membrane domains less challenging.

Acknowledgments

STED images were obtained at the ICFO's Super-Resolution Light Nanoscopy Facility, SLN@ICFO. We acknowledge support from the Spanish Ministry of Science and Innovation (MAT2011-22887), Generalitat de Catalunya (2009 SGR 597), the European Commission (FP7-ICT-2011-7, under grant agreement No. 288263), the HFSP (grant RGP0027/2012), and the ICFONEST fellowship program, a Marie Curie Co-funding of Regional, National, and International Programmes (COFUND) action of the European Commission.

References

Allen, J. A., Halverson-Tamboli, R. A., & Rasenick, M. M. (2007). Lipid raft microdomains and neurotransmitter signalling. *Nature Reviews. Neuroscience, 8*(2), 128–140.

Bader, A. N., Hoetzl, S., Hofman, E. G., Voortman, J., van Bergen en Henegouwen, P. M., van Meer, G., et al. (2011). Homo-FRET imaging as a tool to quantify protein and lipid clustering. *Chemphyschem: A European Journal of Chemical Physics and Physical Chemistry, 12*(3), 475–483.

Bader, A. N., Hofman, E. G., Voortman, J., van Bergen en Henegouwen, P. M., & Gerritsen, H. C. (2009). Homo-FRET imaging enables quantification of protein cluster sizes with subcellular resolution. *Biophysical Journal, 97*(9), 2613–2622.

Betzig, E., Patterson, G. H., Sougrat, R., Lindwasser, O. W., Olenych, S., Bonifacino, J. S., et al. (2006). Imaging intracellular fluorescent proteins at nanometer resolution. *Science, 313*(5793), 1642–1645.

Bournazos, S., Hart, S. P., Chamberlain, L. H., Glennie, M. J., & Dransfield, I. (2009). Association of FcgammaRIIa (CD32a) with lipid rafts regulates ligand binding activity. *Journal of Immunology, 182*(12), 8026–8036.

Brown, D. A. (2006). Lipid rafts, detergent-resistant membranes, and raft targeting signals. *Physiology, 21*, 430–439.

Cahuzac, N., Baum, W., Kirkin, V., Conchonaud, F., Wawrezinieck, L., Marguet, D., et al. (2006). Fas ligand is localized to membrane rafts, where it displays increased cell death-inducing activity. *Blood, 107*(6), 2384–2391.

Castro, B. M., Silva, L. C., Fedorov, A., de Almeida, R. F., & Prieto, M. (2009). Cholesterol-rich fluid membranes solubilize ceramide domains: Implications for the structure and dynamics of mammalian intracellular and plasma membranes. *The Journal of Biological Chemistry, 284*(34), 22978–22987.

Chini, B., & Parenti, M. (2004). G-protein coupled receptors in lipid rafts and caveolae: How, when and why do they go there? *Journal of Molecular Endocrinology, 32*(2), 325–338.

de Bakker, B. I., Bodnar, A., van Dijk, E. M., Vamosi, G., Damjanovich, S., Waldmann, T. A., et al. (2008). Nanometer-scale organization of the alpha subunits of the receptors for IL2 and IL15 in human T lymphoma cells. *Journal of Cell Science, 121*(Pt 5), 627–633.

de Bakker, B. I., de Lange, F., Cambi, A., Korterik, J. P., van Dijk, E. M., van Hulst, N. F., et al. (2007). Nanoscale organization of the pathogen receptor DC-SIGN mapped by single-molecule high-resolution fluorescence microscopy. *Chemphyschem: A European Journal of Chemical Physics and Physical Chemistry*, *8*(10), 1473–1480.

Dunn, K. W., Kamocka, M. M., & McDonald, J. H. (2011). A practical guide to evaluating colocalization in biological microscopy. *American Journal of Physiology—Cell Physiology*, *300*(4), C723–C742.

Eggeling, C., Ringemann, C., Medda, R., Schwarzmann, G., Sandhoff, K., Polyakova, S., et al. (2009). Direct observation of the nanoscale dynamics of membrane lipids in a living cell. *Nature*, *457*(7233), 1159–1162.

Gimpl, G., & Gehrig-Burger, K. (2007). Cholesterol reporter molecules. *Bioscience Reports*, *27*(6), 335–358.

Goswami, D., Gowrishankar, K., Bilgrami, S., Ghosh, S., Raghupathy, R., Chadda, R., et al. (2008). Nanoclusters of GPI-anchored proteins are formed by cortical actin-driven activity. *Cell*, *135*(6), 1085–1097.

Hao, M., Mukherjee, S., & Maxfield, F. R. (2001). Cholesterol depletion induces large scale domain segregation in living cell membranes. *Proceedings of the National Academy of Sciences of the United States of America*, *98*(23), 13072–13077.

Hess, S. T., Girirajan, T. P., & Mason, M. D. (2006). Ultra-high resolution imaging by fluorescence photoactivation localization microscopy. *Biophysical Journal*, *91*(11), 4258–4272.

Im, J. S., Arora, P., Bricard, G., Molano, A., Venkataswamy, M. M., Baine, I., et al. (2009). Kinetics and cellular site of glycolipid loading control the outcome of natural killer T cell activation. *Immunity*, *30*(6), 888–898.

Jares-Erijman, E. A., & Jovin, T. M. (2003). FRET imaging. *Nature Biotechnology*, *21*(11), 1387–1395.

Klar, T. A., & Hell, S. W. (1999). Subdiffraction resolution in far-field fluorescence microscopy. *Optics Letters*, *24*(14), 954–956.

Kusumi, A., Fujiwara, T. K., Chadda, R., Xie, M., Tsunoyama, T. A., Kalay, Z., et al. (2012). Dynamic organizing principles of the plasma membrane that regulate signal transduction: Commemorating the fortieth anniversary of Singer and Nicolson's fluid-mosaic model. *Annual Review of Cell and Developmental Biology*, *28*, 215–250.

Lange, Y. (1992). Tracking cell cholesterol with cholesterol oxidase. *Journal of Lipid Research*, *33*(3), 315–321.

Le Lay, S., Li, Q., Proschogo, N., Rodriguez, M., Gunaratnam, K., Cartland, S., et al. (2009). Caveolin-1-dependent and -independent membrane domains. *Journal of Lipid Research*, *50*(8), 1609–1620.

Lenne, P. F., Wawrezinieck, L., Conchonaud, F., Wurtz, O., Boned, A., Guo, X. J., et al. (2006). Dynamic molecular confinement in the plasma membrane by microdomains and the cytoskeleton meshwork. *The EMBO Journal*, *25*(14), 3245–3256.

Levental, I., Lingwood, D., Grzybek, M., Coskun, U., & Simons, K. (2010). Palmitoylation regulates raft affinity for the majority of integral raft proteins. *Proceedings of the National Academy of Sciences of the United States of America*, *107*(51), 22050–22054.

Lingwood, D., Ries, J., Schwille, P., & Simons, K. (2008). Plasma membranes are poised for activation of raft phase coalescence at physiological temperature. *Proceedings of the National Academy of Sciences of the United States of America*, *105*(29), 10005–10010.

Manzo, C., Torreno-Pina, J. A., Joosten, B., Reinieren-Beeren, I., Gualda, E. J., Loza-Alvarez, P., et al. (2012). The neck region of the C-type lectin DC-SIGN regulates its surface spatiotemporal organization and virus-binding capacity on antigen-presenting cells. *The Journal of Biological Chemistry, 287*(46), 38946–38955.

Manzo, C., van Zanten, T. S., & Garcia-Parajo, M. F. (2011). Nanoscale fluorescence correlation spectroscopy on intact living cell membranes with NSOM probes. *Biophysical Journal, 100*(2), L8–L10.

Mivelle, M., van Zanten, T. S., Neumann, L., van Hulst, N. F., & Garcia-Parajo, M. F. (2012). Ultrabright bowtie nanoaperture antenna probes studied by single molecule fluorescence. *Nano Letters, 12*(11), 5972–5978.

Mollinedo, F., & Gajate, C. (2006). Fas/CD95 death receptor and lipid rafts: New targets for apoptosis-directed cancer therapy. *Drug Resistance Updates: Reviews and Commentaries in Antimicrobial and Anticancer Chemotherapy, 9*(1–2), 51–73.

Mueller, V., Ringemann, C., Honigmann, A., Schwarzmann, G., Medda, R., Leutenegger, M., et al. (2011). STED nanoscopy reveals molecular details of cholesterol- and cytoskeleton-modulated lipid interactions in living cells. *Biophysical Journal, 101*(7), 1651–1660.

Munro, S. (2003). Lipid rafts: Elusive or illusive? *Cell, 115*(4), 377–388.

Owen, D. M., Rentero, C., Rossy, J., Magenau, A., Williamson, D., Rodriguez, M., et al. (2010). PALM imaging and cluster analysis of protein heterogeneity at the cell surface. *Journal of Biophotonics, 3*(7), 446–454.

Owen, D. M., Williamson, D. J., Magenau, A., & Gaus, K. (2012). Sub-resolution lipid domains exist in the plasma membrane and regulate protein diffusion and distribution. *Nature Communications, 3*, 1256.

Pike, L. J. (2006). Rafts defined: A report on the Keystone symposium on lipid rafts and cell function. *Journal of Lipid Research, 47*(7), 1597–1598.

Rao, M., & Mayor, S. (2005). Use of Forster's resonance energy transfer microscopy to study lipid rafts. *Biochimica et Biophysica Acta, 1746*(3), 221–233.

Ripley, B. D. (1977). Modelling spatial patterns (with discussion). *Journal of the Royal Statistical Society B, 39*, 172–212.

Rittweger, E., Han, K. Y., Irvine, S. E., Eggeling, C., & Hell, S. W. (2009). STED microscopy reveals crystal colour centres with nanometric resolution. *Nature Photonics, 3*(3), 144–147.

Rossy, J., Owen, D. M., Williamson, D. J., Yang, Z., & Gaus, K. (2013). Conformational states of the kinase Lck regulate clustering in early T cell signaling. *Nature Immunology, 14*(1), 82–89.

Runnels, L. W., & Scarlata, S. F. (1995). Theory and application of fluorescence homotransfer to melittin oligomerization. *Biophysical Journal, 69*(4), 1569–1583.

Rust, M. J., Bates, M., & Zhuang, X. (2006). Sub-diffraction-limit imaging by stochastic optical reconstruction microscopy (STORM). *Nature Methods, 3*(10), 793–795.

Sengupta, P., Jovanovic-Talisman, T., Skoko, D., Renz, M., Veatch, S. L., & Lippincott-Schwartz, J. (2011). Probing protein heterogeneity in the plasma membrane using PALM and pair correlation analysis. *Nature Methods, 8*(11), 969–975.

Sharma, P., Varma, R., Sarasij, R. C., Ira, Gousset, K., Krishnamoorthy, G., et al. (2004). Nanoscale organization of multiple GPI-anchored proteins in living cell membranes. *Cell, 116*(4), 577–589.

Sieber, J. J., Willig, K. I., Kutzner, C., Gerding-Reimers, C., Harke, B., Donnert, G., et al. (2007). Anatomy and dynamics of a supramolecular membrane protein cluster. *Science, 317*(5841), 1072–1076.

Simons, K., & Gerl, M. J. (2010). Revitalizing membrane rafts: New tools and insights. *Nature Reviews. Molecular Cell Biology, 11*(10), 688–699.

Suzuki, K. G. (2012). Lipid rafts generate digital-like signal transduction in cell plasma membranes. *Biotechnology Journal, 7*(6), 753–761.

Thompson, R. E., Larson, D. R., & Webb, W. W. (2002). Precise nanometer localization analysis for individual fluorescent probes. *Biophysical Journal, 82*(5), 2775–2783.

van Zanten, T. S., Cambi, A., Koopman, M., Joosten, B., Figdor, C. G., & Garcia-Parajo, M. F. (2009). Hotspots of GPI-anchored proteins and integrin nanoclusters function as nucleation sites for cell adhesion. *Proceedings of the National Academy of Sciences of the United States of America, 106*(44), 18557–18562.

van Zanten, T. S., Gomez, J., Manzo, C., Cambi, A., Buceta, J., Reigada, R., et al. (2010). Direct mapping of nanoscale compositional connectivity on intact cell membranes. *Proceedings of the National Academy of Sciences of the United States of America, 107*(35), 15437–15442.

Varma, R., & Mayor, S. (1998). GPI-anchored proteins are organized in submicron domains at the cell surface. *Nature, 394*(6695), 798–801.

Westphal, V., Rizzoli, S. O., Lauterbach, M. A., Kamin, D., Jahn, R., & Hell, S. W. (2008). Video-rate far-field optical nanoscopy dissects synaptic vesicle movement. *Science, 320*(5873), 246–249.

Zidovetzki, R., & Levitan, I. (2007). Use of cyclodextrins to manipulate plasma membrane cholesterol content: Evidence, misconceptions and control strategies. *Biochimica et Biophysica Acta, 1768*(6), 1311–1324.

CHAPTER 7

Serotonin Type 4 Receptor Dimers

Sylvie Claeysen[*,†,‡], Romain Donneger[*,†,‡], Patrizia Giannoni[*,†,‡], Florence Gaven[*,†,‡], and Lucie P. Pellissier[*,†,‡]

[*]CNRS, UMR-5203, Institut de Génomique Fonctionnelle, Montpellier, France
[†]Inserm, U661, Montpellier, France
[‡]Universités de Montpellier 1 & 2, UMR-5203, Montpellier, France

CHAPTER OUTLINE

Introduction 124
7.1 Materials 125
7.2 Methods 127
    7.2.1 Cell Transfection 127
    7.2.2 Detection of 5-HT$_4$R Dimersx by Western Blot 127
        7.2.2.1 Cell Lysate and Membrane Preparation 127
        7.2.2.2 Solubilization and Deglycosylation of the Samples 129
        7.2.2.3 Western Blot and Detection of the 5-HT$_4$ Receptor Dimers ... 129
    7.2.3 Detection of 5-HT$_4$R Dimers by Coimmunoprecipitation 131
    7.2.4 Analysis of 5-HT$_4$R Dimers by Immunofluorescence 131
    7.2.5 Analysis of 5-HT$_4$R Dimers by TR-FRET 134
7.3 Discussion 136
Summary 137
Acknowledgments 137
References 138

Abstract

Numerous class A G protein-coupled receptors and especially biogenic amine receptors have been reported to form homodimers. Indeed, the dimerization process might occur for all the metabotropic serotonergic receptors. Moreover, dimerization appears to be essential for the function of serotonin type 2C (5-HT$_{2C}$) and type

4 (5-HT$_4$) receptors and required to obtain full receptor activity. Several techniques have been developed to analyze dimer formation and properties. Due to our involvement in deciphering 5-HT$_4$R transduction mechanisms, we improved and set up new procedures to study 5-HT$_4$R dimers, by classical methods or modern tools. This chapter presents detailed protocols to detect 5-HT$_4$R dimers by Western blotting and coimmunoprecipitation, including the optimizations that we routinely carry out. We developed an innovative method to achieve functional visualization of 5-HT$_4$R dimers by immunofluorescence, taking advantage of the 5-HT$_4$-RASSL (receptor activated solely by synthetic ligand) mutant that was engineered in the laboratory. Finally, we adapted the powerful time-resolved FRET technology to assess a relative quantification of dimer formation and affinity.

INTRODUCTION

Serotonin type 4 receptors (5-HT$_4$Rs) belong to the extended family of serotonin receptors, which counts 15 different types of receptors involved in a wide range of physiological processes (Berger, Gray, & Roth, 2009). All, except 5-HT$_3$ receptor that is an ionic channel, are G protein-coupled receptors (GPCRs) activating G$_s$-, G$_i$-, or G$_q$-dependent pathways and G protein-independent signaling cascades (Barnes & Sharp, 1999; Bockaert, Claeysen, Becamel, Dumuis, & Marin, 2006; Millan, Marin, Bockaert, & Mannoury la Cour, 2008). Homodimerization of serotonin receptors has been described for the 5-HT$_{1A}$ (Gorinski et al., 2012), 5-HT$_{1B/D}$ (Lee et al., 2000), 5-HT$_{2A}$ (Lukasiewicz, Faron-Gorecka, Kedracka-Krok, & Dziedzicka-Wasylewska, 2011), 5-HT$_{2C}$ (Herrick-Davis, Grinde, & Mazurkiewicz, 2004), 5-HT$_4$ (Berthouze et al., 2005), and 5-HT$_7$ (Renner et al., 2012), suggesting that all metabotropic serotonergic receptors form constitutive homodimers. This dimerization process is essential for receptor function. Indeed, the full activity of 5-HT$_{2C}$ and 5-HT$_4$ receptors has been obtained with the binding of two molecules of ligand and one G protein per dimer (Herrick-Davis, Grinde, Harrigan, & Mazurkiewicz, 2005; Pellissier et al., 2011).

Our team has been involved in the first pharmacological description of 5-HT$_4$ receptors (Dumuis, Bouhelal, Sebben, Cory, & Bockaert, 1988), in the cloning of some splice variants (Claeysen, Sebben, Becamel, Bockaert, & Dumuis, 1999; Claeysen, Sebben, Journot, Bockaert, & Dumuis, 1996) and in the characterization of several original signaling pathways following the activation of these receptors (Bockaert, Claeysen, Compan, & Dumuis, 2011). For years, we have developed methods and tools to detect 5-HT$_4$ receptor dimers that we present in the succeeding text. Ranging from classical to more sophisticated methods, we intend to provide here the hints and tips that facilitate 5-HT$_4$R dimerization study.

We will describe the following procedures:

1. Cell transfection
2. Detection of 5-HT$_4$R dimers by Western blot

3. Detection of 5-HT$_4$R dimers by coimmunoprecipitation
4. Analysis of 5-HT$_4$R dimers by immunofluorescence
5. Analysis of 5-HT$_4$R dimers by time-resolved Förster resonance energy transfer (TR-FRET)

7.1 MATERIALS

1. PBS: phosphate-buffered saline, pH 7.4, LONZA, DPBS, 10× without Ca^{2+} and Mg^{2+}, #BE17-515F.
2. Trypsine–EDTA solution 1×, Life Technologies, #2530096.
3. EP1×: electroporation mix composed of 50 mM K$_2$HPO$_4$, 20 mM CH$_3$CO$_2$K, 20 mM KOH, and 26.7 mM MgSO$_4$, in water. Adjust the pH to 7.4 with acetic acid.
4. 0.4 cm electroporation cuvettes, Eurogentec, #CE-0004-50.
5. Falcon 100 or 150 mm cell culture dishes, cluster of 12, 24, or 96 wells, 12 mL Falcon tubes (BD Biosciences).
6. Microtubes: 1.5 mL capacity.
7. DMEM-10% DFBS: DMEM (DMEM 4.5 g/L glucose with L-glutamine, LONZA, #BE12-604F) supplemented with 10% dialyzed fetal bovine serum (DFBS) (Lonza, #14-810F).
8. Myc-tagged and HA-tagged 5-HT$_4$ receptors cDNA in plasmid suitable for expression in mammalian cells (e.g., pRK5 or pcDNA3.1). Epitopes are located at the N-terminus of the receptor. The signal peptide (SP) from the metabotropic glutamate receptor type 5 is added before HA tag whereas no SP is necessary with Myc-tag.
9. Empty plasmid cDNA used as carrier and control (e.g., pRK5 or pcDNA3.1).
10. Rubber policeman: 34 mm wide rubber scrapper "Model K" on Saint-Gobain, Verneret Plastic catalog, #V101029, ROGO-SAMPAIC (France), or AUXILAB S.L (Spain).
11. Potter homogenizer: 1 mL capacity PYREX® Potter tissue homogenizer, Corning, #7725T-1.
12. Protease inhibitors: cOmplete, EDTA-free Protease Inhibitor Cocktail Tablets, Roche Applied Science, #04693132001.
13. Tris Lysis Buffer: 10 mM Tris–HCl, pH=7.4; 2 mM EDTA; protease inhibitors.
14. Bradford reagent: Quick Start Bradford 1× Dye Reagent, Bio-Rad, #500-0205.
15. Solubilization buffer: 50 mM NaHPO$_4$/NaH$_2$PO$_4$, pH=7.2; 1 mM EDTA; 1% SDS or 10 mM CHAPS; protease inhibitors.
16. Deglycosylation buffer: 50 mM NaHPO$_4$/NaH$_2$PO$_4$, pH=7.2; 10 mM EDTA; 1% SDS or 10 mM CHAPS; protease inhibitors.
17. N-glycosidase F: PNGase F, New England Biolabs, #P0704S.
18. Laemmli buffer 4×: 200 mM Tris/HCl, pH=6.8; 8% SDS; 40% glycerol; 20% β-mercaptoethanol; bromophenol blue.

19. TBST: TBS (20 mM Tris/HCl, pH=7.4; 150 mM NaCl) and 0.25% Tween-20.
20. TBST–milk: TBST, 5% milk.
21. TBST–BSA: TBST, 5% BSA (Albumin from bovine serum, Sigma, #A2153).
22. Ms anti-Myc antibody: mouse anti-c-Myc antibody, Sigma-Aldrich 9E10, #M4439.
23. Rb anti-Myc antibody: rabbit anti-c-Myc antibody, Santa Cruz Biotechnology, sc789, #D1008.
24. Ms anti-HA antibody: mouse anti-HA antibody, Covance, MMS-101P, #E11AF0013.
25. Rb anti-HA antibody: rabbit anti-HA antibody, Life Technologies, #71-5500.
26. Antimouse antibody conjugated with HRP: ECL Mouse IgG, HRP-linked whole Aβ, GE Healthcare; #NA931.
27. GAR—red: Alexa Fluor® 594 Goat Antirabbit IgG, Life Technologies, #A-11012.
28. GAM—green: Alexa Fluor® 488 Goat Antimouse IgG, Life Technologies, #A-11001.
29. Pierce ECL Western Blotting Substrate: Thermo Scientific, #32209.
30. DSP: dithiobis[succinimidyl propionate], Thermo Scientific, #22586. Fist dilution of DSP at 25 mM in DMSO, then final dilution at 1.25 mM in PBS.
31. Phosphatase inhibitors: 10 mM NaF, 2 mM Na^+ vanadate, and 1 mM Na^+ pyrophosphate.
32. DDM: n-dodecyl β-D-maltoside, Sigma, #D5172.
33. DDM lysis buffer: for 10 ml of buffer, 40 mg of DDM, 20 mM HEPES, 150 mM NaCl, 1% NP40, 10% glycerol, phosphatase inhibitors, and half a tablet of protease inhibitors, adjust the volume with water.
34. Anti-HA/agarose beads: mouse anti-HA-tag monoclonal antibody, agarose-conjugated, Sigma, #DMAB8895.
35. Glass slides and coverslips (∅ 18 mm).
36. PORN 1×: poly-L-ornithine hydrobromide (10 mg/L in PBS), Sigma, #P3655.
37. PFA: paraformaldehyde 16% in water, Euromedex, #RT 15710-S.
38. PBS–glycine: PBS, 0.1 M glycine (Sigma, #G7126).
39. PBS–Triton: PBS, 0.05% Triton X-100 (Sigma, #T8532).
40. PBS–gelatin: PBS, 0.2% gelatin (Sigma, #G9391).
41. 5-HT: 5-hydroxytryptamine (serotonin creatinine sulfate monohydrate, Sigma, #H7752).
42. BIMU 8: 2,3-dihydro-N-[(3-endo)-8-methyl-8-azabicyclo[3.2.1]oct-3-yl]-3-(1-methylethyl)-2-oxo-1H-benzimidazole-1-carboxamide hydrochloride, Tocris Bioscience, #4374.
43. VECTASHIELD: VECTASHIELD Mounting Medium, Vector Laboratories, #H-1000.
44. HBS: 20 mM HEPES, 150 mM NaCl, 4.2 mM KCl, 0.9 mM $CaCl_2$, 0.5 mM $MgCl_2$, 0.1% glucose, and 0.1% BSA.
45. HBS-KF: HSB, 200 mM KF.

46. Anti-HA-K: Eu^{3+} cryptate-conjugated mouse monoclonal antibody anti-HA, Cisbio Bioassays, #610HAKLA.
47. Anti-HA-d2: d2-conjugated mouse monoclonal antibody anti-HA, Cisbio Bioassays, #610HADAA.
48. Anti-FLAG M2-K: Eu^{3+} cryptate-conjugated mouse monoclonal antibody anti-FLAG, Cisbio Bioassays, #61FG2KLA.
49. Anti-FLAG M2-d2: d2-conjugated mouse monoclonal antibody anti-FLAG, Cisbio Bioassays, #61FG2DLA.

7.2 METHODS

7.2.1 Cell transfection

Our protocols are based on transient transfection of COS-7 cells or HEK293 cells by electroporation as described in Claeysen et al. (1996). Wash cells at 70% confluence once in PBS, trypsinize them, and, after centrifugation, resuspend them in EP1 × buffer with 25–500 ng of epitope-tagged receptor cDNA and 15 μg empty plasmid that acts as carrier. Transfer 300 μL of cell suspension (10^7 cells) to a 0.4 cm electroporation cuvette and pulse the cells using a Gene Pulser apparatus (settings: 950 μF and 280 V or 270 V for COS-7 or HEK293, respectively). Quickly after the shock, dilute cells in DMEM (10^7 cells/mL) containing 10% DFBS and plate them on 100 or 150 mm Falcon cell culture dishes or into appropriated clusters. 24 h posttransfection, process the cells to study 5-HT_4R dimers.

7.2.2 Detection of 5-HT_4R dimersx by Western blot

Dimerization of 5-HT_4 receptors can be evaluated by Western blotting in denaturing conditions. Indeed 5-HT_4R dimers form with high affinity and resist to detergents. Four mouse splice variants of the 5-HT_4 receptor have been described that differ in length and composition of their C-terminus: 5-$HT_{4(a)}$, $_{(b)}$, $_{(e)}$, and $_{(f)}$ with 387, 388, 371, and 363 amino acids, respectively (Claeysen et al., 1999; Fig. 7.1A). We used the difference in length of these variants to show that they can interact with each other. We also used a truncated 5-HT_4 receptor at the residue 327: $\Delta 327$ that is devoid of the C-terminus of the receptor (Fig. 7.1A).

7.2.2.1 Cell lysate and membrane preparation

Dimerization of 5-HT_4 receptors is analyzed on membrane receptor preparations. Plate four electroporations of COS-7 cells (4×10^7 cells) in two 150 mm Falcon dishes, for each condition. Use 500 ng of Myc-tagged 5-HT_4R cDNA per 10^7 cells in single transfection assays and 250 ng of each receptor per 10^7 cells in cotransfection experiments. 24 h posttransfection, wash the cells with ice-cold PBS, then add 5 mL of cold PBS per 150 mm plate, and scrape the cells on ice with a rubber policeman. Transfer the content of two dishes in one 12 mL Falcon tube. After centrifugation for 5 min at 2400 g and at 4 °C, resuspend each cell pellet in 500 μL of Tris

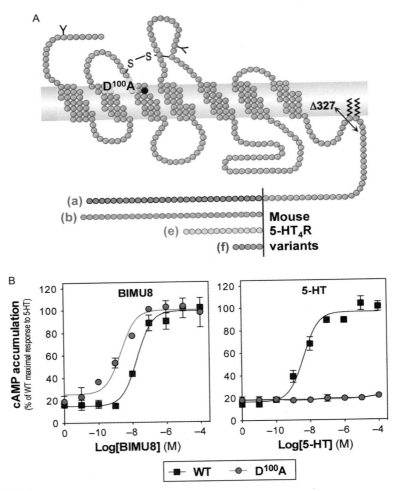

FIGURE 7.1

Mouse 5-HT$_4$ variants and key mutants receptors used. (A) Scheme of the mouse 5-HT$_4$ receptor showing the four splice variants differing in length and composition after a common splicing site (vertical bar). Position of Δ327 mutant, devoid of the C-terminal domain of the receptor, is depicted by a double-head arrow. D^{100}A mutation is indicated with a dark-blue plain circle. (B) Pharmacological profile of the wild-type 5-HT$_4$ receptor (WT) and of the D^{100}A mutant. Both receptors are activated by the full agonist BIMU8, whereas only the WT receptor is capable to respond to the endogenous ligand serotonin (5-HT). 5-HT is unable to bind into D^{100}A mutant. (For interpretation of the references to color in this figure legend, the reader is referred to the online version of this chapter.)

Lysis Buffer and transfer in 1 mL capacity Potter homogenizer. After 20 up and down regular and gentle moves of the pestle in the glass mortar, transfer the cell homogenates in 1.5 mL microtubes, and pellet the membranes by 20 min centrifugation at 43,000 g and at 4 °C. Resuspend the membrane pellet in 200 µL of Tris Lysis Buffer and quantify the protein levels (Bradford reagent) twice with 5 µL of sample. Aliquot the membrane preparations (200 µg/aliquot), freeze in liquid nitrogen, and store at −80 °C for further use.

7.2.2.2 Solubilization and deglycosylation of the samples

In transfected cells, 5-HT$_4$ receptors exist in many glycosylated forms, resulting in smear bands on SDS-PAGE electrophoresis. To circumvent this problem, we add a deglycosylation step of the receptor preparation prior to gel electrophoresis.

Thaw and pellet 400 µg of each sample by 20 min centrifugation at 43,000 g and at 4 °C. To solubilize the membrane proteins, resuspend the pellet in 200 µL of solubilization buffer, and incubate 2 h on a rotating wheel in a cold room at 4 °C. Then, pellet the remaining cell fragments by 20 min centrifugation at 43,000 g at 4 °C and collect the supernatant. Adjust the EDTA concentration in the sample buffer to increase it to 10 mM EDTA (deglycosylation buffer), add 2 µL (1000 units) of N-glycosidase F, and incubate the tubes overnight at 37 °C. After adding 67 µl of Laemmli buffer 4× to the sample, load 25 µg of proteins of each sample on an acrylamide gel.

7.2.2.3 Western blot and detection of the 5-HT$_4$ receptor dimers

Load the samples on Tris/HCl gels (8% or 10% acrylamide/bisacrylamide) in denaturing conditions (SDS). To achieve a good separation of the different dimer bands, use 20 cm glass plate systems (settings, stacking 1 h at 100 V and separation 5 h at 200 V). Transfer the proteins on nitrocellulose using wet electrophoretic system (settings, 30 V, overnight at 4 °C). After transfer, saturate the nitrocellulose membrane for 1 h in TBST–milk, then rinse it with TBST and incubate overnight with the anti-Myc antibody diluted 1/1000 in TBST–milk, under gentle agitation at 4 °C. Wash the membrane six times, 5 min, in TBST, and then incubate it for 1 h with the secondary antibody (e.g., antimouse antibody conjugated with HRP, 1/4000) diluted in TBST–milk, under gentle agitation at room temperature. Wash the membrane six times, 5 min, in TBST, and then reveal the bands using a chemiluminescent kit according to manufacturer instructions (e.g., Pierce ECL Western Blotting Substrate).

Using this technique, we are capable to detect monomers and dimers form for each 5-HT$_4$ receptor variant (Fig. 7.2A). By cotransfection of a "long" variant with a "short" one, for example, variant (a) with variant (e), we can discriminate the dimer formation of (a) and (e) protomers (Fig. 7.2B). 5-HT$_4$ receptor dimers can also be detected using a receptor devoid of its C-terminal domain (Fig. 7.2C) and this truncated receptor is capable to associate with (a) or (b) variant (Fig. 7.2D).

To resume, the key steps in this technique are (1) to start with a membrane preparation of proteins, (2) to apply a deglycosylation step to the samples, and (3) to use long-separation gels.

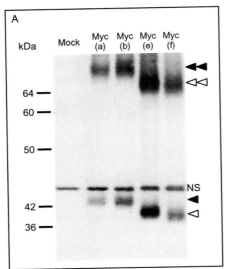
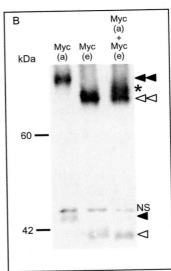
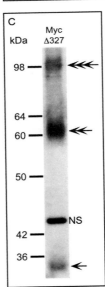
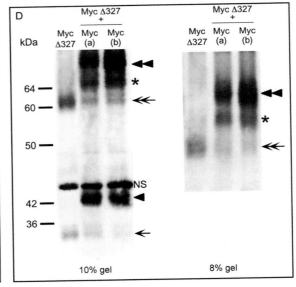

FIGURE 7.2

Analysis of 5-HT$_4$ receptor dimers by Western blot. All receptors used, N-terminally tagged with c-Myc epitope, have been transiently expressed in COS-7 cells and processed as described in the main text. Transfected variant or mutant receptors are indicated above the blots. (A) Transfection of each variant separately show the formation of dimers and monomers of (a), (b), (e), or (f) variants. (B) When variants (a) and (e) are cotransfected (right line), a band of an intermediate size between (a) and (e) homodimer bands appears that corresponds to the formation of a dimer between (a) and (e) protomers (indicated with an

7.2.3 Detection of 5-HT$_4$R dimers by coimmunoprecipitation

Coimmunoprecipitation is a classical technique to demonstrate interactions between proteins. We describe below our protocol to coimmunoprecipitate 5-HT$_4$ receptors.

Transfect 10^7 HEK293 cells with Myc-tagged and HA-tagged 5-HT$_4$ receptors, alone or in combination (500 and 300 ng of each construct, respectively, as HA-tagged receptor is expressed more easily), and seed the cells in 150 mm Falcon petri dish in DMEM-10% DFBS. 24 h posttransfection, apply 15–20 mL of the cross-linking agent DSP in PBS for 30 min at 37 °C. Stop the reaction by washing the cells twice with PBS. Then, add 1 mL of DDM Lysis Buffer per dish and scrape the cells on ice using a rubber policeman. Collect cell lysates in microtubes. Incubate samples for 1 h at 4 °C on a rotating wheel. Centrifuge the samples at 20,000 g for 15 min at 4 °C and keep the supernatants containing solubilized proteins. Quantify the protein concentration (e.g., Bradford reagent). 1 mg of solubilized proteins should then be incubated overnight at 4 °C with anti-HA/agarose beads on a rotating wheel. Centrifuge at 5000 g for 1 min at 4 °C to pellet the beads and remove the supernatant. Wash the beads three times using 1 mL of DDM Lysis Buffer. Resuspend the beads in 40 μL of Laemmli Buffer 1× to elute the immunoprecipitated proteins. Load 40 μL of each sample on 12% acrylamide/bisacrylamide gels, resolve proteins by classical SDS-PAGE gel electrophoresis, and detect them by Western blotting (e.g., use mini-protean and Trans-Blot SD semi-dry transfer cell, Bio-Rad). The immunoblotting protocol described earlier is used for the detection of the receptor bands. Anti-Myc or anti-HA antibodies are diluted in diluted TBST–BSA at 1/1000 (Ms Anti-Myc), 1/400 (Rb Anti-Myc), or 1/500 (Ms anti-HA), respectively. Typical results obtained using this protocol have been shown in Pellissier et al. (2011).

7.2.4 Analysis of 5-HT$_4$R dimers by immunofluorescence

The pharmacological properties of the D^{100}A-5-HT$_4$ mutant receptors (D^{100}A) constitute a great advantage to visualize 5-HT$_4$R dimers by immunofluorescence. The Asp 100 in 5-HT$_4$R transmembrane domain 3 ($D^{3.32}$ in Ballesteros–Weinstein

---

asterisk). (C) Transfection of the ∆327 mutant devoid of its C-terminal domain shows the presence of monomers, dimers, and higher molecular species with a size compatible with "trimers." (D) Cotransfections of ∆327 and (a) or (b) receptors show the capability of ∆327 mutant to interact with (a) or (b) variant by the presence of an intermediate-size band between ∆327 and (a) or (b) dimer bands corresponding to a dimer composed of ∆327 and (a) or (b) protomers (indicated with an asterisk). Legend: plain triangles, monomers of "long" variants (a) and (b); open triangles, monomers of "short" variants (e) and (e); double triangles, dimers of "long" variants (plain) or "short" variants (open); arrows, monomers of ∆327 mutant; double arrows, ∆327 dimers; triple arrows, ∆327 "trimers"; NS, nonspecific band.

nomenclature (Ballesteros & Weinstein, 1995)) corresponds to an aspartate residue that is well conserved in GPCRs responding to biogenic amines and involved in their binding site (Strader et al., 1991). $D^{100}A$ point mutation suppresses the lateral side chain of this aspartate and, consequently, serotonin is unable to bind and activate the mutated 5-HT$_4$ receptor (Fig. 7.1B; Claeysen, Joubert, Sebben, Bockaert, & Dumuis, 2003). However, this receptor remains fully activable by highly selective 5-HT$_4$ R synthetic agonists, such as BIMU 8 (Fig. 7.1B), thus belonging to the RASSL family (receptor activated solely by synthetic ligands) (Conklin et al., 2008). 5-HT$_4$ receptors are rapidly desensitized and internalized by endocytosis after activation by an agonist. Upon 5-HT exposure, $D^{100}A$-5-HT$_4$ receptors stay at the plasma membrane, whereas wild-type (WT) 5-HT$_4$ receptors enter inside the cell. However, if we form dimers between WT and $D^{100}A$ protomers, these molecular complexes can undergo endocytosis, providing a way to separate and to visualize $D^{100}A$ dimers from the other populations of 5-HT$_4$ receptors expressed in the cell.

Prior to cell transfection, place glass coverslips in 12-well clusters and incubate them with 300 μL/well of PORN 1× for minimum 30 min at 37 °C. Wash the slides twice with 1 mL PBS before seeding the cells. Transfect HEK297 cells (10^7 cells) with 300 ng of HA- or Myc-tagged 5-HT$_4$ receptors, alone or in combination. Resuspend them in 20 mL of DMEM-DFBS and plate 1 mL/well (500,000 cell/well). 24 h later, replace the cell culture medium by DMEM alone. 48 h posttransfection, place the clusters at 4 °C in a cold room for 15 min. Remove the medium and incubate for 90 min at 4 °C with the 500 μL of the primary antibodies (Rb anti-HA, 1/300, and/or Ms anti-Myc, 1/400) diluted in cold DMEM. Wash twice with cold DMEM. Place the clusters back at 37 °C in the cell culture incubator under routine parameters for 15 min. Remove cell medium and add 1 mL of stimulation medium (5-HT or BIMU8 and 10^{-5} M in DMEM) or DMEM alone as control for 30 min at 37 °C. Add 145 μL of PFA 16% in each well (2% final concentration) and fix the cells 10 min at 37 °C. Wash the cells three times for 10 min at room temperature with PBS–glycine. Then, permeabilize the cell with 500 μL of PBS–Triton for 5 min at room temperature. Wash three times for 5, 10, and then 15 min with PBS–gelatin and incubate with secondary antibodies coupled to fluorophores (e.g., GAR red, 1/1000, and GAM green, 1/1000) for 1 h at room temperature in the dark. Wash three times for 5, 10, and then 15 min with PBS and then mount the coverslips on glass slide using VECTASHIELD. Image the samples using a confocal microscope. A typical experiment is presented in Fig. 7.3.

Under basal conditions, WT-5-HT$_4$ receptors and $D^{100}A$-5-HT$_4$ receptors expressed alone or coexpressed were predominantly located at the cell surface (Fig. 7.3A, D, G, J, and M). A marginal constitutive internalization, reflecting the constitutive activity of 5-HT$_4$ receptors, was detected for both WT and mutant receptors (e.g., some dotted labeling in Fig. 7.3A). Upon activation with an agonist (5-HT or BIMU8), the WT-5-HT$_4$ receptors were internalized as shown by both an internal dotted labeling and a decrease in cell surface labeling (Fig. 7.3B and C). In the presence of 5-HT, the $D^{100}A$ mutant remained at the cell surface (Fig. 7.3E), whereas it internalized in the presence of BIMU8 (Fig. 7.3F). However, 5-HT, which does not

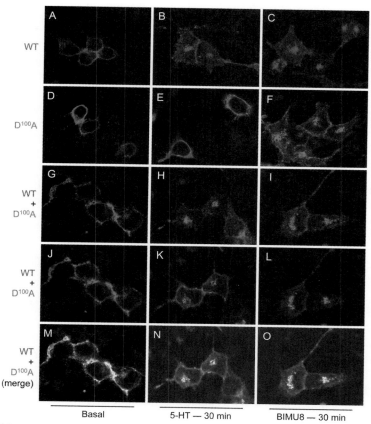

FIGURE 7.3

Functional evidence of 5-HT$_4$R dimerization at the cell surface. Myc-5-HT$_4$R WT (in red, A, B, and C), RhoTag-5-HT$_4$R-D^{100}A (in green, D, E, F), or both (G–O) were transiently expressed in HEK-293 cells and visualized by immunofluorescence confocal microscopy. Cells are fixed in basal conditions (A, D, G, J, and M) or after a 30 min exposure with 10 μM of either 5-HT (B, E, H, K, and N) or BIMU8 (C, F, I, L, and O). (See color plate.)

bind to D^{100}A, induced an internalization of this mutant receptor, when the D^{100}A mutant was coexpressed with WT receptor (Fig. 7.3K and N). Indeed, 5-HT induced the internalization of WT/WT dimers (Fig. 7.3H and N, red dots) and of WT/D^{100}A dimers (Fig. 7.3H, K, and N, yellow dots), whereas D^{100}A/D^{100}A dimers remained at the cell surface (Fig. 7.3K and N, green dots). All types of complexes internalized after a 30 min exposure with BIMU8 (Fig. 7.3I, L, and O). The simplest explanation of these results implicates the existence of a WT/D^{100}A dimer. When the WT monomer of the WT/D^{100}A dimer was occupied by 5-HT, both WT and D^{100}A protomers were internalized.

7.2.5 Analysis of 5-HT$_4$R dimers by TR-FRET

TR-FRET technology provides an easy way to detect the existence of 5-HT$_4$ dimers and to examine their propensity to form heterodimers with other GPCRs.

Transfect COS-7 cells with the appropriate plasmids (N-terminally tagged with HA or FLAG epitopes) and seed them in 96-well plates (100,000 cells/well). Prepare 12 wells per condition. 24 h after transfection, wash the cells with HBS and incubate them at 4 °C for 24 h with the appropriate fluorescent anti-FLAG or anti-HA antibodies diluted in HBS-KF (KF is added to avoid quenching of europium cryptate). In half of the wells, add 50 μL of 4 nM anti-FLAG M2-K and 50 μL of 10 nM anti-HA-d2. In the other half of the wells, add 50 μL of 4 nM anti-FLAG M2-K and 50 μL of HBS-KF. Quantification of FRET signals is performed by homogenous time-resolved fluorescence (HTRF®) settings (Maurel et al., 2004) on appropriate apparatus (see http://www.htrf.com/readers for compatible readers). Express the results as the specific signal over background, Delta F, as described in Maurel et al. (2004).

Figure 7.4 presents the different type of experiments that can be routinely performed using this simple and convenient technology. HA or FLAG-tagged GB$_1$ and GB$_2$ GABA$_B$ receptor subunits are classically used as positive controls of constitutive dimerization. You have to verify that 5-HT$_4$ and GABA$_B$ receptors are expressed at the cell surface in similar amounts by ELISA quantification for example (Fig. 7.4A). In these conditions, the TR-FRET signal detected when we coexpress HA- and FLAG-5-HT$_4$R (Fig. 7.4B) represents only 30% of the signal obtained for GABA$_B$R heterodimers. Indeed, GABA$_B$ receptors expressed at the cell surface are obligatory heterodimers, whereas HA-5-HT$_4$R monomers could associate with either HA-5-HT$_4$ or FLAG-5-HT$_4$ receptors. Therefore, HA-5-HT$_4$R/FLAG-5-HT$_4$R dimers, which are the only couples producing FRET, represent only half of the real amount of dimers at the cell surface. Thus, the real signal for 5-HT$_4$R dimers can be assumed to be around 60% of the GABA$_B$R FRET signal.

By maintaining a constant density of HA-5-HT$_4$Rs (donors, use anti-HA-K antibodies) and increasing the density of FLAG-5-HT$_4$R or FLAG-GB$_2$ (acceptors, use anti-FLAG M2-d2), saturating FRET curves are obtained for WT 5-HT$_4$ receptor, whereas the signal between 5-HT$_4$R and GB$_2$ is linear and unsaturable (Fig. 7.4C). These types of results indicate that 5-HT$_4$R dimerization is specific, whereas the low 5-HT$_4$R/GB$_2$ signal reflects a collisional and nonspecific contact.

Competitions experiments can also be performed to assess specificity of 5-HT$_4$R dimerization. Coexpress constant amounts of HA- and FLAG-5-HT$_4$R with increasing amounts of competing GPCRs, N-terminally tagged with Myc epitope: GABA$_{B2}$R, serotonin type 7 receptor (5-HT$_7$R), or β$_2$-adrenergic receptor (β$_2$-AR), in Fig. 7.4D. As shown by its corresponding horizontal data-fitting line, GABA$_{B2}$R (GB$_2$) is unable to compete with 5-HT$_4$R dimerization (Fig. 7.4D). 5-HT$_7$ receptors compete with the formation of 5-HT$_4$R homodimers at relatively high and probably not physiological concentrations. Interestingly, Myc-β$_2$-AR strongly reduces the FRET signal as efficiently as Myc-5-HT$_4$R and competed with

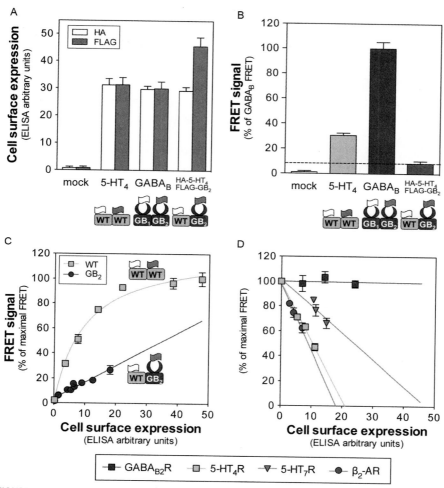

FIGURE 7.4

Study of 5-HT$_4$R dimerization using TR-FRET technology. (A) COS-7 cells were transiently transfected with plasmids encoding epitope-tagged 5-HT$_4$R (250 ng/10^7 cells) and/or GABA$_B$R (1000 ng/10^7 cells). Cell surface expression of 5-HT$_4$R and GABA$_B$R expressed alone or in combination was assessed by ELISA using anti-HA (in white) or anti-FLAG (in black) antibodies in nonpermeabilized, transfected cells as described in Barthet et al. (2005). GB$_1$, GABA$_{B1}$R; GB$_2$, GABA$_{B2}$R. (B) TR-FRET between donor and acceptor fluorophore-labeled antibodies directed against the HA and FLAG tags, respectively, placed at the N-terminus of 5-HT$_4$R and GABA$_B$R as exemplified underneath the graph. Tagged GABA$_B$ receptor subunits GB$_1$ and GB$_2$ were used as a positive control of constitutive dimerization. (C) Saturation FRET experiments. A constant amount of WT HA-5-HT$_4$R (donor) was coexpressed with increasing amounts of FLAG-tagged WT 5-HT$_4$R or GB$_2$ (acceptors). The FRET signal was plotted as a function of cell surface expression of the FLAG-tagged receptors

Continued

HA- and FLAG-5-HT$_4$ receptors for dimer formation, indicating the possible existence of heterodimers between 5-HT$_4$ and β$_2$-adrenergic receptors.

7.3 DISCUSSION

In this chapter, we first described protocols that are classically used in biochemistry of GPCRs such as Western blotting and coimmunoprecipitation. This part is important for us, as detecting 5-HT$_4$ receptors on gel from transient expression in cell lines is a tricky point. Indeed, when this receptor is overexpressed it appears in numerous bands, corresponding to different glycosylation states and maturation. The receptor populations that have reached the plasma membrane are more homogenous in term of sugar maturation, thus starting from a membrane preparation improve the results. One alternative could be to use kits to purify membrane proteins such as Qproteome Plasma Membrane Protein Kit (Qiagen, #37601), which gives good results in our hands, is less time-consuming but more expensive. Adding a deglycosylation step was also a plus to clarify the migration profile of 5-HT$_4$ receptors. The challenge was to find a replacement to the N-Glycosidase F from Roche that was discontinued. This enzyme was efficient in numerous buffers and the actual replacement enzymes (Peptide-N-Glycosidase F, PNGase F) from different suppliers are not as effective. Migration on long gels is another trick to separate more easily complexes of similar size that helps us to show the possibility of heteromerization between 5-HT$_4$R splice variants.

We then described a method to functionally assess the formation of dimers by cross desensitization of the RASSL-5-HT$_4$ mutant that is unable to bind 5-HT (Claeysen et al., 2003). In the presence of WT 5-HT$_4$ receptor and upon 5-HT exposure, the $D^{100}A$ mutant is cointernalized with the WT receptor, thus providing a nice way to show $D^{100}A$/WT dimers. A key step in this procedure is the cells to incubate with the primary antibodies prior to fixation. This procedure avoids the small permeabilization that could be induced by paraformaldehyde. Antibodies in contact with intact cells will label only plasma membrane receptors and not receptors that are in the synthesis pathway. Consequently, the dimers that are immunolabeled and endocyted originate without doubt from the plasma membrane. Interestingly, our cross desensitization assay using the $D^{100}A$ mutant and the WT 5-HT$_4$ receptor could also be used to assess whether a particular 5-HT$_4$ mutant named "M" is able to

FIGURE 7.4—Cont'd determined by ELISA. (D) Competition FRET experiments. A constant amount of WT HA-5-HT$_4$R and FLAG-5-HT$_4$R was expressed and the FRET corresponding to their association was determined in the presence of increasing amounts of Myc-tagged GPCRs belonging to different classes. The FRET signal was plotted as a function of cell surface expression of the Myc-tagged receptors determined by ELISA. Challenger GPCRs, 5-HT$_7$R and 5-HT$_4$R, serotonin receptor subtypes 7 and 4, respectively; β$_2$-AR, β$_2$-adrenergic receptor; GABA$_{B2}$R, subunit of the GABA$_B$ receptor that reaches the cell surface alone (GB$_2$). (For color version of this figure, the reader is referred to the online version of this chapter.)

dimerize or not. If the M mutant that has to be tested retains a classic internalization profile upon 5-HT exposure, it will be coexpressed with the $D^{100}A$ mutant. Upon 5-HT exposure, if RASSL protomers appear to be internalized, one can conclude that the M mutant is capable to dimerize with the $D^{100}A$ protomers. This technique could thus be used as a dimerization-screening test. 5-HT_4 mutants that have been described to disrupt dimer formation (Berthouze et al., 2007) have to be assayed in our procedure to validate this hypothesis. Moreover, due to the high conservation of an aspartate in the third transmembrane domain of biogenic amine receptors (Strader et al., 1991), corresponding to the D^{100} or $D^{3.32}$ position, the point mutation into alanine can easily be transposed in metabotropic serotonergic receptors (Kristiansen et al., 2000) as well as other class A GPCR such as melanocortin-4 receptors (Srinivasan, Santiago, Lubrano, Vaisse, & Conklin, 2007) or histamine H1 receptors (Bakker et al., 2004), conferring similar RASSL properties to the mutated receptors. Consequently, this dimerization-screening test could be extended to other GPCR structurally related to the 5-HT_4 receptors.

We finally described a protocol to achieve relative quantification of 5-HT_4R dimer formation using TR-FRET. Using this method, we can also assess heteromerization with other GPCRs by competition assay. This technology is simple and reliable and can be improved using covalent binding of the acceptor and donor fluorophores to a small suicide enzyme inserted N-terminally in the GPCRs instead of the classical tags (Comps-Agrar et al., 2011). Future developments will aim to use permeant substrates of these enzymes to provide intracellular labeling of the dimers and follow their trafficking.

SUMMARY

Dimerization process is essential for 5-HT_4 receptor function. We provide here methods to analyze 5-HT_4R dimer formation ranging from classical Western blotting and coimmunoprecipitation protocols to cross internalization screening assay and TR-FRET measurements. We intend to describe in details the experimental procedures with the key points necessary to achieve precise and accurate studies regarding 5-HT_4 receptor dimerization.

Acknowledgments

This work was supported by grants from Centre National de la Recherche Scientifique (CNRS), Institut National de la Santé et de la Recherche Médicale (INSERM), Ministère Français de la Recherche (ANR Blanc-2006-0087-02—"GPCR dimers"), and Université de Montpellier.

cAMP quantification, FRET measurements, and ELISA were carried out using the ARPEGE Pharmacology Screening Interactome facility at the Institute of Functional Genomics (Montpellier, France).

References

Bakker, R. A., Dees, G., Carrillo, J. J., Booth, R. G., Lopez-Gimenez, J. F., Milligan, G., et al. (2004). Domain swapping in the human histamine H1 receptor. *Journal of Pharmacology and Experimental Therapeutics*, *311*(1), 131–138. http://dx.doi.org/10.1124/jpet.104.067041.

Ballesteros, J. A., & Weinstein, H. (1995). Integrated methods for the construction of three-dimensional models and computational probing of structure–function relations in G protein-coupled receptors. *Methods in Neurosciences*, *25*, 366–428.

Barnes, N. M., & Sharp, T. (1999). A review of central 5-HT receptors and their function. *Neuropharmacology*, *38*(8), 1083–1152.

Barthet, G., Gaven, F., Framery, B., Shinjo, K., Nakamura, T., Claeysen, S., et al. (2005). Uncoupling and endocytosis of 5-hydroxytryptamine 4 receptors. Distinct molecular events with different GRK2 requirements. *Journal of Biological Chemistry*, *280*(30), 27924–27934.

Berger, M., Gray, J. A., & Roth, B. L. (2009). The expanded biology of serotonin. *Annual Review of Medicine*, *60*, 355–366. http://dx.doi.org/10.1146/annurev.med.60.042307.110802.

Berthouze, M., Ayoub, M., Russo, O., Rivail, L., Sicsic, S., Fischmeister, R., et al. (2005). Constitutive dimerization of human serotonin 5-HT4 receptors in living cells. *FEBS Letters*, *579*(14), 2973–2980.

Berthouze, M., Rivail, L., Lucas, A., Ayoub, M. A., Russo, O., Sicsic, S., et al. (2007). Two transmembrane Cys residues are involved in 5-HT4 receptor dimerization. *Biochemical and Biophysical Research Communications*, *356*(3), 642–647.

Bockaert, J., Claeysen, S., Becamel, C., Dumuis, A., & Marin, P. (2006). Neuronal 5-HT metabotropic receptors: Fine-tuning of their structure, signaling, and roles in synaptic modulation. *Cell and Tissue Research*, *326*(2), 553–572. http://dx.doi.org/10.1007/s00441-006-0286-1.

Bockaert, J., Claeysen, S., Compan, V., & Dumuis, A. (2011). 5-HT(4) receptors, a place in the sun: Act two. *Current Opinion in Pharmacology*, *11*(1), 87–93. http://dx.doi.org/10.1016/j.coph.2011.01.012.

Claeysen, S., Joubert, L., Sebben, M., Bockaert, J., & Dumuis, A. (2003). A single mutation in the 5-HT4 receptor (5-HT4-R D100(3.32)A) generates a Gs-coupled receptor activated exclusively by synthetic ligands (RASSL). *Journal of Biological Chemistry*, *278*(2), 699–702.

Claeysen, S., Sebben, M., Becamel, C., Bockaert, J., & Dumuis, A. (1999). Novel brain-specific 5-HT4 receptor splice variants show marked constitutive activity: Role of the C-terminal intracellular domain. *Molecular Pharmacology*, *55*(5), 910–920.

Claeysen, S., Sebben, M., Journot, L., Bockaert, J., & Dumuis, A. (1996). Cloning, expression and pharmacology of the mouse 5-HT(4L) receptor. *FEBS Letters*, *398*(1), 19–25.

Comps-Agrar, L., Maurel, D., Rondard, P., Pin, J. P., Trinquet, E., & Prezeau, L. (2011). Cell-surface protein–protein interaction analysis with time-resolved FRET and snap-tag technologies: Application to G protein-coupled receptor oligomerization. *Methods in Molecular Biology*, *756*, 201–214. http://dx.doi.org/10.1007/978-1-61779-160-4_10.

Conklin, B. R., Hsiao, E. C., Claeysen, S., Dumuis, A., Srinivasan, S., Forsayeth, J. R., et al. (2008). Engineering GPCR signaling pathways with RASSLs. *Nature Methods*, *5*(8), 673–678.

Dumuis, A., Bouhelal, R., Sebben, M., Cory, R., & Bockaert, J. (1988). A nonclassical 5-hydroxytryptamine receptor positively coupled with adenylate cyclase in the central nervous system. *Molecular Pharmacology*, *34*(6), 880–887.

Gorinski, N., Kowalsman, N., Renner, U., Wirth, A., Reinartz, M. T., Seifert, R., et al. (2012). Computational and experimental analysis of the transmembrane domain 4/5 dimerization interface of the serotonin 5-HT(1A) receptor. *Molecular Pharmacology*, *82*(3), 448–463. http://dx.doi.org/10.1124/mol.112.079137.

Herrick-Davis, K., Grinde, E., Harrigan, T. J., & Mazurkiewicz, J. E. (2005). Inhibition of serotonin 5-hydroxytryptamine2c receptor function through heterodimerization: Receptor dimers bind two molecules of ligand and one G-protein. *Journal of Biological Chemistry*, *280*(48), 40144–40151.

Herrick-Davis, K., Grinde, E., & Mazurkiewicz, J. E. (2004). Biochemical and biophysical characterization of serotonin 5-HT2C receptor homodimers on the plasma membrane of living cells. *Biochemistry*, *43*(44), 13963–13971. http://dx.doi.org/10.1021/bi048398p.

Kristiansen, K., Kroeze, W. K., Willins, D. L., Gelber, E. I., Savage, J. E., Glennon, R. A., et al. (2000). A highly conserved aspartic acid (Asp-155) anchors the terminal amine moiety of tryptamines and is involved in membrane targeting of the 5-HT(2A) serotonin receptor but does not participate in activation via a "salt-bridge disruption" mechanism. *Journal of Pharmacology and Experimental Therapeutics*, *293*(3), 735–746.

Lee, S. P., Xie, Z., Varghese, G., Nguyen, T., O'Dowd, B. F., & George, S. R. (2000). Oligomerization of dopamine and serotonin receptors. *Neuropsychopharmacology*, *23*(4 Suppl), S32–S40. http://dx.doi.org/10.1016/S0893-133X(00)00155-X.

Lukasiewicz, S., Faron-Gorecka, A., Kedracka-Krok, S., & Dziedzicka-Wasylewska, M. (2011). Effect of clozapine on the dimerization of serotonin 5-HT(2A) receptor and its genetic variant 5-HT(2A)H425Y with dopamine D(2) receptor. *European Journal of Pharmacology*, *659*(2–3), 114–123. http://dx.doi.org/10.1016/j.ejphar.2011.03.038.

Maurel, D., Kniazeff, J., Mathis, G., Trinquet, E., Pin, J. P., & Ansanay, H. (2004). Cell surface detection of membrane protein interaction with homogeneous time-resolved fluorescence resonance energy transfer technology. *Analytical Biochemistry*, *329*(2), 253–262.

Millan, M. J., Marin, P., Bockaert, J., & Mannoury la Cour, C. (2008). Signaling at G-protein-coupled serotonin receptors: Recent advances and future research directions. *Trends in Pharmacological Sciences*, *29*(9), 454–464. http://dx.doi.org/10.1016/j.tips.2008.06.007.

Pellissier, L. P., Barthet, G., Gaven, F., Cassier, E., Trinquet, E., Pin, J. P., et al. (2011). G protein activation by serotonin type 4 receptor dimers: Evidence that turning on two protomers is more efficient. *Journal of Biological Chemistry*, *286*(12), 9985–9997. http://dx.doi.org/10.1074/jbc.M110.201939.

Renner, U., Zeug, A., Woehler, A., Niebert, M., Dityatev, A., Dityateva, G., et al. (2012). Heterodimerization of serotonin receptors 5-HT1A and 5-HT7 differentially regulates receptor signalling and trafficking. *Journal of Cell Science*, *125*(Pt 10), 2486–2499. http://dx.doi.org/10.1242/jcs.101337.

Srinivasan, S., Santiago, P., Lubrano, C., Vaisse, C., & Conklin, B. R. (2007). Engineering the melanocortin-4 receptor to control constitutive and ligand-mediated G(S) signaling in vivo. *PLoS One*, *2*(7), e668. http://dx.doi.org/10.1371/journal.pone.0000668.

Strader, C. D., Gaffney, T., Sugg, E. E., Candelore, M. R., Keys, R., Patchett, A. A., et al. (1991). Allele-specific activation of genetically engineered receptors. *Journal of Biological Chemistry*, *266*(1), 5–8.

CHAPTER

Bioluminescence Resonance Energy Transfer Methods to Study G Protein-Coupled Receptor–Receptor Tyrosine Kinase Heteroreceptor Complexes

8

Dasiel O. Borroto-Escuela*, Marc Flajolet[†], Luigi F. Agnati[‡], Paul Greengard[†], and Kjell Fuxe*

*Department of Neuroscience, Karolinska Institutet, Stockholm, Sweden
[†]Laboratory of Molecular and Cellular Neuroscience, The Rockefeller University, New York, USA
[‡]IRCCS Lido, Venice, Italy

CHAPTER OUTLINE

Introduction 142
8.1 The Method Principle 143
8.2 Setting Up a BRET Assay 146
8.3 Methods and Detailed Protocols 148
    8.3.1 Protocol 1: Saturation Assay 151
    8.3.2 Protocol 2: Competition Assays 155
    8.3.3 Protocol 3: Kinetics and Dose–response Assays 156
8.4 Discussion and Note on Critical Parameters 159
Acknowledgments 161
References 162

Abstract

A large body of evidence indicates that G protein-coupled receptors (GPCRs) and receptor tyrosine kinases (RTKs) can form heteroreceptor complexes. In these complexes, the signaling from each interacting protomer is modulated to produce an integrated and therefore novel response upon agonist(s) activation. In the GPCR–RTK heteroreceptor complexes, GPCRs can activate RTK in the absence of added growth factor through the use of RTK signaling molecules. This integrative phenomenon is reciprocal and can place also RTK signaling downstream of GPCR. Formation of either stable or transient complexes by these two important classes of membrane receptors is involved in regulating all aspects of receptor function, from ligand binding to signal transduction, trafficking, desensitization, and downregulation among others. Functional phenomena can be modulated with conformation-specific inhibitors that stabilize defined GPCR states to abrogate both GPCR agonist- and growth factor-stimulated cell responses or by means of small interfering heteroreceptor complex interface peptides. The bioluminescence resonance energy transfer (BRET) technology has emerged as a powerful method to study the structure of heteroreceptor complexes closely associated with the study of receptor–receptor interactions in such complexes. In this chapter, we provide an overview of different $BRET^2$ assays that can be used to study the structure of GPCR–RTK heteroreceptor complexes and their functions. Various experimental designs for optimization of these experiments are also described.

INTRODUCTION

It is now several years since a general dogma established that growth-promoting activity of many G protein-coupled receptor (GPCR) ligands involves activation of receptor tyrosine kinases (RTKs) and their downstream signaling cascades (Luttrell, Daaka, & Lefkowitz, 1999). Such observations led to the emergence of the "transactivation" concept, which refers to the activation of RTK signaling pathways by GPCR ligands; see (Fuxe et al., 2007). However, over the past few years, this concept appears to have become increasingly complex since pharmacological, biochemical, and biophysical studies provided evidence for an engagement of GPCR signaling molecules (e.g., heterotrimeric G proteins and arrestins) in signal transduction generated by various RTK subtypes, which reveals a bidirectional cross communication between RTKs and GPCRs (Dalle et al., 2002; Lin, Daaka, & Lefkowitz, 1998; Pyne & Pyne, 2011).

Perhaps, an even more interesting concept relates to the apparent ability of different GPCRs to associate with RTKs and forming large receptor complexes (GPCR–RTK heteroreceptor complexes) (Borroto-Escuela, Romero-Fernandez, Mudo, et al., 2012; Flajolet et al., 2008; Fuxe et al., 2010). Such a higher molecular organization brings new alternatives in term of physiological functions for each of these two receptor families, but it also increases the possibilities regarding therapeutic targeting (Alderton et al., 2001; Delcourt, Bockaert, & Marin, 2007; Pyne & Pyne, 2011; Waters et al., 2006). Biochemical and/or pharmacological properties reported for GPCR–RTK heteroreceptor complexes are often distinct from those of the corresponding protomers (Borroto-Escuela, Romero-Fernandez, Mudo, et al., 2012; Delcourt et al., 2007; Pyne & Pyne, 2011; Waters et al., 2006). Thus, heteromerization represents an important mechanism for modulating and integrating the physiological functions of GPCRs and RTKs. This can take place at different stages of the receptor's life: biosynthesis, ligand binding, G protein activation, desensitization, internalization, and degradation. Also GPCR–RTK heteroreceptor complexes allow flexibility in terms of programming of spatially controlled signaling pathways and/or increasing signaling gain (Fuxe et al., 2010).

The existence of conformation-specific GPCRs present in the GPCR–RTK heteroreceptor complexes may offer unique opportunities to modulate pathophysiology driven by growth factors. It is likely that long-term and global inhibition of RTK in associated diseases will not be well tolerated due to the role of RTK signaling in normal physiology (Andrae, Gallini, & Betsholtz, 2008; Takeuchi & Ito, 2011). Therefore, GPCR–RTK partnership may lead to novel therapeutic strategies based on specific blockade of RTK signaling via GPCR transinhibition by antagonists and/or inverse agonists of the corresponding GPCR (Pyne & Pyne, 2011). Such strategies are less likely to produce side effects than approaches based on direct RTK inhibition. This could potentially extend the repertoire of biological actions of a large number of compounds and drugs that bind to GPCRs. Combinations of RTK inhibitors and GPCR-specific ligands that reduce G protein or β-arrestin function may be an effective way of abrogating signaling from these heteroreceptor complexes.

In order to progress in our understanding of the functional role of GPCR–RTK heteroreceptor complexes as well as their potential role as therapeutic targets, we need to continuously look for methods and technologies that can be used to demonstrate and evaluate such heteroreceptor complexes and their receptor–receptor interactions. The bioluminescence resonance energy transfer (BRET) technology has emerged as a powerful and straightforward biophysical technique for studying receptor heteromers and heteroreceptor complexes and their receptor–receptor interactions. The present study focuses on recent work illustrating the power of BRET^2 for the study of GPCR–RTK interactions, using A2A-FGFR1 (Flajolet et al., 2008) and 5-HT1A-FGFR1 (Borroto-Escuela et al., 2013; Borroto-Escuela, Romero-Fernandez, Mudo, et al., 2012) heteroreceptor complexes as examples. We highlight the current ways in which the BRET-based methodology is being used to establish the existence of GPCR–RTK heteroreceptor complexes and their specificity. Furthermore, we describe how BRET may be used to establish the involvement of a bidirectional cross-communication mechanism between GPCRs and RTKs.

8.1 THE METHOD PRINCIPLE

BRET is a natural phenomenon found in some marine species (for instance, in the sea pansy, Renilla reniformis), resulting from the nonradiative energy transfer between a luminescent donor (Renilla luciferase—Rluc) and a fluorescent acceptor protein (green fluorescent protein—GFP). Originally developed to study the interactions of circadian clock proteins in bacteria (Xu, Piston, & Johnson, 1999), BRET has subsequently been applied to study receptor–receptor interactions in living cells. It has become a powerful method to assess such interactions at the molecular level. The method is based on the principle of Förster resonance energy transfer, which postulates that the efficacy of energy transfer between a donor and an acceptor is inversely proportional to the sixth power of the distance between them.

When studying receptor–receptor interactions, one protomer is fused to the donor (Rluc) and the other protomer to the acceptor (fluorescent protein). If the two fusion protomers interact and the distance between the energy donor and acceptor is <10 nm, a resonance energy transfer occurs and the emission signal from the acceptor protein can be detected. However, the energy transfer process not only depends on the distance between donor and acceptor but also relies on the overlap of the emission spectrum of the donor with the excitation spectrum of the acceptor and the relative spatial orientation of donor and acceptor. Therefore, the absence of BRET signal between two receptors does not necessarily mean that these receptors do not interact with each other. Thus, all such factors must be taken into account. On the other hand, increasing the acceptor/donor ratio may lead to a detectable but nonspecific BRET signal. This can be illustrated in BRET saturation assays, where a fixed amount of donor is coexpressed with increasing amount of the acceptor. Nonspecific BRET signals tend to increase linearly with increasing acceptor concentrations. In contrast, a progressive increase of the BRET signal in a hyperbolic manner represents the complete saturation of all donors with acceptor molecules and a specific BRET signal (Marullo & Bouvier, 2007; Pfleger & Eidne, 2006).

Over the last years, different generations of BRET (BRET^1, BRET^2, eBRET^2, BRET^3, and QD-BRET) have been developed, depending on the type of enzyme substrate and the nature of donor/acceptor pairs (Kocan, See, Seeber, Eidne, & Pfleger, 2008; Pfleger et al., 2006; Xing, So, Koh, Sinclair, & Rao, 2008). As a result, the nomenclature for describing each of the BRET forms has not followed a unique rigorous pattern. The original BRET method using coelenterazine h (benzyl-coelenterazine) as substrate is nowadays called BRET^1 (Marullo & Bouvier, 2007). In BRET^1, the maximal emission of Rluc is observed at 480 nm, a wavelength that is appropriate for excitation of a yellow fluorescent protein (enhanced YFP: EYFP), which subsequently reemits light at 530 nm. Several other variants of the YFP with identical excitatory properties are YFP topaz, YFP citrine, YFP venus, and YPet (Bacart, Corbel, Jockers, Bach, & Couturier, 2008). BRET^1 is characterized by strong signals and long lifetime, making it one of the most suitable approaches for BRET saturation assays. Changes in the Rluc substrate used resulted in the second generation of BRET methods, the BRET^2. In the BRET^2, the bisdeoxycoelenterazine (DeepBlueC^{TM}) or didehydrocoelenterazine (coelenterazine-400a) is used as Rluc substrate, resulting in different donor emission spectra that shift the maximal light emission of Rluc to 395 nm. Appropriate acceptors are GFP2 and the GFP10 with excitation and emission maxima of 400 and 510 nm, respectively. BRET^2 in comparison to BRET^1 has an improved separation of donor and acceptor emission peaks. This makes this form a better choice for screening assays where high signal-to-noise ratios are required. However, a clear limitation of BRET^2 is the low light emission (DeepBlueC^{TM} as a substrate leads to up to 300 times less light emission) and the short lifetime. More recently, the introduction of a mutated version of the Rluc enzyme, the Rluc8 variant (Rluc mutant containing eight amino acid substitutions), leads to an ~5–30-fold increase in the original $BRET^2$ signal, which results in a new BRET form, the so-called enhanced $BRET^2$ ($eBRET^2$) (Kocan et al., 2008). In the $eBRET^2$, we can combine the advantage of greater spectral resolution of the original $BRET^2$ with a higher quantum yield when using Rluc8. The introduction of a third Rluc substrate, called EnduRen, the protected form of the coelenterazine, resulted in a form of $BRET^1$, known as extended BRET (eBRET) (Pfleger et al., 2006). This enables real-time monitoring of receptor–receptor interactions for extended periods of time and provides stability over time that is logistically advantageous for high throughput screening applications. The eBRET name can result in a nomenclature confusion with the previous name $eBRET^2$.

A similar confusion arises with the introduction of the $BRET^3$. For some authors, $BRET^3$ refers to a BRET form that results from a combination of a red-shifted fluorophore (e.g., mOrange) with Rluc8, using as substrate EnduRen, although the system was validated with coelenterazine h (De, Ray, Loening, & Gambhir, 2009). However, for other authors (Bacart et al., 2008), $BRET^3$ refers to a donor/acceptor pair formed by a firefly luciferase (from Photinus pyralis) and acceptors whose excitation peaks overlap with the emitted light at 565 nm (for instance, the 24-kDa DsRed fluorescent protein; peptides labeled with Cy3 or Cy3.5) using as a substrate the D-luciferin developed by Gammon, Villalobos, Roshal, Samrakandi, and Piwnica-Worms (2009). The firefly luciferase in $BRET^3$ shows lower cellular autofluorescence at the emission wavelength (565 nm) and a more sustained light emission by firefly luciferase compared to Rluc. However, disadvantages are weak signals and overlap between donor and acceptor emission peaks. In addition, the tendency of DsRed to oligomerize has to be considered in these BRET experiments for proper data analysis. It is therefore highly recommended to use instead a DsRed-monomeric variant (Clontech, USA).

Finally, a new BRET version has been introduced: the quantum dot-BRET (QD-BRET) (Bacart et al., 2008; Xing et al., 2008). QDs are semiconductor nanocrystals excited at any wavelength ranging from UV to 530 nm, and their light emission wavelength, which depends on their diameter, can cover the spectrum from blue to near infrared. They are then suitable energy acceptors for $BRET^1$- and $BRET^2$-based assays as their broad interval of excitation wavelengths overlaps the currently used luciferase-emitted light. It should be mentioned that based on the work of Medintz and Mattoussi, it seems that QDs are not so good as energy acceptor but rather used so far as energy donor (Medintz & Mattoussi, 2009). So far, QDs have only been used in $BRET^1$-based assays, where QDs, directly linked to Rluc, were injected into mice and energy transfer monitored in the presence of coelenterazine h (Xing et al., 2008). Even though not really useable at this time, the self-illuminating feature of QD-BRET makes imaging technically possible and could be optimized in the future to work in conditions where the generation of photon is limited such as in tissues. The emission peaks are clearly separated, which makes QD-BRET ideal for screening applications. But its major disadvantages are the large size of QD molecules, ranging from 1.5 to 6 nm, and the fact that QDs are semiconductor nanocrystals. Thus, it is not a genetically coded protein that can be synthesized by living cells and must therefore be added.

Several of these variants of the BRET technology have or can be utilized to provide evidence for receptor–receptor interactions in living cells. In line with the huge potential of BRET technology, bimolecular fluorescence complementation (Cabello et al., 2009) and bimolecular luminescence complementation (see Vidi, Ejendal, Przybyla, & Watts, 2011) have recently been developed and combined with BRET to study more complex receptor–receptor interactions in higher-order receptor oligomers. Several studies continue to improve the potential use of other luciferases and new coelenterazine derivatives have been developed with brighter (ViviRenTM) or extended (EnduRenTM) light emission. Furthermore, the development of the BRET^3 opens the door to the characterization of receptor–receptor interaction *in vivo*.

8.2 SETTING UP A BRET ASSAY

Performing a BRET assay to investigate a potential GPCR–RTK interaction involves several steps:

1. *Selection and generation of the donor/acceptor and substrate combination.* Generation of the two proteins of interest genetically fused with either donor luciferase protein or acceptor fluorescent protein depending of the chosen BRET approach (see Table 8.1) at either the N- or the C-terminus. For a membrane-linked receptor, it is more intuitive to fuse the donor/acceptor protein to the intracellular C-terminal tail of the receptor. A large proportion of GPCRs and RTKs require the presence of a N-terminal signal peptide for correct cell surface expression. This signal is susceptible to proteolytic cleavage and therefore precludes the use of the N-terminal position for the fusion. To avoid to lose the N-terminal moiety after cleavage of the peptide signal, the only alternative is to insert the cDNA of the fluorescent protein downstream of the peptide signal. While this has been done successfully with smaller tags (e.g., myc and HA), it does not seem to be as practical with larger cDNAs.

2. *Design and validation of the receptor of interest, including suitable controls.* For a given BRET assay, the most adequate control protein has to be determined since the inclusion of a negative/positive control is crucial to determine the specificity of the studied interaction. As a positive control, one can generate a double-fused chimeric protein where the donor and acceptor proteins have been linked together. Also, when studying GPCR–RTK interactions, the use of the corresponding homodimeric pair of receptors can be considered as positive control(s). Thus, for both GPCRs and RTKs, it has been extensively documented that they can exist as homodimers. In the case of RTKs, this process takes place upon agonist-induced receptor activation. As negative controls, receptors presenting a similar topology and subcellular localization as the receptor of

Table 8.1 Summary of the Different BRET Methods

Method	Donor	Acceptor	Spectral Properties (Emission, nm) Donor	Spectral Properties (Emission, nm) Acceptor	Substrate
BRET[1]	Rluc	EYFP	480	530	Coelenterazine h
Extended BRET	Rluc	EYFP	480	530	EnduRen
BRET[2]	Rluc	GFP2	395	510	DeepBlueC™/ coelenterazine 400a
Enhanced BRET[2]	Rluc8	GFP2	395	510	DeepBlueC™/ coelenterazine 400a
BRET[3]	Firefly	DsRed	565	583	D-Luciferin and cofactors
BRET[3]	Rluc/ Rluc8	mOrange	480	564	EnduRen/ coelenterazine h
QD-BRET	Rluc	QD	480	605	Coelenterazine h

Over the last years, different generations of BRET (BRET[1], BRET[2], eBRET[2], BRET[3], and QD-BRET) have been developed, depending on the type of enzyme substrate and the nature of donor/acceptor pairs. As a result, the nomenclature for describing each of the BRET forms has not followed a unique rigorous pattern.

interest should be considered. With the development of molecular dynamic simulation of receptor–receptor interactions, it is nowadays possible to predict the receptor interface. Interesting interaction hot spot(s) can be identified and mutated to serve as negative control(s) (Borroto-Escuela, Romero-Fernandez, et al., 2010).

3. *Selection of the cell system and coexpression of the two BRET fusion proteins at a relevant physiological level of expression.* Transient transfection (e.g., calcium phosphate, Lipofectamine, and FuGENE HD) or virus-based transfection systems have been used successfully. Common mammalian cells (e.g., HEK293T, COS-7, and CHO) and primary neuronal culture cells are compatible with BRET assays. Depending on the type of BRET assay, different ratios of donor/acceptor should be used/tested (see Points 1 in Sections 3.1 and 3.2). For some assays, we may consider the use of bicistronic vectors, which may guarantee a more homogeneous receptor ratio expression level.

4. *Detection of the BRET signal and design of an appropriate assay.* The BRET signal can be measured from adherent cells, cell suspensions, subcellular fractions, purified proteins, and also culture medium in the case of secreted proteins using a white plate to avoid light absorption and in parallel a black plate to record total fluorescence values. The biological material (50–100 µl) is then dispatched into the 96-well plate and 10–20 µl substrate is added (5 µM final

concentration for coelenterazine). BRET signals are measured immediately after substrate addition using a BRET reader capable of measuring light emitted at donor and acceptor wavelengths in a quasi-simultaneous manner. The use of the BRET method to study GPCR–RTK interactions goes further than the simple measurement of energy transfer between the donor and acceptor pair together with the use of proper positive and negative controls. If a BRET signal is observed, the proper cellular localization and function of the fusion proteins have to be verified. If the functional validation step is successful, the BRET signal can be studied further to evaluate its specificity. In fact, several assays have been developed with the aim of providing evidence for the existence of specific receptor–receptor interactions and to discriminate between genuine physical interactions between the GPCR and the RTK versus a random collision due to overexpression of the receptor pairs. Therefore, in addition to the classical negative control used, BRET saturation, competition, and ligand-promoted (kinetic and dose–response) assays have been developed. In addition, each of these assays can shed light on the stoichiometry, conformation, and dynamic changes that may take place in GPCR–RTK heteroreceptor complexes.

5. *Analysis of the BRET signal.* The BRET signal or BRET ratio is defined as the light signal of the acceptor emission relative to the light signal of the donor emission. This proportion is corrected for the background signal due to the overlap of donor emission at the acceptor wavelength, always determined in parallel for cells expressing the donor alone. Then, the BRET net value is calculated by subtracting this BRET background ratio from the BRET ratio obtained in cells coexpressing the two partners. The amount of donor can be estimated from the maximal luciferase values measured separately after the BRET reading. The amount of acceptor has to be determined in an independent reading by recording fluorescence values of the acceptor pair in a fluorometer (using a black microplate to avoid light scattering). The calculated acceptor/donor ratio can be used to compare different experiments. To obtain acceptor/donor ratios that correspond to real protein quantities, luciferase and fluorescence values have to be converted into protein amounts using independently established standard correlation curves with real protein quantities (e.g., determined in radioligand binding experiments for each receptor fusion protein).

8.3 METHODS AND DETAILED PROTOCOLS

In recent years, a number of excellent papers have well documented and provided detailed discussions on different forms and approximations to the BRET assay methodology and its straightforward use to study receptor–receptor interactions mainly focused on GPCR–GPCR interactions (Ayoub & Pfleger, 2010; Hamdan, Percherancier, Breton, & Bouvier, 2006; Marullo & Bouvier, 2007). However, in this field, works describing in detail the potential use of BRET methodology to study GPCR–RTK heteroreceptor complexes and their dynamics are still missing.

Consequently, we have used here the A2A-FGFR1 and 5-HT1A-FGFR1 heteroreceptor complexes as examples to illustrate the potential use of different BRET^2 assay formats to monitor GPCR–RTK interactions in living cells.

We chose the BRET^2 variant assay with Rluc8 as a donor and GFP2 as an acceptor with improved spectral separation of the donor and acceptor emission peaks. This implies less bleed through at the acceptor emission maximum and lower background. Also, the rationale for the use of mutated Rluc (Rluc8, GenBank: EF446136) instead of the nonmutated Rluc is mainly based on the fact that Rluc8 gives a higher quantum yield compared to the nonmutated Rluc. Therefore, we do not need to overexpress the luciferase fusion receptor that is detrimental for interaction specificity.

Performing a BRET assay to investigate a potential GPCR–RTK interaction and, in more general terms, any receptor–receptor interaction involves several steps:

1. *Generation and validation of BRET fusion constructs.* The two receptors of interest must be genetically fused to either *Renilla* luciferase (Rluc) or GFP variants at either the N- or C-terminus. The choice of the fusion protein (N- vs. C-terminus) depends on the nature of the studied proteins. For RTK and GPCR, it is, for instance, more advisable to fuse the donor and acceptor protein to the C-terminal tail (see Point 1 in Section 2). It is important to consider preexisting information concerning specific domains of the receptor of interest, like, for example, the tyrosine kinase domains in RTK and their potential binding sites to regulatory/adaptor proteins. Also, posttranslational modification sites, like palmitoylation at the C-terminal of some class A GPCR, can induce a restricted conformation of the C-terminal tail of the receptor, and longer polylinker spacers should be used in order to unmask the donor/acceptor protein and give it a proper orientation. Finally, it is important to ensure that the insertion of the donor/acceptor protein does not interfere with the proper folding of the receptor and insurance of a correct functionality and localization deserves a particular attention. Listed in the succeeding text are the five categories of constructs generated for this study: (1) BRET acceptor constructs: 5-HT1A-GFP2 and A2A-GFP2 for expression of both the 5-HT1A and A2A receptors, C-terminally tagged with GFP2 (PerkinElmer, Sweden). (2) BRET donor constructs: FGFR1–Rluc8 and 5-HT1A-Rluc8 for expression of the FGFR1 and 5-HT1A, C-terminally tagged with Rluc8. (3) Positive control: BRET fusion construct for expression of the Rluc8–GFP2 fusion protein. (4) Negative control BRET constructs: pcDNA3.1-Rluc8 and pGFP2 for expression of Rluc8 and GFP2 alone, respectively. (5) Specificity controls: 5-HT2A–GFP2 and D2mutantR–Rluc8 for expression of other GPCRs were C-terminally tagged with either Rluc8 or GFP2. Each fusion construct was tested for detectable luminescence (using a luminometer following addition of coelenterazine 400a or DeepBlueC™ substrate) or fluorescence (using a fluorometer following direct laser excitation). The fusion receptor was also validated with respect to their function by means of radioligand binding assay and/or gene reporter assays to ensure that the addition of the donor or acceptor molecule has not altered ligand affinity/efficacy/

potency. The use of confocal microscopy is highly recommended to ensure correct protein localization.

2. *Cell transfection and coexpression of the BRET-fused receptors in mammalian cells.* For BRET experiments, we have mostly used human embryonic kidney cells (HEK293T-27). Independent of the kind of BRET assay to be conducted, it is always recommended to perform two sets of transfections: (a) transfection of one set of cells only with donor-fused construct, which will allow to correct for the luciferase signal, and (b) cotransfection of another set of cells with both the cDNAs coding for the donor- and acceptor-fused constructs, which correspond to the BRET pair for the signal measurement. Dependent on the BRET assay to be performed, different or constant donor/acceptor plasmid ratios must be used (see Points 1 in Sections 3.1 and 3.2, respectively).

3. *Harvesting the transfected cells.* $BRET^2$ measurement can be performed in two different ways: (a) 24 h after transfection cells are detached using trypsin-EDTA or Versene (for adherent cell layer). Cells are then microcentrifuged 5 min at $300 \times g$ at room temperature, the supernatant discarded, and the cells washed and resuspended in HEPES-buffered DMEM without phenol red. 40–100 µl of cells is distributed per well of a 96-well white cell culture plate and maintained at 37 °C with 5% CO_2 in a humidified incubator for a further 24 h to allow attachment before $BRET^2$ recording. (b) A second approach consists in detaching the cells 48 h after transfection and microcentrifuge for 5 min at $300 \times g$ at room temperature. The cell pellets are washed once with 1 ml PBS containing 0.5 mM $MgCl_2$, PBS removed, and the cells resuspended in 1 ml of PBS containing 0.5 mM $MgCl_2$ and 0.1% glucose. The cell number in the cell suspension is determined either by measuring the OD600 or by protein concentration (e.g., Bicinchoninic Acid Assay (BCA) method). Equal amounts of cell suspension are distributed per well of a 96-well white cell culture plate and then proceeded to $BRET^2$ measurement.

4. *BRET measurement.* $BRET^2$ measurements are performed under temperature-controlled conditions to obtain reproducible results. $BRET^2$ is initiated by adding the luciferase substrate coelenterazine 400a (5 µM final concentration) and detected using a luminometer (e.g., POLARstar Optima plate reader; BMG Labtechnologies, Offenburg, Germany) that allows the sequential integration of the signals detected with two filter settings 410 ± 80 and 515 ± 30 nm when using coelenterazine400a as substrate, Rluc8 as donor, and GFP2 as acceptor. Light from each well is measured simultaneously through each filter. $BRET^2$, where the rapid decay in the emitted luminescence is faster than in $BRET^1$, presents some limitations and the interval time for measurement must be taken into account. BMG POLARstar is equipped with two online injectors that can deliver compounds and substrate while simultaneously measuring luminescence at two different wavelengths. This significantly increases the reading time per plate. Alternatively, the substrate can be added manually to a maximum of 12 wells at a time, followed by the $BRET^2$ readings. However, this can become easily inconvenient when large numbers of samples are considered.

5. *Calculating BRET ratio.* Data are then represented as a normalized $BRET^2$ ratio, which is defined as the BRET ratio for coexpressed Rluc and GFP2 constructs normalized against the BRET ratio found for the Rluc expression construct alone in the same experiment:

$$BRET^2 \text{ ratio} = [(GFP^2 \text{ emission at } 515 \pm 30 \text{ nm})/(Rluc \text{ emission } 410 \pm 80 \text{ nm})] - cf$$

The correction factor, cf, corresponds to (emission at 515 ± 30 nm)/(emission at 410 ± 80 nm) found with the receptor-*R*luc construct expressed alone in the same experiment.

BRET-based studies of receptor–receptor interactions are particularly prone to false-positive signals and require multiple controls. $BRET^2$ saturation and competition assays have been developed to extend the information obtained from basic $BRET^2$ experiments toward a more quantitative and detailed analysis of the $BRET^2$ signal.

8.3.1 Protocol 1: saturation assay

In titration or saturation $BRET^2$ experiments, cells are transfected with a constant amount of $BRET^2$-donor in the presence or absence of increasing amounts of the acceptor. Theoretically, for any specific interaction between the receptor–donor and receptor–acceptor fusions, the $BRET^2$ ratio increases hyperbolically as a function of increasing GFP/Rluc value, to reach an asymptote (saturation) when all donor molecules are associated with acceptors ($BRET_{max}$, Fig. 8.1). By contrast, in the case of nonspecific interactions (bystander BRET), a quasi-linear plot is expected or eventually reaches a plateau for higher values of receptor density. Nevertheless, the saturation curve should be independent of the total expression level of receptors and the $BRET^2$ configuration used. However, $BRET_{max}$ values cannot be used as a quantitative measure of the relative number of homo/heteromers formed for each combination because they not only are a function of the dimer numbers but also depend on the distance between the energy transfer partners and their relative orientation within the receptor complex. $BRET^2$ saturation curves have been particularly used with the aim to establish the oligomeric order of receptor complexes, as well as the proportion of receptors engaged in dimers or oligomers. They are also used to determine whether ligand-induced $BRET^2$ signals depend on conformational changes or association/dissociation of interacting receptors. Also, saturation assay has been used to compare the relative affinity of receptors for each other and their probability to form a complex, the so-called $BRET_{50}$, which represents the acceptor/donor ratio giving 50% of the maximal signal (Fig. 8.1). The $BRET_{50}$ value is often compared between the two different homomer subtypes and their corresponding heteromer. Many GPCR–GPCR heteromers show no difference in the relative affinity between their receptor homomers and their specific heteromers. Furthermore, neither $BRET_{max}$ nor $BRET_{50}$ values may be modified following the agonist activation of the heteromer, consistent with the general consensus that GPCR homo- and heteromerization is often constitutive. In contrast, GPCR–RTK heteroreceptor complexes

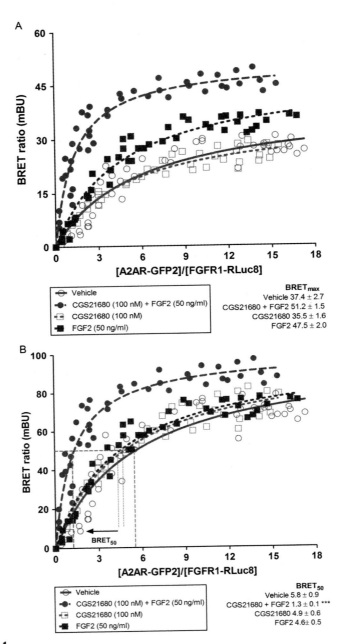

FIGURE 8.1

BRET[2] saturation assay shows specific A2AR and FGFR1 interaction in HEK293T cells. (A) The existence of an A2AR/FGFR1 heteroreceptor complexes and their agonist regulation by CGS21680 and/or FGF2 have been validated using quantitative BRET[2] saturation show more dynamic features where agonist treatment markedly affects $BRET_{max}$, $BRET_{50}$, or both these values (Fig. 8.1).

Details are shown in the succeeding text for typical $BRET^2$ assays for detection and analysis of GPCR–RTK (A2A-FGFR1) heteroreceptor complexes using either adherent cells or cells in suspension.

1. Several independent transfections should be performed using a constant amount of cDNA coding for the BRET donor (10–100 ng, FGFR1-Rluc8) and increasing quantities of cDNA coding for the BRET acceptor (i.e., 0, 10, 20, 50, 100, 200, 300, 500, and 1000 ng; A2A-GFP2) and sufficient "empty" vector (such as pCDNA3.1 or any other cloning vector) to bring the total amount of cDNA in the transfection to 1000 ng/well in a six-well plate.
2. 24 h (preferred option (a), see Point 3 in Section 3) or 48 h (preferred option (b), see Point 3 in Section 3) after transfection cells are washed, detached, and distributed into white opaque 96-well microplates: 40–100 μl per well, incubated as described in the succeeding text (Point 3 in Section 3.1) or moved directly to the BRET ratio measurements.
3. The cells are preincubated in the absence or presence of agonist/antagonist drugs if ligand-induced $BRET^2$ signals will be analyzed. Otherwise, proceed directly to BRET ratio measurements.
4. $BRET^2$ ratio measurements are performed after adding coelenterazine 400a diluted in HBSS or PBS $CaCl_2/MgCl_2$ to each well in order to reach a final concentration of 5 μM. $BRET^2$ ratio readings are preformed using a lumino/fluorometer that allows sequential integration of luminescence signals detected with two filter settings (see Point 4 in Section 3). The specific $BRET^2$ ratio is calculated by subtracting from the mean $BRET^2$ ratio value above the

---

curve assay in HEK293T27 cells cotransfected with a constant amount of FGFR1-Rluc8 plasmid and increasing amount of the A2AR-GFP2 plasmid. In the current analysis, the amount of each receptor effectively expressed in transfected cells was monitored for each individual experiment by correlating both total luminescence and total fluorescence to the number of receptor-binding sites (biochemical binding analysis) in permeabilized cells The linear regression equations derived from these data were thus used to convert fluorescence and luminescence values into femtomoles/mg protein of receptor in order to obtain accurate values. Cells were preincubated 10 min with vehicle CGS21680 (100 nM) or FGF-2 (50 ng/ml) or with both CGS21680 and FGF-2 (100 nM and 50 ng/ml, respectively). The A2AR/FGFR1 curve fitted better to a saturation curve than to a linear regression, F test ($P<0.001$). Data are means±s.e.m. ($n=5$). The $BRET_{max}$ values were significantly enhanced by combined, CGS21680 and FGF-2 treatment alone versus vehicle or CGS21680 treatment alone and FGF2 treatment alone versus vehicle or CGS21680 treatment alone ($P<0.01$). (B) The $BRET_{50}$ values were significantly reduced by combined, CGS21680 and FGF-2 treatment alone versus vehicle ($P<0.001$). (For color version of this figure, the reader is referred to the online version of this chapter.)

background BRET^2 ratio, which corresponds to the signal obtained with cells expressing the BRET donor alone (see Point 5 in Section 3).

5. In saturation assays, specific BRET^2 ratio values are plotted as a function of the GFP2/Rluc fusion protein ratio. Therefore, the total amount of luminescence (BRET donor amount) and fluorescence (BRET acceptor amount) must be determined for each transfection. It is important that the BRET donor levels are relatively constant throughout the experiment. In case of significant variation (difference of 20% or more from the average value), the corresponding points should be excluded from the final plot or the experiment repeated again. To quantify the amount of BRET acceptor in each well, the fluorescence is measured at 510 nm after external excitation at 410 nm in black 96-well microplates. Background fluorescence is obtained by determining fluorescence in wells containing untransfected cells (GFP2 zero (0) value) and subtracted from the fluorescence values measured in cells expressing increasing amounts of BRET acceptor (GFP2) to obtain the specific GFP2 values. Often when distributing cell suspensions into white opaque 96-well microplates (see Point 2 in Section 3.1), a parallel and similar procedure is performed in black 96-well microplates in order to quantify the amount of the acceptor by fluorescence measurements. Otherwise, cells are detached from spare wells of the white 96-well plate using 100 μl PBS-EDTA or Versene, washed twice with PBS, and collected by centrifugation for 5 min at $300 \times g$ at RT. The pellet is resuspended in 150–200 μl PBS and transferred to a black 96-well plate for fluorescence measurements.

6. GFP2-GFP20/Rluc8 fusion protein ratio is calculated for each data point. Depending on the application, it may be necessary to convert luminescence and fluorescence values into absolute amounts of interacting partners using standard curves correlating luminescence and fluorescence signals with amounts of proteins (see Point 7 in Section 3.1).

7. Correlations are assessed between fluorescence or luminescence values and receptor expression levels in BRET^2 experiments. When appropriate radioligands are available, determination of receptor expression levels in BRET^2 assays can be relevant to ascertain that the expression level of fusion proteins falls within the physiological range. It can also be useful to determine the true acceptor/donor ratio in BRET^2 saturation experiments. Luminescence and fluorescence levels of several receptor–donors and receptor–acceptors are found to be linearly correlated with receptor densities (Borroto-Escuela, Garcia-Negredo, Garriga, Fuxe, & Ciruela, 2010). This correlation is an intrinsic characteristic of each fusion protein, and therefore, correlation curves have to be established for each construct. Cells are transfected (a reasonable number of different independent transfections should be performed) with different quantities of either the BRET donor or the BRET acceptor fusion protein plasmid. Then, the luciferase activity (for the BRET donor receptor) and fluorescence values (for the BRET acceptor receptor) are determined as described earlier. In parallel, saturation radioligand binding experiments are

performed using appropriate assay conditions for each receptor. The number of radioligand binding sites is then plotted against fluorescence or luminescence values determined in the same sample; and a linear correlation is expected and obtained. These standard curves can be used to transform fluorescence and luminescence values into f_{mol} of receptor. Thus, the fluorescence/luminescence ratios can be transformed into (receptor-GFP^2)/(receptor-Rluc8) ratios, which allows to determine accurate BRET$_{max}$ and BRET$_{50}$ values.
8. BRET ratio values from Point 4 in Section 3.1 can be plotted as a function of the (GFP2-GFP0)/Rluc8 ratio (Point 6 in Section 3.1) or (receptor-GFP^2)/(receptor-RLuc) ratio as described in Point 7 in Section 3.1. Data are fitted using a nonlinear regression equation assuming a single binding site (GraphPad Prism) and BRET$_{max}$ and BRET$_{50}$ values can be determined (Fig. 8.1).

8.3.2 Protocol 2: competition assays

BRET^2 displacement experiments can also be performed to shed some light on the specific nature of a given receptor–receptor interaction. In a BRET^2 competition or displacement assay, the BRET^2 ratio is measured at a fixed ratio of donor and acceptor in the presence of increasing concentrations of a nontagged native partner. Over the last few years, in order to test the ability of a GPCR to form heteromers and to further investigate the specificity of BRET signals, competition experiments have often been carried out. It has been demonstrated that the use of an untagged receptor X or Y coexpressed with the receptor X-Rluc and receptor Y-GFP2 decreases the BRET ratio signal as a consequence of the ability of the untagged receptor to interact with one or both fusion proteins and compete for the complementary BRET fusion protein. In a GPCR–RTK heteroreceptor complex, a true interacting partner (e.g., an excess of one of them or a competitive interacting receptor) would be also capable of reducing the BRET signal of the receptor complex, whereas a noninteracting partner would not. However, in spite of this theoretically reasonable and until now well-supported concept that reflects true receptor–receptor interactions (at least for several GPCR–GPCR interactions), this does not necessarily represent an unequivocal interpretation. It could be that, for instance, in the case of GPCR–RTK interaction analysis, the results will not be in line with the expected idea that a true interacting partner would be capable of reducing the BRET signal of the receptor complex. GPCRs are seven helix transmembrane receptors and RTKs are single-helix transmembrane receptors. The specificity and diversity of the interface interaction between these two classes of receptor families may allow for a wider plasticity or diversity of the receptor–receptor interaction. Instead of observing a reduction of the BRET ratio upon untagged receptor coexpression, we can still observe a similar or even increased BRET ratio. This behavior could be the result of a reorganization/reconfiguration of the heteroreceptor complex, where more than two receptors could be accommodated, sharing different receptor–receptor interface interactions.

1. Before running a BRET^2 displacement assay, it is highly recommended to conduct a classical BRET^2 saturation experiment as described in Section 3.1. Using around 50–60% of the (GFP2-GFP20)/Rluc8 BRET$_{max}$ values in the saturation experiment will give a better idea of the amount of donor and acceptor needed for transfection in the BRET competition experiments.
2. Perform several independent transfections using a constant ratio amount of cDNA coding for the BRET^2 donor and acceptor (as selected in Point 1 in Section 3.2). Increasing amounts of the untagged receptor are used together with sufficient "empty" vector (such as pCDNA3.1 or any other cloning vector) to keep equal total amount of cDNA/well.
3. Incubate the cells and measure the BRET^2 ratio (see Point 4 in Section 3).
4. Calculate BRET^2 ratios (see Point 5 in Section 3) and plot them as a function of the expression of native receptor determined by binding experiments or as a function of the total untagged cDNA amount employed (Fig. 8.1).

8.3.3 Protocol 3: kinetics and dose–response assays

In addition to monitoring constitutive receptor–receptor interactions, BRET^2 can be used to measure ligand-dependent induction of facilitatory or inhibitory receptor–receptor interactions and also to follow the kinetics of these interactions in real time. Indeed, ligand modulation of the BRET^2 signal may also prove the specificity of the examined interaction. Until now, for most GPCR heteromers studied by BRET, homomerization and heteromerization appear to occur constitutively and independently of receptor activation state. However, the dependence of the energy transfer process on the distance between the donor and acceptor fused to the receptors and on their relative orientation suggests that movements within receptor complexes as a consequence of ligand-induced conformational changes may result in detectable changes in FRET/BRET signals. This does not necessarily mean changes in the association–dissociation between the receptors.

Many interesting examples where specific ligand-induced conformational changes translate into changes in BRET^2 signal were observed in most studies on RTK–RTK interactions and on GPCR–RTK heteroreceptor complexes (Romero-Fernandez et al., 2011). This illustrates the usefulness of the BRET^2 technology as a readout to characterize GPCR–RTK heteroreceptor complexes and their pharmacological features. This is in addition to the more specific outcome of opening up novel avenues for GPCR–RTK drug discovery.

As an example, this protocol describes the use of BRET^2 to measure the kinetics and dose–responses of the agonist-promoted FGFR1 activation (homodimerization) and the effects of combined and single agonist treatment with 8-OH-DPAT and FGF2 on FGFR1 homodimerization in cells containing FGFR1-5-HT1A heteroreceptor complexes (Fig. 8.2). For this type of experiments where we intend to test how the GPCR–RTK heteroreceptor complexes may induce modulation of the ligand-dependent BRET^2 RTK homodimerization signal, cells are cotransfected at a fixed ratio with a plasmid coding for the BRET donor (RTK-Rluc8) in the presence of a single concentration of the plasmid encoding the BRET acceptor (RTK-GFP2)

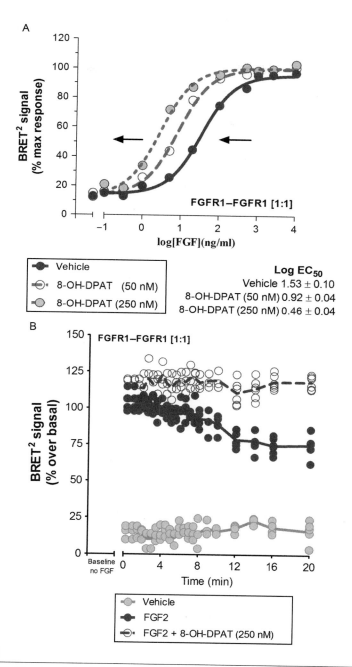

FIGURE 8.2

Effects of combined and single treatment with 8-OH-DPAT and FGF2 on FGFR1 homodimerization in HEK293T27 cells containing FGFR1-5-HT1A heteroreceptor complexes. The modulatory effect of 5-HT1A agonist 8-OH-DPAT was studied on the FGF2

Continued

and the plasmid encoding for the WT-GPCR protomer. BRET measurements are performed over time (up to 20–30 min) after incubation with a RTK-specific agonist or upon combined agonist treatment (RTK and GPCR agonist).

1. HEK-293T cells are prepared and transfected using a constant ratio amount of cDNA coding for the BRET RTK-donor and RTK-acceptor. This ratio should be selected from a previously performed BRET saturation assay. Otherwise, different plasmid DNA concentrations should be tested to determine the optimal ratio of receptor–donor to receptor–acceptor that gives the highest $BRET^2$ signal following receptor activation. Also, a constant amount of the untagged GPCR protomer and sufficient "empty" vector (such as pCDNA3.1 or any other cloning vector) will be used to maintain equal total amount of cDNA per well during transfection.
2. The next day, cells are harvested and dispensed in 50–100 μl of cell suspension containing 20,000 to 30,000 cells into a white opaque 96-microplate in HEPES-buffered DMEM without phenol red.
3. *For agonist-induced dose–response BRET:* 24 h after plating cells are incubated with different concentrations of the selected receptor agonist or PBS alone (basal control condition) for a fixed time. Then, coelenterazine 400a solution at a final concentration of 5 μM is added and the $BRET^2$ signal is measured (Fig. 8.2). The agonist and substrate can be added manually or injected if the plate reader is equipped with built-in online injectors. *For kinetic BRET experiments:* cells are incubated for different time periods with a fixed concentration of the selected receptor agonist or PBS alone (basal control condition). Then, coelenterazine 400a solution at a final concentration of 5 μM is added and the $BRET^2$ signal is measured. We highly recommend when using

FIGURE 8.2—Cont'd induced FGFR1/FGFR1 homodimer formation by means of $BRET^2$ analysis. HEK293T27 cells were transiently cotransfected at a constant ratio (1:1:1) with 5-HT1A, FGFR1-Rluc8, and FGFR1-GFP2. (A) A concentration–response curve with FGF-2 was performed on the development of the $BRET^2$ signal from the FGFR1 homodimer in the HEK293T27 cells. The cells were transiently cotransfected at a constant ratio (1:1:1) of 5-HT1A, FGFR1-Rluc8, and FGFR1-GFP2 and treated with the agonist ligands for 5 min before $BRET^2$ measurement. Treatment with 8-OH-DPAT (with two different concentrations: 50 and 250 nM) shifted the curves of the $BRET^2$ signal to the left, which indicate an enhanced potency of combined treatment with FGF2 and the 5-HT1A agonist versus FGF-2 treatment alone to promote FGFR1 homodimer formation. (B) The kinetics of the FGFR1-Rluc/FGFR1-GFP2 interaction after FGF2 treatment and its modulation by 8-OHDPAT was also studied in transiently transfected HEK293T27 cells using the $BRET^2$ assay to study the FGFR1 homodimer over a period of 20 min. FGF-2 and the combined FGF-2 and 8-OH-DPAT treatments showed no clear-cut changes of the $BRET^2$ value over the 8 min period. However, the combined treatment had a weak tendency to increase the $BRET^2$ signal over time, whereas the FGF2 alone treatment had a markedly tendency to decrease the $BRET_2$ signal over time. (For color version of this figure, the reader is referred to the online version of this chapter.)

BRET^2 assay to incubate first the cells with the selected agonist at different times ranging in the order of minutes (if it is possible) and then read all together instead of performing a more quick kinetic analysis with continuous measurements in view of the fast kinetic delay of substrate when coelenterazine 400a is used (Fig. 8.2).
4. Calculation and interpretation of the BRET^2 signal. In agonist-induced BRET^2 signaling, the BRET^2 ratio is defined as *ligand-promoted BRET^2 ratio = BRET ratio (in presence of a ligand)—BRET ratio (in absence of a ligand or in presence of PBS)*. Plot the BRET data against the logarithm of ligand concentration and analyze it by nonlinear curve fitting (sigmoidal dose–response) using GraphPad PRISM, which will allow to obtain the concentration eliciting a half-maximal response (EC50 value). When analyzing the kinetic BRET^2 data, the BRET^2 signal can be plotted against time to produce kinetic profiles, examples of which are shown in Fig. 8.2. Changes in the time-course profile and apparent association (or dissociation) rate constants can be calculated from such data.

8.4 DISCUSSION AND NOTE ON CRITICAL PARAMETERS

In spite of the increased understanding of the role of GPCR transactivation by RTK ligands and vice versa and their existence as heteroreceptor complexes, only few examples have until now been validated using the BRET^2 methodology (Borroto-Escuela, Romero-Fernandez, Mudo, et al., 2012), most likely due to the particularly difficult nature of such receptor interactions. We believe that a well-controlled and carefully analyzed BRET assay has a great potential to identified and/or study GPCR–RTK heteroreceptor complexes (Fig. 8.3). It also seems to have a high value in investigating the mechanism of action and the pharmacological properties of drugs acting on these important therapeutic targets. The present work gives a step-by-step description of the BRET^2 methodology using specific examples, confronting the pros and cons of various protocols.

BRET^2 assays have been used to study GPCR heteromerization for about a decade now, and as the field has matured, we have a better understanding of the underlying technological limitations and an improved ability to interpret BRET data. The method is particularly suited not only for the molecular characterization of GPCR–RTK heteroreceptor complexes but also for the screening of new compounds that bind specifically to these complexes, with the ultimate goal of identifying novel therapeutic drugs.

Progressive improvements and diversification of the BRET-based assay place this technology among the most powerful methods for the study of GPCR–RTK heteroreceptor complexes and their conformational changes in living cells. Also, new perspectives of this methodology could include the monitoring of these interactions at the subcellular level to unravel the molecular mechanisms underlying GPCR–RTK heteroreceptor complex transactivation/transinhibition processes, internalization, and signaling.

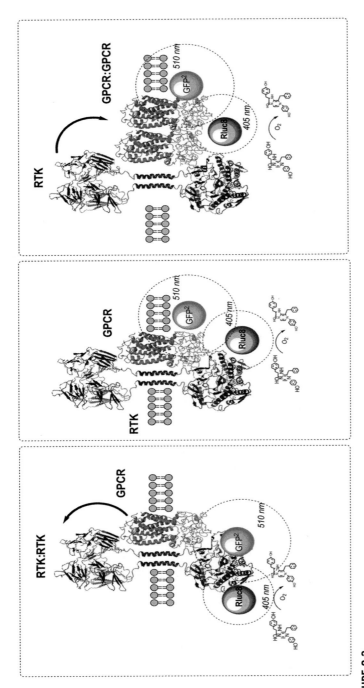

FIGURE 8.3

Different BRET^2 strategy assays to study GPCR–RTK heteroreceptor complexes dynamic. BRET^2 assays seem to be particularly suited for the molecular characterization and study of GPCR–RTK heteroreceptor complexes. Using different combination of receptor–donor/acceptor-fused receptor, it could be possible to unravel the molecular mechanisms underlying GPCR–RTK heteroreceptor complex transactivation/transinhibition processes, internalization, and signaling. (center) For instance, it could be possible to study the existence of the GPCR–RTK heteroreceptor complexes and the pharmacological properties of drugs acting on these important therapeutic targets using as a BRET pairs, the GPCR-fused receptor as a donor, and the RTK-fused receptor as acceptor. (left) Furthermore, the effects of combined and single treatment with GPCR and RTK agonist on the RTK homodimerization in cells containing the GPCR–RTK heteroreceptor complexes could be feasible. (right) In addition, the effects of combined treatment with the GPCR and RTK agonist on GPCR receptor homodimerization in cells containing the GPCR–RTK heteroreceptor complexes could be possible. Each of these different BRET approach could bring new light in the conformation changes and dynamic process that can take place upon agonist treatment in the GPRC–RTK heteroreceptor complexes. (For color version of this figure, the reader is referred to the online version of this chapter.)

However, as any other methodologies, it has its limitations and drawbacks and is constantly being improved with regard to experimental design, instrumentation, and reagents. For example, in some receptor–receptor interaction studies, because the efficiency of energy transfer is tightly dependent on proper orientation of the donor and acceptor dipoles, conformational states of the fusion proteins may fix the dipoles into a geometry that is unfavorable for energy transfer. Thus, two receptors may form heteroreceptor complexes without producing any significant BRET signal. Thus, a negative result with BRET does not necessarily mean absence of the heteroreceptor complex. Furthermore, the luminescent and fluorescent protein fused to the candidate protomers may affect their interaction and alter their subcellular localization, protein folding, and receptor function. Moreover, as with any technique that involves transient transfection, overexpression of the protomers may bring misleading results. Therefore, such drawbacks limit its use today and lead to a demand for proper controls. It is now easier to keep the receptor protomer expression levels within physiologically relevant ranges, thanks to the improvement of donor/acceptor protein quantum yield (e.g., the new mutated Rluc8). On the other hand, $BRET^2$ presents various advantages compared to standard biochemical procedures to study receptor–receptor interactions that require cell-invasive processes such as solubilization and coimmunoprecipitation, which includes even the previously developed FRET technique. $BRET^2$, as opposed to FRET, does not require the excitation of the donor with an external light source. Therefore, it does not suffer from problems usually associated with autofluorescence, light scattering, photobleaching, and/or photoisomerization of the donor moiety, which results in an overall improved signal-to-noise ratio when compared to earlier versions of the resonance energy transfer technologies. Also, the absence of contamination of the light output by the incident light results in a very low background in $BRET^2$ assays, thereby permitting the detection of small changes in the $BRET^2$ signal as compared to FRET. Also, the $BRET^2$ signal is a ratiometric measurement, which can help to reduce data variability caused by fluctuations in light output due to variations in assay volume, cell types, number of cells per well, and/or signal decay across a plate. Finally, the coelenterazine derivative DeepBlueC or coelenterazine 400a used in $BRET^2$ is membrane permeable and nontoxic, which makes $BRET^2$ an ideal assay technology for live cell assays. DeepBlueC penetrates the cell membrane in seconds to activate Rluc8 emission. With respect to receptor–receptor interaction mechanisms and signaling research, its capabilities have now significantly reached beyond studying GPCRs. The $BRET^2$ assay technology is used successfully for a wide range of assay types including GPCR-beta-arrestin assay for monitoring GPCR activity, tyrosine kinase receptor activation, Ca^{2+} and cAMP detection, apoptosis assay, kinase activity, and protease activity.

Acknowledgments

This work has been supported by the Swedish Medical Research Council (04X-715), Telethon TV3's La Marató Foundation 2008, and Hjärnfonden to KF; by grants from the Swedish Royal Academy of Sciences (Stiftelsen B. von Beskows Fond and Stiftelsen Hierta-Retzius stipendiefond) and Karolinska Institutets Forskningsstiftelser 2011 and 2012 to D.O.B-E; and by grants from the National Institutes of Health (Grant DA10044 and MH090963 to PG) and US Army Medical Research contract (W81XWH-10-1-0691 to MF).

References

Alderton, F., Rakhit, S., Kong, K. C., Palmer, T., Sambi, B., Pyne, S., et al. (2001). Tethering of the platelet-derived growth factor beta receptor to G-protein-coupled receptors. A novel platform for integrative signaling by these receptor classes in mammalian cells. *The Journal of Biological Chemistry*, 276(30), 28578–28585.

Andrae, J., Gallini, R., & Betsholtz, C. (2008). Role of platelet-derived growth factors in physiology and medicine. *Genes & Development*, 22(10), 1276–1312.

Ayoub, M. A., & Pfleger, K. D. (2010). Recent advances in bioluminescence resonance energy transfer technologies to study GPCR heteromerization. *Current Opinion in Pharmacology*, 10(1), 44–52.

Bacart, J., Corbel, C., Jockers, R., Bach, S., & Couturier, C. (2008). The BRET technology and its application to screening assays. *Biotechnology Journal*, 3(3), 311–324.

Borroto-Escuela, D. O., Garcia-Negredo, G., Garriga, P., Fuxe, K., & Ciruela, F. (2010). The M(5) muscarinic acetylcholine receptor third intracellular loop regulates receptor function and oligomerization. *Biochimica et Biophysica Acta*, 1803(7), 813–825.

Borroto-Escuela, D. O., Romero-Fernandez, W., Garriga, P., Ciruela, F., Narvaez, M., Tarakanov, A. O., et al. (2013). G protein-coupled receptor heterodimerization in the brain. *Methods in Enzymology*, 521, 281–294.

Borroto-Escuela, D. O., Romero-Fernandez, W., Mudo, G., Perez-Alea, M., Ciruela, F., Tarakanov, A. O., et al. (2012). Fibroblast growth factor receptor 1–5-hydroxytryptamine 1A heteroreceptor complexes and their enhancement of hippocampal plasticity. *Biological Psychiatry*, 71(1), 84–91.

Borroto-Escuela, D. O., Romero-Fernandez, W., Tarakanov, A. O., Gomez-Soler, M., Corrales, F., Marcellino, D., et al. (2010). Characterization of the A2AR-D2R interface: Focus on the role of the C-terminal tail and the transmembrane helices. *Biochemical and Biophysical Research Communications*, 402(4), 801–807.

Cabello, N., Gandia, J., Bertarelli, D. C., Watanabe, M., Lluis, C., Franco, R., et al. (2009). Metabotropic glutamate type 5, dopamine D2 and adenosine A2a receptors form higher-order oligomers in living cells. *Journal of Neurochemistry*, 109(5), 1497–1507.

Dalle, S., Imamura, T., Rose, D. W., Worrall, D. S., Ugi, S., Hupfeld, C. J., et al. (2002). Insulin induces heterologous desensitization of G-protein-coupled receptor and insulin-like growth factor I signaling by downregulating beta-arrestin-1. *Molecular and Cellular Biology*, 22(17), 6272–6285.

De, A., Ray, P., Loening, A. M., & Gambhir, S. S. (2009). BRET3: A red-shifted bioluminescence resonance energy transfer (BRET)-based integrated platform for imaging protein–protein interactions from single live cells and living animals. *FASEB Journal: Official Publication of the Federation of American Societies for Experimental Biology*, 23(8), 2702–2709.

Delcourt, N., Bockaert, J., & Marin, P. (2007). GPCR-jacking: From a new route in RTK signalling to a new concept in GPCR activation. *Trends in Pharmacological Sciences*, 28(12), 602–607.

Flajolet, M., Wang, Z., Futter, M., Shen, W., Nuangchamnong, N., Bendor, J., et al. (2008). FGF acts as a co-transmitter through adenosine A(2A) receptor to regulate synaptic plasticity. *Nature Neuroscience, 11*(12), 1402–1409.

Fuxe, K., Dahlstrom, A., Hoistad, M., Marcellino, D., Jansson, A., Rivera, A., et al. (2007). From the Golgi–Cajal mapping to the transmitter-based characterization of the neuronal networks leading to two modes of brain communication: Wiring and volume transmission. *Brain Research Reviews, 55*(1), 17–54.

Fuxe, K., Marcellino, D., Borroto-Escuela, D. O., Frankowska, M., Ferraro, L., Guidolin, D., et al. (2010). The changing world of G protein-coupled receptors: From monomers to dimers and receptor mosaics with allosteric receptor-receptor interactions. *Journal of Receptor and Signal Transduction Research, 30*(5), 272–283.

Gammon, S. T., Villalobos, V. M., Roshal, M., Samrakandi, M., & Piwnica-Worms, D. (2009). Rational design of novel red-shifted BRET pairs: Platforms for real-time single-chain protease biosensors. *Biotechnology Progress, 25*(2), 559–569.

Hamdan, F. F., Percherancier, Y., Breton, B., & Bouvier, M. (2006). Monitoring protein–protein interactions in living cells by bioluminescence resonance energy transfer (BRET). *Current Protocols in Neuroscience/Editorial Board. Current Protocols in Neuroscience/Editorial Board, Jacqueline N Crawley [et al.],* Chapter 5: Unit 5 23.

Kocan, M., See, H. B., Seeber, R. M., Eidne, K. A., & Pfleger, K. D. (2008). Demonstration of improvements to the bioluminescence resonance energy transfer (BRET) technology for the monitoring of G protein-coupled receptors in live cells. *Journal of Biomolecular Screening, 13*(9), 888–898.

Lin, F. T., Daaka, Y., & Lefkowitz, R. J. (1998). Beta-arrestins regulate mitogenic signaling and clathrin-mediated endocytosis of the insulin-like growth factor I receptor. *The Journal of Biological Chemistry, 273*(48), 31640–31643.

Luttrell, L. M., Daaka, Y., & Lefkowitz, R. J. (1999). Regulation of tyrosine kinase cascades by G-protein-coupled receptors. *Current Opinion in Cell Biology, 11*(2), 177–183.

Marullo, S., & Bouvier, M. (2007). Resonance energy transfer approaches in molecular pharmacology and beyond. *Trends in Pharmacological Sciences, 28*(8), 362–365.

Medintz, I. L., & Mattoussi, H. (2009). Quantum dot-based resonance energy transfer and its growing application in biology. *Physical Chemistry Chemical Physics, 11*(1), 17–45.

Pfleger, K. D., Dromey, J. R., Dalrymple, M. B., Lim, E. M., Thomas, W. G., & Eidne, K. A. (2006). Extended bioluminescence resonance energy transfer (eBRET) for monitoring prolonged protein–protein interactions in live cells. *Cellular Signalling, 18*(10), 1664–1670.

Pfleger, K. D., & Eidne, K. A. (2006). Illuminating insights into protein–protein interactions using bioluminescence resonance energy transfer (BRET). *Nature Methods, 3*(3), 165–174.

Pyne, N. J., & Pyne, S. (2011). Receptor tyrosine kinase-G-protein-coupled receptor signalling platforms: Out of the shadow? *Trends in Pharmacological Sciences, 32*(8), 443–450.

Romero-Fernandez, W., Borroto-Escuela, D. O., Tarakanov, A. O., Mudo, G., Narvaez, M., Perez-Alea, M., et al. (2011). Agonist-induced formation of FGFR1 homodimers and signaling differ among members of the FGF family. *Biochemical and Biophysical Research Communications, 409*(4), 764–768.

Takeuchi, K., & Ito, F. (2011). Receptor tyrosine kinases and targeted cancer therapeutics. *Biological & Pharmaceutical Bulletin, 34*(12), 1774–1780.

Vidi, P. A., Ejendal, K. F., Przybyla, J. A., & Watts, V. J. (2011). Fluorescent protein complementation assays: New tools to study G protein-coupled receptor oligomerization and GPCR-mediated signaling. *Molecular and Cellular Endocrinology, 331*(2), 185–193.

Waters, C. M., Long, J., Gorshkova, I., Fujiwara, Y., Connell, M., Belmonte, K. E., et al. (2006). Cell migration activated by platelet-derived growth factor receptor is blocked by an inverse agonist of the sphingosine 1-phosphate receptor-1. *FASEB Journal: Official Publication of the Federation of American Societies for Experimental Biology, 20*(3), 509–511.

Xing, Y., So, M. K., Koh, A. L., Sinclair, R., & Rao, J. (2008). Improved QD-BRET conjugates for detection and imaging. *Biochemical and Biophysical Research Communications, 372*(3), 388–394.

Xu, Y., Piston, D. W., & Johnson, C. H. (1999). A bioluminescence resonance energy transfer (BRET) system: Application to interacting circadian clock proteins. *Proceedings of the National Academy of Sciences of the United States of America, 96*(1), 151–156.

CHAPTER 9

A Simple Method to Detect Allostery in GPCR Dimers

Eugénie Goupil*, Stéphane A. Laporte*,†, and Terence E. Hébert*

*Department of Pharmacology, McGill University, Montréal, Québec, Canada
†Department of Medicine, McGill University, Montréal, Québec, Canada

CHAPTER OUTLINE

Introduction and Rationale 166
9.1 Materials 169
9.2 Methods 170
    9.2.1 Cell Culture and Transfection 170
    9.2.2 Radioligand Binding and Dissociation Kinetics 170
    9.2.3 Analysis and Interpretation of Results 172
9.3 Discussion 174
Summary 176
Acknowledgments 176
References 176

Abstract

G protein-coupled receptors (GPCRs) represent one of the largest families of cell surface receptors as key targets for pharmacological manipulation. G proteins have long been recognized as allosteric modulators of GPCR ligand binding. More recently, small molecule allosteric modulators have now been widely characterized for a number of GPCRs, and some are now used clinically. Many studies have also underscored the importance of GPCR dimerization or higher-order oligomerization in the control of the physiological responses they modulate. Thus, allosterism can also, between monomer equivalents in the context of a dimer, oligomer, or receptor mosaic, influence signaling pathways downstream. It therefore becomes essential to characterize both small molecule allosteric ligands and allosteric interactions between receptors modulated by canonical orthosteric ligands, in a pathway-specific manner. Here, we describe a simple, radioligand-binding method, which is designed to probe for allosteric modulation mediated by any GPCR interactor, from small molecules to interacting proteins. It can also detect allosteric asymmetries within a

GPCR heterodimer, via orthosteric or allosteric ligands. This assay measures time-dependent ligand occupancy of radiolabeled orthosteric or (with adaptations) allosteric ligands as modulated by either small molecules or receptor dimer partners bound or unbound with their own ligands.

INTRODUCTION AND RATIONALE

Heterotrimeric G proteins remain the best-characterized allosteric modulator of G protein-coupled receptors (GPCRs), through their capacity to modulate ligand-binding affinity. Early models described the effects of G protein recruitment to a ligand-bound receptor as the "ternary complex model" (De Lean, Stadel, & Lefkowitz, 1980; Limbird, 2004). This analysis is also relevant when the heterotrimeric G protein is substituted by an allosteric ligand, which differentially modulates the activated receptor bound by orthosteric agonist (May, Leach, Sexton, & Christopoulos, 2007). By binding to topographically distinct binding sites with respect to the orthosteric binding site, allosteric ligands have rapidly become a viable alternative to modulate the selectivity and functionality of receptors and more recently ligand-directed signaling (Christopoulos, 2002; Kenakin, 2010; Kenakin & Miller, 2010; Urwyler, 2011). Allosteric modulators, as such, have been around a long time but were initially described as noncompetitive antagonists (Litschig et al., 1999; Varney et al., 1999). Such noncompetitive antagonism was often referred as insurmountable, because the receptor–antagonist complex was a new entity in its own right, distinct from the agonist–receptor complex (Vauquelin, Van Liefde, Birzbier, & Vanderheyden, 2002). Allosteric modulators can affect GPCRs at multiple levels: (1) the binding of orthosteric ligand to the receptor, (2) the transmission of ligand-binding information to other parts of the receptor, or (3) the signaling downstream of receptor activation (Fig. 9.1A and B). First, allosteric ligands were shown to modulate the affinity of orthosteric drugs for their binding sites. Cooperativity (also denoted as the "α" factor) between the two binding sites can be neutral (no effect or $\alpha = 1$), positive (leftward shift of the binding curve or $\alpha > 1$), or negative (rightward shift of the binding curve or $\alpha < 1$). Interestingly, allosteric effects are saturable, that is, when all the allosteric sites are occupied, no further allosteric modulation is observed. Moreover, the regulation of the different signaling modes downstream of a given receptor can be achieved by modulating either the efficacy or potency of the response in question, with or without modulation of the binding affinity of the orthosteric ligand, or the coupling efficacy between the G protein and the receptor. For all the modes of modulation described in the preceding text, an allosteric ligand can therefore be a positive allosteric modulator (PAM) or a negative allosteric modulator (NAM). One of the major characteristics of PAMs and NAMs is their relative inability to trigger GPCR-induced responses in the absence of the orthosteric ligand. There are, however, ago-allosteric ligands or allosteric agonists, which behave like agonists but through binding sites distinct from orthosteric ligands. Finally, there are also synthetic ligands that can simultaneously bind to both

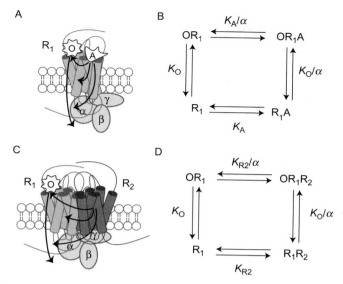

FIGURE 9.1

The many faces of allosteric modulation of GPCRs. (A) Allosteric ligands of GPCRs (denoted "A" on the scheme) can regulate orthosteric (O) ligand binding, the transmission of ligand-induced conformational information to other parts of the receptor or the signaling downstream of receptor activation. (B) The first representation of the allosteric ternary complex model of drug action involves a GPCR, "R_1" being bound by an orthosteric ligand "O," with an affinity constant, "K_O," generating the "OR_1" complex, leading to a cellular stimulus downstream. Then, a third species, the allosteric ligand "A," can bind the "R_1O" complex, with a specific affinity "K_A" and a binding cooperativity value of α, which denotes the magnitude and direction of the allosteric effect on ligand-binding affinity ($\alpha=1$, no effect, $\alpha>1$ positive allostery, $\alpha<$ negative allostery). (C) Dimeric receptor partners (as denoted "R_2" in the scheme) of GPCRs can also be considered as allosteric modulators, in the same way as an allosteric ligand acts on a GPCR monomer or homodimer. (D) The allosteric ternary complex model applied to a dimer partner.

allosteric and orthosteric binding sites, a subclass of the so-called bitopic ligands (Valant, Sexton, & Christopoulos, 2009). Bitopic ligands, in and of themselves, are a very interesting class of molecules, which may combine a number of different pharmacophores, comprising allosteric and orthosteric ligands on the same receptor or between receptor equivalents in homo- and heterooligomeric GPCRs (Kamal & Jockers, 2009; Keov, Sexton, & Christopoulos, 2011).

Allosteric modulation of class A GPCRs is thought to occur via regions outside the heptahelical transmembrane domain, on the extracellular surface of the receptor, which are less conserved and thus providing a basis for receptor-specific action (Peeters et al., 2011). This type of modulation often stabilizes (or induces) specific receptor conformations, altering coupling to distinct effector pathways. The first example of such a regulator was identified for the neurokinin NK2 receptor. When bound to its endogenous ligand, neurokinin A (NKA), NK2 can adopt distinct and sequential conformations, stabilized by two high-affinity binding sites (Palanche et al., 2001). The first conformation, A1L, is thought to facilitate rapid dissociation of NKA from the receptor followed by Gαq-induced calcium mobilization, whereas the second conformation, A2L, results in slower ligand dissociation kinetics and leads to Gαs-induced cAMP production. An allosteric ligand for NK2R, LPI805, preferentially stabilized the A1L conformation, diminishing the intensity of Gαs-mediated cAMP accumulation (Maillet et al., 2007). In a follow-up study, the authors were able to generate derivatives of the original allosteric ligand, which generated distinct selectivity profiles, acting as a PAM for calcium signaling and an NAM for cAMP production (Valant, Maillet, et al., 2009). Another example of an allosteric modulator directly regulating coupling between the receptor and the G protein in a biased manner was shown for the prostaglandin F2α (PGF2α) receptor (FP). We developed a small molecule peptide mimic, PDC113.824, derived originally from the sequence of the second extracellular loop of FP (Goupil et al., 2010). The peptide itself was initially characterized as a tocolytic in a mouse model of preterm labor (Peri et al., 2002). PDC113.824 was demonstrated to stabilize a specific conformation of FP receptor, where Gαq-induced PKC–ERK1/ERK2 activation was potentiated and Gα12-induced, Rho-mediated cytoskeletal rearrangement was inhibited following PGF2α stimulation (Goupil et al., 2010). Interestingly, neither effector pathway was modulated by PDC113.824 alone. However, basal levels of GTPγ[^{35}S] incorporation were altered by PDC113.824 for both Gαq and Gα12, suggesting that the allosteric ligand recognized two distinct preformed, receptor/G protein complexes. This observation has implications regarding the actual molecular targets of biased and/or allosteric ligands. Moreover, the repercussions of such biased signaling were manifested *ex vivo* and *in vivo* by inhibition of myometrial contraction and lipopolysaccharide- or PGF2α-induced preterm labor in mice, respectively, in the presence of PDC113.824. Interestingly, as for LPI805, the dissociation kinetics of [^{3}H]-PGF2α from FP were more rapid in the presence of PDC113.824, and Gαq coupling was enhanced. These studies suggest that allosteric modulators, which enhance GTP binding to G proteins in the absence of orthosteric ligand, may lead to alterations in orthosteric ligand affinity for the receptor, leading to specific G protein-dependent signaling patterns (see Fig. 9.2).

GPCRs were long considered to be monomeric entities; however, it has become clear in recent years that most if not all GPCRs can form dimers and possibly higher-order structures (see Bulenger, Marullo, & Bouvier, 2005; Hébert & Bouvier, 1998; Kniazeff, Prezeau, Rondard, Pin, & Goudet, 2011; Milligan, 2009; Prinster, Hague, & Hall, 2005 for review). Although some GPCRs can function in a monomeric state (Whorton et al., 2007, 2008), dimerization seems critical in regulating all aspects of GPCR function (receptor trafficking to and from the cell surface, ligand binding, G protein coupling, and downstream signaling, reviewed in Terrillon & Bouvier, 2004). While allosteric and biased ligands are useful ways to control GPCR signaling more selectively (Fig. 9.1A and B), the potential relevance of these ligands to individual GPCRs is both increased and complicated by the existence of receptor homo- and heterooligomers (Fig. 9.1C and D). The notion that GPCRs can modulate each another allosterically in the context of a heterodimer, as the G protein does with

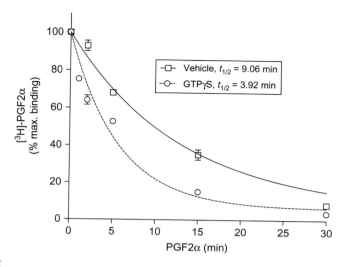

FIGURE 9.2

Allosteric modulation of orthosteric ligand binding to FP by activated G protein. Dissociation kinetics of [^{3}H]-prostaglandin F2α (PGF2α) binding from, FP, in isolated membranes in the presence of a vehicle or 10 μM GTPγS (a nonhydrolyzable form of GTP which activates G proteins). Data are representative of three independent experiments.

respect to receptor ligand-binding affinity in a receptor monomer, can both be addressed with the simple technique described in the following text.

Induced changes in the dissociation kinetics of orthosteric ligands from their cognate GPCRs are a hallmark of allosteric modulators of affinity (Christopoulos & Kenakin, 2002; Kostenis & Mohr, 1996; Lanzafame & Christopoulos, 2004; Proska & Tucek, 1994; Vauquelin, Van Liefde, & Vanderheyden, 2002). This would be true whether the allosteric modulator is a small molecule, a G protein or effector partner (Fig. 9.1A and B), or even another receptor partner in the context of a GPCR homo- or heterodimer (Fig. 9.1C and D). Accordingly, a competitive antagonist (e.g., acting on the orthosteric site) would in principal affect the dissociation rate of a radiolabeled orthosteric ligand in the same way that the unlabeled orthosteric ligand would, while an allosteric modulator, even with relatively low affinity (acting as either PAM or NAM), could promote a conformational change in the receptor that would lead to an alteration in the dissociation kinetics of the orthosteric ligand–receptor complex. Here, we describe a simple assay to measure these effects on ligand dissociation kinetics, which can be performed in native cells with endogenous receptors.

9.1 MATERIALS

1. Phosphate-buffered saline (PBS), pH 7.4
2. EDTA: 0.5 M pH 8.0
3. Homogenization buffer: 10 mM Tris–HCl pH 7.4, 5 mM EDTA

4. Binding buffer: 50 mM Tris–Cl pH 7.4, 10 mM $MgCl_2$, 100 mM NaCl
5. Bovine serum albumin (BSA, Sigma-Aldrich)
6. Wash buffer: 50 mM Tris–Cl pH 7.4 (pH must be adjusted within a temperature between 4 and 10 °C)
7. Protease inhibitors: phenanthroline, pepstatin, leupeptin, and aprotinin (Sigma-Aldrich)
8. Radioligand at levels sufficient to occupy 20% of receptors, [^{3}H]-PGF2α (Perkin Elmer, 150–240 Ci/mmol)
9. GTPγS (Sigma)
10. SARSTEDT tubes for binding: 5 ml, 75 × 12 mm, PS (cat. #55.476), and caps (cat. #65.809)
11. Fisher glass fiber circles (2.4 cm, cat. #09-804-24C)
12. 50 ml PLASTIBRAND pipette tip (cat. #Z332798) and repeater pipette
13. Millipore 1225 sampling manifold or equivalent (cat. #xx2702550)
14. Scintillation tubes and scintillation liquid for tritiated ligands
15. β-scintillation counter for quantification of [^{3}H]-ligand binding

9.2 METHODS

9.2.1 Cell culture and transfection

1. Day 1: HEK 293 cells (or other cell types such as vascular smooth muscle cells) are cultured in MEM or DMEM containing L-glutamine, 10% FBS, and antibiotics, at a density of 500,000 cells (quantity may vary depending on transfection method) per 10 cm dish. Cell lines are grown at 37 °C in 5% CO_2.
   a. To verify the effects of the G protein or an allosteric ligand on radioligand binding of the receptor of interest, only one condition, with the receptor expressed alone, needs to be considered.
   b. To verify the effects of a receptor partner on radioligand binding of the receptor of interest, two conditions are necessary: the receptor of interest expressed alone or expressed with the receptor partner.
2. Day 2: cells are transfected with the desired receptor(s) (1.5–2 μg), using the calcium phosphate method (other transfection methods can be used). Use an empty vector to assure similar DNA quantity in each transfection condition. Alternatively, cells stably expressing receptor(s) of interest can be used.
3. Day 3: change media.
4. Day 4: perform binding experiment (see Section 9.2.2).

9.2.2 Radioligand binding and dissociation kinetics

1. At room temperature, rinse the plates with 5 ml PBS.
2. To detach the cells, add 2 ml of PBS-2 mM EDTA per plate and wait for 5 min.
3. Gently scrape and collect the cells. Put the cells on ice.

4. Dose protein content of whole cells using 35–50 μl of the PBS–EDTA-resuspended cells.
5. Pellet the cells by centrifugation (500 × g) and resuspend the cells in binding buffer containing 0.2% BSA and protease inhibitors to have a concentration of 2 μg/μl (final quantity of cells needed will be 100 μg, may vary depending on the cell type and transfection efficiency and can be predetermined for optimal binding). Alternatively, membranes (especially for the effect of the activated G proteins on radioligand binding to the receptor) can be purified from the cells by homogenizing them with a Teflon potter (by doing 20 strokes on ice), followed by a 30 min centrifugation at 40,000 × g at 4 °C and used for this experiment, instead of whole cells. The quantity of membrane to use for optimal binding has to be predetermined.
6. Prepare the radioligands: Use enough radioligand to occupy no more than 20% of the receptors at equilibrium in a final volume of 100 μl binding buffer. This represents ~50,000–100,000 cpm of [^{3}H]-PGF2α at specific activity of 150–240 Ci/mmol, which has an affinity for FP in the 4–6 nM range. To calculate nonspecific binding, cells are incubated with the same quantity of radioligand and an excess (1000 × the K_i) of cold ligand (see next step).
7. Prepare the binding reaction in the 5 ml SARSTEDT tubes (in duplicate or triplicate), for different times (e.g., 0, 1, 2, 5, 15, and 30 min. In this example, the "0" time point is the total binding, and the nonspecific binding control. Add in the following order, at room temperature (for a final volume of 400 μl):
   a. 220 μl of binding buffer
   b. 50 μl of cold 8 × ligand (for nonspecific binding) or buffer (for total binding)
   c. 50 μl of vehicle (control) or cold 8 × allosteric ligand (to assess the effect of this ligand on the orthosteric radioligand binding to the receptor of interest) or 80 μM GTPγS (to have 10 μM final concentration, to assess the effect of the activated G protein(s) on the orthosteric radioligand binding to the receptor of interest)
   d. 80 μl of radioligand
   e. 50 μl of cells (100 μg protein) to start the binding reaction
   f. Leave the reaction 1 h to reach binding equilibrium (or longer if necessary, to assess the time of binding equilibrium, association kinetics can also be assessed)
8. During this time, equilibrate glass fiber circles in binding buffer containing 0.2% BSA (w/v).
9. After radioligand binding to the receptor has reach equilibrium, start the dissociation kinetics experiment by adding 50 μl of cold orthosteric 10 × ligand, starting with the latest time point (in this example, 30 min) to the shortest.
10. Before the end of the time course experiment, prepare the manifold of the cell harvester for the filtration of the binding reaction by putting the equilibrated glass fiber circles on the manifold and rinse them once with 2 ml of cold 50 mM Tris–HCl, pH 7.4.

11. At specified times, stop the binding reaction by adding 2 ml of cold 50 mM Tris–HCl, pH 7.4 to each binding reaction and immediately filter it on the glass fiber circles placed on the manifold.
12. Rinse the filters twice with 2 ml of cold 50 mM Tris–HCl, pH 7.4.
13. Add the glass fiber circle to SARSTEDT tubes or scintillation tubes (then add 5 ml of scintillation liquid) to prepare for reading.
14. Assess receptor-bound radioactivity with the appropriate counter.

9.2.3 Analysis and interpretation of results

Analysis of dissociation kinetic data can easily be performed using commercial software, such as GraphPad 5 Prism (or higher versions). Prior to importing data into GraphPad, specific binding needs to be calculated by subtracting nonspecific binding from total binding. Moreover, reporting the data as fold over the binding at dissociation kinetic time "0" (i.e., at binding equilibrium, without no cold ligand), or basal, will permit normalization between experiments. Once the binding data is analyzed, it can be plotted as y-axis against dissociation time (in this example, 0, 2, 5, 15, and 30 min). Data points can be further analyzed by nonlinear regression, using a "dissociation-one-phase exponential decay" fit by constraining both the initial binding (Y0) to 100 and maximum dissociation at (e.g., 30 min time) or nonspecific binding (NS) at infinite time to an average common values. The time at which half of the ligand has dissociated, or half-life ($t_{1/2}$), is calculated from $\ln(2)/K$, where K represents the dissociation rate constant (K_{off}) and is an indicator of the affinity between the ligand and its receptor. Once the nonlinear regression analysis is performed, statistical comparisons (extra sum-of-square F test) between the best-fit values can be evaluated assuming the null hypothesis that K is the same for all data sets. Rejection of the null hypothesis (i.e., K differs for each data sets) is thus interpreted as allosteric modulation of the radioligand binding on receptors as compared to control conditions.

Representative examples of the effects GTPγS and PDC113.824 (an allosteric modulator of PGF2α receptor FP) on [^{3}H]-PGF2α binding to FP are shown in Figs. 9.2 and 9.3. Dissociation rates of [^{3}H]-PGF2α binding from FP were increased by almost threefold (K differed for each data sets, $P<0.001$) in the presence of GTPγS, suggesting a negative allosteric modulation of FP binding to its ligand by the functional uncoupling of the G protein to the receptor. Similarly, dissociation of [^{3}H]-PGF2α binding to FP is increased by the allosteric modulator PDC ($t_{1/2}$ of 2.55 min vs. 1.55, $P<0.05$). On the other hand, positive allosteric modulation of ligand binding to its receptor can be detected in the context of a receptor heterodimer, as shown in Fig. 9.4, where dissociation rate of [^{3}H]-PGF2α binding to FP increases ($t_{1/2}$ of 3.40 vs. 5.13 min, $P<0.001$) in the presence of the angiotensin II (Ang II) type I receptor (AT1R). Finally, this technique can also be used in an endogenous context, where overexpression of receptors is not required. When the affinity of a receptor for its ligand is increased by the presence of the receptor partner, $t_{1/2}$ will increase compared to the receptor of interest expressed alone. The opposite is also true. In both cases, if allosteric communication occurs between receptors, the

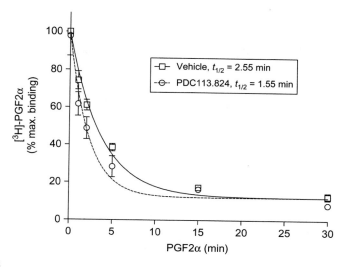

FIGURE 9.3

Allosteric modulation of orthosteric ligand binding to FP by a small molecule allosteric modulator. Dissociation kinetics of [^{3}H]-PGF2α from FP in the presence of a vehicle or 2 μM PDC113.284 (a small molecule allosteric modulator). Data are representative of three independent experiments.

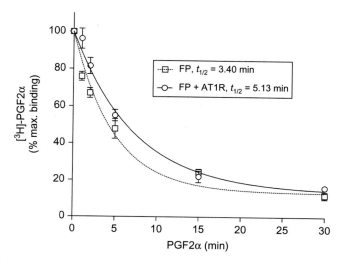

FIGURE 9.4

Allosteric modulation of orthosteric ligand binding to FP via dimerization with another GPCR, AT1R. Dissociation kinetics of (A) [^{3}H]-PGF2α binding from cells expressing FP alone or in the presence of the angiotensin II (Ang II) type I receptor (AT1R). Data are representative of four independent experiments.

comparison of $t_{1/2}$ values can validate these interactions. However, it is possible that no differences in $t_{1/2}$ values would be detected. Differences in the inhibition constant K_i might also be observed.

9.3 DISCUSSION

Here, we provide a simple method that can be added to the armamentarium to characterize allosteric interactions in GPCRs, in both heterologous or homologous expression systems as measured with many different types of potential allosteric modulators, including ligands, G protein, or dimer partners. The technique is not limited to GPCRs and should have general applicability for other receptors as well. The advantages and difficulties of the approach are summarized in Box 9.1.

Box 9.1
Advantages and Pitfalls of This Method

Advantages
- Our method provides a simple way to assess the presence of an allosteric receptor dimer or other allosteric effects with either known or new ligands or interacting proteins. Especially when performed using isolated membranes, a change in dissociation kinetics of the labeled orthosteric ligand can only be explained by an allosteric interaction between receptors.
- This technique is relatively easy to perform and could be used to simply detect the formation of a dimer, as immunoprecipitation of GPCRs can be a tedious process due to the instability of dimers in cell lysates or the poor specificity of antibodies directed against them.
- The technique can be used with endogenous receptors and native cells or primary tissues. No special equipment is required for cellular imaging. Validation of such allosteric interactions between receptors would, however, require depletion of one or both of the endogenous partners (e.g., siRNA or shRNA).
- Alternatively to adding an excess of unlabeled competitor (preferably the same as the orthosteric labeled one), in the reaction to promote the dissociation, one can also dilute the reaction medium (more than 10 times) to promote dissociation at different time point.

Pitfalls
- Our technique requires radioligands, which are often expensive, and certain risks are inherent in their use. However, high-affinity fluorescent ligands might work equally well.
- It is preferable to use the same orthosteric unlabeled ligand to displace the labeled one, in order to avoid the possibility that allosteric modulation influences binding of these two ligands differently. If not possible, the dilution method for assessing the kinetics of dissociation of the labeled ligand should be used.
- Allosterism or asymmetry in binding affinities may not necessarily transmitted to the cell in the same way; that is, a change in dissociation kinetics does not necessarily mean a similar change in signaling downstream of the receptor.
- False negatives might be observed; no change in half-time of ligand dissociation does not necessarily mean that there is no allosteric modulation; it may simply suggest another mechanism of action. It is advisable to examine receptor binding and signaling allostery from multiple vantage points.

Given the recent excitement about allosteric modulation of GPCRs, we will focus our discussion here and further on allosteric interactions in the context of GPCR dimers. Allosteric modulation within a receptor homodimer was elegantly demonstrated for the class A dopamine D_2 receptor (D_2R) (Han, Moreira, Urizar, Weinstein, & Javitch, 2009). In that study, the authors used different combinations of free receptor and receptor/G protein fusions to demonstrate these allosteric interactions. Importantly, it was shown that one protomer actually provided a transactivating signal to the fused G protein of the other protomer when the former was occupied by an agonist. Agonist occupation of the second protomer actually dampened signaling, likely through a mechanism involving negative cooperativity which had previously been demonstrated using hormone desorption experiments in other GPCRs (Guan et al., 2009, 2010; Urizar et al., 2005). Interestingly, as for the class C mGluR (Hlavackova et al., 2005), binding of an agonist to the first protomer of the dimer in conjunction with inverse agonist binding to the second protomer leads to the highest efficacy (Han et al., 2009). Perhaps most intriguingly, this study showed that, as in Class C GPCRs, class A homodimers may be arranged in an asymmetrical fashion with respect to the G protein. These findings need to be reexamined in the context of GPCR heterodimers as they have potentially important implications.

The assembly of asymmetric heterodimers or heterooligomers implies that allosteric machines may be constructed in a cell that respond to a single ligand in terms of signaling output but could be allosterically regulated by ligands binding to different heterodimer partners. If receptor/G protein complexes are in fact preassembled, prior to reaching the plasma membrane (reviewed in Dupré & Hébert, 2006; Dupré, Robitaille, Rebois, & Hébert, 2009), then different orientations of these machines might be constructed by reversing the specific asymmetric arrangement described above. Thus, two distinct, allosterically regulated receptors that respond as a single signaling unit, despite being a receptor heterodimer, may be regulated in distinct and cell-specific ways depending on how they are arranged with respect to each other. Thus, the formation of heterodimers could also lead to the formation of new signaling pathways, as demonstrated with D_1R/D_2R heterodimeric receptor complex. When expressed individually, these receptors do not couple to $G\alpha q$. However, when coexpressed, they were able to stimulate this pathway (Lee et al., 2004; Rashid et al., 2007). Taken together, these findings reveal the capacity of individual protomers to interpret and bias signals delivered to GPCRs and transmit it into the cell in a myriad of new ways. These notions will need to be accommodated in screens for biased and allosteric ligands in future.

It is likely that the dimer is the minimal unit of GPCR organization and that oligomers exist for most receptors. As shown in Fig. 9.1C and D, each GPCR protomer of a dimer is able to modulate its own conformation, when bound and when interacting with the other protomer of the complex. Each ligand might induce a specific conformation responsible for functional selectivity of signaling observed downstream. Several reports now discuss the notion of "receptor mosaics" that would each have specific functions and could be allosterically regulated by a number of unique signaling partners resident in any particular mosaic. Emerging imaging techniques such as resonance energy transfer (RET) or protein fragment complementation assays (PCA) are helping us understand the stoichiometry of these complexes (Ciruela, Vilardaga, & Fernandez-Duenas, 2010; Pétrin & Hébert, 2011; Vidi, Przybyla, Hu, & Watts, 2010). The diversity of responses induced by GPCRs is dependent not only on different types of ligands but also on the arrangement of GPCR protomers within larger oligomeric complexes. More importantly, the formation of oligomers can explain the signaling diversity and the results obtained from ligand-binding studies on native receptors expressed in tissue. Such heterooligomers, once fully characterized, may lead to the development of drugs selective for a given pathway in a given cell type, with fewer undesirable effects. These experiments can reveal other ligand-binding complexities in two ways: first, by examining additional ligands for the receptor partner and, second, by examining other dimer partners in the same context. The use of siRNA in endogenous context could reveal allostery with endogenous receptors.

SUMMARY

The measurement of ligand dissociation kinetics remains a unique way to determine allosteric effects of small ligands, protein partners, or receptor dimerization, including the symmetry of such interactions at the binding level. Since our capacity to use other techniques (immunoprecipitation, fluorescence or bioluminescence resonance energy transfer, etc.) to detect dimers is also limited to overexpression systems, using radioligand binding kinetics may help detect such interactions between receptors in endogenous systems using whole cells or purified membranes.

Acknowledgments

This study was supported by a Canadian Institutes of Health Research (CIHR) Team Grant in GPCR Allosteric Regulation [CTiGAR, CTP 79848], in which both T.E.H. and S.A.L are both coinvestigators, and CIHR grants to T.E.H. (CIHR; MOP-36379, MOP-123470) and S.A.L (CIHR, MOP-74603, MOP-123470). T.E.H. is a Chercheur National of the "Fonds de la Recherche en Santé du Québec" (FRSQ). S.A.L. is supported by a "Chercheur Senior" scholarship from FRSQ.

References

Bulenger, S., Marullo, S., & Bouvier, M. (2005). Emerging role of homo- and heterodimerization in G-protein-coupled receptor biosynthesis and maturation. *Trends in Pharmacological Sciences*, 26, 131–137.

Christopoulos, A. (2002). Allosteric binding sites on cell-surface receptors: Novel targets for drug discovery. *Nature Reviews Drug Discovery*, 1, 198–210.

Christopoulos, A., & Kenakin, T. (2002). G protein-coupled receptor allosterism and complexing. *Pharmacological Reviews*, 54, 323–374.

Ciruela, F., Vilardaga, J. P., & Fernandez-Duenas, V. (2010). Lighting up multiprotein complexes: Lessons from GPCR oligomerization. *Trends in Biotechnology*, 28, 407–415.

De Lean, A., Stadel, J. M., & Lefkowitz, R. J. (1980). A ternary complex model explains the agonist-specific binding properties of the adenylate cyclase-coupled β-adrenergic receptor. *The Journal of Biological Chemistry*, *255*, 7108–7117.

Dupré, D. J., & Hébert, T. E. (2006). Biosynthesis and trafficking of seven transmembrane receptor signalling complexes. *Cellular Signalling*, *18*, 1549–1559.

Dupré, D. J., Robitaille, M., Rebois, R. V., & Hébert, T. E. (2009). The role of Gβγ subunits in the organization, assembly, and function of GPCR signaling complexes. *Annual Review of Pharmacology and Toxicology*, *49*, 31–56.

Goupil, E., Tassy, D., Bourguet, C., Quiniou, C., Wisehart, V., Pétrin, D., et al. (2010). A novel biased allosteric compound inhibitor of parturition selectively impedes the prostaglandin F2α-mediated Rho/ROCK signaling pathway. *The Journal of Biological Chemistry*, *285*, 25624–25636.

Guan, R., Feng, X., Wu, X., Zhang, M., Zhang, X., Hébert, T. E., et al. (2009). Bioluminescence resonance energy transfer studies reveal constitutive dimerization of the human lutropin receptor and a lack of correlation between receptor activation and the propensity for dimerization. *The Journal of Biological Chemistry*, *284*, 7483–7494.

Guan, R., Wu, X., Feng, X., Zhang, M., Hébert, T. E., & Segaloff, D. L. (2010). Structural determinants underlying constitutive dimerization of unoccupied human follitropin receptors. *Cellular Signalling*, *22*, 247–256.

Han, Y., Moreira, I. S., Urizar, E., Weinstein, H., & Javitch, J. A. (2009). Allosteric communication between protomers of dopamine class A GPCR dimers modulates activation. *Nature Chemical Biology*, *5*, 688–695.

Hébert, T. E., & Bouvier, M. (1998). Structural and functional aspects of G protein-coupled receptor oligomerization. *Biochemistry and Cell Biology*, *76*, 1–11.

Hlavackova, V., Goudet, C., Kniazeff, J., Zikova, A., Maurel, D., Vol, C., et al. (2005). Evidence for a single heptahelical domain being turned on upon activation of a dimeric GPCR. *The EMBO Journal*, *24*, 499–509.

Kamal, M., & Jockers, R. (2009). Bitopic ligands: All-in-one orthosteric and allosteric. *F1000 Biology Reports*, *1*, 77.

Kenakin, T. (2010). G protein coupled receptors as allosteric proteins and the role of allosteric modulators. *Journal of Receptor and Signal Transduction Research*, *30*, 313–321.

Kenakin, T., & Miller, L. J. (2010). Seven transmembrane receptors as shapeshifting proteins: The impact of allosteric modulation and functional selectivity on new drug discovery. *Pharmacological Reviews*, *62*, 265–304.

Keov, P., Sexton, P. M., & Christopoulos, A. (2011). Allosteric modulation of G protein-coupled receptors: A pharmacological perspective. *Neuropharmacology*, *60*, 24–35.

Kniazeff, J., Prezeau, L., Rondard, P., Pin, J. P., & Goudet, C. (2011). Dimers and beyond: The functional puzzles of class C GPCRs. *Pharmacology & Therapeutics*, *130*, 9–25.

Kostenis, E., & Mohr, K. (1996). Two-point kinetic experiments to quantify allosteric effects on radioligand dissociation. *Trends in Pharmacological Sciences*, *17*, 280–283.

Lanzafame, A., & Christopoulos, A. (2004). Investigation of the interaction of a putative allosteric modulator, N-(2,3-diphenyl-1,2,4-thiadiazole-5-(2H)-ylidene) methanamine hydrobromide (SCH-202676), with M1 muscarinic acetylcholine receptors. *The Journal of Pharmacology and Experimental Therapeutics*, *308*, 830–837.

Lee, S. P., So, C. H., Rashid, A. J., Varghese, G., Cheng, R., Lanca, A. J., et al. (2004). Dopamine D1 and D2 receptor Co-activation generates a novel phospholipase C-mediated calcium signal. *The Journal of Biological Chemistry*, *279*, 35671–35678.

Limbird, L. E. (2004). The receptor concept: A continuing evolution. *Molecular Interventions*, *4*, 326–336.

Litschig, S., Gasparini, F., Rueegg, D., Stoehr, N., Flor, P. J., Vranesic, I., et al. (1999). CPCCOEt, a noncompetitive metabotropic glutamate receptor 1 antagonist, inhibits receptor signaling without affecting glutamate binding. *Molecular Pharmacology*, *55*, 453–461.

Maillet, E. L., Pellegrini, N., Valant, C., Bucher, B., Hibert, M., Bourguignon, J. J., et al. (2007). A novel, conformation-specific allosteric inhibitor of the tachykinin NK2 receptor (NK2R) with functionally selective properties. *The FASEB Journal*, *21*, 2124–2134.

May, L. T., Leach, K., Sexton, P. M., & Christopoulos, A. (2007). Allosteric modulation of G protein-coupled receptors. *Annual Review of Pharmacology and Toxicology*, *47*, 1–51.

Milligan, G. (2009). G protein-coupled receptor hetero-dimerization: Contribution to pharmacology and function. *British Journal of Pharmacology*, *158*, 5–14.

Palanche, T., Ilien, B., Zoffmann, S., Reck, M. P., Bucher, B., Edelstein, S. J., et al. (2001). The neurokinin A receptor activates calcium and cAMP responses through distinct conformational states. *The Journal of Biological Chemistry*, *276*, 34853–34861.

Peeters, M. C., van Westen, G. J., Guo, D., Wisse, L. E., Muller, C. E., Beukers, M. W., et al. (2011). GPCR structure and activation: An essential role for the first extracellular loop in activating the adenosine A2B receptor. *The FASEB Journal*, *25*, 632–643.

Peri, K. G., Quiniou, C., Hou, X., Abran, D., Varma, D. R., Lubell, W. D., et al. (2002). THG113: A novel selective FP antagonist that delays preterm labor. *Seminars in Perinatology*, *26*, 389–397.

Pétrin, D., & Hébert, T. E. (2011). Imaging-based approaches to understanding G protein-coupled receptor signalling complexes. *Methods in Molecular Biology*, *756*, 37–60.

Prinster, S. C., Hague, C., & Hall, R. A. (2005). Heterodimerization of g protein-coupled receptors: Specificity and functional significance. *Pharmacological Reviews*, *57*, 289–298.

Proska, J., & Tucek, S. (1994). Mechanisms of steric and cooperative actions of alcuronium on cardiac muscarinic acetylcholine receptors. *Molecular Pharmacology*, *45*, 709–717.

Rashid, A. J., So, C. H., Kong, M. M., Furtak, T., El-Ghundi, M., Cheng, R., et al. (2007). D1-D2 dopamine receptor heterooligomers with unique pharmacology are coupled to rapid activation of Gq/11 in the striatum. *Proceedings of the National Academy of Sciences of the United States of America*, *104*, 654–659.

Terrillon, S., & Bouvier, M. (2004). Roles of G-protein-coupled receptor dimerization. *EMBO Reports*, *5*, 30–34.

Urizar, E., Montanelli, L., Loy, T., Bonomi, M., Swillens, S., Gales, C., et al. (2005). Glycoprotein hormone receptors: Link between receptor homodimerization and negative cooperativity. *The EMBO Journal*, *24*, 1954–1964.

Urwyler, S. (2011). Allosteric modulation of family C G-protein-coupled receptors: From molecular insights to therapeutic perspectives. *Pharmacological Reviews*, *63*, 59–126.

Valant, C., Maillet, E., Bourguignon, J. J., Bucher, B., Utard, V., Galzi, J. L., et al. (2009). Allosteric functional switch of neurokinin A-mediated signaling at the neurokinin NK2 receptor: Structural exploration. *Journal of Medicinal Chemistry*, *52*, 5999–6011.

Valant, C., Sexton, P. M., & Christopoulos, A. (2009). Orthosteric/allosteric bitopic ligands: Going hybrid at GPCRs. *Molecular Interventions*, *9*, 125–135.

Varney, M. A., Cosford, N. D., Jachec, C., Rao, S. P., Sacaan, A., Lin, F. F., et al. (1999). SIB-1757 and SIB-1893: Selective, noncompetitive antagonists of metabotropic glutamate receptor type 5. *The Journal of Pharmacology and Experimental Therapeutics*, *290*, 170–181.

Vauquelin, G., Van Liefde, I., Birzbier, B. B., & Vanderheyden, P. M. (2002). New insights in insurmountable antagonism. *Fundamental & Clinical Pharmacology, 16*, 263–272.

Vauquelin, G., Van Liefde, I., & Vanderheyden, P. (2002). Models and methods for studying insurmountable antagonism. *Trends in Pharmacological Sciences, 23*, 514–518.

Vidi, P. A., Przybyla, J. A., Hu, C. D., & Watts, V. J. (2010). Visualization of G protein-coupled receptor (GPCR) interactions in living cells using bimolecular fluorescence complementation (BiFC). *Current Protocols in Neuroscience*, Chapter 5, Unit 5, 29.

Whorton, M. R., Bokoch, M. P., Rasmussen, S. G., Huang, B., Zare, R. N., Kobilka, B., et al. (2007). A monomeric G protein-coupled receptor isolated in a high-density lipoprotein particle efficiently activates its G protein. *Proceedings of the National Academy of Sciences of the United States of America, 104*, 7682–7687.

Whorton, M. R., Jastrzebska, B., Park, P. S., Fotiadis, D., Engel, A., Palczewski, K., et al. (2008). Efficient coupling of transducin to monomeric rhodopsin in a phospholipid bilayer. *The Journal of Biological Chemistry, 283*, 4387–4394.

CHAPTER 10

Fluorescence Correlation Spectroscopy and Photon-Counting Histogram Analysis of Receptor–Receptor Interactions

Katharine Herrick-Davis and Joseph E. Mazurkiewicz
Center for Neuropharmacology & Neuroscience, Albany Medical College, Albany, New York, USA

CHAPTER OUTLINE

Introduction	182
10.1 Materials	184
10.2 Methods	185
10.2.1 Sample Preparation	185
10.2.1.1 Choice of Cell Type	185
10.2.1.2 Choice of Fluorescent Tag	185
10.2.1.3 Labeling the Receptor	185
10.2.1.4 Plating Cells	186
10.2.1.5 Transfecting Cells	186
10.2.2 Instrument Setup	186
10.2.2.1 Environment	186
10.2.2.2 Imaging Setup	186
10.2.2.3 FCS Setup	186
10.2.2.4 Instrument Alignment and Calibration	187
10.2.3 Data Acquisition	188
10.2.3.1 FCS Recording	188
10.2.4 Data Analysis	190
10.2.4.1 Diffusion Coefficient	190
10.2.4.2 Molecular Brightness	192
10.3 Discussion	194
Summary	195
Acknowledgments	195
References	195

Abstract

Fluorescence correlation spectroscopy (FCS) performed using a laser scanning confocal microscope is a technique with single-molecule sensitivity that is becoming more accessible to cell biologists. In this chapter, we describe the use of FCS for the analysis of diffusion coefficients and receptor–receptor interactions in live cells in culture. In particular, we describe a protocol to collect fluorescence fluctuation data from fluorescence-tagged receptors as they diffuse into an out of a small laser-illuminated observation volume using a commercially available system such as the Zeiss ConfoCor 3 or LSM-780 microscope. Autocorrelation analysis of the fluctuations in fluorescence intensity provides information about the diffusion time and number of fluorescent molecules in the observation volume. A photon-counting histogram can be used to examine the relationship between fluorescence intensity and the number of fluorescent molecules to estimate the average molecular brightness of the sample. Since molecular brightness is directly proportional to the number of fluorescent molecules, it can be used to monitor receptor–receptor interactions and to decode the number of receptor monomers present in an oligomeric complex.

INTRODUCTION

Fluorescence correlation spectroscopy (FCS) is a single-molecule detection technique that measures the fluorescence fluctuations of molecules diffusing through a well-defined volume. Introduced over 40 years ago by Magde, Elson, and Webb (1972), one of the initial applications of FCS was to analyze the interaction of ethidium bromide with DNA in solution to measure diffusion and chemical reaction kinetics. The application of FCS to address such questions in live cells, especially the measurement of diffusion of cell surface proteins in biological membranes, was limited by the lack of sufficiently sensitive instrumentation, stable lasers for excitation and a means to reduce the volume in which the measurements were made. Most of these concerns were addressed in the early 1990s with the adoption of the confocal microscope for FCS measurements, providing a sensitive method for monitoring protein dynamics in living cells (reviewed in Elson, 2013).

The main advantage of FCS over other currently used techniques for monitoring receptor interactions is that it provides real-time information about the two-dimensional dynamics of single molecules diffusing within a plasma membrane with diffraction-limited spatial and sub-microsecond temporal resolution. In addition, the most accurate FCS measurements are made in samples with very low protein expression levels, making this technique ideal for monitoring receptors at physiological expression levels. Confocal microscopy-based FCS experiments are performed by focusing a laser beam into a small diffraction-limited spot (0.3 μm) using a high numerical aperture objective to create an observation volume on the order of 0.5×10^{-15} L (Fig. 10.1).

Introduction

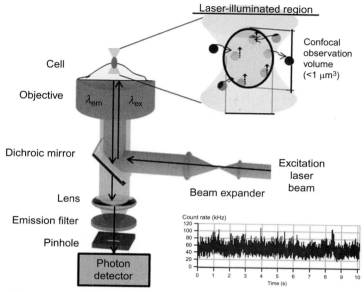

FIGURE 10.1

FCS setup. A laser and a high numerical aperture objective are used to create a very small observation volume focused on the upper plasma membrane of a cell. As fluorescent molecules pass through the observation volume and are excited by the laser, the emitted fluorescence is captured by the objective and is focused through a pinhole onto a photon detector, which records the fluctuations in fluorescence over time.

As the fluorescence-tagged receptors pass through the laser-illuminated observation volume, the fluctuations in fluorescence intensity are recorded in real time by a photon-counting detector. Autocorrelation analysis depicts the fluctuations in fluorescence intensity as a function of particle number and diffusion time (Müller, Chen, & Gratton, 2003). FCS has been used to monitor diffusion and ligand binding for ion channels, tyrosine kinase receptors, and G-protein-coupled receptors (reviewed in Briddon & Hill, 2008; Briddon, Hern, & Hill, 2010; Jameson, James, & Albanesi, 2013).

Fluorescence fluctuation data recorded during an FCS experiment can be used to generate a photon-counting histogram (PCH), providing quantitative information about the number of fluorescent molecules and the number of photon counts per molecule. PCH analysis uses a 3D Gaussian approximation of the laser beam profile and Poisson statistics to predict what the molecular brightness of the fluorescent particle would be when it is at the center of the observation volume (Chen, Muller, So, & Gratton, 1999). Since the molecular brightness of a cluster is directly proportional to the number of fluorescent molecules present in the cluster, molecular brightness can be used to monitor receptor–receptor interactions and to determine the oligomer size of protein complexes. Molecular brightness analysis has been used to explore the oligomeric status of nuclear retinoid X receptors (Chen, Wei, & Muller, 2003), epidermal growth factor receptors (Saffarian, Li, Elson, & Pike, 2007), urokinase plasminogen activator receptors (Malengo et al., 2008), and biogenic amine receptors (Herrick-Davis, Grinde, Cowan, & Mazurkiewicz, 2013; Herrick-Davis, Grinde, Lindsley, Cowan, & Mazurkiewicz, 2012).

This chapter focuses on the application of FCS and PCH for monitoring receptor–receptor interactions. The methods described are for a confocal microscope setup using one photon excitation. Using a commercially available confocal microscope fully equipped for FCS data collection and analysis is the most straightforward approach. A homebuilt FCS system is an option for those who are well acquainted with biophysical techniques (Müller et al., 2003). The first commercial FCS instrument was introduced in 1997 by Zeiss and called the ConfoCor 2 and, in 2000, was combined with a Zeiss LSM 510 as a single integrated instrument platform (Weisshart, Jungel, & Briddon, 2004). Since that time, all the major microscope manufacturers, Nikon, Olympus, and Leica, offer FCS/LSM systems. The methods and data presented in this chapter are based upon our experience with a Zeiss ConfoCor 3 avalanche photodiode detection module mounted on an LSM-510 microscope and with the Zeiss LSM-780 microscope integrated with a gallium arsenide phosphide photon detector.

10.1 MATERIALS

1. Confocal laser scanning microscope (Zeiss, Nikon, Lecia, Olympus)
   The instrument should be equipped with a stable excitation source (argon ion, HeNe laser), a 40 × 1.2 NA C-apochromat water immersion objective, avalanche photodiode or gallium arsenide phosphide photon-counting detector, and autocorrelation analysis software (Zeiss Aim or Zen; Origin; MATLAB).
2. Cells
   HEK293 or CHO for transfection or primary cultures expressing native receptor of interest, along with the appropriate culture medium and culture dishes for growing cells.
3. Fluorescent tag
   A GFP variant attached to the receptor (most common), a fluorescent ligand, or a fluorescent monovalent antibody.
4. Plasmids
   Monomeric (CD-86) and dimeric (CD-28) plasma membrane proteins with GFP or YFP attached to the C-terminus are good controls for molecular brightness analysis of plasma membrane receptors. Monomeric GFP or YFP, expressed in the cell cytoplasm, for adjusting the pinhole alignment prior to the start of each experiment.
5. Transfection reagents
   Lipofectamine reagent, calcium phosphate, or electroporation.
6. Chambers for plating cells
   Use 25 mm glass coverslips (thickness no. 1.0 or 1.5) inserted into a six-well plate or use 35–50 mm dishes with glass bottoms (MatTek). If using coverslips, obtain a

viewing chamber to hold the coverslip (35 mm Attofluor cell chamber, Invitrogen) and ultra fine tweezers for mounting coverslips into the viewing chamber.

7. Substrate to coat glass coverslips for cell adhesion
   Examples include poly-D-lysine (HEK293 or CHO cells), laminin (neuronal cells), fibronectin/collagen (smooth muscle cells), or purchase precoated MatTek dishes.
8. A dilute solution (20 nM) of rhodamine 6G, fluorescein (pH > 7.5), or GFP for instrument calibration. HPLC-grade water for dilution of dye and for placing on the $40\times$ water immersion objective
9. Isotonic solution
   For washing cells and performing FCS experiments, use phosphate-buffered saline, Krebs' ringer solution, or phenol red-free minimal essential media (MEM, CellGro). Add 10 mM HEPES to the solution to keep pH constant during FCS recording. Avoid substances with fluorescence such as phenol.
10. Ethanol and lens cleaning solution
    Clean the bottom of the coverslip or MatTek dish immediately prior to the FCS experiment. Use ethanol on a Kimwipe to clean glass and wipe dry, and then use a lens cleaning solution and wipe dry.

10.2 METHODS

10.2.1 Sample preparation

10.2.1.1 Choice of cell type

HEK293 and CHO cells are cell types that work nicely for FCS experiments. Primary cell cultures, such as neurons or epithelial cells, endogenously expressing the receptor of interest may be used. However, this requires a suitable fluorescent tag capable of labeling native receptors in an exact 1:1 stoichiometry. Membrane stability is important as movement of the membrane within the observation volume during an FCS recording will directly impact the diffusion time and molecular brightness.

10.2.1.2 Choice of fluorescent tag

The ideal fluorescent tag has a large quantum yield (brightness), has a good photostability, and is monomeric. The higher the quantum yield, the better the signal to noise ratio. Photostability is important since photobleaching will decrease fluorescence intensity, producing an artificially fast diffusion time as bleaching will mimic the disappearance of the fluorescence signal from the observation volume. GFP variants, such as eGFP and eYFP, and the newer variants are suitable for FCS. Others, such as CFP, mCherry, and dsRed, have low quantum yield and have the potential to form aggregates (dsRed).

10.2.1.3 Labeling the receptor

Plasmids containing GFP variants are useful for attaching fluorescent labels to the C-terminus of the receptor of interest. This is a commonly employed method to ensure an exact 1:1 labeling of the receptor with the fluorescent probe, which is critical for molecular brightness analysis of receptor–receptor interactions. Fluorescent ligands have been used to label receptors for monitoring diffusion and binding kinetics (Briddon et al., 2010) but would require very slow dissociation kinetics in order to be useful for molecular brightness analysis. Alternatively, a monoclonal, monovalent Fab, recognizing the native conformation of the receptor, could be used provided each Fab has exactly one fluorescent tag.

10.2.1.4 Plating cells
Seed cells at 5×10^5 cells per 25 mm coverslip or MatTek dish (coated with appropriate substrate) 12–18 h prior to transfection or following electroporation (40% confluence). Note that the type of substrate will affect cell attachment and overall cell shape, which has important implications for making FCS recordings from the plasma membrane.

10.2.1.5 Transfecting cells
Lipofectamine, calcium phosphate, and electroporation are common transfection methods. End the transfection in phenol red-free culture medium, as phenol has autofluorescence. Wait 24 h prior to the start of the FCS experiment to allow time for the receptors to reach the plasma membrane. Fluorescent receptors in the ER/Golgi or in vesicles can enter the observation volume complicating interpretation of plasma membrane FCS results. If the endogenous ligand for the chosen receptor is present in serum, use media with dialyzed serum to minimize receptor internalization.

10.2.2 Instrument setup
10.2.2.1 Environment
The room should be maintained at constant temperature during FCS recording (68–70 °F). Detector sensitivity can decrease at high room temperatures, and changes in temperature surrounding the sample can affect diffusion time.

10.2.2.2 Imaging setup
Turn on the instrument, allow 30 min for the laser to stabilize, and establish the proper settings for the fluorescent probe. Choose the appropriate laser line, dichroic mirror, and set the appropriate emission range to capture as much of the emission spectrum as possible. Select the fastest scan speed using 12 bits with 514×514 resolution, high detector gain (1100), pinhole of 1 airy unit, and low laser intensity (1%) to minimize photobleaching.

10.2.2.3 FCS setup
Choose the appropriate laser line, dichroic mirror or main bean splitter and set the appropriate emission range to capture as much of the emission spectrum as possible. Use a pinhole diameter of 1 airy unit and the lowest laser setting that gives good signal to noise with minimal photobleaching. Note that laser intensity will depend on the

sensitivity of the instrument (on the Zeiss ConfoCor 3, we use 1%, and on the LSM-780, we use 0.1%) and must be determined experimentally. Select a correlation bin time of 0.2 μs, PCH bin time of 10 μs, and data acquisition time as 10 s for 10 consecutive repetitions (runs). For proper signal statistics, the acquisition time should be ~1000-fold over the diffusion time of the protein being studied. For membrane receptors, with diffusion times on the order of 10–100 ms, a total acquisition time of 100 s is sufficient. It is essential to use the same settings for all FCS recordings.

10.2.2.4 Instrument alignment and calibration

Align the light beam path and pinhole, and determine the dimensions of the observation volume.

a. Clean the lens of the 40× objective and adjust the collar to match coverslip thickness (0.17 mm for no.1.5 and 0.15 mm for no.1.0 coverslip thickness). Place a drop of HPLC-grade (ultrapure, low fluorescence) water on the objective lens.
b. Prepare a 20 nM dilution of calibration dye, such as rhodamine 6G, fluorescein (pH > 7.5), or GFP. Place a 100 μl drop on a coverslip or MatTek dish. Be aware that sample evaporation will concentrate the sample and introduce error into the calibration process.
c. Place the holder with dye solution on the microscope stage and raise the objective until the water just contacts the surface of the coverslip. Adjust the focus upward until the focal plane of the laser is in the middle of the dye solution.
d. Using the imaging and FCS settings established in the preceding text (for the fluorescent probe used to label your receptor of interest, not the calibration dye), capture an image of the dye solution. In the imaging setup menu, set the zoom to 3 and capture another image. Locate the "region of interest" marker and position the marker in the center of the screen. Perform pinhole adjustments for the X and Y planes.
e. Begin an FCS recording and adjust the laser setting to give a count rate of approximately 50 kHz. Begin an FCS recording for 10 consecutive 10 s intervals. Repeat. Fit the data to a 3D model for Brownian diffusion and perform an autocorrelation analysis. The autocorrelation curve represents the time-dependent decay in the fluorescence signal. The midpoint of the curve provides a measure of τ_D (the average dwell time of the fluorescent particles in the observation volume). Use τ_D to calculate the radius of the observation volume (ω_0) as in Eq. (10.1):

$$\omega_0 = \sqrt{4D\tau_D} \qquad (10.1)$$

where D is the known diffusion coefficient of the calibration dye (rhodamine 6G, or fluorescein, 400 μm^2/s; GFP in solution, 87 μm^2/s). The theoretical value of ω_0 is calculated as $0.61 \times \lambda/\text{NA}$ where λ is the laser wavelength and NA is the numerical aperture of the objective. The number of molecules in the observation volume (N) is calculated as in Eq. (10.2):

$$N = \frac{1}{G(0)} \cdot \gamma \qquad (10.2)$$

where $G(0)$ is the amplitude of the autocorrelation curve (y-intercept) and γ is the point spread function describing the shape of the observation volume (for calibration dye, use $\gamma = 0.35$ for a 3D Gaussian confocal observation volume). Commercially available software designed for FCS analysis will automatically display $G(0)$, the number of molecules, the count rate, and the diffusion time in a table. Use the number of molecules (N) to calculate the confocal volume (V) as in Eq. (10.3):

$$V = \frac{N}{C \cdot Na} \qquad (10.3)$$

where C is the concentration of calibration dye (20 nM) and Na is Avogadro's number. Finally, calculate the structural parameter (s, the ratio of the axial length to radial width of the observation volume) as in Eq. (10.4):

$$s = \frac{V}{\pi \omega_0^2} \qquad (10.4)$$

Successful instrument alignment and calibration will produce a structural parameter between 3 and 8.

10.2.3 Data acquisition

10.2.3.1 FCS recording

Always perform a pinhole adjustment using a sample of the fluorescent probe chosen to label the receptor of interest before beginning an FCS recording session. Begin and end each session with control samples and perform FCS recordings on the isotonic viewing solution and in the middle of untransfected cells to establish autofluorescence levels.

a. Prepare a coverslip of control cells transfected with plasmid containing the chosen fluorescent probe (e.g., monomeric GFP or YFP expressed in the cytosol) by washing three times using an isotonic solution such as PBS, Krebs' ringer, or phenol red-free MEM. Place the coverslip in a viewing chamber and add 1 ml of room temperature isotonic solution buffered with 10 mM HEPES. Place a drop of HPLC-grade water on the 40× objective, place sample on the microscope stage, and raise the objective until the waterdrop touches the coverslip. Working quickly, illuminate the sample and focus upward until the cells are visible, and then turn off. It is important to minimize illumination and viewing of the cells to minimize photobleaching.

b. Collect an image of the cells and display on the computer screen. Choose a cell of low to medium brightness (avoid bright cells) and adjust the zoom setting to 3. Working quickly, use fast continuous scanning to position the cell in the center of the screen. For FCS measurements in the cytosol, place the region of interest marker near the center of the cell, just to one side of the nucleus where the cell is likely to be tallest, and fill the observation volume in the axial direction. Perform a pinhole adjustment in X and Y and note the settings.

c. To establish the proper laser power setting, begin an FCS recording with low intensity (0.1–1%). Note the extent of photobleaching that occurs during the recording. Repeat the process on different cells testing different laser powers and monitor the extent of photobleaching. Select the lowest laser power setting that gives minimal photobleaching (no leftward shift in the autocorrelation curve) but still gives a good signal from the fluorescent control sample and low background fluorescence. Use this laser setting for all subsequent experiments.

d. Perform an FCS recording (for 10×10 s intervals) in the cytosol of five different cells, saving the data files after each run for subsequent analysis at a later time. If the software allows, monitor the counts per molecule during the recording to get an idea of the molecular brightness of the control sample.

e. If the receptor of interest is a plasma membrane receptor, prepare a coverslip with control cells expressing a known plasma membrane monomeric or dimeric control such as CD-86 or CD-28, respectively. Quickly capture an image of the cells (Fig. 10.2).

Choose a cell of low to medium brightness (count rate of 50–250 kHz), and adjust the zoom setting to 3. Working quickly, use fast continuous scanning to position the cell in the center of the screen and focus on the upper plasma membrane. Mark a region on the upper plasma membrane (over the nucleus) and begin an FCS recording. During the first 10 s interval there will be a dramatic photobleaching of the nonmobile fraction of receptors that will appear as a rapid decline in the count rate. During the subsequent 10 s intervals, it will be necessary to adjust the focal plane of the sample

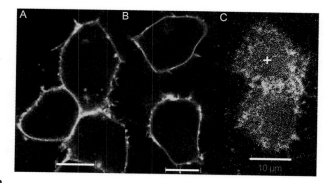

FIGURE 10.2

Confocal image of YFP-tagged receptors in the plasma membrane of HEK293 cells. (A) Beta-adrenergic receptors; (B) CD-28 receptors; (C) upper plasma membrane of two transfected cells with + marking the region where an FCS recording was made. Scale bar = 10 μm. (For color version of this figure, the reader is referred to the online version of this chapter.)

up and down (along the z-axis) to get the plasma membrane positioned in the center of the observation volume. Monitor the counts per molecule while making small adjustments, in 0.1 μm steps, along the z-axis. Find the position in z corresponding to the maximal counts per molecule, and then make a 10 by 10 s FCS recording from that position. Save the data file and move to a different region of the coverslip, select a new cell, and begin again. Each time, adjust the sample focus along the z-axis to find the optimal positioning of the membrane in the observation volume. Make recordings from at least 5 to 10 different cells. Repeat the process for the tagged receptors of interest and end the session with another control sample. Repeat the process with freshly prepared cells on two additional test days.

10.2.4 Data analysis

Commercially available systems can be purchased with autocorrelation analysis software. If using a homebuilt system, software such as Origin or MATLAB can be used.

10.2.4.1 Diffusion coefficient

As the fluorescence-tagged receptors pass through the laser-illuminated observation volume, the fluctuations in fluorescence intensity are recorded in real time by the photon-counting detector, and a fluorescence intensity trace for the observation period is generated (Fig. 10.3A). During the first two 10 s intervals of the 100 s FCS recording time, photobleaching of the immobile fraction of plasma membrane receptors occurs. For G-protein-coupled receptors, this typically represents 40–50% of the receptor pool. Data analysis is performed on the mobile fraction of receptors monitored in the third through tenth 10 s FCS recording intervals (runs 3–10). Autocorrelation analysis of the fluorescence signal is performed as in Eq. (10.5):

$$G(\tau) = \frac{\langle \delta F(t) \cdot \delta F(t+\tau) \rangle}{\langle F(t) \rangle^2} \quad (10.5)$$

where $G(\tau)$ is the ⟨time average⟩ of the change in fluorescence fluctuation intensity (δF) at some time point (t) and at a time interval later ($t+\tau$) divided by the square of the average fluorescence intensity. Autocorrelation analysis of the fluorescence intensity trace is performed using a nonlinear least-squares fitting routine that graphically represents the autocorrelation function $G(\tau)$ on the ordinate and diffusion time on the abscissa (Fig. 10.3B).

The rate at which the fluorescence-tagged receptor diffuses within the plasma membrane is reported as the average dwell time (τ_D) within the observation volume and is calculated from the midpoint of the autocorrelation decay curve. For autocorrelation analysis, select a 2D model for plasma membrane receptors or a 3D model for cytosolic receptors. Most autocorrelation curves will have a minimum of two components, a very fast component (<1 ms) related to the photophysical properties of the fluorescent probe, and a slower component representing the diffusion of the fluorescence-tagged receptor. Begin by fitting the data to a two-component model. The examples shown in Fig. 10.3B are best fit by a two-component, 2D model with

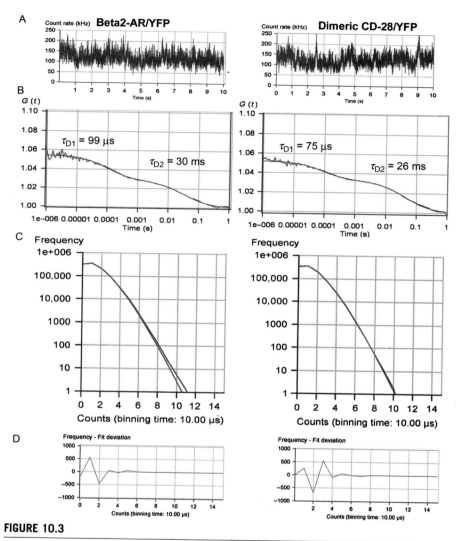

FIGURE 10.3

Sample FCS recordings for YFP-tagged beta2-adrenergic and dimeric CD-28 receptors diffusing within the plasma membrane of HEK293 cells. (A) fluorescence intensity trace; (B) autocorrelation analysis of the fluorescence intensity trace; (C) photon-counting histogram; (D) residuals of the curve fit. (For color version of this figure, the reader is referred to the online version of this chapter.)

a very fast component characteristic of the photophysical properties of YFP (50–100 μs) and a slower component representing the translational diffusion of the receptor within the plasma membrane. Once the diffusion time (τ_D) for the receptor has been determined, then the diffusion coefficient (D) can be calculated using

Eq. (10.1) as described in the preceding text. Diffusion rates for different types of plasma membrane receptors are summarized in Briddon & Hill, 2008.

10.2.4.2 Molecular brightness

The average photon count rate (k) of a given sample is determined by the number of fluorescent molecules (N) and their molecular brightness (ε) as in Eq. (10.6):

$$k = N \cdot \varepsilon \qquad (10.6)$$

Thus, dividing the photon count rate (in kHz) by the number of molecules (N) yields the molecular brightness (ε) of the sample expressed as counts per second per molecule (CPSM). Note that the absolute numerical values of molecular brightness will vary depending on whether a 2D or 3D model is used to calculate N in Eq. (10.2). This is due to the different numerical values of γ used for 2D samples (where $\gamma = 0.5$) versus 3D samples (where $\gamma = 0.35$). Even though the surface of a cell may be 3D, the plasma membrane is only 5 nm thick and doesn't fill the observation volume in the axial direction (~ 1 μm). Thus, a plasma membrane is 2D with respect to the FCS observation volume. However, as long as the same model is used for control and unknown samples, the choice of 2D versus 3D for molecular brightness analysis of plasma membrane receptors becomes less critical. This is illustrated in the example provided in the succeeding text:

a. FCS: FCS software packages automatically calculate the counts per molecule (molecular brightness) by dividing the photon count rate by the number of molecules. A 2D analysis of the FCS data presented in Fig. 10.3 is performed by dividing the average count rate obtained from the fluorescence intensity trace in Fig. 10.3A (130 kHz) by the number of molecules calculated as in Eq. (10.2) where $G(0)$ is the amplitude (y-intercept) of the autocorrelation curve shown in Fig. 10.3B ($N = 10$). This yields a molecular brightness of 13,000 CPSM. In this example, the beta2-adrenergic receptor and the dimeric CD-28 control have similar molecular brightness, indicating that beta2-adrenergic receptors form homodimers. Membrane receptor expression level can be determined by dividing the number of fluorescence-tagged receptors in the observation volume (N, calculated using Eq. (10.2) where $\gamma = 0.5$) by the area of the observation volume calculated as $\pi \omega_0^2$ (where ω_0 is the radius of the observation volume determined experimentally using Eq. (10.1)).

b. PCH: Software packages with PCH modules are available commercially. PCH analysis uses a 3D Gaussian approximation of the laser beam profile and Poisson statistics to predict what the molecular brightness of the fluorescent particle would be when it is at the center of the observation volume (Chen et al., 1999). For PCH analysis, choose cells with a count rate of 50–250 kHz and use a bin time of 10 μs yielding 0.5–2.5 counts per bin (a good range for molecular brightness analysis). If studying membrane receptors, photobleaching of the immobile fraction will occur during the first two 10 s runs of the 100 s observation period. Therefore, molecular brightness values from runs 3–10 are averaged.

Segments that show large spikes or drifts in fluorescence intensity (due to cell movement) are excluded from the analysis. Since a constant intensity light source produces a photon count distribution that follows Poisson statistics, as fluorescent molecules enter and diffuse through the nonhomogenously illuminated observation volume, the fluctuations in fluorescence intensity result in a broadening of the Poisson distribution. This super-Poisson characteristic is observed in the tail of PCH curve.

Sample histograms are shown in Fig. 10.3C. The shape of the histogram is a function of the number of fluorescent molecules and their molecular brightness. To generate a histogram, each 10 s fluorescence intensity trace is broken down into 1 million 10 μs intervals or bins. The number of 10 μs bins is plotted on the y-axis and photon counts on the x-axis, revealing the number of bins that register 1,2,3, . . .,n photon counts during one 10 s observation period. Both histograms show an average of 1.25 counts per 10 μs bin time, which equals 125,000 counts per second. Dividing by the average number of molecules (calculated from the amplitude of the autocorrelation curves in Fig. 10.3B using Eq. (10.2) with a 3D model, where $N=7$) yields an average molecular brightness of 17,857 CPSM. Histograms for monomeric YFP (not shown) yield a molecular brightness of 8750 CPSM. In this manner, comparison of molecular brightness values with known controls can be used to determine receptor–receptor interactions. Even though the 2D FCS and 3D PCH models yielded different numerical values, both predicted a dimeric structure for the beta2-adrenergic receptor.

The residuals of the curve fit (Fig. 10.3D) provide a measure of how well the data fit the selected model. In the example provided, the data were fit to a one-component model for a homogenous population of fluorescence-tagged receptors and yielded reduced chi-square values equal to unity. The residuals of the curve fit are <2 standard deviations and are randomly distributed about zero, indicating that the data are a good fit for the selected model. When the residuals are systematic, non-uniformly distributed about zero, the data are not adequately described by the selected model.

PCH provides an estimate of the average molecular brightness of all fluorescent species present in the sample (Müller et al., 2000). Thus, if the receptor being studied exists as a mixture of species (monomers/dimers/tetramers), then the observed molecular brightness value would be an average based on the monomer/dimer/tetramer composition of the sample. Initially, PCH data should be fit to a one-component model with concentration and molecular brightness set as free and the first-order correction fixed at zero. Reduced chi-square values <3 (for runs 3–10 where photobleaching should be minimal) indicate a good fit to the selected model. If greater than three, two-component and three-component analyses are performed. In this case, molecular brightness values are fixed in multiples of established control values for monomers/dimers. Inspection of the residuals of the curve fit and reduced chi-square analysis are used to determine the goodness of fit to the selected model.

10.3 DISCUSSION

The most important considerations in performing FCS molecular brightness experiments in living cells are instrument alignment and calibration, choice of controls, fluorescence labeling of the receptor, and proper positioning of the sample in the observation volume. These topics are reviewed in more detail in the succeeding text.

Begin every FCS recording session by performing a pinhole adjustment using a control sample of the fluorescent probe used to label the receptor of interest to ensure proper alignment of the light path. Control samples, such as GPF or YFP expressed in the cytosol or a fluorescent dye in solution, provide a means to monitor day-to-day variability in instrument performance through reproducibility of molecular brightness values at a given laser setting. Likewise, end each session with additional control samples to monitor instrument stability throughout the session. If studying plasma membrane receptors, suitable plasma membrane controls are essential. Monomeric and dimeric receptors, such as CD-86 and CD-28, are a good choice for "decoding" the oligomeric composition of unknown samples by molecular brightness analysis. Be sure that the selected control and receptor of interest have their fluorescent tags in similar position and orientation with respect to the plasma membrane. Avoid the use of GPI-anchored or short myristoyl/palmitoyl chains to target GFP variants to the plasma membrane as they can enhance clustering within microdomains (Zacharias, Violin, Newton, & Tsein, 2002). We have tested such constructs and found them to be several times brighter than other monomeric controls. The use of the A206K/L221K mutations to eliminate GFP aggregation can be used, but receptor levels studied in FCS experiments are well below the concentration at which aggregation occurs (Zacharias et al., 2002). It is critical to use the lowest laser power possible while still maintaining a good signal to noise ratio, as photobleaching will result in artificially fast diffusion times and decreased molecular brightness.

A GFP variant attached to the C-terminus of the receptor is the most common method for receptor labeling. It has the advantage of ensuring an exact 1:1 labeling of the receptor, critical for molecular brightness analysis. A fluorescent ligand can be used if it has very slow dissociation kinetics. However, if the receptor of interest forms dimers, for example, experiments must be performed to confirm that the ligand binds with slow dissociation kinetics to both protomers of the dimer. A case in point is risperidone, which binds in a pseudo-irreversible manner to only one protomer of the 5-HT_7 receptor homodimer (Teitler, Toohey, Knight, Klein, & Smith, 2010). Alternatively, a monoclonal, monovalent Fab fragment recognizing the native conformation of the receptor could be used provided each Fab has exactly one fluorescent tag. A singly labeled Fab, such as Fab cloned into a GFP expression plasmid, would provide the ultimate tag as it could be used to label native receptors endogenously expressed in primary cultures.

The most critical parameter in molecular brightness analysis is proper positioning of the sample within the observation volume. The observation volume is not illuminated homogenously, and the detected photon counts decrease as a fluorescent molecule moves away from the center of the observation volume. Since a plasma membrane is approximately 1/200th of the axial dimension, optimal positioning within the center of the observation volume is critical for accurate molecular brightness determination of fluorescence-tagged membrane receptors. Positioning of the plasma membrane in the center of the observation volume is best achieved by scanning the sample along the z-axis while simultaneously monitoring the photon counts per molecule. The z position corresponding to the maximal photon counts per molecule is selected for FCS recording. FCS measurements generally are made on the upper plasma membrane, directly above the cell nucleus. We have discovered several factors that influence the shape of the cell and thus the ability to achieve optimal membrane position within the observation volume including the cell type, the receptor being expressed, and the choice of substrate for cell adhesion. Control samples should most closely approximate the conditions used for the receptor being studied.

SUMMARY

During the past four decades, fluorescence fluctuation spectroscopy has evolved into a highly sensitive method for monitoring protein dynamics in living cells. Laser scanning confocal microscopes fully equipped with sensitive photon-counting detectors and FCS analysis software have greatly expanded the application of FCS to address current issues in cell biology research at the single-molecule level. In this chapter, we reviewed a basic method for monitoring receptor–receptor interactions using FCS and molecular brightness analysis. Once a technique used predominantly by biophysicists and chemists, commercial availability of sensitive instrumentation and step-by-step protocols for performing FCS experiments are making the FCS method more accessible to researchers in general.

Acknowledgments

Funding for this work is provided by NIMH R21MH086796 to KHD.

References

Briddon, S. J., Hern, J. A., & Hill, S. J. (2010). Use of fluorescence correlation spectroscopy to study the diffusion of G protein-coupled receptors. In D. R. Poyner & M. Wheatley (Eds.), *G protein coupled receptors* (pp. 169–195). Chichester: Wiley & Sons.

Briddon, S. J., & Hill, S. J. (2008). Pharmacology under the microscope: The use of fluorescence correlation spectroscopy to determine properties of ligand–receptor complexes. *Trends in Pharmacological Sciences, 28*, 637–645.

Chen, Y., Muller, J. D., So, P. T., & Gratton, E. (1999). The photon counting histogram in fluorescence fluctuation spectroscopy. *Biophysical Journal, 77*, 553–567.

Chen, Y., Wei, L., & Muller, J. D. (2003). Probing protein oligomerization in living cells with fluorescence fluctuation spectroscopy. *Proceedings of the National Academy of Sciences, 100*, 15492–15497.

Elson, E. L. (2013). 40 years of FCS: How it all began. *Methods in Enzymology*, *518*, 1–10.

Herrick-Davis, K., Grinde, E., Cowan, A., & Mazurkiewicz, J. E. (2013). Fluorescence correlation spectroscopy analysis of serotonin, adrenergic, muscarinic and dopamine receptor dimerization: the oligomer number puzzle. *Molecular Pharmacology*, *84*, 1–13.

Herrick-Davis, K., Grinde, E., Lindsley, T., Cowan, A., & Mazurkiewicz, J. E. (2012). Oligomer size of the serotonin 5-HT$_{2C}$ receptor revealed by fluorescence correlation spectroscopy with photon counting histogram analysis: Evidence for homodimers without monomers or tetramers. *The Journal of Biological Chemistry*, *287*, 23604–23614.

Jameson, D. M., James, N. G., & Albanesi, J. P. (2013). Fluorescence fluctuation spectroscopy approaches to the study of receptors in live cells. *Methods in Enzymology*, *519*, 87–113.

Magde, D., Elson, E., & Webb, W. W. (1972). Thermodynamic fluctuations in a reacting system: Measurement by fluorescence correlation spectroscopy. *Physical Review Letters*, *29*, 705–708.

Malengo, G., Andolfo, A., Sidenius, N., Gratton, E., Zamai, M., & Caiolfa, V. R. (2008). Fluorescence correlation spectroscopy and photon counting histogram on membrane proteins. *Journal of Biomedical Optics*, *13*, 031215.

Müller, J. D., Chen, Y., & Gratton, E. (2000). Resolving heterogeneity on the single molecular level with the photon-counting histogram. *Biophysical Journal*, *78*, 474–486.

Müller, J. D., Chen, Y., & Gratton, E. (2003). Fluorescence correlation spectroscopy. *Methods in Enzymology*, *361*, 69–92.

Saffarian, S., Li, Y., Elson, E. L., & Pike, L. J. (2007). Oligomerization of the EGF receptor investigated by live cell fluorescence intensity distribution analysis. *Biophysical Journal*, *93*, 1021–1031.

Teitler, M., Toohey, N., Knight, J. A., Klein, M. T., & Smith, C. (2010). Clozapine and other competitive antagonists reactivate risperidone-inactivated h5-HT7 receptors: Radioligand binding and functional evidence for GPCR homodimer protomer interactions. *Psychopharmacology*, *212*, 687–697.

Weisshart, K., Jungel, V., & Briddon, S. J. (2004). The LSM 510 META—ConfoCor 2 system: An integrated imaging and spectroscopic platform for single-molecule detection. *Current Pharmaceutical Biotechnology*, *5*, 135–154.

Zacharias, D. A., Violin, J. D., Newton, A. C., & Tsein, R. Y. (2002). Partitioning of lipid-modified monomeric GFPs into membrane microdomains of live cells. *Science*, *296*, 913–916.

CHAPTER

Monitoring Receptor Oligomerization by Line-Scan Fluorescence Cross-Correlation Spectroscopy

11

Mark A. Hink and Marten Postma

Section of Molecular Cytology, van Leeuwenhoek Centre for Advanced Microscopy (LCAM), Swammerdam Institute for Life Sciences (SILS), University of Amsterdam, The Netherlands

CHAPTER OUTLINE

Introduction 198
11.1 Materials 199
11.1.1 Solutions, DNA Constructs, and Cells 199
11.1.2 Fluorescent Proteins 200
11.1.3 Microscope 200
11.2 Methods 202
11.2.1 Growth and Transfection of HeLa Cells 202
11.2.2 Microscope Calibration 203
11.2.3 Fluorescence Fluctuation Measurements 204
11.2.4 Auto- and Cross-Correlation Analysis 205
11.3 Discussion 209
Acknowledgments 210
References 210

Abstract

Membrane-localized receptor proteins are involved in many signaling cascades, and diffusion and oligomerization are key processes controlling their activity. In order to study these processes in living cells, fluorescence fluctuation spectroscopy techniques have been developed that allow the quantification of concentration levels, diffusion rates, and interactions between fluorescently labeled receptor proteins at the nanomolar concentration level. This chapter presents a brief introduction to the technique and a protocol to measure and quantify the diffusion and oligomerization of

Methods in Cell Biology, Volume 117

ISSN 0091-679X
http://dx.doi.org/10.1016/B978-0-12-408143-7.00011-6

human histamine1 receptor complexes in living HeLa cells using line-scanning fluorescence cross-correlation spectroscopy.

INTRODUCTION

Diffusion and oligomerization are essential processes controlling the activity of numerous membrane-localized receptor proteins that are involved in signaling cascades. Various biochemical and biophysical approaches have been used to quantify these processes *in vitro* using artificial or reconstituted membrane systems. Recently, techniques have been developed to study these in living cells, and in this chapter, the application of fluorescence fluctuation techniques will be discussed.

Fluorescence correlation spectroscopy (FCS) is the most well-known member of the family of fluorescence fluctuation spectroscopy (FFS) techniques analyzing temporal changes of the fluorescence intensity and relating these fluctuations to physical parameters of the observed fluorescent molecules. The high sensitivity of the technique allows the detection of single molecules and obtaining information about concentration levels, diffusion rates, and interactions with other molecules (Haustein & Schwille, 2007). Using a confocal microscope and illumination with a focused laser beam in combination with a pinhole, fluorescence is detected from a subfemtoliter volume element ($\sim 1\ \mu m^3$). Photons emitted by the fluorescent molecules present in this element pass through a pinhole and are detected by a sensitive detector. During the measurement, molecules will, due to their (Brownian) motion, move into and out of the volume element and emit photons in a burst-type manner. The observed fluctuations can be used to determine the average time a molecule requires passing through the volume element. Due to the high signal-to-noise ratio that could be achieved, measurements can be performed at the single-molecule level. In practice, FFS operates in the nanomolar concentration range, which is close to the physiological expression levels of many membrane receptors.

The mobility of a membrane receptor is related, among other factors, to the size of the molecule. Therefore, it is possible to examine if receptors are moving freely or are part of a large, more slowly moving complex. Meseth, Wohland, Rigler, and Vogel (1999) examined the resolving power of FCS to distinguish between different sized molecules. In case of an unchanged fluorescence yield upon binding, the diffusion coefficients of the free and complexed form have to differ at least 1.6-fold, corresponding to a fourfold mass increase for globular particles, which is required to discriminate both species without prior knowledge of the system. This will not suffice to discriminate receptor monomers from dimers, which involves a doubling of the mass. However, the distribution of the detected fluorescence intensities can still give information about dimerization, since dimers are twice as bright as monomers, as is exploited in photon counting histogram (PCH) (Chen, Johnson, Macdonald, Wu, & Müller, 2010; Hink, Shah, Russinova, de Vries, & Visser, 2008; Herrick-Davis, Grinde, Lindsley, Cowan, & Mazurkiewicz, 2012; Chapter 10) or using number and brightness (N&B) analysis (Crosby et al., 2013; Digman, Dalal, Horwitz, & Gratton, 2008).

An alternative approach to monitor oligomerization has been developed by Schwille, Meyer-Almes, and Rigler (1997) who applied dual-color fluorescence cross-correlation spectroscopy (FCCS). In these studies, each of the two types of interacting molecules is tagged by a spectrally different fluorescent group, for example, a green or a red fluorescent dye. The different fluorescent dyes can be excited either with different lasers or with the same laser. The emission light is split into two different detectors so that the fluorescence of the two fluorophores can be monitored over time simultaneously. In contrast to single-color FCS, the discriminating factor is not the increase of the molecular mass or increased brightness upon complex formation but the simultaneous detection of fluorescence bursts in both detection channels. Although most FCCS studies published use two dyes to monitor interactions between two types of molecules, also higher-order complexes could be monitored using more than two fluorophores (Heinze, Jahnz, & Schwille, 2004; Shcherbakova, Hink, Joosen, Gadella, & Verkhusha, 2012).

Classical FCS techniques as described earlier are based on point illumination by parking the laser beam at a preferred location within the cell, like the plasma membrane, and illuminate this point during the data acquisition. However, membrane movement or fluorophore photobleaching, caused by a low mobility of membrane receptors or their complexes, often results in artifacts in the FCS curves. Photobleaching can be reduced by exciting the molecules less frequently, for example, by moving the laser beam, thus scanning lines or even whole images. FFS approaches that combine temporal and spatial information are referred to as image correlation spectroscopy techniques (Kolin & Wiseman, 2007). One of these techniques, line FCS, has the advantage that movement of the cell and or the plasma membrane during the measurement can easily be corrected for after acquisition and that the obtained diffusion time is independent on the z-focus (Ries, Yu, Burkhardt, Brand, & Schwille, 2009).

In this chapter, we describe the use of dual-color line-scan FCCS in order to study the dimerization and diffusion of fluorescently labeled human histamine1 receptors (H_1R) in nonstimulated living HeLa cells. Typically, these experiments are complemented by PCH or N&B analysis to monitor the stoichiometry of the complexes being formed. The application of PCH is being discussed in Chapter 10 in this book.

11.1 MATERIALS

11.1.1 Solutions, DNA constructs, and cells

1. Phosphate buffered saline (PBS): 137 mM NaCl, 2.7 mM KCl, 10 mM sodium phosphate dibasic, 2 mM potassium phosphate monobasic (pH 7.4).
2. Microscopy medium: 20 mM HEPES (pH = 7.4), 137 mM NaCl, 5.4 mM KCl, 1.8 mM $CaCl_2$, 0.8 mM $MgCl_2$, and 20 mM glucose.
3. Purified mTurquoise2 and sYFP2 in PBS pH 7.4. The fluorescent proteins (FPs) are purified using the IMPACT (New England Biolabs) system, which utilizes inducible self-cleavage activity of an intein splicing element to separate the FP from the affinity tag. Transformed bacteria are collected by

centrifugation and resuspended in 50 mM PBS. Cells are lysed by passage through a French pressure cell. Soluble protein is obtained after centrifugation at 20,000 g for 30 min at 4 °C. The protein binds to the chitin beads and after extensive wash with PBS, the FPs are eluted from the column by incubating the beads overnight in 50 mM dithiothreitol. Protein purity is checked on SDS-PAGE.

4. Solution of purified N1 plasmid DNA encoding for the protein fusions mTurquoise2-p63-sYFP2, human H_1R-mTurquoise2 or H_1R-sYFP2.
5. DMEM growth medium or phenol-free DMEM growth medium (Gibco) supplemented with GlutaMAX (Gibco).
6. Live cell incubator at 37 °C with 5% CO_2.
7. Lipofectamine and Opti-MEM transfection agents (Invitrogen).
8. Sterile circular (ø 24 mm, size 1) coverslips (Menzel-Gläser).
9. Attofluor coverslip holder (Invitrogen).
10. Six-well container (Greiner).
11. 96-well microtiter plates with borosilicate bottom (Whatman).
12. Confocal laser scanning microscope FV1000 (Olympus) equipped with lasers and MPD detectors as described in detail in the succeeding text.
13. FFS data processor 2.3 software package (Scientific Software Technologies Center).
14. Ptu converter 0.40 (van Leeuwenhoek Centre for Advanced Microscopy).
15. Matlab 7.0 (MathWorks).

11.1.2 Fluorescent proteins

Because FCS is a method that relies on the analysis of relative fluorescence fluctuations, the average concentration of fluorescent molecules should be kept low, typically between 1 nM and 1 µM. While the molecules pass through the detection volume, as many photons as possible should be detected. As shown by Koppel (1974), the molecular brightness of the fluorophore is a key parameter to obtain high-quality FCS data. Since our work focuses on studying cellular signaling within living cells, genetically encoded FPs are the most suitable dyes. The two FPs selected on the basis of highest molecular brightness, minimal absorbance spectrum overlap, fast chromophore maturation, and good photostability are mTurquoise2 (mTq2) (Goedhart et al., 2012) and sYFP2 (Kremers, Goedhart, van Munster, & Gadella, 2006).

11.1.3 Microscope

Data are acquired using an Olympus IX81 inverted microscope with a Fluoview 1000 scan and confocal detection head, controlled by FV3.1 software (Olympus). The 514 nm line of a continuous wave Ar^+ laser (Omnichrome) is combined with a pulsing laser diode of 440 nm (PicoQuant) operated at a rate of 20 MHz, as controlled by a Sepia II laser driver unit (PicoQuant). The light intensity is attenuated 10 times by a neutral density filter and guided via a D440/514 primary dichroic mirror (Chroma)

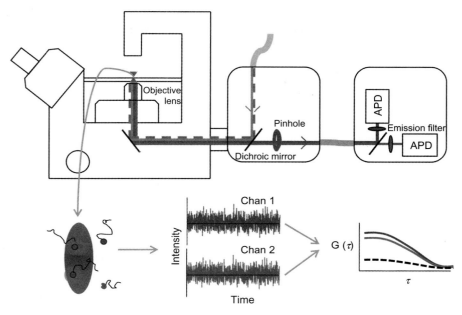

FIGURE 11.1

FCCS setup. Light from a pulsed 440 nm diode and a continuous 514 nm Ar^+ laser are focused into the sample using highly corrected objective lens with a large numerical aperture. A line is scanned perpendicular to the membrane at millisecond intervals. The fluorescence is guided via a pinhole into the detection unit where a dichroic mirror and emission filters separate the light into two sensitive detectors. Diffusion of monomeric and dimeric fluorescent receptors through the detection volumes causes intensity fluctuations over time that can be analyzed by fitting the auto- and/or cross-correlation curves. (See color plate.)

through a silicone oil immersion 60 × UPlanSApo objective (NA 1.3) into the sample. Although not discussed in this chapter, total internal reflection fluorescence (TIRF) combined with FCS could be useful for measuring membrane receptors as well, since TIRF will only excite molecules within a region of a few hundred nanometers above the cover glass, thereby reducing the size of the observation volume and background contributions from the cell cytoplasm (Hassler et al., 2005). A critical part of the equipment is the objective lens, especially since multicolor analysis is being used here. To minimize detection of volume shape distortions within the cells, a silicone immersion objective with high numerical aperture (N.A.) is used, which approaches the refractive index of the cellular environment closely, resulting in high-quality intracellular FCS curves (Hink, 2012).

Cellular samples grown on round coverslips are stored in Attofluor sample holders. The emission light is guided via a size-adjustable pinhole, set at 120 μm, through the Olympus detection box to the fiber output channel. The optical fiber is coupled to a custom-made detection box (PicoQuant) containing PDM avalanche photodiodes (MPD). The emission light is split by a 515 nm dichroic mirror and guided into one of the two MPDs where the light is filtered by a 440/45 or 525/45 emission filter (Chroma). The photon arrival times are recorded by a PicoHarp 300 time-correlated single-photon counting system (PicoQuant), controlled by SymPhoTime 64 1.5 software (PicoQuant).

The contribution of cross talk, the fluorescence of one type of dye detected in the "other" detector, will give rise to false-positive cross-correlation. This artifact will be prevented by using pulsed interleaved excitation (PIE) (Müller, Zaychikov, Bräuchle, & Lamb, 2005). The absorbance spectra of most fluorophores are relatively narrow, and therefore, it is possible to select a pair of dyes such that each dye can be excited exclusively by a specific laser line. When two pulsed laser units are emitting in an alternating mode at a switching time much faster than the residence time of the labeled receptor in the detection volume, one will obtain the fluorescence photons of both dyes in different time periods. Here, a semi-PIE approach is used where the CFP laser is pulsing in combination with a continuous wave YFP laser. By time gating the detected fluorescence, the cross-talk photons from mTurquoise2 in the YFP detection channel can be omitted in the calculation of the auto- and cross-correlation curves. YFP fluorescence in the CFP detection channel can be neglected (Fig. 11.1).

11.2 METHODS

In the following section, a protocol is presented that can be used to measure dual-color FCCS in living cells using line-scan acquisition. Here, the homodimerization of mTq2- and sYFP2-labeled H1 receptors, produced in HeLa cells, is studied but the protocols can easily be adapted to monitor hetero-oligomerization. Protocols that describe point FCCS and PCH measurements are published elsewhere (Hink, in preparation; Hink, de Vries, & Visser, 2011; Chapter 10).

11.2.1 Growth and transfection of HeLa cells

HeLa cells are grown in DMEM medium supplemented with GlutaMAX in a live cell incubator. One day before transfection, the cells are transferred from the culture flask to sterile circular coverslips and stored in a six-well container, at a confluency of $\sim$60% ($\sim$5.0 $\times$ 10^5 cells). Six hours after the cell transfer, the growth medium is replaced by phenol-free DMEM medium, thereby lowering the autofluorescence intensity. Cells are transfected using Lipofectamine in Opti-MEM according to the manufacturer's protocol (Invitrogen). Since the H1 DNA constructs used here contain a CMV high expression promoter and FFS requires low concentration levels, the amount of DNA used in the transfection procedure is only 1 ng per construct. The FFS experiments are carried out half a day after transfection.

11.2.2 Microscope calibration

The microscope is placed in an air-conditioned room to ensure temperature stability, and the calibration samples are stored in glass-bottomed 96-well plate. In order to reduce the amount of detected background light, a protective blackened aluminum cover is placed on top of the sample during the measurements. The microscope and lasers are turned on roughly 30 min before starting the measurements in order to stabilize. The mTurquoise2-labeled samples will be excited with the 440 nm laser line pulsing at 20 MHz and the sYFP2 samples are excited using the 514 nm line of an Ar^+ laser.

To optimize the microscope settings, the fluorescence of a $\sim$100 nM solution of purified mTurquoise2 in PBS is measured at a laser power of $\sim$12 kW cm^{-2} in the focal plane. The correction collar of the objective lens is adjusted to the position such that the highest fluorescence count rate is observed in one of the two detection channels. In order to prevent that scattered excitation light or fluorescence from adsorbed probe at the bottom of the sample holder is optimized, instead of the fluorescence signal from the sample of interest, the focus should be set at least 10 μm into the solution. Furthermore, the position of the pinhole is optimized by moving the X-, Y-, and, if available, the Z-position of the pinhole that yields the highest detected fluorescence of mTq2. The lens position in front of the APD is optimized as well in a similar way.

The purified mTq2 or sYFP2 samples are measured, using their corresponding laser lines at 12 (440 nm) and 9 kW cm^{-2} (514 nm). The acquisition time, typically 2 min, was adjusted such that the resulting correlation curves are smooth in the decaying part of the curve. The intensity traces are imported into the FFS data processor 2.3 software, autocorrelated, and fitted to a model (Eq. 11.1) including terms for triplet state kinetics and three-dimensional Brownian diffusion (Skakun et al., 2005). Because of the relatively slow scanning speed and the oversampling (pixel size of $\sim$100 nm) in the line-FCCS experiments, the y- and z-dimensions of the detection volumes can be estimated from point FCS measurements. In order to estimate the size of the cross-correlation detection volume, the mTq2 signal, as acquired in both detectors, is cross-correlated, using the relative large number of mTq2 cross-talk photons that are detected in the YFP channel.

$$G(\tau) = 1 + \frac{\gamma}{N} \cdot \left(\frac{1 - T + Te^{(-\tau/\tau_{TRIP})}}{(1-T)} \cdot \frac{1}{\sqrt{1 + \frac{\tau}{\tau_{dif,x}}}\sqrt{1 + \frac{\tau}{\tau_{dif,y}}}\sqrt{1 + \frac{\tau}{\tau_{dif,z}}}} \right) \quad (11.1)$$

The autocorrelation function, $G(\tau)$, contains a parameter N, which corresponds to the average number of fluorescent particles in the detection volume. τ_{dif} denotes the average diffusion time of the molecules, T denotes the fraction of molecules present in the dark state, and τ_{TRIP} is the average time a molecule resides in the dark state. Parameter γ represents the shape factor of the observation volume and equals 0.3535 for a 3D Gaussian or 1.0 for a cylindrical-shaped observation volume. From the autocorrelation fits a shape factor, a, describing the ratio between the axial (ω_z) and lateral (ω_{xy}) e^{-2} radii of the detection volume ($a = \omega_z/\omega_{xy}$) is obtained. Although the shape of the detection volumes is close to a 3D Gaussian in "ideal conditions," the optical aberrations introduced by measuring inside the living cell allow to approximate the detection volumes by a cylindrical shape. The amplitude of the autocorrelation function, $G(0)$, equals γ/N where γ corresponds to the shape of the detection volume. In all the equations used in this chapter, cylindrical detection volumes are assumed with $\gamma = 1$.

To validate the correct calibration of the microscope, the obtained fitting parameters are checked. For 3D diffusion in an open cylindrical volume, as measured here, the diffusion times in x- and y-direction are equal and the diffusion time in the z-direction along the optical axis is given by

$$\tau_{\text{dif},z} = \frac{\omega_z^2}{\omega_{xy}^2} \times \tau_{\text{dif},x} \tag{11.2}$$

The diffusion times $\tau_{\text{dif},x}$ for mTq2 or sYFP2 are in the order of 180 µs and the shape factors a range from 4 to 15. The observed brightness values vary between 8 and 12 kHz per molecule. Note that these values will be different for other microscope systems. In general, the higher the brightness per particle, the better the quality of the obtained data will be. The system has to be calibrated such that the highest possible brightness is reached without photobleaching and/or saturation artifacts (Gregor, Patra, & Enderlein, 2005).

By solving Eqs. (11.3) and (11.4), using the translational diffusion coefficient (D) of 90 µm^2 s^{-1} for both mTurquoise as sYFP2 in buffer, the dimensions of the "cyan" and "yellow" detection volumes, V, can be calculated by approximating these as cylinders:

$$\tau_{\text{dif},x} = \frac{\omega_{xy}^2}{4D} \tag{11.3}$$

$$V = 2 \cdot \pi \cdot a \cdot \omega_{xy}^3 \tag{11.4}$$

11.2.3 Fluorescence fluctuation measurements

A coverslip with the transfected HeLa cells is sealed in an Attofluor cell chamber (Invitrogen) submerged in microscopy medium. Note that growth media should not be used since this will acidify when incubated in the absence of 5% CO_2 during the measurement.

The effective cross-correlation observation volume is estimated measuring cross-correlation using a positive control, for example, cells producing a fusion protein of mTq2-p63-sYFP2. To select transfected cells with FFS-compatible expression levels, the sample is imaged by the confocal microscope using standard imaging settings for the PMT detector sensitivity while the pinhole is completely open. The cells that are more fluorescent at the membrane than the mock-transfected control cells, but do not give rise to detector saturation in the image, are selected for FFS measurements. For cross-correlation measurements, only cells are selected that produce both the mTq2 and sYFP2 fusion proteins. The laser power is set not higher than 2.5 kW cm^{-2} for the 440 nm laser line and 2.0 kW cm^{-2} for the 514 nm laser line to prevent photobleaching and cellular damage and reduce photophysical effects. Photobleaching and probe saturation lead to significant distortions of the correlation curves, which will compromise the analysis. An overview xy-image is obtained at the equatorial plane of the cell such that the fluorescent plasma membrane is located in the middle (Fig. 11.2). In order to make use of the highest scanning rate of the laser beam, that is, along the x-axis, the membrane should be visible as a vertical stripe in the image. A horizontal line of 16 or 32 pixels is selected for measurements, intersecting the plasma membrane at the center of this line. The zoom factor is adjusted to obtain pixel sizes of approximately 90×90 nm, and the line is scanned 32,766 times at a rate of 1.7 ms per line. This measurement is repeated three times to improve photon statistics.

Finally, the samples of interest are being measured. All raw fluorescence intensity files are saved for processing. If desired, cross-correlation could be measured in a negative control experiment, which consists of mutants of the two sample proteins lacking the possibility to interact with each other, for example, due to point mutations or domain deletions. Alternatively, HeLa cells are cotransfected with separate constructs encoding for membrane-targeted mTq2 or sYFP2.

11.2.4 Auto- and cross-correlation analysis

Each raw data set is converted into a single kymograph tif image using the in-house-built Ptu converter software. Only photons are exported from the CFP detector that arrived within the first 25 n after the 440 nm laser pulse at $t=0$ ns. For the YFP channel, only the photons between $t=25$ ns and $t=50$ ns are exported. The image is loaded into Matlab and processed using in-house-written scripts according to the following protocol.

First, the data are corrected for movement of the plasma membrane along the x-axis during the acquisition. Thereto, a moving averaged box of 20 lines is fitted to an equation describing a Gaussian profile summed with a step function to take into account intracellular fluorescence. Each individual line is translated such that the peak fluorescence intensity in the box, which corresponds to the fluorescent plasma membrane, will be located at the center of the line. From this aligned carpet, the five central pixels per line are summed to generate a single intensity value per line. When the intensity data set contains large intensity spikes (>10 times the intensity standard deviation around the mean intensity), significant signal drift, or photobleaching ($>20\%$ intensity loss per minute), the data set is discarded for further analysis. Stable sections of the two intensity traces, one for each detection channel, are auto- and

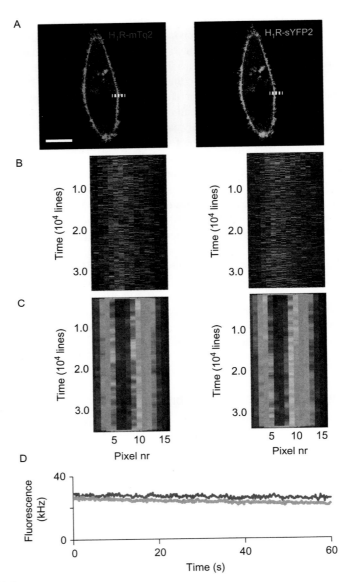

FIGURE 11.2

Working scheme for the line-FCCS data acquisition. Averaged confocal image (A) at the equatorial plane of a HeLa cell cotransfected with H_1R-mTurquoise2 and H_1R-sYFP2 shows membrane-localized fluorescence (scale bar, 5 μm). A line of 16 pixels (dotted line) perpendicular to the plasma membrane was selected for data acquisition. The intensity peaks in the time-pixel kymographs (B), corresponding to the plasma membrane, were aligned to correct for membrane movement during the measurement (C). For each line, the intensities in a box of 2 pixels around the center were summed. The resulting intensity traces (D) were auto- and cross-correlated. (See color plate.)

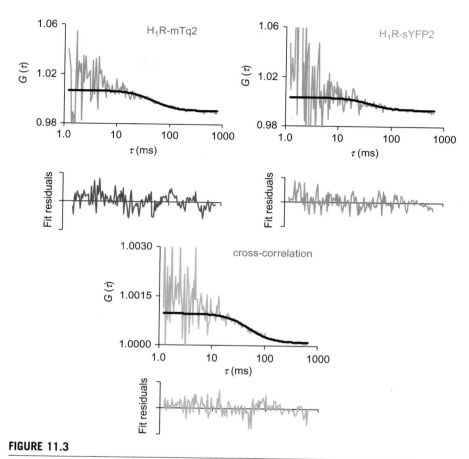

FIGURE 11.3

Typical auto- and cross-correlation curves for HeLa cells cotransfected with H_1R-mTurquoise2 and H_1R-sYFP2. The curves were fitted (black lines) to models including terms for diffusion, dark-state kinetics, and the presence of monomers and dimers. (For color version of this figure, the reader is referred to the online version of this chapter.)

cross-correlated using multi-tau algorithms (Wahl, Gregor, Patting, & Enderlein, 2003) resulting in the two autocorrelation curves $G_{CC}(\tau)$ (for the CFP channel) and $G_{YY}(\tau)$ (for the YFP channel) and the cross-correlation curve $G_{CY}(\tau)$. Figure 11.3 shows some typical experimental curves.

The two autocorrelation and cross-correlation curves are fitted according to Eq. (11.5). Since we define the direction of line scanning to be the x-axis, so perpendicular to the membrane, diffusion along the x-axis can be neglected due to the limited thickness of the plasma membrane (3–10 nm) in relation to the e^{-2} radius of the detection volume (~250 nm).

$$G_{a,b}(\tau) = 1 + \left(1 - \frac{F_{\text{background},a}}{F_{\text{total},a}}\right) \cdot \left(1 - \frac{F_{\text{background},b}}{F_{\text{total},b}}\right) \cdot G_{\text{dynamics}} + \text{Offset}$$

$$\text{where } G_{\text{dynamics}} = \left(G_{a,b}(0) \cdot \frac{1 - T + Te^{(-\tau/\tau_T)}}{(1-T)} \sum_m \Phi_m \cdot \frac{1}{\sqrt{1 + \frac{\tau}{\tau_{\text{dif}}}} \sqrt{1 + \frac{\tau}{\tau_{\text{dif}}} \cdot \left(\frac{\omega_{xy}}{\omega_z}\right)^2}} \right)$$

$$\text{with } \Phi_m = \frac{\left(\eta_m^2 Y_m\right)}{\left(\sum_m \eta_m Y_m\right)^2} \tag{11.5}$$

Here, the first two factors between brackets correct for the background fluorescence in the corresponding detection channel ($F_{\text{background},x}$). The value of this parameter is determined by averaging the fluorescence intensity in the plasma membrane in mock-transfected cells. In most cases, this correction can be neglected since typically $F_{\text{background}} < 0.01 \times F_{\text{total}}$.

Each molecular species, m, contributes to the autocorrelation curve function according to its fraction (Y_m) and molecular brightness (η_m). This information is obtained from PCH or N&B analysis. The ratio ω_{xy}/ω_z is fixed to the value obtained from the calibration measurement for each channel. An offset can be included in the fit to correct for the effect of small and slow intensity drifts in the selected data trace.

The amplitude of the fitted correlation curves, $G_{a,b}(0)$, contains the information of the number of labeled particles. $G_{CY}(0)$ scales to the number of doubly labeled particles N_{CY}, as observed in the cross-correlation observation volume, V_{CY}. Equations (11.6)–(11.8) describe the auto- and cross-correlation amplitudes, when corrected for the presence of monomers and dimers. In addition, one assumes the probability of forming CC-, CY-, or YY-labeled dimers is equal. The absence of higher-order complexes is validated analyzing the same data sets by PCH confirming the presence of molecules having a similar and an almost double (1.92 × and 1.89 ×, respectively) molecular brightness compared to (monomeric) membrane-targeted mTq2 or sYFP2.

$$G_{CC}(0) = 1 + \frac{1}{\left(N_C + N_{CY} + (1.92)^2 N_{CC}\right)} \tag{11.6}$$

$$G_{YY}(0) = 1 + \frac{1}{\left(N_Y + N_{CY} + (1.89)^2 N_{YY}\right)} \tag{11.7}$$

$$G_{CY}(0) = 1 + \frac{N_{CY}}{(N_C + N_{CY} + N_{CC})(N_Y + N_{CY} + N_{YY})} \tag{11.8}$$

where N_C and N_Y are the number of monomeric receptors labeled with mTq2 and sYFP2, respectively. N_{CC}, N_{YY}, and N_{CY} correspond to the number of dimers with a single or both types of FP labels.

For this experiment, we assume a double probability of observing a CY dimer over a CC or YY dimer, and there is an equal production of mTq2- and sYFP2-labeled receptors.

$$N_{CY} = 0.5 \cdot N_{CC} = 0.5 \cdot N_{YY} \tag{11.9}$$

Equations (11.5)–(11.9) are linked and fitted globally including a weight factor to correct for the different detection volume sizes, as obtained from the calibration measurements. In addition, the number of complexes is corrected for the fraction of overlap between the cyan and yellow detection volume as is estimated from the positive control. From these numbers, the dissociation constant, K_D (Mäder et al, 2007), for H_1R dimerization is calculated. An example of our experimental values is presented in Table 11.1. Due to the different expression levels among the cells, caused by the transient transfection procedure, the interaction values are presented as percentage complex in each detection channel and as apparent dissociation constant assuming the absence of nonfluorescent and endogenous protein.

11.3 DISCUSSION

Quantitative measurements using FCCS require precise calibration measurements because of the different detection volume sizes; the potential existence of nonmature dark FPs and photobleaching can affect the results. Dimerization could induce a

Table 11.1 Line-FCCS Analysis of HeLa Cells Cotransfected with H_1R-mTq2 and H_1R-sYFP2 and the Control Samples

Construct(s)	Diffusion Time τ_x (ms)	% Interaction (N_{CY}/N_{Xtotal})	Apparent K_D (nM)	n (−)
H_1R-mTq2	30±8	0±0	n.a.	3
H_1R-sYFP2	37±11	0±0	n.a.	3
H_1R-mTq2 + H_1R-sYFP2	32±9 39±12	12±4 18±5	646±91	5
mTq2-p63 sYFP2 (fusion, pos.)	24±10 28±13	99±2 98±1	n.a	3
CAAX-mTq2 + CAAX-sYFP2 (neg.)	1.2±0.7 1.1±0.7	1±1 0±0	3512±3984	2

The data are analyzed using Eqs. (11.5)–(11.9) and the obtained parameters are presented for the detection channel corresponding to the FP in the construct. The errors presented are SEM. n.a., nonapplicable.

strong energy transfer between the FPs, which will result in a reduced molecular brightness of the donor, in this case mTq2. This will lead to a more complicated contribution of the CY dimer to the correlation curves. However, fluorescence lifetime analysis showed a relatively small reduction of the fluorescence lifetime ($<10\%$) for the cotransfected (H_1R-mTq2 + H_1R-sYFP2) cells with respect to H_1R-mTq2-transfected cells, and therefore, the FRET effect can be neglected.

In addition, the presence of endogenous, nonfluorescent H_1R can affect the estimated K_D values. Western blotting showed a relative low contribution of endogenous H_1R in resting HeLa cells in comparison to the average H_1R-FP protein level after successful transfection. However, when studies would be performed where cells are stimulated by the histamine ligand, a significant increase of the endogenous H_1R protein levels is expected (Das et al., 2007). In this situation, the fraction of dimers where both monomers are fused to a FP will be lowered. This can be solved by using siRNA strategies to lower the concentration of endogenous protein or the data can be corrected as described by Foo, Naredi-Rainer, Lamb, Ahmed, and Wohland (2012).

In this experiment, a single type of protein, H_1R, is genetically fused to mTq2 or sYFP2. Both plasmids, which are almost identical in code and size, are cotransfected in HeLa cells. FCS experiments have shown that in this case, the production level of both proteins is equal, validating Eq. (11.9). However, when this equation has to be adapted for experiments with two different kinds of proteins/constructs, a similar and preferably known ratio of proteins is desired. We found that IRES and viral 2A peptides can be used to coexpress proteins at a fixed ratio at the single-cell level (Goedhart et al., 2011) that is very helpful for multicolor FCCS experiments as described in this chapter.

Acknowledgments

This work was supported by Middelgroot (834.09.003) and Echo (711.011.018) investment grants and a VIDI fellowship awarded to M.P. (864.09.015) from the Netherlands Organisation for Scientific Research (NWO). We thank Kevin Crosby and Kobus van Unen for sharing the DNA constructs.

References

Chen, Y., Johnson, J., Macdonald, P., Wu, B., & Müller, J. D. (2010). Observing protein interactions and their stoichiometry in living cells by brightness analysis of fluorescence fluctuation experiments. *Methods in Enzymology*, *472*, 345–363.

Crosby, K. C., Postma, M., Hink, M. A., Zeelenberg, C. H. C., Adjobo-Hermans, M. J., & Gadella, T. W. J. (2013). Quantitative analysis of self-association and mobility of the Ca^{2+} binding protein annexin A4 in cells. *Biophysical Journal*, *104*, 1875–1885.

Das, A. K., Yoshimura, S., Mishima, R., Fujimoto, K., Mizuguchi, H., Dev, S., et al. (2007). Stimulation of histamine H1 receptor up-regulates histamine H1 receptor itself through activation of receptor gene transcription. *Journal of Pharmacological Sciences, 103,* 374–382.

Digman, M. A., Dalal, R., Horwitz, A. F., & Gratton, E. (2008). Mapping the number of molecules and brightness in the laser scanning microscope. *Biophysical Journal, 94,* 2320–2332.

Foo, Y. H., Naredi-Rainer, N., Lamb, D. C., Ahmed, S., & Wohland, T. (2012). Factors affecting the quantification of biomolecular interactions by fluorescence cross-correlation spectroscopy. *Biophysical Journal, 102,* 1174–1183.

Goedhart, J., van Weeren, L., Adjobo-Hermans, M. J. W., Elzenaar, I., Hink, M. A., & Gadella, T. W. J., Jr. (2011). Quantitative co-expression of proteins at the single cell level—Application to a multimeric FRET sensor. *PloS One, 6,* e27321.

Goedhart, J., von Stetten, D., Noirclerc-Savoye, M., Lelimousin, M., Joosen, L., Hink, M. A., et al. (2012). Structure-guided evolution of cyan fluorescent proteins towards a quantum yield of 93%. *Nature Communications, 3,* 751–759.

Gregor, I., Patra, D., & Enderlein, J. (2005). Optical saturation in fluorescence correlation spectroscopy under continuous-wave and pulsed excitation. *ChemPhysChem, 6,* 164–170.

Hassler, K., Leutenegger, M., Rigler, P., Rao, R., Rigler, R., Giisch, M., et al. (2005). Total internal reflection fluorescence correlation spectroscopy (TIR-FCS) with low background and high count-rate per molecule. *Optics Express, 13,* 7415–7423.

Haustein, E., & Schwille, P. (2007). Fluorescence correlation spectroscopy: Novel variations of an established technique. *Annual Review of Biophysics and Biophysical Chemistry, 36,* 151–169.

Heinze, K. G., Jahnz, M., & Schwille, P. (2004). Triple-color coincidence analysis: One step further in following higher order molecular complex formation. *Biophysical Journal, 86,* 506–516.

Herrick-Davis, K., Grinde, E., Lindsley, T., Cowan, A., & Mazurkiewicz, J. E. (2012). Oligomer size of the serotonin 5-HT2C receptor revealed by fluorescence correlation spectroscopy with photon counting histogram analysis: Evidence for homodimers without monomers or tetramers. *The Journal of Biological Chemistry, 287,* 23604–23614.

Hink, M. A. (2012). Single-molecule microscopy using silicone oil immersion objective lenses. *The Biomedical Scientist, 2012,* 83–85.

Hink, M. A. (in preparation) Fluorescence correlation spectroscopy. In P. Verveer (Vol. Ed.). *Methods in molecular biology. Advanced fluorescence microscopy: Methods and protocols.* Amsterdam: Springer.

Hink, M. A., de Vries, S. C., & Visser, A J W G (2011). Fluorescence fluctuation analysis of receptor kinase dimerization. In *Plant kinases. Methods and protocols.* In N. Dissmeyer & A. Schnittger (Eds.), *Methods in molecular biology,* Vol. 779, (pp. 199–216). Dordrecht: Humana Press.

Hink, M. A., Shah, K., Russinova, E., de Vries, S. C., & Visser, A J W G (2008). Fluorescence fluctuation analysis of AtSERK and BRI1 oligomerization. *Biophysical Journal, 94,* 1052–1062.

Kolin, D. L., & Wiseman, P. W. (2007). Advances in image correlation spectroscopy: Measuring number densities, aggregation states, and dynamics of fluorescently labeled macromolecules in cells. *Cell Biochemistry and Biophysics, 49,* 141–164.

Koppel, D. E. (1974). Statistical accuracy in fluorescence correlation spectroscopy. *Physical Review A, 10,* 1938–1945.

Kremers, G. J., Goedhart, J., van Munster, E. B., & Gadella, T. W. J., Jr. (2006). Cyan and yellow super fluorescent proteins with improved brightness, protein folding, and FRET Förster radius. *Biochemistry, 45,* 6570–6580.

Mäder, C. I., Hink, M. A., Kinkhabwala, A., Mayr, R., Bastiaens, P. I., & Knop, M. (2007). Spatial regulation of Fus3 MAP kinase activity through a reaction–diffusion mechanism in yeast pheromone signaling. *Nature Cell Biology*, *9*, 1319–1326.

Meseth, U., Wohland, T., Rigler, R., & Vogel, H. (1999). Resolution of fluorescence correlation measurements. *Biophysical Journal*, *76*, 1619–1631.

Müller, B. K., Zaychikov, E., Bräuchle, C., & Lamb, D. C. (2005). Pulsed interleaved excitation. *Biophysical Journal*, *89*, 3508–3522.

Ries, J., Yu, S. R., Burkhardt, M., Brand, M., & Schwille, P. (2009). Modular scanning FCS quantifies receptor–ligand interactions in living multicellular organisms. *Nature Methods*, *6*, 643–645.

Schwille, P., Meyer-Almes, F. J., & Rigler, R. (1997). Dual-color fluorescence cross-correlation spectroscopy for multicomponent diffusional analysis in solution. *Biophysical Journal*, *72*, 1878–1886.

Shcherbakova, D., Hink, M. A., Joosen, L., Gadella, T. W. J., Jr., & Verkhusha, V. V. (2012). An orange fluorescent protein with a large Stokes shift for single-excitation multicolor FCCS and FRET imaging. *Journal of the American Chemical Society*, *134*, 7913–7923.

Skakun, V. V., Hink, M. A., Digris, A. V., Engel, R., Novikov, E. G., Apanasovich, V. V., et al. (2005). Global analysis of fluorescence fluctuation data. *European Biophysics Journal*, *34*, 323–334.

Wahl, M., Gregor, I., Patting, M., & Enderlein, J. (2003). Fast calculation of fluorescence correlation data with asynchronous time-correlated single-photon counting. *Optics Express*, *11*, 3583–3591.

CHAPTER 12

Biochemical Assay of G Protein-Coupled Receptor Oligomerization: Adenosine A_1 and Thromboxane A_2 Receptors Form the Novel Functional Hetero-oligomer

Natsumi Mizuno*, Tokiko Suzuki[†], Yu Kishimoto*, and Noriyasu Hirasawa*

Department of Pharmacotherapy of Life-style Related Disease, Graduate School of Pharmaceutical Sciences, Tohoku University, Sendai, Japan
[†]*Department of Cellular Signaling, Graduate School of Pharmaceutical Sciences, Tohoku University, Sendai, Japan*

CHAPTER OUTLINE

Introduction 214
12.1 Detection of GPCR Oligomerization 216
    12.1.1 Coimmunoprecipitation 216
        12.1.1.1 Coimmunoprecipitation of the GPCR Oligomer 216
        12.1.1.2 Protocol for Coimmunoprecipitation 216
        12.1.1.3 Coimmunoprecipitation Results 218
    12.1.2 $BRET^2$ Assay 218
        12.1.2.1 $BRET^2$ Assay to Analyze GPCR Oligomerization 218
        12.1.2.2 $BRET^2$ Saturation Assay 219
        12.1.2.3 Competitive $BRET^2$ Assay 219
12.2 Measurement of the Signaling Pathway 220
    12.2.1 Cyclic AMP Assay 220
    12.2.2 Extracellular signal-regulated Kinase1/2 Assay 222
    12.2.3 Effect of the Coexpression of Receptors on the Signaling Pathway .. 222
12.3 Materials and Methods 223
    12.3.1 Materials 223
    12.3.2 Construction of Plasmids 223
    12.3.3 Cell Culture and Transfection 223
    12.3.4 Western Blot Analysis 224
Conclusion 224
References 224

Abstract

G protein-coupled receptors (GPCRs) are classified into a family of seven transmembrane receptors. Receptor oligomerization may be the key to the expression and function of these receptors, for example, ligand binding, desensitization, membrane trafficking, and signaling. The accumulation of evidence that GPCRs form an oligomerization with a functional alternation may change the strategy for the discovery of novel drugs targeting GPCRs. Identification of the oligomer is essential to elucidate GPCR oligomerization. GPCR oligomerizations have been demonstrated using various biochemical approaches, which include the coimmunoprecipitation method, fluorescence resonance energy transfer assay, and bioluminescence RET assay. Thus, various assays are useful for the study of GPCR oligomerization, and we should choose the best method to match the purpose. We previously targeted adenosine A_1 and thromboxane A_2 (TP) receptors to form a functionally novel hetero-oligomer, since both receptors function in the same cells. This chapter describes the methods used to detect GPCR oligomerization and alterations in the signaling pathways, principally according to our findings on oligomerization between adenosine A_1 and TPα receptors.

INTRODUCTION

G protein-coupled receptors (GPCRs) are classified into a family of seven transmembrane receptors and are the targets for various clinical drugs in the market today. Increasing evidence has shown that homo- and hetero-oligomerization formed by various GPCRs play crucial roles in GPCR signaling (Milligan, 2009). Receptor oligomerization may be the key to the expression and function of these receptors, for example, ligand binding, desensitization, membrane trafficking, and signaling.

The $GABA_B$ receptor is known to be composed of two isoforms, the $GABA_{B1}$ and $GABA_{B2}$ subunits. Hetero-oligomerization between $GABA_{B1}$ and $GABA_{B2}$ produces the fully functional receptor at the cell surface but is nonfunctional when each monomer is expressed individually (Jones et al., 1998; Kaupmann et al., 1998). Studies on GPCR oligomerization would be valuable in understanding their roles in pathological conditions. Oligomerization may increase the diversity of GPCR signaling.

Identification of the oligomer is essential to elucidate GPCR oligomerization. Recently, predicting the interacting partners among GPCRs has been performed in several studies. Until now, GPCR oligomerizations have been demonstrated using various biochemical approaches.

In the early stages, the coimmunoprecipitation method, which is one of the biochemical methods generally used for GPCR oligomerization, was used to identify the β_2-adrenergic receptor homo-oligomer (Hebert et al., 1996). Coimmunoprecipitation strategies are simple and do not require special equipment. However, this method has several drawbacks. The solubilization of hydrophobic GPCRs can lead to aggregation, which may be mistaken for oligomerization. Furthermore, the solubilization process with detergents may dissociate between GPCRs (Ramsay, Kellett, McVey, Rees, & Milligan, 2002; Salim et al., 2002).

The use of resonance energy transfer (RET) may improve the drawbacks associated with coimmunoprecipitation. The RET technique allowed for the detection of sensitive receptor–receptor interactions in living cells, real-time monitoring, and visualization with microscopy (Ayoub & Pfleger, 2010; Pfleger & Eidne, 2005). Furthermore, RET is valuable for observing receptor–receptor interactions and also ligand–receptor interactions (Albizu et al., 2010). The initial approach for GPCR oligomerization using RET has been the fluorescence RET (FRET) assay (Overton & Blumer, 2000). Receptor–receptor interactions can be monitored by measuring RET between cyan fluorescent protein- and yellow fluorescent protein-fused receptors. Conventionally, RET-based techniques cannot clearly distinguish dimers from high-order oligomers. However, three-chromophore FRET technique, which is one of the RET-based techniques being developed, could extend to this issue (Galperin, Verkhusha, & Sorkin, 2004). Additionally, bioluminescence RET (BRET) is a biophysical technique that represents a powerful tool with which to measure protein–protein interactions in living cells (Albizu et al., 2010; Pfleger & Eidne, 2006). BRET commonly resorts to the use of *Renilla* luciferase (Rluc) as the donor. Rluc transfers energy to the green fluorescent protein (GFP) as the acceptor in the presence of coelenterazine, the substrate of Rluc. FRET monitors the energy transfer between two fluorescent proteins, while BRET can monitor without external excitation because it occurs after the oxidation of the substrate. BRET can avoid some of the problems associated with FRET, for example, photobleaching and the coincidence of the donor and acceptor (Pfleger & Eidne, 2006). Therefore, GPCR oligomerization has been confirmed with the BRET assay in a large number of studies (Pfleger & Eidne, 2005).

Recent studies on GPCR oligomerization have also demonstrated the paradigm of structural studies. Using atomic force microscopy techniques, rhodopsin molecules, which are categorized as class A GPCR, have been visualized as dimers in native membranes. These findings also provided unequivocal evidence that receptor dimers and multimers exist in native tissues (Fotiadis et al., 2003), although their functional relevance to GPCR dimerization remains controversial (Chabre, Deterre, & Antonny, 2009; Chabre & le Maire, 2005; Gurevich & Gurevich, 2008).

Thus, various assays are useful for the study of GPCR oligomerization, and we should choose the best method to match the purpose. On the other side of GPCR oligomerization, it is important to translate *in vitro* observations to the physiologically relevant functions of cells and tissues. However, this is still difficult because GPCRs function in an intricate environment. Diversification of the signaling pathway, a requirement for membrane trafficking and altering the ligand specificity, should be taken into consideration to understand GPCR-mediated cell functions.

Clinical drugs in the market today have been produced based on the theory that GPCR is coupled to a single G protein in 1:1 stoichiometry. GPCR oligomerization can increase the diversity of cellular signaling and is likely to occur *in vivo*. Therefore, the accumulation of evidence that GPCRs form an oligomerization with a functional alternation may change the strategy for the discovery of novel drugs targeting GPCRs.

We previously targeted adenosine A_1 and thromboxane A_2 (TP) receptors to form a functionally novel hetero-oligomer, since both receptors function in the same cells (Dare, Schulte, Karovic, Hammarberg, & Fredholm, 2007; Martinez-Salgado, Garcia-Cenador, Fuentes-Calvo, Macias Nunez, & Lopez-Novoa, 2007; Nakahata, 2008). Adenosine receptors have been subclassified into A_1, A_{2A}, A_{2B}, and A_3 subtypes (Fredholm, IJzerman, Jacobson, Klotz, & Linden, 2001), and human thromboxane A_2 receptors have been divided into two subtypes, TXA_2 receptor α and $TP\alpha$ and TXA_2 receptor β (Hirata et al., 1991). This chapter describes the methods used to detect GPCR oligomerization and alterations in the signaling pathways, principally according to our findings on oligomerization between adenosine A_1 and $TP\alpha$ receptors.

12.1 DETECTION OF GPCR OLIGOMERIZATION
12.1.1 Coimmunoprecipitation
12.1.1.1 Coimmunoprecipitation of the GPCR oligomer

Coimmunoprecipitation has been the most frequently used and effective method for studying GPCR oligomerization (Kroeger, Pfleger, & Eidne, 2003). This technique requires the solubilization of cells expressing the two targeted GPCRs. Importantly, the highly hydrophobic nature of the seven transmembrane domains makes the solubilization of GPCRs difficult. The detergent used must be strong enough to extract GPCRs from cell membranes; however, it sometimes dissociates the oligomers. There is also a risk of the detection of nonspecific complexes produced during this step. It is important to carefully choose the detergent and examine the composition of the solubilizing buffer. Additionally, the absence of coimmunoprecipitation should be confirmed in cells expressing only one receptor of the target or coexpressing the noninteracting partner (Jordan & Devi, 1999). We coimmunoprecipitated homo- and hetero-GPCR oligomers using cells individually coexpressing target GPCRs with distinct tags. In addition, we performed immunoprecipitation using lysates of the cells expressing only one receptor of the target receptors to assess the validity of the method used. We consequently obtained results to suggest that some GPCRs form homo- and/or hetero-oligomers (Suzuki, Namba, Mizuno, & Nakata, 2013; Suzuki, Namba, Tsuga, & Nakata, 2006).

In the following section, we described the protocol for the coimmunoprecipitation of adenosine A_1 and $TP\alpha$ receptors.

12.1.1.2 Protocol for coimmunoprecipitation

Human embryonic kidney 293T (HEK293T) cells were seeded onto 10 cm culture dishes at a density of 3.5×10^6 cells/dish. At 24 h post seeding, cells were transfected with myc-A_1R, HA-tagged $TP\alpha$ receptor (HA-$TP\alpha$), and HA-LPA1R, respectively. The transfected cells were collected by centrifugation at $1900 \times g$ and washed twice with Dulbecco's phosphate-buffered saline (Table 12.1). The cells were disrupted by sonication using a Handy Sonic UR-20P (Tomy Seiko, Tokyo, Japan) in 300 µl of

12.1 Detection of GPCR Oligomerization

Table 12.1 Reagents

Dulbecco's phosphate-buffered saline (DPBS)
137 mM NaCl
2.7 mM KCl
1.5 mM KH_2PO_4
8.1 mM Na_2HPO_4
Adjust pH to 7.4

Lysis buffer
20 mM Tris–HCl pH7.4
1 mM EDTA
150 mM NaCl
1% TritonX-100
1 mM Na_3VO_4
Store at 4 °C

Laemmli sample buffer
75 mM Tris–HCl
2% SDS
10% glycerol
3% 2-mercaptoethanol
0.003% bromophenol blue
Store at −20 °C

$BRET^2$ buffer
DPBS containing
0.1 mg/ml $CaCl_2$
0.1 mg/ml $MgCl_2$
1 mg/ml D-glucose

Tyrode–HEPES solution
137 mM NaCl
2.7 mM KCl
1.0 mM MgCl
1.8 mM $CaCl_2$
10 mM HEPES
5.6 mM glucose
Adjust pH to 7.4
Store at 4 °C

lysis buffer (Table 12.1). After rotating for 3 h at 4 °C, the solution was centrifuged at 17,400 × g for 20 min at 4 °C, and the supernatant was precleared with 30 μl/ml of 50% (v/v) Protein G Sepharose™ 4 Fast Flow (GE Healthcare, Piscataway, NJ) in lysis buffer, followed by centrifugation at 17,400 × g for 10 s to remove nonspecifically bound proteins to Sepharose. Anti-myc 9E10 antibodies (10 μg/ml) were added into the supernatant and rotated for 1 h at 4 °C. Fifty microliters per milliliter of 50% (v/v) Protein G Sepharose™ 4 Fast Flow was then added into the supernatant and rotated for a further 2 h. The mixture was centrifuged, the resulting precipitate was washed twice with 500 μl of lysis buffer, and bound proteins were eluted by incubation with 30 μl of the Laemmli sample buffer (Table 12.1) at room temperature.

12.1.1.3 Coimmunoprecipitation results

We examined using coimmunoprecipitation whether adenosine A_1 and TPα receptors form hetero-oligomers. Using cells cotransfected with myc-A_1R and HA-TPα, we detected myc-A_1R in the complex precipitated with anti-myc antibodies, which suggested the validity of the method used here (Mizuno, Suzuki, Hirasawa, & Nakahata, 2012). Interestingly, HA-TPα was observed in the complex precipitated with anti-myc antibodies only in the cells cotransfected with myc-A_1R and HA-TPα. To confirm that HP-TPα did not form an inappropriate aggregation, we explored the immunoprecipitation using the cells transfected with HA-TPα alone and the cells cotransfected with myc-A_1R and HA-LPA1R, which did not interact with each other. In these cases, we did not detect a specific band in the complex precipitated with anti-myc antibodies, which indicated that HP-TPα and myc-A_1R formed a specific interaction.

12.1.2 BRET² assay
12.1.2.1 BRET² assay to analyze GPCR oligomerization

BRET² is a method to sensitively detect receptor oligomerization in intact living cells. Biophysical assays based on RET are valuable to overcome the issues of artificial aggregation in the coimmunoprecipitation described earlier. A large number of GPCRs have recently been reported to form homo- or hetero-oligomerization using RET (Pfleger & Eidne, 2006).

The efficiency of RET depends on the degree of the spectral overlap and the distance between the donor and acceptor. BRET typically occurs in the 1–10 nm region, which is consistent with the distance in which two molecules can interact with each other (Fig. 12.1). BRET² is an advanced BRET technology that uses Rluc as the

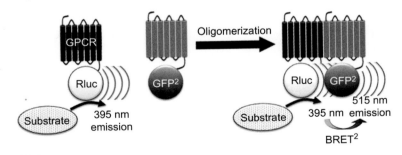

FIGURE 12.1

A schematic representation of the principle of BRET² in GPCR oligomerization. BRET² consists of nonradiative energy transfer between a donor, *Renilla* luciferase (Rluc), and an acceptor, modified green fluorescent protein (GFP²). Excitation of the donor results in the emission of light by the acceptor only if the two molecules are in close proximity (10∼100 Å).

donor, the codon-humanized form of GFP (GFP^2) as the acceptor, and bisdeoxycoelenterazine (DeepBlueC) as the substrate instead of benzyl-coelenterazine, which has often been used in the classic BRET technique (Ayoub & Pfleger, 2010; Pfleger & Eidne, 2006). To apply BRET^2 to the study of GPCR oligomerization, Rluc and GFP^2 must be fused to the GPCRs of interest, and the BRET^2 signal can be measured in the cells expressing the target receptors.

12.1.2.2 BRET^2 saturation assay

The BRET saturation assay is performed using cells coexpressing a fixed amount of the Rluc-fused receptor of interest and increasing amounts of another GFP-fused receptor (Terrillon et al., 2003). Oligomerization induces saturation of the BRET signal according to high concentrations of the GFP-fused receptor, whereas nonspecific interactions cause a continuous rise in BRET signals with increasing amounts of the Rluc-fused receptor in a pseudolinear pattern (Mercier, Salahpour, Angers, Breit, & Bouvier, 2002). We previously examined GPCR oligomerization using the BRET^2 saturation assay with cells expressing the Rluc- or GFP^2-fused adenosine A_1 receptor. Consequently, it became possible to determine the specific BRET^2 signal for homo-oligomerization of the adenosine A_1 receptor in living cells (Suzuki et al., 2009), and this result supported the previous studies that showed indicating homo-oligomerization of the adenosine A_1 receptor using other methods (Ciruela et al., 1995; Yoshioka, Hosoda, Kuroda, & Nakata, 2002).

12.1.2.3 Competitive BRET^2 assay

Competitive BRET^2 can occur when BRET^2 between two partners expressed at a fixed donor–acceptor ratio is inhibited by the increasing coexpression of nonfused partners, while no inhibition was seen with a nonfused noninteracting protein (Marullo & Bouvier, 2007). Therefore, the competitive BRET^2 assay allows distinguished oligomerization to be from random collisions between overexpressing proteins.

We performed a competitive BRET^2 assay in HEK293T cells transiently transfected with Rluc- and GFP^2-fused HA-tagged adenosine A_1 receptor (HA-A_1R) using nonfused HA-TPα. The interaction between adenosine A_1 and TPα receptors was assessed based on the competition of adenosine A_1 receptor homodimerization with the heterodimerization of adenosine A_1 and TPα receptors in living cells.

12.1.2.3.1 Protocol for the competitive BRET^2 assay

HEK293T cells (5×10^5 cells/35 mm dish) were cotransfected with a fixed amount of the HA-A_1R-Rluc and HA-A_1R-GFP^2 plasmids and increasing amounts of the unfused plasmids, HA-A_1R, HA-TPα, or HA-LPA1R (Table 12.2). Nontransfected cells were used as a control. The cells were harvested and suspended in BRET^2 buffer at 48 h after transfection (Table 12.1). Suspended cells were distributed in a white-walled 96-well plate (OptiPlate, Perkin Elmer Life Sciences, Boston, MA) at a density of 1×10^6 cells/well and incubated for 20 min at 37 °C. DeepBlueC (Perkin Elmer Life Sciences) or coelenterazine 400A (Biotium Inc., CA) was then added at a final concentration of

Table 12.2 Concentrations of Plasmids for the Competitive BRET^2 Assay

	HA-A$_1$R-Rluc	HA-A$_1$R-GFP^2	Unfused Receptor Plasmids
HA-A$_1$R-Rluc/HA-A$_1$R-GFP^2/ HA-A$_1$R	1	1	0, 0.4, 0.8, 1.2, 1.6, 2.0
HA-A$_1$R-Rluc/HA-A$_1$R-GFP^2/ HA-TPα	1	1	0, 0.4, 0.8, 1.2, 1.6, 2.0
HA-A$_1$R-Rluc/HA-A$_1$R-GFP^2/ HA-LPA1R	1	1	0, 0.6, 1.2, 1.8, 2.7, 3.6

Plasmids (μg/5 × 10^5 cells)
Adjust the amount of total plasmids using vector plasmids.

5 μM. Assays were conducted immediately using a Fusion α universal microplate analyzer (Perkin Elmer Life Sciences) for the detection of Rluc at 410 nm and GFP^2 at 515 nm. Receptor association calculated as the BRET ratio are emission at 515 nm over emission at 410 nm. The values were collected by subtracting the background signal detected using nontransfected cells in the following equation:

$$\text{BRET}^2 \text{ ratio} = \frac{(515\,\text{nm emission} - \text{control}\,515\,\text{nm emission})}{(410\,\text{nm emission} - \text{control}\,410\,\text{nm emission})}$$

12.1.2.3.2 Results of the competitive BRET^2 assay

The expression of HA-TPα or HA-A$_1$R in cells coexpressing Rluc- and GFP^2-fused HA-A$_1$R decreased the BRET^2 signals for homo-oligomerization between HA-A$_1$R-Rluc and HA-A$_1$R-GFP^2 (Fig. 12.2A and B). In contrast, the expression of HA-LPA1R did not eliminate the BRET^2 signal for A$_1$R receptor homodimerization (Fig. 12.2C). These results suggest that the adenosine A$_1$ receptor selectively formed a homo-oligomer or hetero-oligomer with the TPα receptor. Thus, the formation of hetero-oligomerization could be demonstrated in living cells.

12.2 MEASUREMENT OF THE SIGNALING PATHWAY

12.2.1 Cyclic AMP assay

The pertussis toxin (PTX)-sensitive family of G proteins, G$_{i/o}$ proteins, inhibits cyclic AMP (cAMP) production. The adenosine A$_1$ receptor is mainly coupled to G$_{i/o}$ proteins (Ralevic & Burnstock, 1998). To investigate whether the formation of hetero-oligomerization between adenosine A$_1$ and TPα receptors affects signaling by the adenosine A$_1$ receptor via G$_{i/o}$, we examined intracellular cAMP levels using cells coexpressing both receptors.

The protocol for the cAMP assay was as follows. HEK293T cells were seeded onto 48-well plates at a density of 2×10^4 cells/well. The cells were transfected with

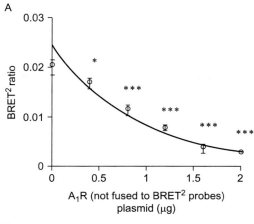
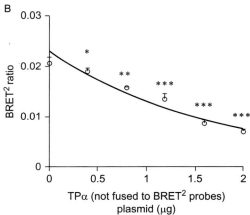
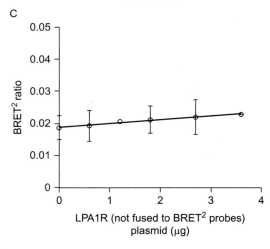

FIGURE 12.2

The competitive BRET^2 assay for GPCR oligomerization. HEK293T cells were cotransfected with a fixed concentration of HA-A$_1$R-Rluc and HA-A$_1$R-GFP^2 plasmids and increasing amounts of HA-A$_1$R (A), HA-TPα (B), or HA-LPA1R (C) plasmids. Data represent the mean ± SEM for three independent experiments. *$P<0.05$, **$P<0.005$, ***$P<0.001$ versus 0 (Dunnett's test).

the 3HA-adenosine A_1 receptor (3HA-A_1R) and/or HA-TPα plasmids 24 h post seeding. After 48 h, the medium was changed to Eagle's minimum essential medium, 20 mM HEPES, and preincubated for 20 min at 37 °C. The cells were incubated with 100 μM of Ro 20-1724 as a phosphodiesterase inhibitor for 15 min with or without receptor antagonists. The cells were then stimulated with a receptor agonist and 100 μM forskolin as an adenylyl cyclase activator and were incubated for another 10 min. Reactions were terminated by adding 2.5% perchloric acid. Acid extracts were mixed with a 1/10 volume of 4.2 N KOH to neutralize the acid, which resulted in the formation of potassium perchlorate as the precipitate. The cAMP in the supernatant was succinylated and determined using a radioimmunoassay kit (Yamasa, Tokyo, Japan) according to the manufacturer's directions. Briefly, the mixture of succinyl cAMP and fixed amount of [^{125}I]cAMP was reacted with anti-cAMP antibodies. After the removal of unbound cAMP by adsorption on dextran-coated charcoal radioactivity, the supernatant was measured with a gamma spectrometer.

12.2.2 Extracellular signal-regulated kinase1/2 assay

One of the key components in GPCR-induced signaling is mitogen-activated protein kinase (MAPK) cascades. It has been reported that extracellular signal-regulated kinase1/2 (ERK1/2), a major kinase of MAPK cascades, is phosphorylated by the individual activation of both the adenosine A_1 and TPα receptors (Dickenson, Blank, & Hill, 1998; Miggin & Kinsella, 2001). We examined whether the coexpression of adenosine A_1 and TPα receptors affected ERK1/2 phosphorylation.

The protocol for the activation of the ERK1/2 assay was as follows. HEK293T cells were seeded onto 12-well plates at a density of 1×10^5 cells/well. The cells were transfected with the 3HA-A_1R and HA-TPα plasmids 24 h post seeding. The cells were washed with a Tyrode–HEPES solution 48 h after transfection (Table 12.1) and were preincubated for 20 min at 37 °C. Stimulation with CPA and/or U46619, which are adenosine A_1 and TPα receptors agonists, respectively, for 10 min at 37 °C, was terminated by aspiration of the medium, and cells were lysed in 150 μl of ice-cold Laemmli sample buffer.

12.2.3 Effect of the coexpression of receptors on the signaling pathway

The coexpression of adenosine A_1 and TPα receptors affected receptor signal transduction. The accumulation of cAMP and activation of ERK1/2 were enhanced by the costimulation of both receptor agonists (Mizuno et al., 2012). These enhancements were only observed in the coexpressing cells and not in the cells expressing either of these receptors alone, which indicated that ligand–receptor-selective intracellular responses were modulated by another ligand–receptor system via receptor oligomerization and/or crosstalk between signals.

12.3 MATERIALS AND METHODS
12.3.1 Materials
Forskolin was purchased from Wako Pure Chemicals (Osaka, Japan). PTX and 4-(3-butoxy-4-methoxybenzyl) imidazolidin-2-one (Ro20-1724) were from Calbiochem (San Diego, CA). 1S-[1α,2α (Z),3α,4α]-7-[3-[[2-[(phenylamino)carbonyl]hydrazine] methyl]-7-oxabicyclo[2.2.1]hept-2-yl]-5-heptenoic acid (SQ29548) and 9,11-dideoxy-9α,11α-epoxymethanoprostaglandin $F_2α$ (U46619) were from Cayman Chemical (Ann Arbor, MI). N^6-cyclopentyladenosine (CPA) and 8-cyclopentyl-1, 3-dipropylxanthine (DPCPX) were from Sigma Aldrich (St. Louis, MO). The anti-HA antibody and FuGENE HD Transfection Reagent were purchased from Roche Applied Science (Mannheim, Germany). HRP-conjugated anti-mouse IgG, Protein G Sepharose^{TM}, and ECL^{TM} Western blotting detection reagent were purchased from GE Healthcare (Piscataway, NJ). The anti-myc 9E10 antibody was purchased from Covance (Berkeley, CA). The anti-adenosine A_1 receptor antibody was from Acris Antibodies GmbH (Hiddenhausen, Germany). The anti-ERK 1/2 antibody, anti-phospho-ERK 1/2 antibody, and HRP-conjugated anti-rabbit IgG were purchased from Cell Signaling Technology (Beverly, MA). HRP-conjugated anti-rat IgG was purchased from Santa Cruz Biotechnology (Santa Cruz, CA). The 3HA-A_1R and HA-lysophosphatidic acid 1 receptor (HA-LPA1R) plasmids were purchased from UMR cDNA Resource Center (Rolla, MO). Other chemicals used were of reagent grade or the highest quality available.

12.3.2 Construction of plasmids
The HA-TPα and HA-A_1R had the hemagglutinin (HA) epitope tag (YPYDVPDYA), and myc-tagged adenosine A_1 receptor (myc-A_1R) had the myc epitope tag (EQKLISEEDL) at the N-terminus of the target receptors, which were constructed as described previously (Sasaki et al., 2007; Suzuki et al., 2006).

We constructed HA-A_1R-Rluc (*Renilla* luciferase) and HA-A_1R-GFP^2 for the BRET^2 assay by amplifying the HA-A_1R sequence, a without stop codon, using sense and antisense primers containing distinct restriction enzyme sites at the 5′ and 3′ ends (Suzuki et al., 2006).

12.3.3 Cell culture and transfection
HEK293T cells were grown in Dulbecco's modified Eagle's medium (Sigma Aldrich, St. Louis, MO) containing 10% fetal calf serum, 50 units/ml penicillin, and 50 μg/ml streptomycin in a humidified incubator with a 5% CO_2 atmosphere at 37 °C. Transfections were done with FuGENE HD Transfection Reagent as described previously (Suzuki et al., 2006). Briefly, the optimal volumes of each plasmid and FuGENE HD Transfection Reagent were added to Opti-MEM1,

respectively, and were immediately mixed. After incubation for 15 min at room temperature, the FuGENE HD Transfection Reagent/DNA mixture was added to the cells. The cells were assayed at 48 h posttransfection.

12.3.4 Western blot analysis

Samples were loaded on a 10% sodium dodecyl sulfate (SDS)-polyacrylamide gel for electrophoresis and transferred to polyvinylidene difluoride membranes. The membranes were blocked with 5% nonfat milk in TBST (10 mM Tris–HCl, 100 mM NaCl, and 0.05% Tween 20, pH 7.4), incubated with the anti-HA antibody (1:2000), anti-adenosine A_1 receptor antibody (1:2000), anti-ERK1/2 antibody (1:2000), or anti-phospho ERK1/2 antibody (1:2000) for 90 min followed by HRP-conjugated anti-rat IgG (1:2000) or anti-rabbit IgG (1:5000) for 90 min at room temperature, and were detected with ECL™ Western blotting detection reagents.

CONCLUSION

In this section, we described the methods to identify GPCR oligomerization, coimmunoprecipitation and $BRET^2$ techniques. Coimmunoprecipitation is the most general biochemical approach used to detect GPCR oligomerization. However, coimmunoprecipitation is limited to declare a specific interaction of GPCRs. Additional methods, such as the $BRET^2$ assay, which can support the weak point of coimmunoprecipitation, are essential for conclusions regarding GPCRs forming oligomers. A large number of studies have also addressed the effect of receptor functions (Panetta & Greenwood, 2008; Szidonya, Cserzo, & Hunyady, 2008). Likewise, we showed that signal transductions were affected using cells coexpressing adenosine A_1 and TPα receptors (Mizuno et al., 2012). However, as not all receptors expressed in the cells form hetero-oligomers, the study of GPCR functions using coexpression cells could not reflect the events induced directly by oligomerization. This is one area that needs to be improved in the future about our studies. The accumulation of evidence on GPCR oligomerization and the accompanying functional effects with various biochemical methods will be significant for the elucidation of physiological mechanisms and novel drug discoveries.

References

Albizu, L., Cottet, M., Kralikova, M., Stoev, S., Seyer, R., Brabet, I., et al. (2010). Time-resolved FRET between GPCR ligands reveals oligomers in native tissues. *Nature Chemical Biology*, *6*, 587–594.

Ayoub, M. A., & Pfleger, K. D. (2010). Recent advances in bioluminescence resonance energy transfer technologies to study GPCR heteromerization. *Current Opinion in Pharmacology, 10*, 44–52.

Chabre, M., Deterre, P., & Antonny, B. (2009). The apparent cooperativity of some GPCRs does not necessarily imply dimerization. *Trends in Pharmacological Sciences, 30*, 182–187.

Chabre, M., & le Maire, M. (2005). Monomeric G-protein-coupled receptor as a functional unit. *Biochemistry, 44*, 9395–9403.

Ciruela, F., Casado, V., Mallol, J., Canela, E. I., Lluis, C., & Franco, R. (1995). Immunological identification of A1 adenosine receptors in brain cortex. *Journal of Neuroscience Research, 42*, 818–828.

Dare, E., Schulte, G., Karovic, O., Hammarberg, C., & Fredholm, B. B. (2007). Modulation of glial cell functions by adenosine receptors. *Physiology & Behavior, 92*, 15–20.

Dickenson, J. M., Blank, J. L., & Hill, S. J. (1998). Human adenosine A1 receptor and P2Y2-purinoceptor-mediated activation of the mitogen-activated protein kinase cascade in transfected CHO cells. *British Journal of Pharmacology, 124*, 1491–1499.

Fotiadis, D., Liang, Y., Filipek, S., Saperstein, D. A., Engel, A., & Palczewski, K. (2003). Atomic-force microscopy: Rhodopsin dimers in native disc membranes. *Nature, 421*, 127–128.

Fredholm, B. B., IJzerman, A. P., Jacobson, K. A., Klotz, K. N., & Linden, J. (2001). International Union of Pharmacology. XXV. Nomenclature and classification of adenosine receptors. *Pharmacological Reviews, 53*, 527–552.

Galperin, E., Verkhusha, V. V., & Sorkin, A. (2004). Three-chromophore FRET microscopy to analyze multiprotein interactions in living cells. *Nature Methods, 1*, 209–217.

Gurevich, V. V., & Gurevich, E. V. (2008). How and why do GPCRs dimerize? *Trends in Pharmacological Sciences, 29*, 234–240.

Hebert, T. E., Moffett, S., Morello, J. P., Loisel, T. P., Bichet, D. G., Barret, C., et al. (1996). A peptide derived from a beta2-adrenergic receptor transmembrane domain inhibits both receptor dimerization and activation. *The Journal of Biological Chemistry, 271*, 16384–16392.

Hirata, M., Hayashi, Y., Ushikubi, F., Yokota, Y., Kageyama, R., Nakanishi, S., et al. (1991). Cloning and expression of cDNA for a human thromboxane A2 receptor. *Nature, 349*, 617–620.

Jones, K. A., Borowsky, B., Tamm, J. A., Craig, D. A., Durkin, M. M., Dai, M., et al. (1998). GABA(B) receptors function as a heteromeric assembly of the subunits GABA(B)R1 and GABA(B)R2. *Nature, 396*, 674–679.

Jordan, B. A., & Devi, L. A. (1999). G-protein-coupled receptor heterodimerization modulates receptor function. *Nature, 399*, 697–700.

Kaupmann, K., Malitschek, B., Schuler, V., Heid, J., Froestl, W., Beck, P., et al. (1998). GABA(B)-receptor subtypes assemble into functional heteromeric complexes. *Nature, 396*, 683–687.

Kroeger, K. M., Pfleger, K. D., & Eidne, K. A. (2003). G-protein coupled receptor oligomerization in neuroendocrine pathways. *Frontiers in Neuroendocrinology, 24*, 254–278.

Martinez-Salgado, C., Garcia-Cenador, B., Fuentes-Calvo, I., Macias Nunez, J. F., & Lopez-Novoa, J. M. (2007). Effect of adenosine in extracellular matrix synthesis in human and rat mesangial cells. *Molecular and Cellular Biochemistry, 305*, 163–169.

Marullo, S., & Bouvier, M. (2007). Resonance energy transfer approaches in molecular pharmacology and beyond. *Trends in Pharmacological Sciences, 28*, 362–365.

Mercier, J. F., Salahpour, A., Angers, S., Breit, A., & Bouvier, M. (2002). Quantitative assessment of beta 1- and beta 2-adrenergic receptor homo- and heterodimerization by bioluminescence resonance energy transfer. *The Journal of Biological Chemistry, 277,* 44925–44931.

Miggin, S. M., & Kinsella, B. T. (2001). Thromboxane A(2) receptor mediated activation of the mitogen activated protein kinase cascades in human uterine smooth muscle cells. *Biochimica et Biophysica Acta, 1539,* 147–162.

Milligan, G. (2009). G protein-coupled receptor hetero-dimerization: Contribution to pharmacology and function. *British Journal of Pharmacology, 158,* 5–14.

Mizuno, N., Suzuki, T., Hirasawa, N., & Nakahata, N. (2012). Hetero-oligomerization between adenosine A(1) and thromboxane A(2) receptors and cellular signal transduction on stimulation with high and low concentrations of agonists for both receptors. *European Journal of Pharmacology, 677,* 5–14.

Nakahata, N. (2008). Thromboxane A2: Physiology/pathophysiology, cellular signal transduction and pharmacology. *Pharmacology & Therapeutics, 118,* 18–35.

Overton, M. C., & Blumer, K. J. (2000). G-protein-coupled receptors function as oligomers in vivo. *Current Biology, 10,* 341–344.

Panetta, R., & Greenwood, M. T. (2008). Physiological relevance of GPCR oligomerization and its impact on drug discovery. *Drug Discovery Today, 13,* 1059–1066.

Pfleger, K. D., & Eidne, K. A. (2005). Monitoring the formation of dynamic G-protein-coupled receptor-protein complexes in living cells. *The Biochemical Journal, 385,* 625–637.

Pfleger, K. D., & Eidne, K. A. (2006). Illuminating insights into protein–protein interactions using bioluminescence resonance energy transfer (BRET). *Nature Methods, 3,* 165–174.

Ralevic, V., & Burnstock, G. (1998). Receptors for purines and pyrimidines. *Pharmacological Reviews, 50,* 413–492.

Ramsay, D., Kellett, E., McVey, M., Rees, S., & Milligan, G. (2002). Homo- and hetero-oligomeric interactions between G-protein-coupled receptors in living cells monitored by two variants of bioluminescence resonance energy transfer (BRET): Hetero-oligomers between receptor subtypes form more efficiently than between less closely related sequences. *The Biochemical Journal, 365,* 429–440.

Salim, K., Fenton, T., Bacha, J., Urien-Rodriguez, H., Bonnert, T., Skynner, H. A., et al. (2002). Oligomerization of G-protein-coupled receptors shown by selective co-immunoprecipitation. *The Journal of Biological Chemistry, 277,* 15482–15485.

Sasaki, M., Sukegawa, J., Miyosawa, K., Yanagisawa, T., Ohkubo, S., & Nakahata, N. (2007). Low expression of cell-surface thromboxane A2 receptor beta-isoform through the negative regulation of its membrane traffic by proteasomes. *Prostaglandins & Other Lipid Mediators, 83,* 237–249.

Suzuki, T., Namba, K., Mizuno, N., & Nakata, H. (2013). Hetero-oligomerization and specificity changes of G protein-coupled purinergic receptors: Novel insight into diversification of signal transduction. *Methods in Enzymology, 521,* 239–257.

Suzuki, T., Namba, K., Tsuga, H., & Nakata, H. (2006). Regulation of pharmacology by hetero-oligomerization between A1 adenosine receptor and P2Y2 receptor. *Biochemical and Biophysical Research Communications, 351,* 559–565.

Suzuki, T., Namba, K., Yamagishi, R., Kaneko, H., Haga, T., & Nakata, H. (2009). A highly conserved tryptophan residue in the fourth transmembrane domain of the A adenosine receptor is essential for ligand binding but not receptor homodimerization. *Journal of Neurochemistry, 110,* 1352–1362.

Szidonya, L., Cserzo, M., & Hunyady, L. (2008). Dimerization and oligomerization of G-protein-coupled receptors: Debated structures with established and emerging functions. *The Journal of Endocrinology, 196*, 435–453.

Terrillon, S., Durroux, T., Mouillac, B., Breit, A., Ayoub, M. A., Taulan, M., et al. (2003). Oxytocin and vasopressin V1a and V2 receptors form constitutive homo- and heterodimers during biosynthesis. *Molecular Endocrinology, 17*, 677–691.

Yoshioka, K., Hosoda, R., Kuroda, Y., & Nakata, H. (2002). Hetero-oligomerization of adenosine A1 receptors with P2Y1 receptors in rat brains. *FEBS Letters, 531*, 299–303.

CHAPTER 13

Oligomerization of Sweet and Bitter Taste Receptors

Christina Kuhn[*] and Wolfgang Meyerhof[†]

[*]Axxam SpA, Bresso (Milan), Italy
[†]Department of Molecular Genetics, German Institute of Human Nutrition Potsdam-Rehbruecke, Nuthetal, Germany

CHAPTER OUTLINE

Introduction and Rationale	230
13.1 Materials	232
13.2 Methods	233
13.2.1 Generation of Expression Constructs	233
13.2.2 Expression Analysis of Receptor Constructs	234
13.2.3 Coimmunoprecipitation	235
13.2.4 Bioluminescence Resonance Energy Transfer	237
13.3 Discussion	240
Summary	241
Acknowledgments	241
References	241

Abstract

The superfamily of G protein-coupled receptors (GPCRs) mediates numerous physiological processes, including neurotransmission, cell differentiation and metabolism, and sensory perception. In recent years, it became evident that these receptors might function not only as monomeric receptors but also as homo- or heteromeric receptor complexes. The family of TAS1R taste receptors are prominent examples of GPCR dimerization as they act as obligate functional heteromers: TAS1R1 and TAS1R3 combine to form an umami taste receptor, while the combination of TAS1R2 and TAS1R3 is a sweet taste receptor. So far, TAS2Rs, a second family of ~25 taste receptors in humans that mediates responses to bitter compounds, have been shown to function on their own, but if they do so as receptor monomers or as homomeric receptors still remains unknown. Using two different experimental approaches, we have recently shown that TAS2Rs can indeed form both homomeric and heteromeric receptor complexes. The employed techniques, coimmunoprecipitations and bioluminescence resonance energy transfer (BRET), are based on different principles and complement each other well and therefore provided compelling evidences for TAS2R oligomerization. Furthermore, we have adapted the protocols to include a number of controls and for higher throughput to accommodate the investigation of a large number of receptors and receptor combinations. Here, we present the protocols in detail.

INTRODUCTION AND RATIONALE

Our sense of taste serves us in our daily lives to advise us on which foods are safe and worth consuming and turns eating into pleasure. On the molecular level, the different taste modalities—sweet, umami, bitter, salty, and sour—are detected by different taste receptors expressed on the tongue, ionotropic receptors and metabotropic receptors. Taste receptors of the TAS1R (also referred to as T1R) family, mediating sweet and umami taste, and those of the TAS2R (or T2R) family, mediating bitter taste, belong to the superfamily of G protein-coupled receptors (GPCRs) (Yarmolinsky, Zuker, & Ryba, 2009). Both families are unrelated, however; the $\sim$25 members of the human TAS2R family have short N-termini and are distantly related to opsins, whereas the three different TAS1Rs are class C GPCRs, have long N-termini, and are related to metabotropic glutamate and $GABA_B$ receptors.

While it was initially common belief that class A (rhodopsin-like) and class B (secretin receptor family) GPCRs act as singular units, over the past 10–15 years, many reports demonstrated that GPCRs might not (only) function as monomeric receptors but that they can also assemble with one another and form receptor dimers or even higher-order oligomers with potentially dramatic and important implications (see, e.g., George, O'Dowd, & Lee, 2002). However, many of these discoveries were made in *in vitro* systems and the functional and physiological significance of these findings remained debated. But at least for some receptors, a functional significance and/or relevance *in vivo* is evident. Recently, *in vivo* functional dimerization could be demonstrated for the luteinizing hormone receptor. Mice with either a binding-deficient mutant receptor or a signaling-deficient mutant receptor have only small gonads and are sterile. In contrast, when mice are heterozygous for both mutants, receptor dimerization will lead to receptor complementation and the phenotype of the mice is rescued; they have normal-sized gonads and are fertile (Rivero-Müller et al., 2009).

Many studies have also established that for class C GPCRs, homo- or hetero-oligomerization is indispensable for receptor function (Pin, Galvez, & Prezeau, 2003). For example, $GABA_B$ receptors are formed by $GABA_B$ R1 and $GABA_B$ R2 class C GPCRs. The assembly of both receptor subunits was shown to take place in neurons *in vivo* and is essential for the targeting of the dimeric $GABA_B$ receptors to the plasma membrane and for ligand binding and signaling (Jones et al., 1998,

Kaupmann et al., 1998). Other class C GPCRs, for example, the metabotropic glutamate receptors and the calcium-sensing receptor, function as homomers.

Indeed, the receptors of the TAS1R family are prime examples for GPCR dimerization. Expression analysis of the receptors in taste buds gave early hints about their possible structure and function: TAS1R3 is coexpressed with either TAS1R1 or TAS1R2, but neither receptor seems to be expressed alone at significant levels (Nelson et al., 2001). Functional analysis of the receptors in heterologous expression systems then showed that the combination of TAS1R3 with TAS1R1 is an umami taste receptor, detecting L-glutamate (human receptors) or a variety of L-amino acids (mouse receptors), while TAS1R2/TAS1R3 serves as a general sweet taste receptor and is activated not only by natural sugars but also by artificial sweeteners and sweet-tasting D-amino acids and proteins (Li et al., 2002, Nelson et al., 2001, 2002). Studies using knockout mice subsequently confirmed these results *in vivo*; both respective subunits of the sweet and umami taste receptor are necessary for functional detection of the molecules and the behavioral attraction of the animals towards sweet and umami tastants (Zhao et al., 2003). Furthermore, actual active dimerization of the receptors could be shown—at least *in vitro*—using two different techniques: coimmunoprecipitation (CoIP) and bioluminescence resonance energy transfer (BRET) (Jiang et al., 2004; Nelson et al., 2002). All of these findings taken together make a very strong case for TAS1Rs as obligate heteromeric receptors, and TAS1R3 remains until now the only known GPCR that was shown to be an essential partner in functionally distinct receptors with essentially nonoverlapping ligand profiles.

Mammals have another family of taste receptors, the TAS2Rs, with ~25 members in humans (Shi, Zhang, Yang, & Zhang, 2003). Many evidences obtained from *in vitro* and *in vivo* studies showed that these receptors are bitter taste receptors that detect potentially toxic substances and induce aversive behaviors towards them (see, e.g., Behrens, Reichling, Batram, Brockhoff, & Meyerhof, 2009; Yarmolinsky et al., 2009). Indeed, most of the receptors of this family have been deorphanized by now. While some receptors seem to be specialized receptors detecting only few substances, many of them show a broad ligand spectrum, a characteristic that might explain how thousands of bitter-tasting substances can be detected with a rather small set of receptors (Meyerhof et al., 2010). The question if also these receptors form protein assemblies remained, however. To answer this question, we investigated TAS2R oligomerization with two different experimental approaches and found that TAS2Rs can indeed form receptor–receptor complexes (Kuhn, Bufe, Batram, & Meyerhof, 2010). Oligomerization seemed to be a general feature of these receptors as we found that 90% of all possible binary receptor combinations showed positive results.

In this chapter, we present the experimental protocols for analysis of oligomerization of TAS1Rs and TAS2Rs in greater detail compared to what we reported in our previously published study (Kuhn et al., 2010). As the number of receptors to be investigated is large (there are 25 different TAS2Rs and 325 different possibilities for combinations of two receptors!), we want to put special emphasis on the adaption of the used methods for higher throughput.

13.1 MATERIALS

INSTRUMENTS

Ultracentrifuge
Luminescent image analyzer
Luminescence plate reader (with emission filters 370–450 and 500–530 nm), we used a FLUOstar OPTIMA (BMG Labtech)
Fluorescence plate reader (with excitation at 488 nm and emission at 515 nm), we used a FLIPR (Molecular Devices)

VECTORS

Mammalian expression vector, for example, pcDNA3 (Invitrogen), pcDNA5/FRT/TO (Invitrogen), and pEAK10 (Edge Biosystems)
BRET vector, pGFP^2-MCS-Rluc(h) (PerkinElmer)

CELL LINES

HEK293T cells (we used a modified cell line stably expressing the Gα protein chimera G$\alpha_{16gust44}$)

CELL CULTURE DISHES AND PLATES

10 cm cell culture dishes, coated with poly-D-lysine
96-Well plates, white, clear-bottomed (Greiner Bio-One); coated with poly-D-lysine
96-Well plates, black, clear-bottomed (Greiner Bio-One); coated with poly-D-lysine

BUFFERS AND SOLUTIONS

PBS (140 mM NaCl, 10 mM Na$_2$HPO$_4$, 2.7 mM KCl, and 1.76 mM KH$_2$PO$_4$, pH 7.4)
C1 buffer (130 mM NaCl, 5 mM KCl, 10 mM HEPES, 2 mM CaCl$_2$, and 1 mM glucose, pH 7.4)
CoIP buffer (120 mM NaCl; 50 mM Tris, pH 8.0; 1 mM EDTA, pH 8.0; 0.5% (v/v) IGEPAL; 1 mM PMSF; 2 μg/ml leupeptin; 2 μg/ml pepstatin A; and 10 μg/ml aprotinin)
2× Laemmli buffer (125 mM Tris, pH 6.8; 4% (w/v) SDS; 20% (v/v) glycerol; and 0.04% (w/v) bromophenol blue) without DTT or mercaptoethanol!
TBS (10 mM Tris, pH 8.0, and 150 mM NaCl)
TBST (10 mM Tris, pH 8.0; 150 mM NaCl; and 0.2% (v/v) Tween)

ANTIBODIES

Rabbit anti-FLAG® antibody (Sigma-Aldrich, catalog #F7425)
Rabbit anti-GFP antibody (Abcam, catalog #ab6556)
Donkey anti-rabbit IgG, coupled to horseradish peroxidase (Amersham/GE Healthcare)
EZview™ Red anti-FLAG® M2 affinity gel (Sigma-Aldrich, catalog #F2426)

OTHER REAGENTS

Transfection reagent, for example, Lipofectamine® 2000 (Invitrogen)
BRET coelenterazine substrate DeepBlueC (PerkinElmer)
Amersham ECL Western blotting detection reagent (GE Healthcare)

13.2 METHODS

13.2.1 Generation of expression constructs

Functional expression of TAS2R receptors in mammalian cell lines requires that the receptors carry an additional sequence at their N-terminus (e.g., the 45 N-terminal amino acids of the rat somatostatin receptor 3, "rsstr3"), which facilitates export of the protein to the plasma membrane (Bufe, Hofmann, Krautwurst, Raguse, & Wolfgang Meyerhof, 2002). TAS1Rs on the other hand do not need any N-terminal modification to support their expression. For CoIP experiments, it is necessary to further include a FLAG-tag at the C-terminus of the receptors, while BRET experiments require constructs encoding for receptor–GFP^2 and receptor–*Renilla reniformis* luciferase (Rluc) fusion proteins.

For easy and fast generation of the receptor constructs, we suggest to first build a cloning cassette in a modular way (Fig. 13.1) in a mammalian expression vector (such as pcDNA5/FRT). Then, the following approach can be followed: First, GFP^2 and Rluc sequences are amplified from the BRET vector by PCR; the oligonucleotides should additionally contain the sequences of the restriction enzymes 3 and 4, thereby adding

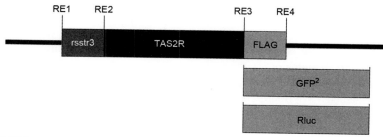

FIGURE 13.1

Modular cloning cassette for taste receptor expression constructs. A cloning cassette for TAS2Rs is generated in a mammalian expression vector, for example, pcDNA5/FRT. Sites for different restriction enzymes (RE) separate the modules from each other and enable easy subcloning of the different parts. rsstr3: sequence of the first 45 amino acids of the rat somatostatin receptor 3, used as N-terminal tag for TAS2Rs to enable functional plasma membrane expression of the receptors; TAS2R: TAS2R coding sequence without stop codon; FLAG: sequence of the FLAG-tag; GFP^2: sequence of a modified green fluorescent protein (GFP), used for BRET and CoIP experiments; Rluc: sequence of the luciferase of *Renilla reniformis*, used for BRET experiments. TAS1R sequences can be cloned into the same cassette using RE1 and RE3, replacing the combined sequences of rsstr3-TAS2R.

these sequences to the 5′ and 3′ ends of the GFP^2 and Rluc sequences, respectively. Next, these fragments are individually cloned into one cassette at the 3′ end of one receptor sequence; the correct sequence and the maintenance of the open reading frame should be confirmed. Because of its small size, the FLAG-tag can be added to the receptor sequence directly by PCR, bypassing this first cloning step. Subsequently, using restriction enzymes 2 and 3, the receptor sequence in the constructs in the preceding text can be replaced with the sequences of all other receptors. Alternatively, restriction enzyme 1 in combination with restriction enzyme 3 can be used to replace the complete sequence of sst-TAS2R with that of another sst-TAS2R sequence or with the sequences of TAS1R2 and TAS1R3, respectively.

13.2.2 Expression analysis of receptor constructs

The finished receptor constructs should be analyzed for their expression—good expression levels are needed for both the CoIP and BRET experiments. Receptor–Rluc protein expression can be easily quantified via luminescence activity of the luciferase moiety; receptor–GFP^2 proteins can by quantified by GFP^2 fluorescence. The constructs are transiently transfected into the desired mammalian cell line (e.g., HEK293T cells) in 96-well plate format. Cells are seeded at ~10% confluence in white, clear-bottomed, poly-D-lysine-coated 96-well plates (for luminescence detection) or black, clear-bottomed, poly-D-lysine-coated 96-well plates (for fluorescence detection) and transfected 48 h later using a lipofection reagent according to the manufacturer's recommendation; the confluence of the cells should be ~50% before transfection. For replicate measurements, cells of three wells should be transfected with the same receptor construct. It is also advantageous for the analysis to include mock-transfected cells as negative control and cells transfected with the BRET vector (expressing a direct fusion protein of GFP^2-Rluc) as a positive control.

The following day, cells are washed twice with C1 buffer using an automated cell washer leaving a volume of 50 μl on the cells. Washing the cells manually bears the risk of cells detaching from the bottom of the plate. GFP^2 fluorescence can immediately be detected in a fluorescence plate reader (such as FLIPR) using excitation at 488 nm (a suboptimal but satisfactory excitation wavelength; the ideal excitation wavelength would be at 395 nm) and emission detection at 515 nm. For measuring luciferase activity, the bottom of the plates is covered with white tape before the last wash. The plates are then read in a luminometer (such as FLUOstar OPTIMA). In this protocol, 20 μl of 17.5 μM DeepBlueC in C1 buffer (5 μM final concentration) is added to one well, and light emission is detected at 370–450 nm before the instrument moves on to the next well.

For both fluorescence and luminescence measurements, the obtained values for the receptor constructs will be much lower compared to the BRET vector, which is to be expected as the receptor fusion proteins are integral membrane proteins, whereas the GFP^2–Rluc fusion protein is a cytosolic protein and can therefore reach much higher expression levels. It is difficult to suggest a general cutoff for unacceptably low expression, but we found that receptor constructs showing expression levels below 35% of the mean expression level will not be suitable for performing CoIP and BRET experiments.

13.2.3 Coimmunoprecipitation

In a first experimental approach to investigate receptor complexes, CoIP experiments are performed (Fig. 13.2). Two different receptors (or the same receptor with two different tags or marker protein moieties) are coexpressed. Then, immunoprecipitation of one receptor is performed using a specific antibody against this receptor or its tag. Subsequently, the precipitates are analyzed by Western blots for the presence of the originally immunoprecipitated receptor and for the presence of the other receptor.

This type of approach had already been performed for *in vitro* analysis of TAS1Rs (Nelson et al., 2002). In these experiments, the authors used one tagged receptor (with HA-tag) and one unmodified receptor as a receptor-specific antibody

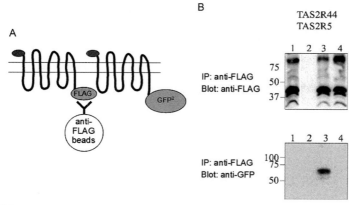

FIGURE 13.2

Detection of physical interactions between taste receptors by coimmunoprecipitation (CoIP). (A) Receptors with a C-terminal FLAG-tag or fusion with GFP^2, respectively, are coexpressed in a mammalian cell line, for example, HEK293T cells. Cells are then lysed; membrane proteins are solubilized and subjected to an immunoprecipitation with anti-FLAG affinity beads. These immunoprecipitates are then separated by SDS-PAGE and analyzed for the presence of the different receptors by Western blots using anti-FLAG and anti-GFP antisera. (B) Sample Western blots of a CoIP experiment with TAS2R44 and TAS2R5. The receptor–GFP^2 protein can only be immunoprecipitated when coexpressed with a FLAG-tagged receptor, which shows physical interactions between the two analyzed receptors. 1, immunoprecipitation (IP) with lysates from cells expressing only TAS2R44-FLAG; 2, IP with lysates from cells expressing TAS2R5-GFP^2; 3, IP with cells coexpressing of TAS2R44-FLAG and TAS2R5-GFP^2; 4, mixed lysates of cells expressing either TAS2R44-FLAG or TAS2R5-GFP^2. IP always with anti-FLAG; upper blot: anti-FLAG, lower blot: anti-GFP.

had been generated for this receptor. For almost all TAS2Rs, however, commercial receptor-specific antibodies are either not available or are not working reliably (Behrens et al., 2012). Therefore, differently tagged or marked receptors will have to be used. We recommend the usage of FLAG-tagged receptors and receptor–GFP^2 fusion proteins, respectively.

To perform the CoIP experiments, cells are seeded at ~20% confluence in 10 cm cell culture dishes coated with poly-D-lysine. The next day, cells are transfected with DNA for the receptor combination to be analyzed (one receptor-FLAG, one receptor–GFP^2 construct) in a 1:1 ratio or with either single receptor construct (performed in duplicate) using a lipofection reagent according to the manufacturer's recommendation. One set of the cells transfected with only one receptor construct serves as control for several purposes: by analyzing the cell lysate, expression of the construct can be confirmed. Second, by using the cell lysates for the immunoprecipitation procedure, one can show that the procedure was performed successfully and with specificity, that is, that only the FLAG-tagged receptor can be pulled down but not the GFP^2-marked receptor expressed on its own. Third, both parts together also help to confirm the specificity of the antibodies used. The second set of cells transfected with a single receptor construct will be mixed later during the cell lysis procedure. This will serve as control to verify that the observed protein–protein interactions did not arise artificially during the purification process but were already present inside the cells.

The following day, cells are prepared for the lysis procedure by putting the cell culture dishes on ice. Cells are washed twice with 10 ml cold PBS before adding 1 ml of ice-cold CoIP buffer; incubate for 5–10 min on ice. Cells and cell remnants should detach easily and be transferred to a glass homogenizer; move the pestle up and down several times to fully lyse the cells and solubilize the membrane proteins. At this step, one set of lysates of cells transfected with a single receptor can be mixed to control for occurrence of protein–protein interactions inside the cells vs. during the procedure. To ensure that all remaining cell debris and membrane fragments are removed from the solubilized proteins, ultracentrifugation at $100{,}000g$ and $4\,°C$ should be performed for 90 min.

In the meantime, the anti-FLAG-agarose beads used for immunoprecipitation of FLAG-tagged proteins can be prepared. Be careful to always keep them at $4\,°C$ or on ice! Resuspend the anti-FLAG affinity gel well by pipetting up and down with a yellow pipette tip whose tip has been cut off and take out the appropriate amount (use 20 μl of agarose bead suspension per sample, which equals 10 μl of packed gel). Add ice-cold CoIP buffer (volume $= 20\times$ volume of packed gel), mix gently, and centrifuge for 30 s at $8200g$, $4\,°C$. Remove the liquid without disturbing the gel pellet and repeat this washing step twice more. After the final washing step, add an appropriate amount of ice-cold CoIP buffer and aliquot the washed anti-FLAG affinity gel on ice into new 1.5 ml tubes.

After the ultracentrifugation is finished, take the supernatant of the cell lysates and put them on ice. Transfer 20 μl of the lysates to new tubes and freeze these samples; they will later be separated by SDS-PAGE and analyzed by Western blots and serve as controls for expression and antibody specificity (see the preceding text).

Add the rest of the supernatants to separate aliquots of the prepared anti-FLAG affinity gel. If necessary, add some more CoIP buffer for a total volume of 1 ml and incubate with gentle shaking at 4 °C overnight.

The next day, wash the anti-FLAG affinity gel at least 3× (ideally 6×) with 300 μl CoIP buffer each. After the final washing step, remove all liquid carefully without disturbing the gel pellet; add 20 μl of 2× Laemmli buffer, mix gently, heat to 95 °C for 5 min, and briefly spin it down. Take all of the supernatant carefully without transferring any of the gel; add DTT to a final concentration of 200 mM to all samples.

The samples, together with the cell lysates kept at −20 °C, can now be analyzed by SDS-PAGE and Western blot. Blots are blocked in 5% dried milk/TBST and then incubated with the primary antibodies (anti-FLAG 1:4000 and anti-GFP 1:1000, respectively) in blocking solution with gentle shaking at 4 °C overnight. The following day, blots are washed 3× with TBST and then incubated with secondary antibody (anti-rabbit HRP 1:5000) for 1 h at room temperature. The blots are washed 3× with TBST and 2× with TBS before adding ECL substrate according to the manufacturer's recommendation. Blots can be developed and bands visualized by standard X-ray film exposure or using a luminescent image analyzer (such as LAS-1000 and with the following settings: exposure time from 30 s to 30 min, aperture 0.85, and integration of 2 × 2 pixels).

The expected protein sizes of TAS2Rs are ~40 kDa for FLAG-tagged receptors and 67 kDa for receptor–GFP^2 fusion proteins. However, multiple bands can often be observed, most likely due to differently glycosylated receptor proteins (Reichling, Meyerhof, & Behrens, 2008). The expected protein sizes of TAS1Rs are ~90 kDa for FLAG-tagged receptors and 120 kDa for receptor–GFP^2 fusion proteins. However, the actual observed protein sizes can be greater as protein glycosylation has also been observed for TAS1Rs (Max et al., 2001).

FLAG-tagged receptors and receptor–GFP^2 fusion proteins should be detectable in all lysates of cells expressing the respective receptor constructs. In immunoprecipitates of cells expressing only one receptor or the mix control, however, only FLAG-tagged receptors should be stained, a result that demonstrates the specificity of the antibodies and the suitability of the protocol. On the other hand, a positive CoIP result is obtained if receptor–GFP^2 fusion proteins can be detected in immunoprecipitates of cells coexpressing this receptor construct together with a FLAG-tagged receptor, a result that is indicative of an assembly of the two respective receptor proteins through protein–protein interactions.

13.2.4 Bioluminescence resonance energy transfer

In the second experimental approach to investigate receptor oligomerization, BRET experiments are carried out (Fig. 13.3). We used the BRET^2 technology (PerkinElmer) with the luciferase of *Renilla reniformis* (Rluc) as BRET donor and a modification of the green fluorescent protein (GFP^2) as BRET acceptor. Both proteins are fused to the C-terminus of the TAS1R and TAS2R receptor proteins.

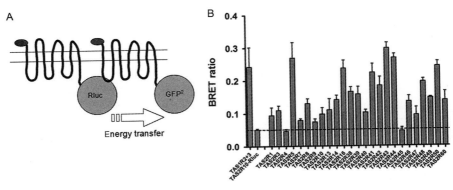

FIGURE 13.3

Analysis of TAS2R homo- and hetero-oligomerization by bioluminescence resonance energy transfer (BRET). (A) Two receptors bearing C-terminal GFP^2 and Rluc moieties, respectively, are coexpressed in a mammalian cell line, for example, HEK293T cells. Administration of the coelenterazine DeepBlueC will lead to light emission from the luciferase at 370–450 nm. If the GFP^2 is brought into close proximity by protein–protein interactions of the respective receptors, it can be excited and light emission at 500–530 nm can be detected. (B) Graph showing BRET results obtained with TAS2R10-Rluc and all other TAS2R-GFP^2. The sweet taste receptor heteromer (TAS1R2-Rluc + TAS1R3-GFP^2) serves as a positive control, TAS2R10-Rluc alone as a negative control (dashed line). Almost all combinations show BRET ratios (light emission at 500–530 nm over light emission at 370–450 nm) higher than the negative control, which is indicative of receptor oligomerization. Only TAS2R10-Rluc with TAS2R4-GFP^2 and TAS2R45-GFP^2, respectively, have lower values. Data represent means and standard deviation from two experiments carried out in duplicate.

When both BRET donor and acceptor molecules are brought in close proximity to each other through protein–protein interactions of the receptors, an energy transfer from the luciferase to the GFP^2 can occur, resulting in detectable fluorescence of GFP^2 (Fig. 13.3).

BRET has been successfully employed for the analysis of dimerization of several GPCRs including TAS1Rs (Jiang et al., 2004). But while the reading was in general being performed in 96-well plates and luminescent plate readers, the cells had been grown and transfected in larger cell culture dishes and only then had the cells been transferred to 96-well plates for BRET assay reading. We have adapted the procedure for higher throughput by directly growing the cells in 96-well plates and performing the BRET assay in the same 96-well plate.

Therefore, modified HEK293T cells are seeded at ~10% confluence in white, clear-bottomed, poly-D-lysine-coated 96-well plates. After 48 h, at ~50% confluence, cells are transfected with the plasmids encoding for the BRET fusion proteins (one receptor–Rluc and one receptor–GFP^2 construct) using a lipofection reagent according to the manufacturer's recommendation. Cotransfection of two plasmids by lipofection works equally well as transfection with calcium phosphate but is easier to perform for many samples. The ratio of the plasmid DNAs should be three parts receptor–GFP^2 and one part receptor–Rluc; this ratio has been optimized to yield high BRET values (due to higher saturation of the BRET donors with BRET acceptors) while still resulting in high enough absolute light emission values to produce reliable and reproducible results. For replicate measurements, cells in 2–3 wells should be transfected with the same receptor combination. An equal number of wells with mock-transfected cells should be included as blanks. Also, cells transfected with only the receptor–Rluc construct alone should be included as negative control; cells transfected with the BRET vector serve as a positive control.

The next day, cells are washed twice with C1 buffer using an automated cell washer leaving a volume of 50 µl on the cells after the last wash. Also, before the last wash, the bottom of the plates is covered with white tape to improve luminescence detection. The plates are then read in a luminometer, which is able to perform a simultaneous reading with two different channels (e.g., FLUOstar OPTIMA). For reading the BRET assay, 20 µl of 17.5 µM DeepBlueC in C1 buffer (5 µM final concentration) is added to one well and light emission is detected simultaneously at 370–450 and 500–530 nm, respectively, for 4–12 s before the instrument moves on to the next well.

To analyze the data, the BRET ratio is calculated as light emission at 500–530 nm (minus the light emission at 500–530 nm of mock-transfected cells) over light emission at 370–450 nm (minus the light emission at 370–450 nm of mock-transfected cells). The negative control of the receptor–Rluc transfected alone is expected to give a BRET ratio of ~0.05, which is due to some residual light emission of the coelenterazine substrate at 500–530 nm. The positive control, cells expressing the direct fusion protein of GFP^2-Rluc, will yield a very high BRET ratio, approaching a value of 1. This is due to very high expression levels of this cytosolic protein and, secondly, as a direct fusion protein, it provides all BRET acceptor moieties with a BRET donor partner, leading to saturation and maximum BRET values. BRET ratios of the receptor combinations will be much lower; they are expected to be in the range of 0.05 to ~0.4. Only values significantly higher than the values from the respective receptor–Rluc alone are considered positive signals, that is, they stem from receptors in close proximity to each other and are indicative of receptor–receptor interactions.

We further recommend to always analyzing both possible receptor combinations, for example, TAS2R16-GFP^2 with TAS2R44-Rluc and TAS2R16-Rluc with TAS2R44-GFP^2. The results may vary from each other. Possible explanations might be differences in expression levels between the receptors; a higher BRET ratio will be obtained when the receptor expressed at a higher level is fused to the BRET acceptor. Alternatively or additionally, the BRET donor and acceptor protein moieties might be folded differently under the respective receptors, bringing them closer together or further apart in the different combinations. For these reasons, we generally assumed that two receptors can interact with each other when at least one of the two combinations yielded a positive BRET signal.

13.3 DISCUSSION

With CoIP and BRET experiments, we and others could show oligomerization of both families of taste receptors, TAS1Rs and TAS2Rs (Jiang et al., 2004, Kuhn et al., 2010, Nelson et al., 2002). The demonstration of receptor complexes could therefore be made with two different and independent techniques; both methods have their advantages and disadvantages.

CoIP is an often-used biochemical technique and can be used for analysis of protein–protein interactions *in vitro* and *in vivo*. When differently tagged or marked receptors are used *in vitro*, homo-oligomerization can be investigated and hetero-oligomerization. However, the use of detergents to lyse cells and solubilize membrane proteins could lead to artificial aggregation of proteins during the process; one point of critique is that one cannot be sure that the protein complexes occurred already in the cells under physiological conditions (Jordan & Devi, 1999, Milligan & Bouvier, 2005). Another critical point is that one has to ensure that all membrane proteins have been fully solubilized and that no intact membrane patches are still present in the preparation as this might lead to false-positive results (Milligan & Bouvier, 2005). To address these points, we used only a mild detergent, incorporated an ultracentrifugation step into our protocol, and included a control with mixed lysates of cells expressing only one receptor. As a result, this procedure is rather long and requires a lot of hands-on time, which makes the protocol only suitable for a small number of samples.

BRET on the other hand is a newer biophysical technique and is adaptable for higher throughput, as shown in this protocol. The BRET method requires the generation of special receptor fusion proteins and can therefore only be employed for *in vitro* studies. Homo- and hetero-oligomerization can be investigated. One definite advantage over CoIP experiments is that the analysis is performed in living cells and under physiological conditions (Pfleger & Eidne, 2005). A point of critique is, however, that BRET experiments do not show, like CoIP analysis, for example, direct protein–protein interactions but only demonstrate close proximity of the investigated receptor proteins.

Both methods together can overcome most of the limitations of the single methods and together provide compelling evidence for oligomerization, like here for TAS1R and TAS2R taste receptors. It is therefore advised to use both methods for oligomerization studies. If only one method can be used, it would be best to consider using CoIPs for investigating only few samples; no special equipment is needed. For many samples, however, it might be better to use BRET. BRET requires a luminometer with dual-emission detection and special filter sets and a coelenterazine as substrate. But it has the advantages of being suitable for a higher throughput and of analyzing proteins in cells under physiological conditions.

CoIPs could, for example, also be used for studies of taste receptor oligomerization *in vivo*. Recently, TAS2R receptor-specific antibodies became commercially available and have already been successfully used to stain receptor proteins in taste tissue (Behrens et al., 2012). These antibodies could theoretically also be used for

TAS2R CoIP studies *in vivo*. However, the availability of adequate amounts of tissue still provides a challenge.

BRET experiments on the other hand could be useful to analyze many different receptor variants or mutants for their ability to form oligomers, for example, regarding the functionality of the heteromeric sweet taste receptor or, for example, regarding the identification of the domain(s) involved in and necessary for oligomerization of TAS1Rs and TAS2Rs, respectively.

SUMMARY

We have here provided a detailed protocol for the investigation of the oligomerization of both TAS1R and TAS2R taste receptor families. Starting with the easy and straightforward generation of receptor constructs needed for the following methods, we also provide a protocol for quick analysis of expression of these constructs in mammalian cell lines. We have then detailed CoIP experiments for the biochemical analysis of protein–protein interactions, incorporating important controls into our protocol. Further, we have also explained the use of BRET for the investigation of receptor oligomerization in living cells and for a high throughput. The protocols provided here should also be easily transferable to studies on oligomerization of other receptors, for example, other rhodopsin-like GPCRs or other class C GPCRs.

Acknowledgments

This work was supported by the German Research Foundation (DFG, Me 1024/2-1, 2-2).

References

Behrens, M., Born, S., Redel, U., Voigt, N., Schuh, V., Raguse, J. D., et al. (2012). Immunohistochemical detection of TAS2R38 protein in human taste cells. *PLoS One, 7*(7), e40304.

Behrens, M., Reichling, C., Batram, C., Brockhoff, A., & Meyerhof, W. (2009). Bitter taste receptors and their cells. *Annals of the New York Academy of Sciences, 1170*, 111–115.

Bufe, B., Hofmann, T., Krautwurst, D., Raguse, J. D., & Wolfgang Meyerhof, W. (2002). The human TAS2R16 receptor mediates bitter taste in response to beta-glucopyranosides. *Nature Genetics, 32*, 397–401.

George, S. R., O'Dowd, B. F., & Lee, S. P. (2002). G-protein-coupled receptor oligomerization and its potential for drug discovery. *Nature Reviews Drug Discovery, 1*, 808–820.

Jiang, P., Ji, Q., Liu, Z., Snyder, L. A., Benard, L. M., Margolskee, R. F., et al. (2004). The cysteine-rich region of T1R3 determines responses to intensely sweet proteins. *Journal of Biological Chemistry, 279*, 45068–45075.

Jones, K. A., Borowsky, B., Tamm, J. A., Craig, D. A., Durkin, M. M., Dai, M., et al. (1998). GABA(B) receptors function as heteromeric assembly of the subunits GABA(B)R1 and GABA(B)R2. *Nature, 396*, 674–679.

Jordan, B. A., & Devi, L. A. (1999). G-protein-coupled receptor heterodimerization modulates receptor function. *Nature, 399*, 697–700.

Kaupmann, K., Malitschek, B., Schuler, V., Heid, J., Froestl, W., Beck, P., et al. (1998). GABA(B)-receptor subtypes assemble into functional heteromeric complexes. *Nature, 396*, 683–687.

Kuhn, C., Bufe, B., Batram, C., & Meyerhof, W. (2010). Oligomerization of TAS2R bitter taste receptors. *Chemical Senses, 35*(5), 395–406.

Li, X., Staszewski, L., Xu, H., Durick, K., Zoller, M., & Adler, E. (2002). Human receptors for sweet and umami taste. *Proceedings of the National Academy of Sciences of the United States of America, 99*, 4692–4696.

Max, M., Shanker, Y. G., Huang, L., Rong, M., Liu, Z., Campagne, F., et al. (2001). Tas1r3, encoding a new candidate taste receptor, is allelic to the sweet responsiveness locus Sac. *Nature Genetics, 28*, 58–63.

Meyerhof, W., Batram, C., Kuhn, C., Brockhoff, A., Chudoba, E., Bufe, B., et al. (2010). The molecular receptive ranges of human TAS2R bitter taste receptors. *Chemical Senses, 35*, 157–170.

Milligan, G., & Bouvier, M. (2005). Methods to monitor the quaternary structure of G protein-coupled receptors. *FEBS Journal, 272*, 2914–2925.

Nelson, G., Chandrashekar, J., Hoon, M. A., Feng, L., Zhao, G., Ryba, N. J., et al. (2002). An amino-acid taste receptor. *Nature, 416*, 199–202.

Nelson, G., Hoon, M. A., Chandrashekar, J., Zhang, Y., Ryba, N. J., & Zuker, C. S. (2001). Mammalian sweet taste receptors. *Cell, 106*, 381–390.

Pfleger, K. D., & Eidne, K. A. (2005). Monitoring the formation of dynamic G-protein-coupled receptor-protein complexes in living cells. *Biochemical Journal, 385*, 625–637.

Pin, J. P., Galvez, T., & Prezeau, L. (2003). Evolution, structure, and activation mechanism of family 3/C G-protein-coupled receptors. *Pharmacology & Therapeutics, 98*, 325–354.

Reichling, C., Meyerhof, W., & Behrens, M. (2008). Functions of human bitter taste receptors depend on N-glycosylation. *Journal of Neurochemistry, 106*(3), 1138–1148.

Rivero-Müller, A., Chou, Y. Y., Ji, I., Lajic, S., Hanyaloglu, A. C., Jonas, K., et al. (2009). Rescue of defective G protein-coupled receptor function in vivo by intermolecular co-operation. *Proceedings of the National Academy of Sciences of the United States of America, 107*, 2319–2324.

Shi, P., Zhang, J., Yang, H., & Zhang, Y. P. (2003). Adaptive diversification of bitter taste receptor genes in mammalian evolution. *Molecular Biology and Evolution, 20*, 805–814.

Yarmolinsky, D. A., Zuker, C. S., & Ryba, N. J. (2009). Common sense about taste: From mammals to insects. *Cell, 139*, 234–244.

Zhao, G. Q., Zhang, Y., Hoon, M. A., Chandrashekar, J., Erlenbach, I., Ryba, N. J., et al. (2003). The receptors for mammalian sweet and umami taste. *Cell, 115*, 255–266.

CHAPTER 14

Analysis of Receptor–Receptor Interaction by Combined Application of FRET and Microscopy

Sonal Prasad, Andre Zeug, and Evgeni Ponimaskin
Cellular Neurophysiology, Center of Physiology, Hannover Medical School, Hannover, Germany

CHAPTER OUTLINE

- Introduction and Rationale 244
- 14.1 Theory 245
  - 14.1.1 Basic Principles of FRET 245
  - 14.1.2 Fluorescent Proteins and FRET 247
  - 14.1.3 Principle Behind Lux-FRET Quantification and its Outcome 248
- 14.2 Materials / Experimental Setups 249
  - 14.2.1 Reagents 249
  - 14.2.2 Devices 251
  - 14.2.3 Reagent Setup 252
    - 14.2.3.1 Preparation of Cell Culture Media 252
    - 14.2.3.2 Buffer A for Cell Lysis 252
    - 14.2.3.3 Buffer B for FRET Imaging and Fluorolog Experiments 252
    - 14.2.3.4 Protease Inhibitors and Phosphate-Buffered Saline Stock Solutions 253
- 14.3 Methods (Protocols and Procedure) 253
  - 14.3.1 Cell Preparation and Transfection Timing 253
    - 14.3.1.1 Culture of Cells 253
    - 14.3.1.2 Reculture and Seeding Cells 253
    - 14.3.1.3 Transfection 253
  - 14.3.2 Cell Lysis for Fluorolog 254
  - 14.3.3 Procedure Part I: Data Acquisition 254
    - 14.3.3.1 Setting Individual Protocols According to Experimental Setup 255
  - 14.3.4 Spectroscopic Measurement at a Fluorescence Spectrometer (Fluorolog 3v.2.2.) 255

14.3.5 FRET Imaging of Receptor–Receptor Interaction for Intact Cells
with Subcellular Spatial Resolution 256
    14.3.5.1 *Zeiss LSM 780* 256
    14.3.5.2 *Spinning Disk Microscope* 257
    14.3.5.3 *TIRF Microscope* 259
14.3.6 Procedure Part II: Data Evaluation 260
    14.3.6.1 *Evaluation Procedure* 261
14.4 Results and Discussion **263**
Acknowledgments **264**
References **264**

Abstract

G protein-coupled receptors (GPCRs) participate in the regulation of many cellular processes and, therefore, represent key targets for pharmacological treatment. The existence of GPCR homo- and heterodimers has become generally accepted, and a growing body of evidence points to the functional importance of oligomeric complexes for the receptor trafficking, receptor activation, and G protein coupling in native tissues. Quantitative molecular microscopy is becoming more and more important to investigate such receptor–receptor interaction in their native environments. Förster resonance energy transfer (FRET) is thereby utilized to aim at investigating the interaction of molecules at distances beyond diffraction-limited spatial resolution. The exact determination of the FRET signals, which are often only fractions of the fluorescence signals, requires extensive experimental effort. Moreover, the correct interpretation of FRET measurements as well as FRET data-based modeling represents an essential challenge in microscopy and biophysics.

In this chapter, we present and discuss variety of acquisition protocols and models based on "linear unmixing FRET" (lux-FRET) to investigate receptor–receptor interaction in living cells with high spatial and temporal resolution. Here, we show how to apply lux-FRET in spectroscopic and different imaging devices, based either on spectral detection or on filter cubes. We focus on detailed description for FRET measurements and analyses based on sophisticated acquisition procedures according to different experimental setups and also provide several examples of biological applications.

INTRODUCTION AND RATIONALE

G protein-coupled receptors (GPCRs) belong to a large and diverse family of integral membrane proteins that participate in the regulation of many cellular processes and, therefore, represent key targets for pharmacological treatment. GPCRs are known to be able to activate multiple downstream signaling modules at the plasma membrane (Gorinski et al., 2012); however, the mechanisms regulating multimodal GPCR-mediated signaling are still poorly understood. Until recently,

GPCRs were assumed to exist and function as monomeric units that interact with corresponding G proteins in a 1:1 stoichiometry. However, biochemical, structural, and functional evidence collected during the last decade indicates that GPCRs can form oligomers. Oligomerization can occur between identical receptor types (homomerization) or between different receptors of the same or different GPCR families (heteromerization). Because signaling processes mediated by GPCR are transient and fast and usually involve multiple proteins, quantitative molecular microscopy is mandatory to study such receptor–receptor interaction and to obtain unprecedented detailed information. Dynamic interaction between receptors is thought to have a key role in regulating most cellular signal transduction pathways.

Förster resonance energy transfer (FRET) is a technique that allows investigation of molecular processes in nanometer resolution. FRET is a nonradiative process that can occur between fluorophores when the energy is transferred from a donor fluorophore to an acceptor fluorophore. There are different fluorescence lifetime and spectral or intensity-based approaches used in microscopy like time- and frequency-domain fluorescence lifetime techniques, acceptor photobleaching, semiquantitative sensitized emission, FRET stoichiometry 3-cube quantification, spectral RET, and the linear unmixing FRET (lux-FRET) (Zeug, Woehler, Neher, & Ponimaskin, 2012). Lux-FRET approach allows us to measure not only the apparent FRET efficiency but also the abundance of total receptor and total donor, as well as their ratio (Wlodarczyk et al., 2008). In this chapter, we show how lux-FRET can be applied in spectroscopic and imaging devices based on spectral detection as well as on filter cubes and describe how to use this method for measuring dynamic receptor–receptor interaction by changes in FRET efficiency and also discuss various imaging protocols and analysis modes.

Using the lux-FRET technique in combination with microscopy, we have previously characterized interaction between serotonin receptors 5-HT$_{1A}$ and 5-HT$_7$ at the single-cell level (Fig. 14.1). However, the lux-FRET method can be applied to investigate complex protein behavior like oligomerization between any proteins of interest. The correct interpretation of FRET measurements as well as FRET data-based modeling represents an essential challenge in microscopy and biophysics.

14.1 THEORY
14.1.1 Basic principles of FRET

The theory of resonance energy transfer was originally developed by Theodor Förster and, in honor of his contribution, has been named after him (Forster, 1946, 1948, 2012). FRET is well suited to investigate protein–protein interaction that occurs between two molecules positioned in close proximity of each other. The fundamental mechanism of FRET involves a donor fluorophore in an excited electronic

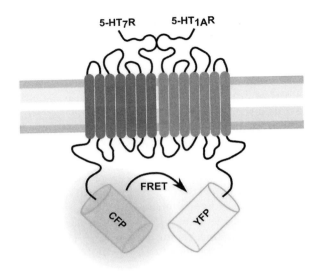

FIGURE 14.1

Heterodimerization of the serotonin receptors 5-HT$_{1A}$ and 5-HT$_7$. The cartoon shows intermolecular interaction between two proteins, one tagged with a cyan FP (donor) and the other tagged with a yellow FP (acceptor). Only when specific interaction occurs (i.e., receptors dimerizes), FRET signal can be measured. (For interpretation of the references to color in this figure legend, the reader is referred to the online version of this chapter.)

state, which may transfer its excitation energy to a nearby acceptor fluorophore (or chromophore) in a nonradiative fashion through long-range dipole–dipole interaction. Basically, there are three major primary conditions required for FRET (see Box and Fig. 14.2).

Primary conditions required for FRET are
- close proximity of donor and acceptor molecules (typically 1–10 nm),
- donor emission spectrum with considerable spectral overlap acceptor excitation spectrum,
- proper donor and acceptor transition dipole moment orientations.

The Förster theory shows that FRET efficiency (E) varies with an inverse sixth power of the distance (r) between the two chromophores:

$$E_{\text{FRET}} = \frac{R_0^6}{r^6 + R_0^6} \tag{14.1}$$

R_0 is the Förster distance where the FRET efficiency is 50% and typical values are between 4 and 6 nm in case of acute FRET pairs. Therefore, it is suitable to employ FRET for protein–protein interaction, because under physiological condition,

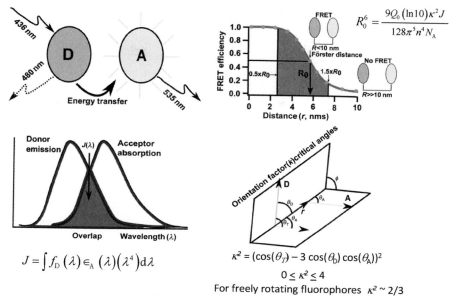

FIGURE 14.2

Principles of FRET. The technique of Förster resonance energy transfer permits determination of the approach between two molecules within several nanometers, a distance sufficiently close for molecular interaction to occur. It is a strategy for detection of protein–protein interaction tagged with fluorescent proteins. For more detailed information, see Ref. Lakowicz (2006). (For color version of this figure, the reader is referred to the online version of this chapter.)

average molecular distance is far above R_0 and FRET could be only seen if interaction occurs. More important, FRET allows us to investigate protein–protein interaction, which occurs far below spatial resolution of optical microscopy.

14.1.2 Fluorescent proteins and FRET

To study protein–protein interaction by FRET, the first step is to tag proteins of interest with either donor or acceptor fluorophore. Several labeling strategies are available for FRET signal detection (Lam et al., 2012). Genetic labeling with fluorescent proteins (FPs) for live-cell FRET imaging represents one of the central labeling strategies. At present, there is a large availability of green fluorescent protein (GFP) variants as possible donor–acceptor pairs that can be used to study quantitative FRET in living cells. Cyan fluorescent protein (CFP) and yellow fluorescent protein (YFP) derivatives are widely used in fluorescence microscopy studies as acute FRET pairs (Piston & Kremers, 2007). For protein–protein interaction, it is important to ensure that the FPs per se do not interact with each other. It is known that original GFP showed intrinsic tendency to dimerize to some extent; hence, special effort was taken to create the so-called monomeric versions of GFP derivatives, which more or less sufficiently suppress the native dimerization tendency.

Since the dimensions of GFPs are already in the order of R_0, proper linker design is very crucial. Hence, appropriate linker length and steric conformation (includes distance and orientation) flexibility is necessary. At the same time, the functionality of receptor is essential, in particular if the analysis of downstream signaling is of interest.

To quantify FRET, not only high FRET efficiency but also sufficient brightness of FPs is important. High quantum yield and superior excitation coefficients of FPs can be achieved by using new optimized CFP and YFP derivatives. High quantum yield of donor and acceptor fluorophores improves quality of FRET investigations by increasing signal-to-noise ratio, which is important for pixel-based FRET quantification in live cells. From the intensity level of the signal, one can estimate the expression level of receptors and also exclude saturation artifacts.

14.1.3 Principle behind lux-FRET quantification and its outcome

There are several methods available to quantify FRET. Applying fluorescence lifetime-based methods, the FRET efficiency E can be determined directly. However, biologically relevant in most types of experimental designs is the fraction of molecules present in FRET state. From fluorescence lifetime techniques, like time-correlated single-photon counting, the fraction of FRET complexes scaled by total donor concentration $f_D \equiv [DA]/[D^t] = A_{\tau_{DA}}/(A_{\tau_{DA}} + A_{\tau_D})$, where D is donor and A is acceptor, can be obtained from the fractional amplitudes of the multiexponential decay $A_{\tau_{DA}}$ and A_{τ_D} (Zeug et al., 2012). For reliable results, a high number of photons are required, which limits the spatiotemporal resolution and makes it rarely suitable for live-cell imaging. Moreover, the fraction of FRET complexes scaled by total acceptor concentration $f_A \equiv [DA]/[A^t]$ cannot be directly obtained from fluorescence lifetime. Therefore, to study protein–protein interaction, we developed lux-FRET (Wlodarczyk et al., 2008); for an overview of spectral FRET approaches, see also Zeug et al. (2012)). Lux-FRET is a spectral quantitative FRET approach based on two sequential excitations and linear unmixing of fluorescence emission by spectral references of donor $F_D^{i,\text{ref}}(\lambda)$ and acceptor $F_A^{i,\text{ref}}(\lambda)$ at the corresponding excitation wavelength i:

$$F^i(\lambda) = \delta^i \cdot F_D^{i,\text{ref}}(\lambda) + \alpha^i \cdot F_A^{i,\text{ref}}(\lambda) \tag{14.2}$$

From the contributions δ^i and α^i from donor and acceptor emission, respectively, the apparent FRET efficiencies Ef_D and Ef_A and the donor mole fraction x_D can be obtained by

$$Ef_D \equiv E\frac{[DA]}{[D^t]} = \frac{\Delta\alpha}{\Delta r \delta^1 + \Delta\alpha} \tag{14.3}$$

$$Ef_A \equiv E\frac{[DA]}{[A^t]} = R_{TC}\frac{\Delta\alpha}{\alpha^2 r^{\text{ex},1} - \alpha^1 r^{\text{ex},2}} \tag{14.4}$$

$$x_D \equiv \frac{[D^t]}{[D^t]+[A^t]} = \frac{1}{1+\dfrac{(\alpha^2 \cdot r^{\text{ex},1} - \alpha^1 \cdot r^{\text{ex},2})}{R_{\text{TC}}(\delta^1 \cdot \Delta r + \Delta \alpha)}} \tag{14.5}$$

Here $\Delta \alpha = \alpha^2 - \alpha^1$ and $\Delta r = r^{\text{ex},2} - r^{\text{ex},1}$. $R_{\text{TC}} = [D^{\text{ref}}]/[A^{\text{ref}}]$ must be obtained by a separate lux-FRET measurement from the so-called a tandem construct, where donor and acceptor are in a known fixed stoichiometry. The ratios of excitation strengths (donor/acceptor) $r^{\text{ex},i}$ at excitation wavelengths i can be obtained from the spectral parameters of the reference measurements by

$$r^{\text{ex},i} \equiv \frac{\varepsilon_D^i}{\varepsilon_A^i} \cdot \frac{[D^{\text{ref}}]}{[A^{\text{ref}}]} = \frac{F_D^{i,\text{ref}}(\lambda)}{e_D(\lambda)} \cdot \frac{e_A(\lambda)}{F_A^{i,\text{ref}}(\lambda)} \cdot \frac{Q_A}{Q_D} \tag{14.6}$$

where $\varepsilon_{D,A}^i$ are the extinction coefficient, $e_{D,A}(\lambda)$ are the characteristic emission spectra, and $Q_{D,A}$ the fluorescence quantum yield of donor and acceptor.

Ef_D is calculated by Eq. (14.3) from experimental quantities, without any additional information. To calculate Ef_A and x_D by Eqs. (14.4) and (14.5), further information from a tandem construct with known donor–acceptor stoichiometry is required. Fluorescence quantum yield of donor and acceptor is also necessary for lux-FRET calculations.

From lux-FRET, we receive the apparent FRET efficiencies Ef_D and Ef_A rather than the true quantum efficiency of the energy transfer E. High E values however help to improve the signal-to-noise ratio of the analysis. The fractions of DA complexes scaled by total donor and acceptor concentration f_D and f_A are of potential interest to study protein–protein interaction. By changing the donor mole fraction x_D, which can be assessed by varying the relative expression level of donor and acceptor and keeping the total concentration constant, we get specific values for apparent FRET efficiencies Ef_D and Ef_A. A model characterizing apparent FRET efficiencies Ef_D and Ef_A as a function of x_D for oligomeric structures has been defined:

$$Ef_D = E\left(1 - x_D^{n-1}\right), \quad Ef_A = E \frac{x_D}{x_D - 1}\left(1 - x_D^{n-1}\right) \tag{14.7}$$

Fitting this model to experimental data allows for the estimation of the true energy transfer efficiency E and also provides information about the oligomerization state n (Meyer et al., 2006). When receptors interact and form dimers, the equations become linear. For hetero-oligomerization, the model gets more complex (Renner et al., 2012).

14.2 MATERIALS / EXPERIMENTAL SETUPS
14.2.1 Reagents
- Mouse N1E-115 neuroblastoma cells from America Type Culture Collection (LGC Promochem)
- Lipofectamine 2000 transfection reagent (Life Technologies, cat.no.11668-019)
- Phenylmethylsulfonyl fluoride (PMSF) (Roth, cat.no.6367.1)

- Triton X-100 solution (Sigma, cat.no.93427)
- D+glucose (Sigma, cat.no.G7021)
- $NaHPO_4$ (Roth, cat.no.P030.1)
- KH_2PO_4 (Roth, cat.no.P018.1)
- Ethanol—99.8% (Roth, cat.no.90654)
- Acetone—99.5% (Roth, cat.no.5025.2)
- Methanol—99% (Roth, cat.no.8388.5)
- Double-distilled water

Cell culture media and reagents
- Dulbecco's modified Eagle's medium—high glucose (DMEM) (Sigma Aldrich, cat.no.D5648-10X)
- Fetal bovine serum (Biochrom, cat.no.S 0415)
- Penicillin–streptomycin (Invitrogen, cat.no.15070-063)
- NaCl (Roth, cat.no.3957.5)
- $NaHCO_3$ (Roth, cat.no.8551.1)
- KCl (Roth, cat.no.6781.1)
- $MgCl_2$ hexahydrate (Roth, cat.no.2189.1)
- $CaCl_2$ dihydrate (Roth, cat.no.5239.1)
- HEPES (Roth, cat.no.9105.3)
- NaOH—32% solution (Roth, cat.no.T197.1)
- Dimethyl sulfoxide (DMSO) (Sigma Aldrich, cat.no.472301)
- Poly-L-lysine (PLL) (Sigma Aldrich, cat.no.P2636-25)
- Opti-MEM reduced-serum medium (1×), liquid (Life Technologies, cat.no.11058021)
- Cell culture petri plates (10 cm) (Nunc, cat.no.150350)
- 6-well cell culture plate (Greiner, cat.no.657160)
- 12-well cell culture plate (Greiner, cat.no.665180)
- 18 mm (thickness 0.177 mm), 50 mm (Thermo Scientific, cat.no.004710482, 004711182), and 24 × 24 mm (Roth, cat.no.H 875) glass coverslips

Amino acids and organic compounds for medium and protease inhibitors
- L-Cystine (Sigma, cat.no.C7352-25G)
- L-Alanine (Sigma, cat.no.A7469-25G)
- L-Aspartic acid (Sigma, cat.no.A4534-100G)
- L-Proline (Sigma, cat.no.P5607-25G)
- L-Glutamic acid monosodium salt hydrate (Sigma, cat.no.G5889-100G)
- L-Asparagine monohydrate (Sigma, cat.no.A4284-100G)
- Leupeptin (Roth, cat.no.CN33.1)
- Chymostatin (Sigma, cat.no.C7268-5MG)
- Antipain (Roth, cat.no.2933.1)
- Pepstatin (Roth, cat.no.2936.1)

14.2.2 Devices

- **Fluorescence spectrometer (Fluorolog 3-2.2, HORIBA Jobin Yvon):**
  - Double-grating emission monochromators, improving the system's stray light rejection.
  - A PMT (950 V) with a high dynamic range ($1-10^7$) that allows better signal-to-noise ratio.
  - A swing-away mirror allows collection of sample luminescence at standard right angle 90° to the excitation beam or front face at 22.5°.
  - XBO lamp (450 W).
  - FluorEssence version 3.0 software for data acquisition used FluorEssence-3.0 as a plug-in for OriginPro software.
  - Quartz cuvette 10 mm light path (Hellma, type no, 101-QS).
- **Inverted confocal laser scanning microscope—Zeiss LSM 780:**
  - Axio Observer.Z1.
  - Laser lines: diode 440 (25 mW), argon 488 (25 mW) and 514 (25 mW), DPSS 561 (20 mW) for CFP–YFP, and GFP–mCherry derivatives.
  - Objectives: C-Apochromat $40 \times /1.2$W DIC III, Plan-Apochromat $63 \times /1.4$ Oil DIC II, and Plan-Apochromat $20 \times /0.8$ DIC II.
  - Dichroics (InVis and VisTwin gate main beam splitter (MBS)): MBS 440 and 458/514 for CFP–YFP and MBS 488/561 for GFP–mCherry FPs.
  - A motorized piezo XY stage (Merzhäuser).
  - Definite focus.
  - HBO lamp (100 W).
  - ZEN 2010 imaging software package.
  - Stage adapter (64–0327) and low-profile open bath chamber (RC-41LP) (Warner Instruments).
- **Upright semiconfocal spinning disk microscope:**
  - Spinning disk (CSUX1-M2N-E Yokogawa Electric Corporation).
  - Laser lines: diode 447 (50 mW) and 515 (50 mW) and DPSS 561 (50 mW).
  - Objectives: $60 \times /1.00$W LUMPlanFLN and $20 \times /0.95$W XLUMPPlanFL (Olympus).
  - OptoSplit II image splitter (CAIRN Research, UK) equipped with filterset I (dichroic 510dcxr, CFP emitter 483/32 nm, and YFP emitter 542/27 nm) and filterset II (dichroic 560dcxr, GFP emitter 537/26 nm, and mCherry emitter 609/54 nm).
  - NanoScanZ-100 µm (Prior Scientific).
  - EMCCD iXon 897 with 95% Q.E and 35fps camera (Andor Technology).
  - HG lamp (100 W).
  - Andor iQ1.10.5 imaging software.
  - Three different kinds of black Teflon-coated inserts (Luigs & Neumann) and 50 mm glass coverslip were glued with hot wax at the bottom of this chamber.

- **TIRF microscope (TILL Photonics):**
  - iMIC platform with extensions for TIRF and FRET measurements (TILL Photonics): iMIC microscope with a polytrope and dichrotome dual emission extension.
  - Laser lines: 440 (50 mW), 488 (50 mW), 515 (40 mW), and 640 (50 mW) from TOPTICA Photonics for CFP–YFP and GFP–mCherry derivatives FRET pairs.
  - α-Plan-Apochromat 63×/1.46 Oil TIRF (Zeiss).
  - Two MBSs MBS405/488/561 and MBS400/514/561/640 (AHF Analysentechnik).
  - Filterset I (dichroic 515dcxr, CFP emitter 482/18 nm, and YFP emitter 514/3 nm) and filterset II (560dcxr, GFP emitter 561/14 nm, and mCherry emitter 655/40 nm).
  - CCD camera for recording simultaneous two emission channels.
  - Oligochrome rapid filter switch device with xenon (150 W).
  - Multichannel mode of the live acquisition imaging software.
  - Custom-built chambers.

14.2.3 Reagent setup

14.2.3.1 Preparation of cell culture media

1. To prepare 1 L of DMEM: First dissolve 13.4 g of powdered DMEM in 100 ml of distilled water. Then, add 1.81 g of HEPES, 2.53 g of $NaHCO_3$ and bring the volume up to 800 ml by adding water. Then, add 10 ml of each amino acid and adjust the pH to 7.4 with NaOH and bring the final volume to 1 l. Sterile filter the medium and store at 4 °C up to 1 month.
2. Take bottle of the DMEM medium (in cooling room) and put in water bath (37 °C).
3. Take one flask of frozen FCS (fetal bovine serum) and one flask of frozen S/P (penicillin–streptomycin) and put as well in water bath.
4. Add thawed 10% FBS and 1% S/P (5 ml) to a 500 ml DMEM medium and mix it.
5. Sterile filter the media and store in a refrigerator (4 °C).

14.2.3.2 Buffer A for cell lysis

Combine 10 mM HEPES, 150 mM NaCl, and 1% Triton. We usually prepare 1 M of HEPES and NaCl stock solution and keep them for 2 months. To prepare 500 ml of buffer A, weigh HEPES and NaCl according to their molecular weight and add 500 ml of double-distilled water. Adjust the pH to 7.4 with 1 M NaOH. Filter the buffer before use. The buffer can be stored at 4 °C and is stable for 2 months.

14.2.3.3 Buffer B for FRET imaging and Fluorolog experiments

Combine 150 mM NaCl, 5 mM KCl, 1 mM $MgCl_2$, 2 mM $CaCl_2$, and 10 mM HEPES. Adjust the pH to 7.4 with 1 M NaOH. Check the osmolality of the DMEM medium and the buffer B. Bring the osmolality of the buffer same as the medium by adding D + glucose, which is generally around 342 mOsmol before starting of FRET measurements.

14.2.3.4 Protease inhibitors and phosphate-buffered saline stock solutions

PMSF inhibitor: Prepare 50 ml of 100 mM stock solution in 70% ethanol and store them in 500 µl aliquots at −20 °C.

> *CLAP100× inhibitor*: Prepare separate stock solutions of each organic compound and store them at −20 °C: leupeptin (10 mg/ml in distilled water), chymostatin (10 mg/ml in DMSO), antipain (10 mg/ml in 70% ethanol), and pepstatin (1 mg/ml in methanol). To prepare 50 ml of solution, mix all the compounds in 50 ml of 70% ethanol and prepare 500 µl of aliquots and sterile filter it before storing at −20 °C. Thaw before use.
> *Amino acids*: Prepare 200 ml stock solution of each amino acid (1.050 g of L-cystine, 0.712 g of L-alanine, 1.060 g of L-aspartic acid, 0.920 g of L-proline, 1.350 g of L-glutamic acid, and 1.351 g of L-asparagine) and store as 10 ml aliquots at −20 °C.
> *10× phosphate-buffered saline (PBS)*: Combine 80 g of NaCl, 14.195 g of $NaHPO_4$, 2.0 g of KCl, and 2.4 g of KH_2PO_4 and dissolve them in 1 l of distilled water. Autoclave the solution and later adjust the pH to 7.4 with 1 M NaOH. Store the stock solution at room temperature for 2 months. Dilute 10 times before using for experiments.

14.3 METHODS (PROTOCOLS AND PROCEDURE)

14.3.1 Cell preparation and transfection timing

14.3.1.1 Culture of cells

Start new culture: Resuspend the frozen cells in 1 ml of DMEM media and add the suspension into a flask containing 19 ml of DMEM media.

14.3.1.2 Reculture and seeding cells

Neuroblastoma cells (N1E-115 cell lines)

1. Put a vial of fresh media DMEM (FCS/PS) into the water bath at 37 °C for 15 min.
2. Pour 6 ml of media per petri plates.
3. Pour the old media out of the plate.
4. Add 5 ml of fresh media into the plate and detach the cells and separate the cells by gently sucking in and out 3–4 times. Avoid forming bubbles.
5. Pour 1.5 ml of the cells into the new petri plates and 500 µl for 6-well and 100 µl for 12-well (including glass coverslips on the bottom for microscopes) plates.
6. Let the cells grow normally for 2 days or one for transfection at 37 °C/5% CO_2.

14.3.1.3 Transfection

1. Calculate the volume of Opti-MEM, Lipofectamine, and vectors (depending on concentration) to be used according to manufacturer's instruction.

2. Change the medium of the cells after 3–4 h from DMEM (FCS) to DMEM (FCS/PS).
3. Let the cells grow at 37 °C/5% CO_2 overnight. Overnight transfection is sufficient for proper expression of plasmids. The cells can be used for spectral and FRET measurements at any time, usually 14–24 h after transfection.

14.3.2 Cell lysis for Fluorolog

1. Keep the cells on ice.
2. Remove the media and wash the cells with 700–1000 μl of PBS.
3. Mix buffer A (200 μl/plate) and two inhibitors (10 μl/ml) (PMSF100 mM and CLAP100×).
4. Add the inhibitors to cells and keep at 4 °C for 15–20 min with shaking.
5. Centrifuge the cells for 15 min at 14,000 rpm at 4 °C.
6. Remove the pellet carefully and transfer the supernatant to a cuvette and bring the volume to 2 ml by adding buffer B and further proceed with experiment (like titration measurements over time).
7. For spectral analysis measurements, lysing of cells is not required. Bring the cells from incubator and remove the media and wash the cells with 1 ml of PBS once. Then, detach the cells from the plate by adding 1 ml of buffer B twice by gently sucking in and out with 1 ml pipette so that the cells are alive for the experiment. Transfer 2 ml volume of cells to cuvette and continue with experiment.

14.3.3 Procedure part I: data acquisition

Quantitative lux-FRET measurement requires acquisition procedure shown in the flowchart (Fig. 14.3). For the spectral separation of donor and acceptor fluorescence contribution, reference spectra of cells transfected with only donor and only acceptor and others must be obtained for each excitation wavelength under identical acquisition settings (i.e., excitation intensity, exposure time, and detector gain) as in the FRET experiment. From that reference measurements, $F_D^{i,\mathrm{ref}}(\lambda)$ and $F_A^{i,\mathrm{ref}}(\lambda)$ are obtained. In case of multicolor experiments, all other spectral components must be obtained as separate measurements in a similar manner. Especially for experiments with a fluorescence spectrometer, measuring the autofluorescence and the water Raman spectrum is required. Accordingly, Eq. (14.2) must be extended for spectral unmixing. Data acquisition of a standard tandem construct is required to determine R_{TC}. Dependent on the experiment, positive and negative control must be also recorded. The real FRET experiment with the sample of interest should then be recorded, best under identical conditions. Take into account that all reference measurements are required for later data evaluation. Thus, it is advisable to prepare some additional reference samples as well as to acquire references and controls before starting extensive sample experiments.

14.3 Methods (Protocols and Procedure)

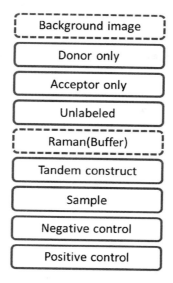

FIGURE 14.3

Flowchart showing steps to acquire various references according to different experimental setups for quantitative lux-FRET measurement. (For color version of this figure, the reader is referred to the online version of this chapter.)

14.3.3.1 Setting individual protocols according to experimental setup

In the following, we describe experimental procedures to investigate the protein–protein interaction using serotonin receptors 5-HT$_{1A}$ and 5-HT$_7$ tagged with monomeric versions of CFP and YFP as an example. However, the same strategy can be applied to any protein of interest.

14.3.4 Spectroscopic measurement at a fluorescence spectrometer (Fluorolog 3v.2.2.)

This system is highly suitable for FRET investigations in titration and time lapse experiments for lysed cells and cell suspensions. Mouse N1E-115 neuroblastoma cells were seeded in 35 mm dishes and cotransfected with plasmid DNAs encoding for 5-HT$_{1A}$–CFP and 5-HT$_7$–YFP receptors. 16 h after transfection, cells were washed once with 1 × PBS to remove residual medium and then resuspended in 2 ml buffer B solution. All measurements were performed in 10 mm pathway quartz cuvettes. We use front-face detection, where the fluorescence is collected from the sample's surface. Typically, fluorescence signals are dwarfed by stray or scattered light from the sample. We set chamber temperature to 25 °C to compare results with microscope experiments. During acquisition, the cell suspension was continuously stirred with a magnetic stirrer. For reference measurements, cells were cotransfected with plasmid encoding a single fluorophore-tagged receptor (only donor and only acceptor).

To maintain total plasmid concentration, an equal amount of empty vector was transfected together with receptor. To estimate the spectral contributions of light scattering and autofluorescence of cells, reference spectra of empty vector transfected cells were used. Background and Raman spectra were unmixed along with references during the data analysis. A CFP–YFP tandem construct with a 1:1 stoichiometry and $E=35\%$ was used to obtain R_{TC} and as positive control. As negative control, we used CFP and YFP, which are expressed in cytosol.

In contrast to FRET microscopy, where we use 440 and 514 nm excitation for CFP–YFP pairs, excitation wavelength ex1:440 and ex2:488 nm was applied in the Fluorolog. At 440 nm, the relative excitation of CFP to YFP is optimal. Ex2 however differs from microscope experiments of optimally 514 nm. To find best excitation wavelengths, we test lux-FRET accuracy and error propagation by using various couples of excitation wavelengths. We found that when CFP and YFP wavelengths are longer than 490 nm it lead to serious unmixing errors since reference spectra of acceptor and cell background have similar line shape above 500 nm and thus can hardly be distinguished.

Acquisition conditions were set to excitation slit width 4 nm (emission was recorded 450–600 nm for ex1 and 498–600 nm for ex2) and emission slit width 2 nm, with step width of 1 nm and integration time 1 s. In case of time lapse experiments, acquisition protocol can be further optimized by sparsely recording second excitation (e.g., only at start and end of time lapse).

14.3.5 FRET imaging of receptor–receptor interaction for intact cells with subcellular spatial resolution

14.3.5.1 Zeiss LSM 780

For pixel-based analysis and in order to achieve high spatial resolution in 3D in detail, we perform the following calibration steps for best confocal overlap at both excitation conditions: (1) MBS to pinhole calibration after thermal setting of 24 h is especially required for CFP–YFP experiments with 440/514 nm excitation since the two laser lines are introduced to the optical path by different "TwinGate" main beam splitters, which could cause x–y-pixel shifts up to 200 nm due to thermal effects. For constant thermal settings, LSM part of the microscope is not switched off on daily basis. (2) Correction collar adjustment to correct for coverslip thickness mainly influences confocality in z-direction and thus influences brightness and leads to spectral aberration at suboptimal settings. (3) Collimator adjustment for 440 nm laser line is done to gain maximum intensity and best spatial overlap with 514 nm excitation in z-direction. Calibration steps 2 and 3 can be performed best in the "fast Z" mode, where x–z images can be recorded sufficiently fast.

After calibrating the system, live-cell FRET experiments were performed. FRET images of N1E-115 cells expressing 5-HT$_{1A}$–CFP and 5-HT$_7$–YFP receptors were acquired using a 40×/1.2 NA water immersion objective at frame size 1024 × 1024, bit depth 16 bit, and pinhole setting of 1–1.5 AU. To obtain best spatial overlap of both excitations, we applied line-wise switching of excitation. This could be achieved with the ZEN 2010 software with two different acquisition protocols: (1) Line-wise excitation switching is allowed only in standard channel mode configuration. Therefore, we defined two tracks, one per excitation wavelength, each track having maximal 10 emission channels ranging from 415 to 735 nm. To cover the appropriate spectral range, a constant binning of two PMT channels to each channel was defined (Fig. 14.4A). Unfortunately, acquisition settings are difficult to define, and unmixing capabilities of the microscope cannot be fully employed due to binning. (2) As an alternative, we designed an acquisition protocol generated by the LIC Macro toolbox from Roland Nitschke (Life Imaging Center, University of Freiburg, Germany), which allows us to use a nice feature called multitrack lambda mode, where one could define lambda tracks according to number of excitations in a single acquisition protocol with earlier-defined scan settings and sequential line-wise switching of the excitation wavelength (Fig. 14.4B). In this mode with two or more subsequential excitations, we can acquire fluorescence signal information from 32 channels per track at single time frame (Fig. 14.4C). This has been recently implemented in the measurement protocol for lux-FRET imaging. LIC Macro enables us to fully employ spectral capabilities of the Zeiss LSM 780. This acquisition protocol with multitrack lambda mode could also be further applied for dual and triple FRET measurements.

General acquisition conditions are as follows: Laser power was limited to 3–5% for each excitation to avoid bleaching over time and through z-scan. Images were acquired over duration up to 10 min with z-stack consisting of 20 slices with optimal interval settings. In case of using GFP and mCherry as a FRET pair, we used 488 and 561 nm as excitation wavelength. Images for reference were acquired from cells expressing only receptor tagged with respective donor and acceptor with abovementioned acquisition settings for a single image.

14.3.5.2 Spinning disk microscope

The custom-tailored spinning disk microscope represents a good compromise between spatiotemporal resolution and light exposure. Although the pixel size of 222×222 nm is far below requirements for best resolution, it allows us to achieve scan frequencies of up to 10fps even under moderate expression levels. Due to the limited spatial resolution, calibration and correction for aberration is not as crucial as at Zeiss LSM.

General acquisition settings were as follows: Andor iQ 1.10.5 was set to fast lambda z-mode, with alternating excitation of 445 and 514 nm laser power of 30% and 5%, respectively; minimal exposure time was set to 50 ms; EMCCD readout was set to 14 bit AD conversion with 512×512 pixels; and camera gain was kept constant to 100. Single-frame acquisition settings were the same for all image series of an experiment. FRET series with z-stack (20 planes with step width of 1 μm) are recorded in frame-wise scanning mode. The acquisition protocol allows a loop pattern with a delay of about 2 s per single z-stack including alternating excitation. No significant bleaching was observed. Measurements were performed with $60 \times$ water immersion objective. Data were analyzed offline in MATLAB. Schematic representation of this setup is shown in Fig. 14.5.

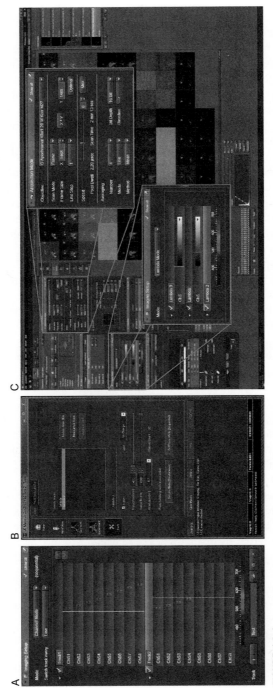

FIGURE 14.4

Lux-FRET acquisition protocols at Zeiss LSM 780. (A) Standard channel mode configuration with defined two excitation tracks with a constant binning of two PMT channels to each channel. (B) Application of the LIC Macro toolbox to improve spectral detection capabilities for lux-FRET. (C) LIC Macro enables us to use a multitrack lambda mode with sequential line-wise switching of alternating excitation wavelength. Sample recording is shown with three lambda track mode. (See color plate.)

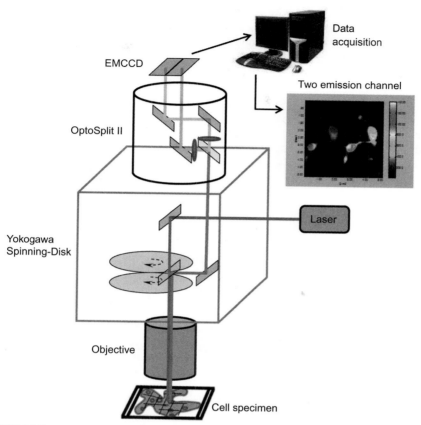

FIGURE 14.5

Schematic representation of spinning disk microscope setup. Pathway of excitation as well as emission light through the setup is shown schematically. Laser excitation light passes spinning disk and Yokogawa dichroic mirror (DM) into the objective exciting the specimen. The emission passes pinhole spinning disk and is reflected by DM and guided to the OptoSplit II that separates the emission with a filterset of dichroic mirror and emitters into two emission channels. This results in two spatially similar, but spectrally different images on both half sides of the EMCCD camera. (See color plate.)

14.3.5.3 TIRF microscope

The extension configuration available at our TIRF microscope allows us to alter the excitation incident angle stepless to switch easily between wide field and TIRF mode. FRET measurements were conducted in two modes: wide field (incident angle 0°) and TIRF (incident angle 75.48°). Due to the difference in signal intensity, images had to be recorded with adapted laser intensities and exposure times. Widefield mode was measured with 30% laser intensity and 50 ms exposure time, and TIRF did require 100% intensity and 500 ms exposure time.

14.3.6 Procedure part II: data evaluation

To determine apparent FRET efficiency for 5-HT$_7$ and the 5-HT$_{1A}$ receptor interaction, we used lux-FRET method described in theory section. The detailed data evaluation steps are illustrated in Fig. 14.6. For the imaging systems, various additional steps are required, since evaluation is pixel-based. In the following, the procedure steps 2–4, 6–8, 14, and 15 are only required for imaging systems.

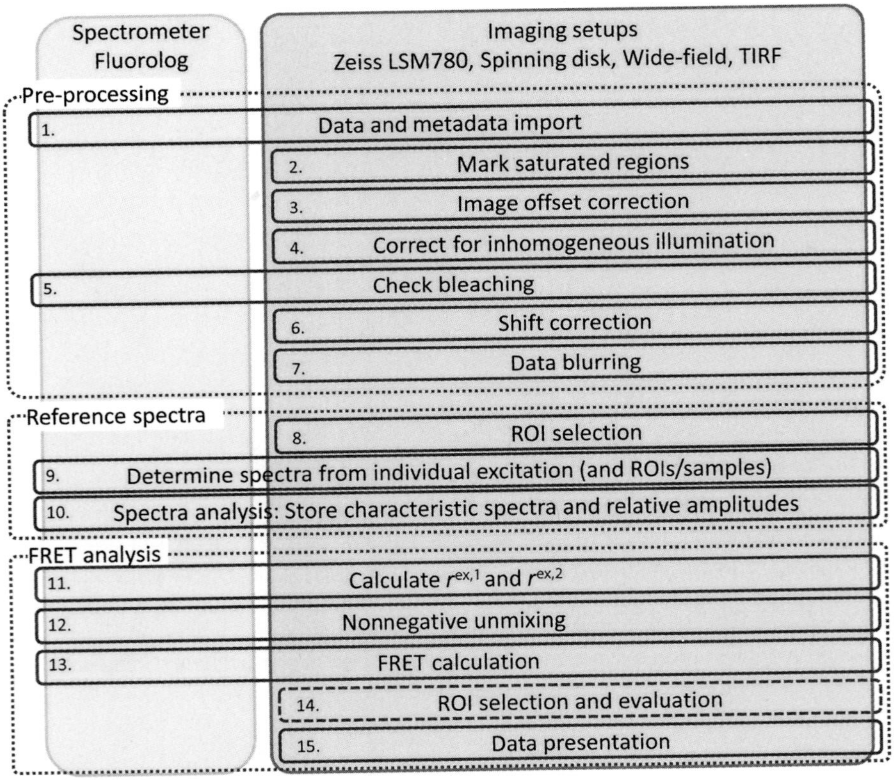

Spectrometer Fluorolog | **Imaging setups** Zeiss LSM780, Spinning disk, Wide-field, TIRF

Pre-processing
1. Data and metadata import
2. Mark saturated regions
3. Image offset correction
4. Correct for inhomogeneous illumination
5. Check bleaching
6. Shift correction
7. Data blurring

Reference spectra
8. ROI selection
9. Determine spectra from individual excitation (and ROIs/samples)
10. Spectra analysis: Store characteristic spectra and relative amplitudes

FRET analysis
11. Calculate $r^{ex,1}$ and $r^{ex,2}$
12. Nonnegative unmixing
13. FRET calculation
14. ROI selection and evaluation
15. Data presentation

FIGURE 14.6

Flowchart describing preprocessing and offline data evaluation of lux-FRET quantification for data sets obtained at different setups. (For color version of this figure, the reader is referred to the online version of this chapter.)

14.3.6.1 Evaluation procedure

Preprocessing **includes steps from 1 to 7**

1. Data import to evaluation software must be achieved individually. In our case, the Fluorolog data were imported in ASCII format, and the microscope images and metadata were imported with self-written routines.
2. Saturated regions of images were indicated and excluded from further processing.
3. Since acquisition devices work with unsigned integer, we introduced an artificial offset while acquisition. Images were offset corrected either done by selecting regions where fluorescence signal was absent or done by using dark images recorded extra with no laser excitation. Values from dark regions were further subtracted from the sample image.
4. The inhomogeneous illumination correction was done according to the specific excitation wavelengths. The inhomogeneous illumination was recorded by fluorescent slides (Chroma Technology). The correction for inhomogeneous illumination is essential for low-quality objectives and large field of views, because the intensity at the peripheral regions of the images can drop to 80%. The intensity drop usually shows an individual characteristic for the used excitation wavelength and the specific emission channels.
5. Since bleaching correction is hardly possible, we preferred to adjust and optimize acquisition conditions to avoid it. This check is also recommended for Fluorolog data.
6. For pixel-based analysis, subpixel shift correction between two excitations for Zeiss LSM and between emission channels caused by imperfect alignment of the OptoSplit II and dichrotome in case of spinning disk and TIRF is necessary. Data shift correction was performed by self-written routines based on image registration and regional image transformation, allowing shifts with a low-order polynomial spatial function.
7. Gauss blurring (standard 0.4 pixel) was applied to soften the image view in order to slightly reduce noise level that could be increased according to imaging conditions.

Reference spectra analysis **(steps 8–10)**

8. From reference images, several regions of interest were selected to define reference spectra for unmixing in further FRET analysis.
9. Reference spectra were determined from ROIs (microscopes) and reference samples (Fluorolog) selected at each excitation. In addition, all acquisition parameters were kept constant along with stored spectra.
10. Spectral analysis was done and the characteristic spectra and relative amplitudes were stored for further FRET calculation (for Fluorolog data as well).

FRET analysis

11. Ratios of excitation strengths $r^{ex,1}$ and $r^{ex,2}$ are calculated for both excitations. These ratios require reference as well as normalized characteristic spectra (see Eq. (14.6)) and provide a link between absorption coefficients and receptor concentrations.
12. Nonnegative unmixing on the basis of $F_D^{i,\text{ref}}(\lambda)$ and $F_A^{i,\text{ref}}(\lambda)$ and all additional contributions was done for both excitations to obtain the excitation wavelength-specific donor and acceptor contributions δ^i and α^i for each pixel. An example of Fluorolog data unmixing is shown in (Fig. 14.7).
13. With the relative donor and acceptor contributions δ^i and α^i, FRET calculation was performed according to Eqs. (14.3)–(14.5). Values and images of Ef_D, Ef_A, and x_D are calculated for Fluorolog and imaging data, respectively. To estimate the error of the quantities, channel intensity information was stored together with the results for further processing. Therewith, weighted mean values for Ef_D, Ef_A, and x_D can be calculated.
14. Selection of region of interest for FRET images is done mainly to distinguish spatial regions or to further evaluate time traces for individual regions.
15. Finally, data presentation is done according to specific biological interest. To check for analysis artifacts, 2D histogram of all FRET parameters versus pixel brightness was generated. FRET parameters should not depend on pixel brightness. Further, from the regions of interest, selective parameters can be evaluated.

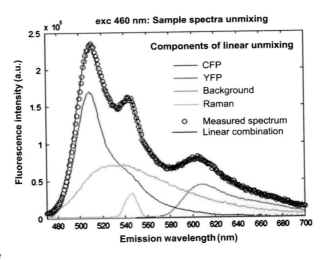

FIGURE 14.7

Principle of spectral linear unmixing. The measured emission spectrum (460 nm excitation) was fitted by a linear combination of reference spectra for YFP, CFP, Raman, and background. The reference spectra were obtained separately. (See color plate.)

14.4 RESULTS AND DISCUSSION

It is now widely accepted that GPCRs can interact and form oligomers, and a growing body of evidence points to the functional importance of oligomeric complexes for receptor trafficking, receptor activation, and G protein coupling in native tissues (Rivero-Muller et al., 2010). The clinical significance of GPCR oligomerization has also become more evident during recent years, leading to identification of receptor oligomers as novel therapeutic targets (Gonzalez-Maeso et al., 2008). We have recently established the spectral quantitative FRET analysis called lux-FRET method to analyze receptor–receptor interaction with high spatial and temporal resolution in live cells. From lux-FRET analysis, we can not only obtain the apparent FRET efficiencies but also receive information about the relative expression levels and can correlate their oligomerization state with the effector response. Lux-FRET in combination with microscopy provides unique possibility to monitor protein–protein interaction in living cells at a higher resolution in real time under physiological conditions. Based on quantitative lux-FRET, we have developed custom-tailored acquisition protocols that are optimized according to biomedical application and are important in quantitative molecular microscopy. Noteworthy, this approach allows to calculate and visualize (Fig. 14.8; Renner et al., 2012) the apparent FRET efficiencies for fractions of donors and acceptors over a wide range of donor molar fraction. Examples of microscopic FRET images measured at LSM 780, spinning disk, Widefield, and TIRF setting with above mentioned defined acquisition protocols are shown in Fig. 14.9. This analysis strategy now enables us to apply this technique further for various biologically interesting single-cell experiments.

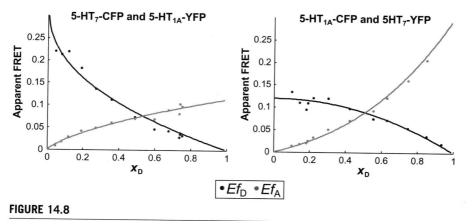

FIGURE 14.8

Dimerization of 5-HT$_{1A}$ and 5-HT$_7$ receptors investigated by lux-FRET. Apparent FRET efficiencies Ef_D (blue) and Ef_A (green) were calculated and are shown as functions of the donor mole fraction x_D for 5-HT$_{1A}$–5-HT$_7$ receptors heteromers. Experimental data were fitted according model for dynamic oligomerization described in Ref. Renner et al. (2012). (See color plate.)

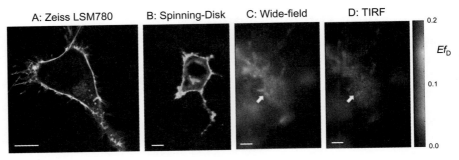

FIGURE 14.9

Interaction of 5-HT$_{1A}$ and 5-HT$_7$ receptors investigated by lux-FRET. Images of apparent FRET efficiency Ef_D in an N1E-115 cell coexpressing 5-HT$_{1A}$–CFP and 5-HT$_7$–YFP receptors were created according to the lux-FRET method measured at (A) Zeiss LSM 780 and (B) spinning disk microscope, and cell coexpressing 5-HT$_{1A}$–YFP and 5-HT$_7$–CFP receptors recorded using the iMIC platform in (C) Wide-field and (D) TIRF mode. Scale bar: 10 μm. (See color plate.)

Acknowledgments

This study was supported by the Federal Ministry of Education and Research of the Federal Republic of Germany (0315690D to S.P. and A.Z.). E.G.P. was supported by the Deutsche Forschungsgemeinschaft (grants PO732 and SFB621/C12).

References

Forster, T. (1946). Energiewanderung und Fluoreszenz. *Die Naturwissenschaften, 33*(6), 166–175. http://dx.doi.org/10.1007/bf00585226.

Forster, T. (1948). Zwischenmolekulare Energiewanderung und Fluoreszenz. *Annalen der Physik, 2*(1–2), 55–75.

Forster, T. (2012). Energy migration and fluorescence. 1946. *Journal of Biomedical Optics, 17*(1), 011002.

Gonzalez-Maeso, J., Ang, R. L., Yuen, T., Chan, P., Weisstaub, N. V., Lopez-Gimenez, J. F., et al. (2008). Identification of a serotonin/glutamate receptor complex implicated in psychosis. *Nature, 452*(7183), 93–97. http://dx.doi.org/10.1038/nature06612.

Gorinski, N., Kowalsman, N., Renner, U., Wirth, A., Reinartz, M. T., Seifert, R., et al. (2012). Computational and experimental analysis of the transmembrane domain 4/5 dimerization interface of the serotonin 5-HT(1A) receptor. *Molecular Pharmacology, 82*(3), 448–463. http://dx.doi.org/10.1124/mol.112.079137.

Lakowicz, J. (2006). *Principles of fluorescence spectroscopy* (3rd ed.). New York, Boston, Dordrecht, London, Moscow: Kluwer Academic/Plenum Publishers.

Lam, A. J., St-Pierre, F., Gong, Y., Marshall, J. D., Cranfill, P. J., Baird, M. A., et al. (2012). Improving FRET dynamic range with bright green and red fluorescent proteins. *Nature Methods, 9*(10), 1005–1012. http://dx.doi.org/10.1038/nmeth.2171.

Meyer, B. H., Segura, J. M., Martinez, K. L., Hovius, R., George, N., Johnsson, K., et al. (2006). FRET imaging reveals that functional neurokinin-1 receptors are monomeric and reside in membrane microdomains of live cells. *Proceedings of the National Academy of Sciences of the United States of America*, *103*(7), 2138–2143. http://dx.doi.org/10.1073/pnas.0507686103.

Piston, D. W., & Kremers, G. J. (2007). Fluorescent protein FRET: The good, the bad and the ugly. *Trends in Biochemical Sciences*, *32*(9), 407–414. http://dx.doi.org/10.1016/j.tibs.2007.08.003.

Renner, U., Zeug, A., Woehler, A., Niebert, M., Dityatev, A., Dityateva, G., et al. (2012). Heterodimerization of serotonin receptors 5-HT1A and 5-HT7 differentially regulates receptor signalling and trafficking. *Journal of Cell Science*, *125*(Pt 10), 2486–2499. http://dx.doi.org/10.1242/jcs.101337.

Rivero-Muller, A., Chou, Y. Y., Ji, I., Lajic, S., Hanyaloglu, A. C., Jonas, K., et al. (2010). Rescue of defective G protein-coupled receptor function in vivo by intermolecular cooperation. *Proceedings of the National Academy of Sciences of the United States of America*, *107*(5), 2319–2324. http://dx.doi.org/10.1073/pnas.0906695106.

Wlodarczyk, J., Woehler, A., Kobe, F., Ponimaskin, E., Zeug, A., & Neher, E. (2008). Analysis of FRET signals in the presence of free donors and acceptors. *Biophysical Journal*, *94*(3), 986–1000. http://dx.doi.org/10.1529/biophysj.107.111773.

Zeug, A., Woehler, A., Neher, E., & Ponimaskin, E. G. (2012). Quantitative intensity-based FRET approaches—A comparative snapshot. *Biophysical Journal*, *103*(9), 1821–1827. http://dx.doi.org/10.1016/j.bpj.2012.09.031.

CHAPTER

Site-Specific Labeling of Genetically Encoded Azido Groups for Multicolor, Single-Molecule Fluorescence Imaging of GPCRs

15

He Tian, Thomas P. Sakmar, and Thomas Huber

Laboratory of Chemical Biology and Signal Transduction, The Rockefeller University, New York, New York, USA

CHAPTER OUTLINE

- 15.1 Purpose 270
- 15.2 Theory 270
  - 15.2.1 Site-Specific Labeling of GPCRs with Fluorophores Suitable for SMD–TIRF 270
  - 15.2.2 Single-Molecule Immunoprecipitation of GPCRs 271
  - 15.2.3 Automated, Multicolor SMD–TIRF Microscopy 272
- 15.3 Protocol 1: Genetically Encoded Azido Groups for Bioorthogonal Conjugation 273
  - 15.3.1 Step 1: Transfection of the HEK293-F Suspension Cells to Express Azido-Group Tagged Opsin 273
    - 15.3.1.1 Materials 273
    - 15.3.1.2 Culture HEK293-F Suspension Cells for Transfection 274
    - 15.3.1.3 Cotransfect the Cells with the Plasmids for Amber Codon Suppression 274
  - 15.3.2 Step 2: Purification of Heterologously Expressed Wild-Type and Azido-Tagged Rhodopsin 275
    - 15.3.2.1 Materials 275
    - 15.3.2.2 Regeneration of Heterologously Expressed Opsin with 11-cis-retinal 275
    - 15.3.2.3 Purification of Rhodopsin 275
    - 15.3.2.4 Characterization of Purified Rhodopsin with UV–Vis Spectroscopy 276

15.4 Protocol 2: Labeling with Fluorescent Probes Using Cyclooctynes 276
    15.4.1 Step 1: Fluorescent Labeling of Azido-Tagged Rhodopsin
        with SpAAC 277
        15.4.1.1 Materials 277
        15.4.1.2 Procedures 277
    15.4.2 Step 2: Characterization of Labeled Receptor Using UV–Vis
        Spectroscopy and In-Gel Fluorescence 278
        15.4.2.1 Material 278
        15.4.2.2 UV–Vis Spectroscopy 278
        15.4.2.3 In-Gel Fluorescence 278
    15.4.3 Step 3: Kinetic Study of the SpAAC 278
        15.4.3.1 Materials for Silver Staining: 278
        15.4.3.2 Prepare Samples of the SpAAC for time-series Analysis 281
        15.4.3.3 In-Gel Fluorescence 281
        15.4.3.4 Silver Staining 281
        15.4.3.5 Data Analysis 281
15.5 Protocol 3: Preparation of Biotinylated Antibodies 282
    15.5.1 Method 1: Biotinylation of Periodate-Oxidized Antibodies with
        Biocytin Hydrazide 282
        15.5.1.1 Materials 282
        15.5.1.2 Solutions and Buffers 283
        15.5.1.3 Procedure 284
    15.5.2 Method 2: Biotinylation of Antibodies with Sulfo-NHS-SS-biotin .. 284
        15.5.2.1 Materials 284
        15.5.2.2 Solutions and Buffers 284
        15.5.2.3 Procedure 285
15.6 Protocol 4: Preparation of Detergent-Solubilized Lipids 285
    15.6.1 Materials 285
    15.6.2 Solutions and Buffers 285
    15.6.3 Procedure 286
15.7 Protocol 5: Single-Molecule Immunoprecipitation on Glass-Bottom Microplates . 287
    15.7.1 Step 1: Cleaning of Glass-Bottom Microtiter Plate 287
        15.7.1.1 Materials 287
        15.7.1.2 Solutions and Buffers 287
        15.7.1.3 Procedure 288
    15.7.2 Step 2: Functionalize Glass Surface with Epoxy Silane 289
        15.7.2.1 Materials 289
        15.7.2.2 Solutions and Buffers 289
        15.7.2.3 Procedure 289
    15.7.3 Step 3: Immobilization of Biotinylated-Bovine Serum Albumin 289
        15.7.3.1 Materials 289
        15.7.3.2 Solutions and Buffers 290
        15.7.3.3 Procedure 290

15.7.4 Step 4: Immobilization of NeutrAvidin 291
        15.7.4.1 Materials 291
        15.7.4.2 Solutions and Buffers.. 291
        15.7.4.3 Procedure 292
15.7.5 Step 5: Immobilization of Biotinylated Antibody.......................... 292
        15.7.5.1 Materials 292
        15.7.5.2 Solutions and Buffers.. 292
        15.7.5.3 Procedure 293

15.8 Protocol 6: Automated Multicolor, Single-Molecule TIRF Microscopy 293
    15.8.1 Step 1: Microscope Optical Setup... 295
    15.8.2 Step 2: Alignment of Laser-to-Fiber, Fiber-to-Fiber Beam Combination System... 298
    15.8.3 Step 3: Alignment of DualView and QuadView Simultaneous Imaging Systems... 299
    15.8.4 Step 4: Calibration of the Multiwell Plate Geometry..................... 299
    15.8.5 Step 5: Adjust TIRF Excitation 300
    15.8.6 Step 6: Sample for Subpixel Alignment of Color Channels............ 300
    15.8.7 Step 7: Alignment of the CRISP Focus-Stabilization (autofocus) System 300
    15.8.8 Step 8: Preparation of Antibleaching and Antiblinking Imaging Buffer System 300
        15.8.8.1 Materials 300
        15.8.8.2 Solutions and Buffers.. 301
        15.8.8.3 Procedure 301
    15.8.9 Step 9: Multiposition, Multicolor, Time-Lapse Imaging of Single Molecules 302
    15.8.10 Step 10: Image Processing for Optimal Detection of Single Molecules... 302

Acknowledgments 302
References 302

Abstract

Heptahelical G protein-coupled receptors (GPCRs) mediate transmembrane signal transduction to facilitate intercellular communication. GPCRs assemble in the membrane bilayer with a variety of cytoplasmic adapter and scaffold proteins to form molecular machines, or "signalosomes," which undergo complex dynamic assembly and disassembly reactions. Despite significant recent advances in structural studies of GPCRs and their associated cytoplasmic components, understanding transmembrane signaling in four dimensions with chemical precision requires new approaches. One promising approach to study allosteric effects involved in signalosome reaction pathways is to use multicolor single-molecule detection (SMD) fluorescence experiments in biochemically defined systems. We describe here the methodological foundation for automated, multicolor, single-molecule fluorescence studies of the structural and compositional dynamics of macromolecular complexes involved in signal transduction. We present a general, simple, and robust method for stoichiometric, site-specific fluorescence labeling of expressed GPCRs. The method is based on bioorthogonal conjugation of a fluorescent reporter group to a genetically encoded azido group introduced into expressed GPCRs using amber codon suppression. We then present a strategy to reconstitute labeled GPCRs in native-like membranes and to tether-oriented samples onto surfaces amenable for interrogation by total internal reflectance fluorescence (TIRF) spectroscopy. We describe how to assemble an automated four-color epifluorescence microscope with SMD–TIRF optics. Finally, we discuss how to adapt engineered samples for high-throughput imaging with the aim of understanding the kinetic relationships between ligand binding and the dynamic regulation of the GPCR signalosome.

15.1 PURPOSE

The protocols described in this chapter provide a foundation for automated, multicolor, single-molecule fluorescence studies of the structural and compositional dynamics of macromolecular complexes involved in signal transduction. In particular, we are interested in the dynamic assembly of signaling complexes ("signalosomes") involving G protein-coupled receptors (GPCRs). We present a general, simple, and robust method for stoichiometric, site-specific fluorescence labeling of GPCRs. The method is based on bioorthogonal conjugation of a fluorescent reporter group to a genetically encoded azido group. Furthermore, we describe high-throughput adapted approaches to single-molecule imaging by total internal reflection fluorescence (TIRF) microscopy.

15.2 THEORY

Single-molecule detection (SMD) fluorescence methods have appeared first in the late 1980s, and there has been almost exponential growth (Joo, Balci, Ishitsuka, Buranachai, & Ha, 2008) of work in this field, ever since. One of the fascinating features of these methods is that detailed kinetic information is accessible from observations of a fluctuating system under equilibrium conditions (Weiss, 1999). Protein–nucleic acid interactions have become the spearhead of biological applications (Roy, Hohng, & Ha, 2008), in part due to favorable kinetics (Blanchard, 2009).

The protocols presented in this chapter address a series of technical challenges in the study of membrane proteins, especially GPCRs.

15.2.1 Site-specific labeling of GPCRs with fluorophores suitable for SMD–TIRF

Using our amber codon suppression technology (Ye et al., 2008), it is possible to genetically encode a reactive *p*-azido-phenylalanine (azF) residue with just a single-site mutation (Naganathan, Ye, Sakmar, & Huber, 2013; Ye, Huber, Vogel, & Sakmar,

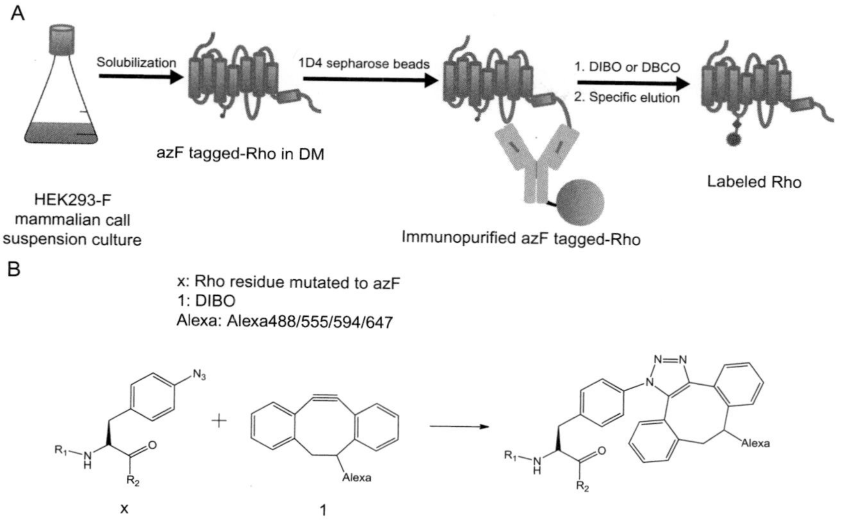

FIGURE 15.1

Labeling scheme of rhodopsin with strain-promoted alkyne–azide cyclooctyne (SpAAC). (A) Labeling scheme of rhodopsin with strain-promoted alkyne–azide cyclooctyne. First p-azido-L-phenylalanine (azF) was incorporated into opsin heterologously expressed in HEK293-F suspension cells using amber codon suppression technology. The cells expressing opsin were harvested and regenerated with 11-cis-retinal. The regenerated cells were solubilized and immunoprecipitated with 1D4 sepharose 2B resin. After the labeling reaction, the unreacted dyes were washed away. The labeled receptor was eluted and recovered in the elution buffer. (B) The dibenzocyclooctyne (DIBO) reagents derivatized with fluorophore were used to generate fluorescently labeled receptor. (For color version of this figure, the reader is referred to the online version of this chapter.)

2009). The site-specific azido group is then conjugated to a fluorophore or any other biophysical probe (Fig. 15.1). The goal here is to introduce a unique amino acid side chain at a site that would not affect function, but would be informative with respect to receptor conformational changes related to structural dynamics of signalosomes. Our method circumvents the fundamental problem in generating site-specific fluorescently labeled proteins based on single-accessible cysteine mutants, that is, the necessity to generate first a cysteine-free protein background by identification of nonessential and accessible cysteines and substitution with alternative amino acids. The problem here is circular, as one would need a sufficiently informative functional assay to detect any unwanted side effects of these mutations, while the purpose of the labeling experiment itself is to establish such an assay.

15.2.2 Single-molecule immunoprecipitation of GPCRs

TIRF methods (Axelrod, Burghardt, & Thompson, 1984) allow selective excitation of fluorophores in the evanescent wave penetrating about 100 nm deep into the aqueous layer beyond the glass–water interface. Capturing a molecule of interest in this spatial

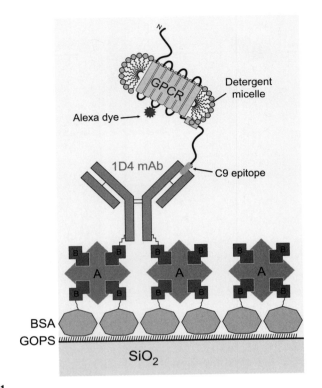

SCHEME 15.1

Single-molecule immunoprecipitation (pull-down) for single-molecule detection (SMD) fluorescence experiments. We developed the following scheme to self-assemble the depicted structure. The glass surface of the chip is first coated with glycidoxy propyl silane (GOPS) to introduce reactive epoxy groups on the glass surface. Biotinylated BSA is then covalently attached to the surface by overnight reaction with the epoxy groups. The biotinylated BSA efficiently captures NeutrAvidin (A), a deglycosylated form of the tetrameric biotin-binding protein avidin with a more neutral isoelectric point, and NeutrAvidin captures a biotinylated antibody (1D4 mAb). 1D4 mAb that recognizes the C-terminal nonapeptide of rhodopsin ("C9-tag"), which we routinely use as purification tags for other GPCRs. The remaining biotin binding sites are saturated with biotin to convert the avidin tetramer in the high-affinity state with fM binding affinity for biotin. The antibody captures the Alexa dye-labeled GPCR by its C-terminal C9-tag (green) from detergent solution. (For interpretation of the references to color in this figure legend, the reader is referred to the online version of this chapter.)

region allows long-term observation of a single molecule (Scheme 15.1). In this way, slow dissociation/association processes can be observed (Huber & Sakmar, 2011).

15.2.3 Automated, multicolor SMD–TIRF microscopy

The number of single-molecule fluorescence techniques is large and steadily growing, but for the study of GPCR signalosomes, a smaller number of methods appear to be especially well suited. They are based on (i) intensity and residence time of fluorescence spots, (ii) dual-color colocalization (>10 nm), and (iii) FRET (3–10 nm). Multicolor methods have the unique advantage to allow simultaneous observation of compositional and conformational dynamics of macromolecular complexes. Automated SMD–TIRF microscopy facilitates systematic studies of biochemical parameters, such as concentration, pH, ionic strength, and osmotic pressure (and molecular crowding). For example, two concentrations have to be systematically varied in the study of three-component (ternary) complexes that are characteristic for the allosteric GPCR signalosomes (Huber & Sakmar, 2011).

15.3 PROTOCOL 1: GENETICALLY ENCODED AZIDO GROUPS FOR BIOORTHOGONAL CONJUGATION

We have established a robust transient transfection protocol for amber codon suppression in HEK293-T adherent cells. However, the adherent cells are cumbersome for scaling up. We have adapted our transient transfection protocol for HEK293-F suspension cells, which greatly reduced our time invested on protein expression. We have incorporated an unnatural amino acid p-azido-L-phenylalanine (azF) into rhodopsin and shown that both wild-type and mutant rhodopsin could be regenerated and purified from suspension cells. UV–Vis spectroscopy can characterize the purified rhodopsin. The expression level of UAA-containing rhodopsin in the suspension cells is comparable to that of the adherent cells.

15.3.1 Step 1: transfection of the HEK293-F suspension cells to express azido-group tagged opsin

15.3.1.1 Materials
FreeStyle™ 293-F Cells (Invitrogen, R790-07)
Gibco® FreeStyle™ 293 Expression Medium (Invitrogen, 12338018)
FreeStyle™ MAX Reagent (Invitrogen, 16447-100)
OptiPRO™ SFM (Invitrogen, 12309-019)
p-azido-L-phenylalanine (Chem-Impex International, 06162, MW=206.1 g/mol)
125-mL polycarbonate, disposable, sterile Erlenmeyer flasks
Millipore Steriflip-GP filter unit, 0.22 μm (Millipore, SCGP00525)
Sterile Eppendorf tube and conical, polycarbonate centrifuge tube (15 and 50 mL)
11-*cis*-retinal (1.4 mM stock solution dissolved in ethanol)
5% CO_2 cell culture incubator (Forma Scientific)
Platform shaker for growing suspension cell culture (New Brunswick Scientific, Innova 2000)
Room temperature tabletop centrifuge
A dark room for handling light-sensitive materials

PLASMIDS FOR TRANSFECTION:
1. pSVB.Yam carrying the gene encoding an amber suppressor tRNA
2. pcDNA.RS carrying the tRNA synthetase for azF

3. pMT4 carrying the synthetic gene encoding bovine rhodopsin (Franke, Sakmar, Oprian, & Khorana, 1988)

E. coli TOP10 (Invitrogen) was used for plasmid propagation and isolation. Plasmid DNA was purified using standard MaxiPrep Kits from Qiagen. The amber mutations were introduced into the rhodopsin gene using QuikChange mutagenesis (Stratagene). Oligonucleotides for mutagenesis were obtained from Fisher Scientific. All plasmid constructs were confirmed by automated DNA sequencing.

15.3.1.2 Culture HEK293-F suspension cells for transfection
- Culture the HEK293-F suspension cells in serum-free Gibco® FreeStyle™ 293 Expression Medium in a 125 mL disposable, sterile, polycarbonate Erlenmeyer flask at 37 °C in 5% CO_2 atmosphere. Shake the cell culture at a constant speed of 125 rpm.
- Use the cells in rapid growth phase for transfection (the density of the culture ranges between 1 and 2×10^6 cells/mL and the viability should be above 90%). Determine the cell culture density and calculate the amount of cells needed for the experiment. A standard transfection experiment is performed in a 30 mL volume and the number of cells to transfect is 3×10^7 (final cell density of 1×10^6 cells/mL). Harvest the correct number of cells in a sterile conical centrifuge tube by centrifuging the suspension culture at $100 \times g$ for 5 min.

15.3.1.3 Cotransfect the cells with the plasmids for amber codon suppression
- Dissolve azF in the FreeStyle™ 293 Expression Medium. As azF is sparingly soluble at neutral pH (<3 mM), we ultrasonicate the medium to accelerate dissolution. The medium should be filtered prior to use, as the unnatural amino acid powder does not come sterile. The medium needs to be warmed up in a 37 °C water bath before adding into the cell culture.
- Resuspend the HEK293-F cells to a density of 10^6 cells/mL in 30 mL culture supplemented with 1 mM azF.
- For amber codon suppression, add 38.6 μg of plasmid DNA (18.4 μg of pMT4.Rho containing the amber codon, 18.4 μg of pSVB.Yam, and 1.84 μg pcDNA.RS) into OptiPro SFM reduced serum medium to make a total volume of 0.6 mL.
- In another sterile tube, dilute 38.6 μL of FreeStyleMax reagent in OptiPro SFM reduced serum medium to a total volume of 0.6 mL.
- Gently add the transfection reagent dilution into the DNA, and incubate the mixture at room temperature for 10 min before adding into the cell culture.
- Harvest the cells transfected with wild-type rhodopsin 48 h posttransfection. For amber codon suppression in rhodopsin, harvest after 96 h, which give the maximal yield for azF-containing rhodopsin. Centrifuge the culture at $100 \times g$ for 5 min and remove the medium. Wash the cells once with DPBS containing protease inhibitor. The total cell number upon harvesting normally ranged from 6×10^7 to 8×10^7.

15.3.2 Step 2: purification of heterologously expressed wild-type and azido-tagged rhodopsin

15.3.2.1 Materials

n-Dodecyl-β-D-maltopyranoside (DM) (Affymetrix, Anatrace, D310, MW = 510.6 g/mol)

C9 peptide for specific elution (NH_2-TETSQVAPA-COOH, synthesized by Bio Basic)

1D4 Sepharose 2B resin (2 mg 1D4 monoclonal antibody/mL resin, prepared by cyanogen bromide activation procedure) (Oprian, Molday, Kaufman, & Khorana, 1987; Knepp, Grunbeck, Banerjee, Sakmar, & Huber, 2011)

SOLUBILIZATION BUFFER: (Sakmar, Franke, & Khorana, 1989)
 1% (w/v) DM
 50 mM HEPES or Tris–HCl, pH 6.8
 100 mM NaCl
 1 mM $CaCl_2$
 With complete protease inhibitor cocktail (Roche)

REACTION AND WASH BUFFER:
 0.1% (w/v) DM in Dulbecco's phosphate-buffered saline (DPBS), pH 7.2

ELUTION BUFFER:
 0.33 mg/mL C9 peptide
 0.1% (w/v) DM
 2 mM phosphate buffer, pH 6.0
 Centrifugation apparatus capable of centrifugation at $100,000 \times g$. We use a Beckman Optima ultracentrifuge with a Beckman TLA100.3 rotor.
 15 mL polypropylene Falcon tube for the solubilization process.
 A deli-refrigerator installed in the dark room.
 Lambda-800 spectrophotometer (PerkinElmer Life Sciences).

15.3.2.2 Regeneration of heterologously expressed opsin with 11-cis-retinal

- Resuspend the harvested cells in DPBS (containing protease inhibitor) at a density of 10^7 cells/mL in a 15 mL conical, polypropylene tube (Falcon).
- Add 11-*cis*-retinal (1.4 mM ethanol solution) in the dark room into the cell suspension to a final concentration of 5 μM. Nutate the suspension at 4 °C overnight.
- Remove the excess 11-*cis*-retinal by spinning down the cells and discard the supernatant fraction. The regenerated cells can be immediately used or stored at −20 °C for several months.

15.3.2.3 Purification of rhodopsin

- In order to prevent photobleaching of rhodopsin, always maintain the regenerated cells and lysates in the dark.

- The regenerated cells were lysed with the solubilization buffer for at least 1 h at 4 °C.
- Centrifuge the cell lysates at $100,000 \times g$ for 30 min at 4 °C to clear the cell lysate.
- Mix the supernatant fraction with 50 or 100 μL 1D4-mAb sepharose 2B and incubated overnight at 4 °C.
- Transfer the resin into a Millipore Ultrafree-MC centrifugal filtering unit (pore size 0.45 μm), which allows for easy removal of wash buffer.
- Wash the resin with the wash buffer for three times (30 min each time) and then once with the elution buffer depleted of the C9 peptide to remove the salt.
- Elute the receptor with 100 μL (no less than the volume of the packed beads) elution buffer.
- Incubate the resin with the elution buffer at 4 °C for at least 1 h.
- Collect the purified receptor in a clean 1.5 mL Eppendorf tube by centrifugation. Two elutions should recover 70–80% of the receptor.
- Add 150 mM NaCl to the sample to give appropriate ionic strength. If no further experiment follows immediately, keep the eluates at −80 °C and thaw on ice before use.

15.3.2.4 Characterization of purified rhodopsin with UV–Vis spectroscopy

- Record the absorption spectrum of rhodopsin on a Lambda-800 spectrophotometer (PerkinElmer Life Sciences) in a cuvette with a 10 mm path length. We recommend using a cuvette suitable for samples of a small volume.
- First obtain the dark spectrum of rhodopsin. Make sure the sample has not been exposed to light source that causes the photoisomerization of 11-*cis*-retinal.
- To acquire the light spectrum, photobleach the receptor by irradiating the sample with a 505 nm LED light source (Thorlabs) and adding 25 mM NH_2OH to cleave the Schiff-base bond between opsin and retinal. Record the spectrum.
- Calculate the difference spectrum (dark spectrum–light spectrum).
- Determine the concentration of rhodopsin based on the intensity of the 500 nm peak in the difference spectrum (assuming an extinction coefficient of $40,600 \ M^{-1}cm^{-1}$ at A_{500}). The yield for heterologously expressed wt rhodopsin normally ranges from 5 to 8 g per 10^7 cells. The yield for amber codon suppression is ~0.5–1 μg per 10^7 cells, but it is dependent on the site of mutation.

15.4 PROTOCOL 2: LABELING WITH FLUORESCENT PROBES USING CYCLOOCTYNES

Strain-promoted azide–alkyne cycloaddition (SpAAC) is chosen to label the azido group incorporated into rhodopsin. This chemistry has the following advantages that make it suitable for our specific purpose: (1) it exhibits good bioorthogonality, that is, reacts specifically to azido group rather than to other reactive groups occurring in the natural amino acids; (2) its fast kinetics enables robust labeling of the protein of interest; (3) and unlike copper-catalyzed azide–alkyne cycloaddition, it does not involve any additional catalyst and thus is far less likely to damage the protein by inducing undesirable radical reactions. So far, a series of cyclooctyne reagents have been developed. We chose dibenzocyclooctyne (DIBO), which has been reported for its superior bioorthogonality. In the first step, we tested this reaction for sites in different regions of rhodopsin (Y102 in the second extracellular loop, M155 in helix IV, and S144 in the second intracellular loop). We found that the resulting stoichiometry is dependent on the site of labeling. We characterized the labeled receptor in two ways: (1) UV–Vis spectroscopy to determine the stoichiometry and (2) in-gel fluorescence to confirm the purity of the labeled product. We chose Rho S144azF, a mutant that expresses at relatively high level (1 μg/10^7 cells) and reacts well with the cyclooctyne reagents (dye-to-protein ratio ≈ 1) to study the kinetics of the reaction and examine the reactivity of DIBO reagents conjugated to different Alexa fluorophores (Alexa 488, Alexa 555, Alexa 594, and Alexa 647). In-gel fluorescence was used to measure the degree of fluorescence labeling, while silver staining was used to quantify the amount of protein.

15.4.1 Step 1: fluorescent labeling of azido-tagged rhodopsin with SpAAC

15.4.1.1 Materials
Dibenzocyclooctyne reagents conjugated to Alexa fluorophores (Invitrogen, dissolved in DMSO as 5 mM stock solution and stored at $-20\ °C$)
Thermomixer to control the temperature and shaking speed (Eppendorf)

15.4.1.2 Procedures
- Immobilize the regenerated rhodopsin (expressed from a 30 mL HEK293-F culture) to 100 μL 1D4-mAb sepharose 2B as described in Section 15.3.2.3.
- Transfer resin into a 1.5 mL Eppendorf tube and wash with the reaction buffer for three times (30 min each time).
- Mix the resin with 200 μL reaction buffer to give a 300 μL slurry. Add appropriate amount of labeling reagent into the slurry. Allow the reaction to proceed at 25 °C in the dark with agitation (shaking speed 900 – 1000 rpm) for desired reaction time.
- Stop the reaction by centrifuging the resin and removing the supernatant fraction.
- Wash the resin with the reaction/wash buffer to deplete the unreacted dyes. Monitor the washing steps by recording the UV–Vis spectra of the wash buffer. Based on our experience, five washes should be sufficient.
- Elute the labeled receptor with the C9 peptide (100 μL elution buffer ×2). Supplement the eluate with 150 mM NaCl.

15.4.2 Step 2: characterization of labeled receptor using UV–Vis spectroscopy and in-gel fluorescence

15.4.2.1 Material
Lambda-800 spectrophotometer (PerkinElmer Life Sciences)
Apparatus for SDS-PAGE electrophoresis
Confocal Typhoon 9400 fluorescence scanner (GE Healthcare, Life Sciences)

15.4.2.2 UV–Vis spectroscopy
The acquisition of the dark, light (photobleached), and difference spectra for labeled rhodopsin is performed as described for unlabeled samples in Section 15.3.2.4. Results for samples labeled at different positions and with different fluorophores are shown in Figs. 15.2 and 15.3. Determine the labeling stoichiometry (dye-to-protein ratio) from the maximal absorption peak of Alexa Fluors (A_{fluor}) in the light spectrum and from the rhodopsin absorption at 500 nm ($A_{Rho, 500}$) in the difference spectrum:

$$\text{Labeling stoichiometry} = \frac{c_{fluor}}{c_{Rho}} = \frac{A_{flour}/\varepsilon_{flour}}{A_{Rho,500}/\varepsilon_{Rho,500}}$$

The extinction coefficients used for the calculation are listed in the succeeding table:

Chromophore	Rho	Alexa 488-DIBO	Alexa 555-DIBO	Alexa 594-DIBO	Alexa 647-DIBO
$\varepsilon/M^{-1}\,cm^{-1}$	40,600	71,000	155,000	92,000	239,000

15.4.2.3 In-gel fluorescence
- Load 50–100 ng of the labeled rhodopsin under reducing conditions to SDS-PAGE gel. There is no boiling treatment of the sample prior to loading so as to reduce protein aggregation.
- Wash the gels briefly in ultrapure water.
- Visualize the gel on a confocal Typhoon 9400 fluorescence scanner. For Alexa 488, 488 nm laser is used as the excitation light and the emission light is filtered with a 520 ± 10-nm band-pass. For Alexa 555, 532 nm laser and 580 ± 15 nm band-pass are used.

15.4.3 Step 3: kinetic study of the SpAAC

15.4.3.1 Materials for silver staining:
Acetic acid
Methanol
Ultrapure water (Milli Q Synthesis)
Silver nitrate (powder)

15.4 Protocol 2: Labeling with Fluorescent Probes Using Cyclooctynes

Citric acid
Sodium hydroxide
30% ammonium hydroxide
38% formaldehyde (formalin)
Magnetic stirrer

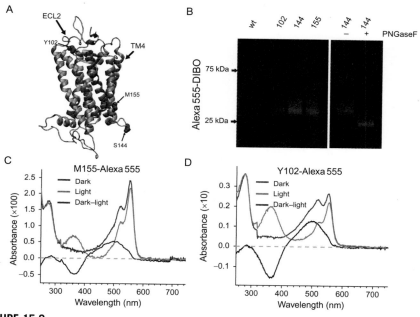

FIGURE 15.2

Fluorescent labeling of rhodopsin with Alexa 555-DIBO. (A) The tested sites are highlighted in red in the crystal structure (PDB 1U19). (B) The corresponding in-gel fluorescence images of the wild-type and azF-containing rhodopsin mutants labeled with Alexa 555-DIBO. For all the three sites tested here, the initial concentration of Alexa 555-DIBO was 50 μM. Following SDS-PAGE electrophoresis, the image of the gel was acquired with a confocal fluorescence scanner using 532 nm laser. The absence of fluorescent band for wt rhodopsin treated under the same condition indicated that this reaction is specific for azF. PNGaseF treatment of Rho S144-Alexa 555 resolved the monomer, dimer, and higher oligomer bands. (C) The dark-state, light-state (after photobleaching), and difference (dark–light) spectra of Rho M155-Alexa 555. The label-to-protein ratio was calculated from the 555 nm absorbance in the green spectrum and the 500 nm absorbance in the difference spectrum (blue). The extinction coefficients used for Alexa 555 and rhodopsin is 155,000 $M^{-1}\,cm^{-1}$ and 40,500 $M^{-1}\,cm^{-1}$, respectively. The dye-to-protein ratio for Rho M155-Alexa 555 was calculated to be 1.00. (D) The dark-state, light-state, and difference spectra of Rho Y102-Alexa 555. The dye-to-protein ratio was calculated to be 0.44. (For interpretation of the references to color in this figure legend, the reader is referred to the online version of this chapter.)

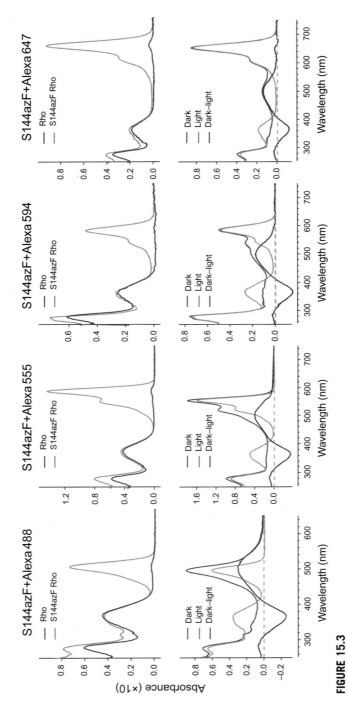

FIGURE 15.3

Fluorescent labeling of azF-containing rhodopsin at site S144 with Alexa Fluor-DIBO. Upper panel: the normalized spectra for Rho S144-Alexa Fluor and Rho wt treated under the same conditions. Lower panel: the dark-state, light-state, and difference UV-visible spectra rhodopsin labeled at S144. The initial concentration of Alexa 488-DIBO is 5 μM, for Alexa 555-DIBO 50 μM, for Alexa 594-DIBO 5 μM, and for Alexa 647-DIBO 20 μM. The extinction coefficients for Alexa 488-DIBO, Alexa 555-DIBO, Alexa 594-DIBO, and Alexa 647-DIBO are assumed to be 71,000 M^{-1} cm^{-1}, 155,000 M^{-1} cm^{-1}, 92,000 M^{-1} cm^{-1}, and 239,000 M^{-1} cm^{-1}. The apparent dye-to-protein ratios for Rho S144-Alexa 488, Rho S144-Alexa 555, Rho S144-Alexa 594, Rho S144-Alexa 647 are 1.13, 1.11, 1.20, and 1.03, respectively. (For color version of this figure, the reader is referred to the online version of this chapter.)

15.4.3.2 Prepare samples of the SpAAC for time-series analysis
- Perform the reaction under the same condition as described in Section 15.4.1. The starting mixture consists of 100 μL resin and 200 μL reaction buffer (a total volume of 300 μL).
- At different time points, take out 30 μL of the resin/buffer mixture and add it into 0.4 mL of precooled reaction/wash buffer in a clean 1.5 mL Eppendorf tube to quench the reaction. Centrifuge the resin to remove the labeling reagents.
- Wash the resin once with 0.4 mL of reaction/wash buffer (30 min).
- Elute the labeled rhodopsin using the elution buffer containing the C9 peptide (15 μL elution buffer ×2).

15.4.3.3 In-gel fluorescence
The in-gel fluorescence experiment is carried out in the same way as described in Section 15.4.2.3. Lacking the knowledge for the exact concentrations of time-series samples, we load 10 μL eluate to the SDS-PAGE gel for each time point. Wash the gel briefly before scanning it on a Typhoon 9400 fluorescence scanner.

15.4.3.4 Silver staining
- After acquiring the in-gel fluorescence image, immediately soak the SDS-PAGE gel with the fixation buffer for at least 1 h with two or three buffer changes or overnight to fix the gel.
- Wash the gel thoroughly washed with ultrapure water for over 1 h with at least three changes to remove all the acetic acid and methanol.
- Meanwhile, prepare the staining solution. Prepare Solution A (dissolve 0.4 g silver nitrate in 2 mL of ultrapure water), and Solution B (add 0.7 mL of 30% ammonium hydroxide to 10.5 mL of 0.36% sodium hydroxide solution). Both solutions should be prepared fresh. Add Solution A dropwise into Solution B under stirring to make the staining solution. The brown precipitate should clear in several seconds. Add 37 mL ultrapure water to the silver-ammonia complex solution to bring the total volume to 50 mL.
- Transfer the gel into a clean container, and stain with the staining solution under gentle agitation.
- After 15 min, rinse the gel with ultrapure water to remove the residue staining solution. Then remove the water and add freshly prepared developing buffer (mix 0.5 mL of 1% citric acid with 50 μL 38% formaldehyde, then add into 100 mL water) to visualize the band. Collect the used staining solution in a waste bottle. Monitor closely the development process to avoid overstaining. When the background begins to turn yellowish, stop the reaction by rinsing the gel with 1% acetic acid.
- Scan the gel on a regular image scanner to acquire the digital data.

15.4.3.5 Data analysis
- Analyze the signal intensity of in-gel fluorescence and silver staining data using ImageJ. As rhodopsin runs as a smear on the SDS-PAGE gel, the entire region of the smear should be included when measuring the overall signal intensity. The normalized signal was calculated as

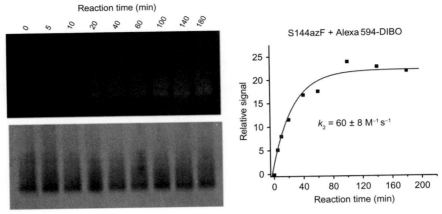

FIGURE 15.4
The kinetic study of the strain-promoted alkyne–azide cycloaddition between Alexa 594-DIBO and azF-tagged rhodopsin. (A) The reaction was performed under identical conditions as described in the experimental section. The initial concentration of Alexa 594-DIBO was 5 μM. For each time point, 10 μL of eluted sample was loaded to the SDS-PAGE gel. The labeled receptor was first visualized by in-gel fluorescence and then silver stained to determine the relative amount of protein. The intensities of the fluorescence signals and the silver staining bands were quantified using ImageJ. The normalized fluorescence intensity ($I = F/S$) was plotted against the reaction time and fitted using pseudo first-order kinetic model. The second-order rate constant was calculated to be $60 \pm 8 \, M^{-1} \, s^{-1}$. (For color version of this figure, the reader is referred to the online version of this chapter.)

$$\text{Relative signal} = \frac{I_{\text{fluorescence}}}{I_{\text{silver}}}$$

The relative signal was plotted against the reaction time to calculate the second-order rate constant (Fig. 15.4).

15.5 PROTOCOL 3: PREPARATION OF BIOTINYLATED ANTIBODIES

15.5.1 Method 1: biotinylation of periodate-oxidized antibodies with biocytin hydrazide

15.5.1.1 Materials
Millipore Steriflip
Amicon Ultra (Ultracel 100 K, regenerated cellulose 100 kDa MWCO, 15 mL device)

Zeba 5 mL, 7 kDa MWCO spin columns
Slide-A-Lyzer 10 kDa MWCO 0.5–3 mL dialysis unit
Sodium periodate (MW = 213.89 g/mol)
Sodium sulfite (MW = 126.04 g/mol)
Sodium cyanoborohydride (MW = 62.84 g/mol)
Biocytin hydrazide (Pierce 28020; MW = 368.20 g/mol)
BupH phosphate 7.2 (Pierce 28372)
Glycerol
1D4 mAb (10 mg/mL in PBS with 10% glycerol 0.02% NaN_3, stored at $-80\ °C$)

15.5.1.2 Solutions and buffers
10 × PBS:
Dissolve one pack BupH phosphate in purified water to give 50 mL total volume (1000 mM Na phosphate pH 7.2, 1500 mM NaCl). Filter with 0.2 μm Steriflip filtration unit.

PBS:
5 mL 10 × PBS.
45 mL purified water.
Filter and degas with 0.2 μm Steriflip filtration unit.

10% NaN_3:
100 mg sodium azide
0.9 mL purified water

350 mM $NaIO_4$:
75 mg sodium metaperiodate.
1 mL purified water.
Prepare solution directly before use and protect from light.

400 mM Na_2SO_3:
50 mg sodium sulfite.
1 mL purified water.
Prepare directly before use.

50 mM BIOCYTIN HYDRAZIDE:
18.41 mg biocytin hydrazide
1 mL DMSO

300 mM CYANOBOROHYRIDE:
18.9 mg sodium cyanoborohydride.
1 mL PBS.
Prepare directly before use.

15.5.1.3 Procedure
- Buffer exchange and concentration of antibody. Dilute 2 mL (20 mg) 1D4 mAb with 12 mL PBS. Concentrate solution with centrifugal ultrafiltration device (100 kDa Amicon Ultra-15, at RT with $2000 \times g$) down to 0.4 mL volume. Repeat dilution and concentration two more times. Test the concentration of the concentrated sample by UV–Vis spectroscopy. Typically, more than 85% protein is recovered.
- Periodate oxidation of antibody. Dilute antibody concentrate to 20 mg/mL with PBS. Add 1 part 200 mM $NaIO_4$ (final concentration 10 mM) to 35 parts of the antibody solution, mix well, and incubate protected from light for 15 min at RT.
- Stop reaction with 50 µL 400 mM Na_2SO_3 (final concentration 20 mM).
- Add 100 µL 50 mM biocytin hydrazide (final concentration 5 mM) and mix on a rotating wheel for 2 h at RT.
- Cool sample on ice and add 100 µL 300 mM cyanoborohydride (final concentration 30 mM). Incubate on ice for 40 min.
- Dialyze sample in 10 kDa Slide-A-Lyzer against prechilled 500 mL PBS supplemented with 55 mL glycerol + 1.11 mL 10% NaN_3 (final concentration 0.02%) at 4 °C. Exchange dialysis buffer twice every 4–8 h.
- Determine the concentration of the dialyzed sample by UV–Vis spectroscopy. Typically, we find better than 50% overall yield. The sample can be stored at 4 °C for up to 3 months without significant loss of activity.

15.5.2 Method 2: biotinylation of antibodies with sulfo-NHS-SS-biotin

15.5.2.1 Materials
Disposable column, 10 mL (Pierce 29924)
Sephadex G50, fine (GE) 50% slurry in 20% ethanol
2 mL 10 mg/mL 1D4 mAb (in PBS with 10% glycerol 0.02% NaN_3, stored at -80 °C)
Sulfo-NHS-SS-biotin (Pierce 21328; MW = 606.69 g/mol)

15.5.2.2 Solutions and buffers
10 mM SULFO-NHS-SS-BIOTIN:
1 mg sulfo-NHS-SS-biotin.
164 µL purified water.
Prepare immediately before use.

78% GLYCEROL:
10 mL purified water.
35 mL glycerol.
Adjust volume to 45 mL with water.
Filter with a 0.2 µm Steriflip unit and store at -20 °C.

10% NaN₃:
  100 mg sodium azide
  0.9 mL purified water

15.5.2.3 Procedure
- Biotinylation of antibody with amino-reactive sulfo-NHS-SS-biotin. Supplement 2 mL 10 mg/mL 1D4 mAb with 200 µL 10× PBS. Add 150 µL 10 mM sulfo-NHS-SS-biotin (12-fold molar excess over protein); mix for 2 h at RT.
- Equilibrate a disposable plastic column with 10 mL packed Sephadex G50 with 30 mL PBS. Load reaction mixture. Elute with PBS and collect 1 mL fractions. Determine the protein concentration in the fractions by UV–Vis spectroscopy.

Combine the peak fractions. Supplement 1 mL sample with 148 µL 78% glycerol (final concentration 10%) and 2.3 µL 10% NaN₃ (final concentration 0.02%) and filter with 0.22 µm syringe filter. Snap freeze 200 µL aliquots in liquid nitrogen and store at −80 °C.

15.6 PROTOCOL 4: PREPARATION OF DETERGENT-SOLUBILIZED LIPIDS

15.6.1 Materials
  CHAPS (3-[(3-cholamidopropyl)-dimethylammonio]-1-propane sulfonate, anagrade, MW=614.9 g/mol; Affymetrix)
  DM (n-dodecyl-β-D-maltopyranoside Anagrade, MW=510.6 g/mol; Affymetrix)
  CHS (cholesteryl hemisuccinate (Tris salt), MW=607.9 g/mol; Affymetrix)
  DOPC (1,2-dioleoyl-*sn*-glycero-3-phosphocholine, MW=786.1 g/mol; Avanti Polar Lipids)
  DOPS (1,2-dioleoyl-*sn*-glycero-3-phospho-L-serine (sodium salt), MW=810.0 g/mol; Avanti Polar Lipids)

15.6.2 Solutions and buffers
78% GLYCEROL:
  10 mL purified water.
  35 mL glycerol.
  Adjust volume to 45 mL with water.
  Filter with 0.2 µm Steriflip and store at −20 °C.

2 M (NH₄)₂SO₄:
  13.2 g ammonium sulfate.
  35 mL purified water.
  Adjust volume to 50 mL with water.
  Filter with 0.2 µm Steriflip and store at −20 °C.

1 M TRIS–HCl BUFFER PH 7.0:
  7.28 g Tris–HCl salt.
  0.47 g Tris base.
  Dissolve with purified water and adjust volume to 50 mL.
  Filter with 0.2 μm Steriflip and store at $-20\,°C$.

20% CHAPS:
  5 g CHAPS.
  20 mL purified water.
  Dissolve by mixing on a motorized wheel (~30 min).
  Filter with 0.2 μm Steriflip and store at $-20\,°C$.

10% DM:
  5 g DM.
  45 mL purified water.
  Dissolve by mixing on a motorized wheel (~30 min).
  Filter with 0.2 μm Steriflip and store at $-20\,°C$.

15.6.3 Procedure

- Disperse 590 mg CHS-Tris salt (=472 mg CHS) in 10 mL 20% CHAPS (final concentrations 4.65% CHS and 19.7% CHAPS) on a magnetic stirrer. The opalescent solution is subjected to multiple freeze–thaw cycles until optically clear. The clear solution is filtered with a 0.2 μm Steriflip unit and store at $-20\,°C$.
- Determine the weight of the clean and dry 50 mL glass round flasks using an analytical balance.
- Transfer the 4 mL 25 mg/mL DOPC solution in chloroform to a 50 mL round flask using a clean glass syringe. Evaporate the solvent under a steam of argon or nitrogen under a chemical fume hood. Rotate the flask to generate an evenly thin lipid film. Remove the last traces of chloroform by putting the container on a vacuum system overnight. Determine the amount of lipid by weighing the flask and subtraction of the empty weight.
- Add 10 mL 10% DM to the flask with 105.5 mg dried DOPC (1.06% final concentration) and disperse the lipid using a water bath sonicator. The resulting opalescent solution is cleared by multiple freeze–thaw cycles. The clear solution is filtered with a 0.2 μm syringe filter using a glass syringe (or methanol washed disposable plastic syringe) and stored at $-20\,°C$.
- Determine the weight of the clean and dry 15 mL glass round flasks using an analytical balance.
- Transfer the content of two glass vials with each 25 mg DOPS solution in chloroform to a 15 mL round flask using a glass Pasteur pipette. Evaporate the solvent under a steam of argon or nitrogen under a chemical fume hood. Rotate the flask to generate an evenly thin lipid film. Remove the last traces of

chloroform by putting the container on a vacuum system overnight. Determine the amount of lipid by weighing the flask and subtraction of the empty weight.
- Add 4.8 mL 10% DM to the flask with 57.6 mg dried DOPS (1.2% final concentration) and disperse the lipid using a water bath sonicator. The resulting opalescent solution is cleared by one or two freeze–thaw cycles. The clear solution is filtered with a 0.2 μm syringe filter using a glass syringe (or methanol washed disposable plastic syringe) and stored at −20 °C.
- Preparation of buffer N. Mix 850 μL DOPC (1.06% in 10% DM), 330 μL DOPS (1.2% in 10% DM), 460 μL 10% DM, 80 μL 20% CHAPS, 750 μL CHS (4.65% in 19.7% CHAPS), 6.43 mL 78% glycerol, 2.5 mL 2 M $(NH_4)_2SO_4$, and 1 mL 1 M Tris–HCl buffer, pH 7.0. Add purified water to 50 mL final volume. Filter with 0.2 μm Steriflip and store aliquots at −20 °C. Use this buffer for solubilization and purification of GPCRs, such as C–C chemokine receptor 5 (CCR5) (Knepp, Grunbeck, Banerjee, Sakmar, & Huber, 2011).
- Prepare buffer N2 by substituting Tris buffer with 50 mM HEPES–NaOH buffer pH 7.5 and supplement with 2 mg/mL BSA.

15.7 PROTOCOL 5: SINGLE-MOLECULE IMMUNOPRECIPITATION ON GLASS-BOTTOM MICROPLATES

15.7.1 Step 1: cleaning of glass-bottom microtiter plate

15.7.1.1 Materials

384-well, PS SensoPlate, Small Volume, LoBase, Glass Bottom (Greiner 788896) 384-well microplate
Dimethyl sulfoxide (DMSO, ACS grade, Fisher)
Ethanol USP 95%
Hydrochloric acid (37% HCl, ACS grade, Fisher)
Purified and filtered water (Milli Q Synthesis)
Plastic bottle with sprayer
Filtration unit (e.g., Millipore Steritop-GP)
Ultrasonic cleaning bath (Bransonic)
Automated pipetting system (epMotion 5070, Eppendorf) with 50 μL multichannel tool (TM-50) or manual multichannel pipette (1–20 μL, Pipetman)
Benchtop centrifuge with microplate swinging bucket rotor
Whatman filter paper
Polypropylene ziplock plastic bag (11 × 16 cm)

15.7.1.2 Solutions and buffers
70% FILTERED ETHANOL:
365 mL 95% ethanol.
Add purified water to 500 mL.

Filter with 0.45 µm vacuum filtration unit.
Transfer to plastic spray bottle.

1 N HCl:
918 mL purified water.
82 mL 37% HCl (add acid slowly to water!).
Filter with 0.2 µm nylon filter.

15.7.1.3 Procedure

- Fill the low volume, 384-well glass-bottom microplate with 20 µL DMSO per well.
- In a chemical fume hood, set up an ultrasonic water bath with a surface large enough to fit a microplate (e.g., 15 × 15 cm).
- Carefully place the plate (without lid) on the water surface of the ultrasonic water bath.
- Turn on the sonicator for 10 min. A fine mist of DMSO will appear above the plate. Do not cover the bath with a lid, as DMSO might be corrosive to it. After sonication, wipe the bath with a wet paper towel to remove any DMSO that might have settled from the mist.
- Remove the microplate from the surface.
- Over a sink, shake off the DMSO from the wells.
- Spray the plate evenly with 70% ethanol to wash the wells. Shake off the liquid into the sink and repeat this ethanol rinse three more times.
- Place a regular 384-well plate upside-down on the bench.
- Add two sheets of Whatman filter paper cut in 11 × 7.5 cm pieces.
- Place the ethanol washed plate upside-down onto the filter papers.
- Load this sandwich in a microplate swinging bucket and balance the rotor.
- Centrifuge 5 min at 500 × g.
- Inspect a representative sample of the wells using a microscope. These glass-bottom microplates have variable amounts of glue residues in the wells that sometimes require several rounds of sonication with DMSO to obtain clean wells. Repeat the earlier steps up to three times.
- Fill the plate with 20 µL 1 N HCl per well.
- Incubate at RT for 30 min.
- Empty wells by shaking off the acid into the sink. (Use appropriate personal protective equipment, especially goggles!)
- Spray the plate evenly with 70% ethanol to wash the wells. Shake off the liquid into the sink and repeat this ethanol rinse three more times.
- Assemble stack with filter papers as the preceding text and centrifuge plate.
- Remove plate from stack and cover it with its lid.
- The cleaned plate may be stored in a sealed propylene bag for up to 30 days or used immediately for the next step.

15.7.2 Step 2: functionalize glass surface with epoxy silane
15.7.2.1 Materials
GOPS, 3-glycidyloxypropyl trimethoxysilane (Aldrich 440167, Dow Z-6040)
Acetic acid, glacial (ACS grade, Fisher)
Ethanol, absolute (200 proof, USP grade)
Purified and filtered water (Milli-Q Synthesis, Millipore)
Steriflip-GP, 0.22 µm polyethersulfone (PES) membrane (Millipore)

15.7.2.2 Solutions and buffers
GOPS SOLUTION:
500 µL acetic acid (1% final concentration, resulting pH ~4.5).
2.5 mL purified water (5% final concentration).
46 mL absolute ethanol (92% final concentration).
1 mL GOPS (aspirate from Sure/Seal bottle using 22-gauge needle and methanol-cleaned and air-dried plastic syringe).
Mix on wheel for 15 min.
Filter and degas with Steriflip and use immediately.

15.7.2.3 Procedure
- Load 20 µL GOPS solution per well of the freshly cleaned 384-well glass-bottom microplate.
- Incubate 60 min at RT.
- Over a sink, shake off the GOPS solution from the wells.
- Spray the plate evenly with 70% ethanol to wash the wells. Shake off the liquid into the sink and repeat this ethanol rinse three more times.
- Place a regular 384-well plate upside-down on the bench.
- Add two sheets of Whatman filter paper cut in 11×7.5 cm pieces.
- Place the ethanol washed plate upside-down onto the filter papers.
- Load this sandwich in a microplate swinging bucket and balance the rotor.
- Centrifuge 5 min at $500 \times g$.
- Remove plate from stack and cover it with its lid.
- Transfer plate in the vacuum chamber of a SpeedVac and keep it overnight under high vacuum of a dual-stage oil pump (less than 0.05 Torr/6.7 Pa pressure).
- The silanized plate should be used within a day but may be stored in a sealed plastic bag for a few weeks.

15.7.3 Step 3: immobilization of biotinylated-bovine serum albumin
15.7.3.1 Materials
Biotin-LC-BSA (Pierce 29130, nominal 25 mg, 8 biotin/BSA)
Purified and filtered water (Milli-Q Synthesis, Millipore)
Sodium tetraborate decahydrate (Sigma B9876, MW = 381.4 g/mol)
Tris base (Trizma base, Sigma 93362 , MW = 121.14 g/mol)
BSA, fatty acid-free (Sigma A8606)

15.7.3.2 Solutions and buffers

0.2 M BORATE:
   0.954 g sodium tetraborate decahydrate (M=381.4, Sigma B9876).
   Dissolve with purified water and adjust to 50 mL.
   The solution has pH 9.2.
   Filter with Steriflip.

1 M TRIS BASE:
   6.057 g Tris base.
   Dissolve with purified water and adjust to 50 mL.
   Filter with Steriflip.

20 mg/mL BIOTINYL-BSA:
   30 mg biotin-LC-BSA.
   1.5 mL purified water.
   Aliquot (250 µL) and store at −80 °C.

5% BSA:
   2.5 g BSA, fatty acid-free.
   47.5 mL purified water.
   Filter with Steriflip.
   Aliquot (1 mL per vial) and store at −80 °C.

COATING SOLUTION:
   250 µL 20 mg/mL biotinyl-BSA (final concentration 1 mg/mL).
   10 µL 5% BSA (final concentration 0.1 mg/mL).
   2.24 mL H_2O.
   2.5 mL 0.2 M borate.
   1.85 µL 10% DM.
   Filter with Steriflip.

BLOCKING SOLUTION:
   300 µL 1 M Tris base.
   2.7 mL purified water.
   3 mL 0.2 M borate.
   Filter with Steriflip.

15.7.3.3 Procedure
- Load 10 µL coating solution per well to columns 2–23 of the silanized microplate. (On epMotion, use multidispense, change tips before aspiration and water liquid-type settings.)
- Centrifuge 5 min at $500 \times g$.
- Seal plate in a ziplock plastic bag with a wet filter paper.
- Incubate first 90 min at RT and then overnight at 4 °C.

- Next day, aspirate wells manually using a very fine plastic tip placing on a single position close to the side of the well bottom. It is important not to scratch the surface in the center of the well.
- Load 10 μL blocking solution per well to columns 2–23.
- Centrifuge 5 min at $500 \times g$.
- Incubate 30 min at RT.
- Aspirate wells.
- Load 20 μL purified water per well. (On epMotion, use multidispense, change tips after command and water liquid-type settings.)
- Aspirate wells.
- Load 20 μL purified water per well.
- Aspirate wells and dry plate in a laminar flow (tissue culture) hood for 2 h.
- Store plate in a plastic bag at 4 °C (may be stored for a few weeks).

15.7.4 Step 4: immobilization of NeutrAvidin

15.7.4.1 Materials
NeutrAvidin (Pierce 31000)

15.7.4.2 Solutions and buffers
10 mg/mL NEUTRAVIDIN:
  10 mg NeutrAvidin.
  990 μL purified water.
  Aliquot (50 μL per vial) and store at -80 °C.

0.37% DM:
  10 μL 10% DM
  260 μL purified water

PRIMING SOLUTION:
  10 μL 0.37% DM
  990 μL PBS

NEUTRAVIDIN SOLUTION (0.06 mg/mL):
  6 μL 10 mg/mL NeutrAvidin
  40 μL 5% BSA.
  10 μL 0.37% DM.
  944 μL PBS.
  Centrifuge 10 min at $12{,}000 \times g$.

BLOCKING SOLUTION:
  40 μL 5% BSA
  10 μL 0.37% DM
  944 μL PBS

15.7.4.3 Procedure
- Select block of wells used for one experiment, for example, 4 columns and 12 rows (wells C3–N14). Mark the lid to keep track of used wells.
- Load 10 μL priming solution to each well.
- Aspirate wells.
- Load 10 μL NeutrAvidin solution to each well. (On epMotion with TS-50 single channel tool, use multidispense, change tip before each aspiration and water liquid-type settings.)
- Inspect wells for incomplete wetting and centrifuge if necessary.
- Incubate 15 min at RT.
- Aspirate wells.
- Load 10 μL blocking solution to each well.
- Incubate 10 min at RT.
- Proceed with next step, or store at 4 °C for up to a week.

15.7.5 Step 5: immobilization of biotinylated antibody
15.7.5.1 Materials
D-Biotin (Invitrogen B-1595, MW = 244.31 g/mol)
1.0 M KOH (volumetric, Fluka)

15.7.5.2 Solutions and buffers
0.2 M BIOTIN:
48.9 mg D-biotin.
951 μL 1.0 M KOH.
Aliquot (50 μL) and store at −80 °C.

1 mM BIOTIN:
5 μL 0.2 M biotin
995 μL PBS

BIOTINYLATED ANTIBODY SOLUTION:
10 μg biotinylated 1D4 (e.g., 1.84 μL 5.42 mg/mL 1D4-ss-biotin)
40 μL 5% BSA
10 μL 0.37% DM
950 μL PBS

BIOTIN SOLUTION:
3 μL 1 mM biotin (3 μM final concentration)
40 μL 5% BSA
10 μL 0.37% DM
947 μL PBS

BLOCKING SOLUTION:
40 μL 5% BSA
10 μL 0.37% DM
950 μL PBS

15.7.5.3 Procedure

- Load 10 μL biotinylated antibody solution to each well. (On epMotion with TS-50 single channel tool, use multidispense, change tip before each aspiration and water liquid-type settings.)
- Inspect wells for incomplete wetting and centrifuge if necessary.
- Incubate 15 min at RT.
- Aspirate wells.
- Load 10 μL biotin solution to each well.
- Incubate 10 min at RT.
- Aspirate wells.
- Load 10 μL blocking solution to each well.
- Incubate 10 min at RT.

15.8 PROTOCOL 6: AUTOMATED MULTICOLOR, SINGLE-MOLECULE TIRF MICROSCOPY

The block diagram in Scheme 15.2 shows the components in our customized SMD–TIRF microscopy workstation based on a Zeiss AxioVert 200M inverted microscope. The system enables automatic time-lapse, multichannel, and multiposition single-molecule experiments in a 384-well, small-volume, glass-bottom microplate. The microplate is in contact with a $100\times/1.45$NA oil immersion TIRF objective (Fig. 15.5). Multiple positions in a single well may be sampled to increase the statistics of the single-molecule observables. Multiple wells facilitate systematic variation of biochemical parameters and routine inclusion of proper controls and standards.

Brightness and photostability are two important properties of fluorescent labels employed in SMD methods. The chemical environment can influence these properties. For example, atmospheric oxygen contributes to irreversible, light-induced reactions of fluorophores, leading to photobleaching products. We employ a reducing and oxidizing system (ROXS) in combination with an enzymatic oxygen scavenger to minimize photobleaching and blinking of our fluorophores (Vogelsang et al., 2008). The oxygen scavenger generates a gradient of the partial oxygen pressure (pO_2) from high values at the buffer–air interface to an oxygen-depleted environment close to the glass-bottom where the laser excitation beam generates an approximately 100 nm deep, 0.2 mm diameter evanescent wave excitation volume (Fig. 15.5). An oxygen-depleted environment dramatically enhances the photostability of cyanine dyes, such as Alexa 555 and 647 (Scheme 15.3).

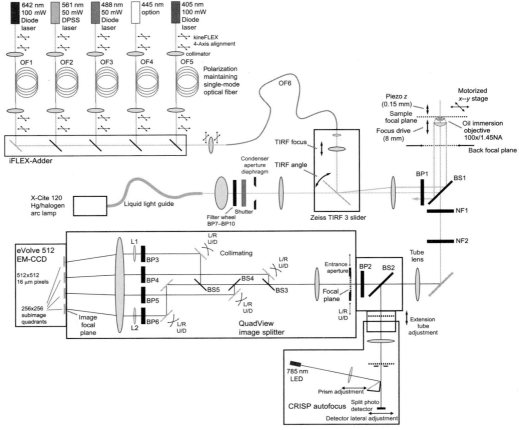

SCHEME 15.2

Schematic of the 4-color SMD–TIRF microscope with alternating-laser excitation (ALEX). The 561 nm frequency-doubled, diode-pumped, solid-state (DPSS) laser and the 405, 488, and 642 nm diode lasers are controlled by TTL signals generated by an Arduino microcontroller. The lasers are combined into a single polarization-maintaining optical fiber using the kineFLEX/iFLEX-Adder system from Qioptiq. The kineFLEX fiber is connected to the Zeiss TIRF3 slider. A quad band-pass filter (BP1) with quad band-pass dichroic mirror (BS1) direct the laser beam to the Zeiss α-Plan-Fluar 100×/1.45NA oil immersion TIRF objective. In the emission light path, a dual-notch filter (NP1) and single-notch filter (NP2) block residual laser light. The CRISP autofocus unit is equipped with a customized 785 nm LED. The dichroic mirror (BS2) directs the infrared beam of the CRISP unit back towards the objective, where the glass–water interface of the sample reflects the beam. The split photo detector of the CRIPS unit monitors the position of the reflected beam to control the Piezo Z drive in a feedback loop that locks the sample in focus. The quad band-pass filter (BP2) eliminates any residual light from the LED and Lasers. The QuadView image splitter separates the emission light path in four different wavelength windows that are defined by the dichroic mirrors (BS3–BS5) and band-pass filters (BP3–BP6). The correction lenses (L1 and L2) reduce chromatic aberration of the long wavelength emission channels. An electron-multiplying charge-coupled device (EM CCD) camera simultaneously acquires the four subimages generated by the QuadView optic. The EM CCD exposure is synchronized with the TTL-controlled lasers by the Arduino microcontroller. A motorized x–y stage enables automated single-molecule experiments on multiwell microtiter plates. (For color version of this figure, the reader is referred to the online version of this chapter.)

15.8 Protocol 6: Automated Single-Molecule Microscopy

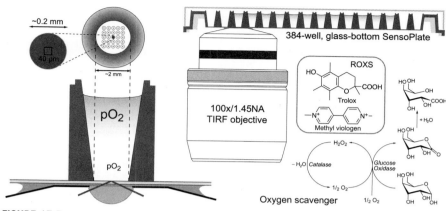

FIGURE 15.5

A motorized *x–y* stage increases throughput and facilitate experiments in high-density microplates. The scheme shows in-scale the well geometry of the 384-well, small-volume, glass-bottom microplate (SensoPlate, Greiner). The total internal reflection of the laser beam (red) excites molecules in a 0.2 mm diameter spot. A 40 μm square field of view is imaged by the QuadView optics on the EM CCD camera. Trolox and methyl viologen form a reducing and oxidizing system (ROXS) that in combination with an enzymatic oxygen scavenger improves the photostability of fluorophores. (For interpretation of the references to color in this figure legend, the reader is referred to the online version of this chapter.)

The combination of laser lines and optical filters in our system can be used to image the popular fluorophores, Alexa 488, 555, 594, and 647 (Fig. 15.6). We have used DIBO derivatives of these dyes to label the GPCR rhodopsin for the three-color, single-molecule fluorescence images shown in Fig. 15.7.

15.8.1 Step 1: microscope optical setup

The following optical components are used in Scheme 15.2:

SEMROCK:
  FF01-390/40-25 (BP7)
  FF02-482/18-25 (BP8)
  FF01-563/9-25 (BP9)
  FF01-640/14-25 (BP10)
  FF01-390/482/563/640-25 (BP1)
  Di01-R405/488/561/635-25×36 (BS1)
  FF484-FDi01-12.5D (BS3)
  FF560-FDi01-12.5D (BS4)
  FF640-FDi01-12.5D (BS5)
  FF01-446/523/600/677-25 (BP2)
  FF01-445/20-25 (BP4)

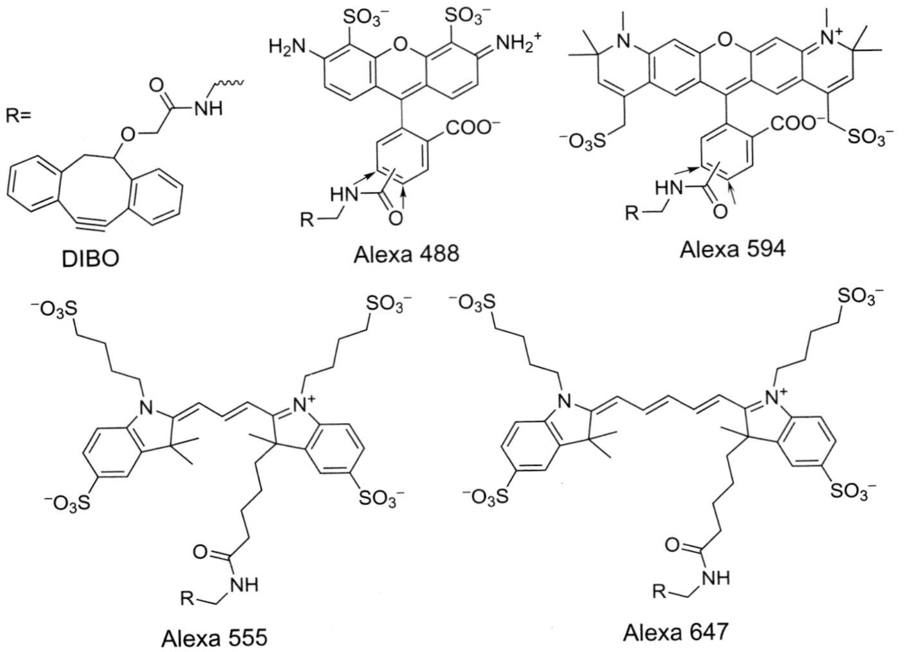

SCHEME 15.3

Fluorescent labels for strain-promoted azide–alkyne cycloaddition (SpAAC). The structure of Alexa 555 is inferred from Alexa 647 (Cordes, Maiser, Steinhauer, Schermelleh, & Tinnefeld, 2011). Alexa 488 and 594 are rhodamines, whereas Alexa 555 and 647 are cyanines.

    FF01-525/30-25 (BP5)
    FF01-605/15-25 (BP3)
    FF01-676/29-25 (BP6)

CHROMA:
    ZET405/488/635NF EM (NF1)
    ZET561NF EM (NF2)

EDMUND OPTICS:
    750 nm dichroic shortpass filter (25×36, #69-219)

APPLIED SCIENTIFIC INSTRUMENTATION:
    CRISP Autofocus (with 780 nm LED)
    PZ-2000-XY-Piezo Z system (closed loop XY stage with rotary encoders and 150 μm Piezo Z top plate)

PHOTOMETRICS:
    Evolve 512 eXcelon EMCCD
    QuadView Imager

15.8 Protocol 6: Automated Single-Molecule Microscopy

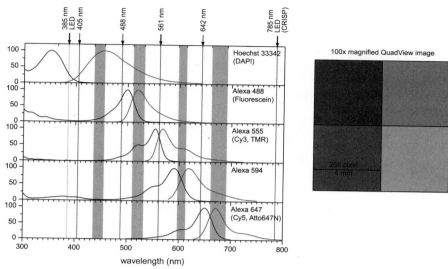

FIGURE 15.6

Selection of optimal laser lines, fluorophores, and band-pass filter combinations for multicolor single-molecule detection fluorescence experiments. Fluorophores for SMD experiments should show high absorption and quantum yield, low frequency of intersystem crossing, and high photostability. An attractive combination of lasers and fluorescent dyes for triple-color SMD–TIRF experiments is shown. In addition, epifluorescence excitation by a 385 nm LED facilitates localization of cell stained with nuclear stains (DAPI or Hoechst 33342). The 405 nm laser line enables efficient excitation of single quantum dot fluorescent nanocrystals matched to one or more emission bands. The spectra of Alexa 555 and Alexa 594 dyes illustrate the trade-off to match the 561 nm excitation or to match the emission band around 605 nm. (For color version of this figure, the reader is referred to the online version of this chapter.)

QIOPTIQ:
   kineFLEX-P-2-S-488-640-0.7-FCP-P2
   kineFLEX-P-1-S-405-0.7-0.7-P2
   kineFLEX-P-1-S-445-0.7-0.7-P2
   kineFLEX-P-1-S-488-0.7-0.7-P2
   kineFLEX-P-1-S-532-0.7-0.7-P2
   kineFLEX-P-1-S-640-0.7-0.7-P2
   4 laserPLATE CUBE (with kineMATIX-P2)
   iFLEX-Adder-RGBB(445)V

ZEISS:
   Zeiss α-Plan-Fluar 100×/1.45NA TIRF objective
   TIRF Slider 405–640 nm, manual (4236839010000)
   TIRF Adjusting Aid Set (4236849010000)

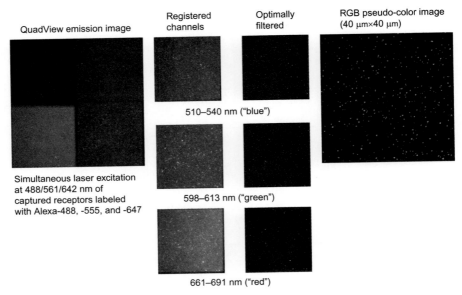

FIGURE 15.7

Raw image showing two classes of single-fluorophore spots. One side of the quadratic image corresponds to 40 μm in the sample plane. A false color image of the same data after subtraction of static background from "hot" pixels of the CCD sensor and after application of a 2D convolution with a spot-enhancing filter that was shown to be the optimal detector of a Gaussian-like spot in $1/\omega^2$ noise (Sage, Neumann, Hediger, Gasser, & Unser, 2005). (For color version of this figure, the reader is referred to the online version of this chapter.)

> Signal Distribution Box (with Trigger-Board)
> AxioVision 4 (with modules Mark and Find 2, Fast Acquisition, ASI Scanning Stage XYZ, Driver for Roper)
> AxioCam MRm, cooled CCD
> AxioVert 200M inverted Microscope

15.8.2 Step 2: alignment of laser-to-fiber, fiber-to-fiber beam combination system

- Turn on the first laser and set the output power to a low setting (1–5 mW).
- Measure the optical power with a laser power meter or a photodiode (DET-100A, Thorlabs) connected to a voltmeter.
- Verify polarization of the laser using a linear polarizer (Thorlabs). Lasers might have vertical, horizontal, or 45° orientation of the polarization axis.
- Check the orientation of the notch for the polarization pin of the polarization-maintaining collimator beam probe on the kineFLEX 4-axis adjuster. If necessary, change the orientation to match that of the laser. LaserPLATE holders for the cube diode laser modules (Coherent) allow vertical and horizontal

mounting of the kineFLEX 4-axis adjuster. 45°-orientated lasers, such as the 50 mW, 532 nm DPSS laser from CrystaLasers, require a custom machined holder.
- Use the kineFLEX coarse alignment tool in two orientations to adjust the 4-axis alignment screws and find the optical axis of the laser beam.
- Replace the alignment tool with the kineFLEX collimator of the fiber matching the laser wavelength.
- Monitor the optical power on the other end of this fiber and perform a fine adjustment of the 4-axis alignment screws to maximize the power throughput. Typically, we obtain better than 50% power throughput.
- Connect the fiber to the correct input port of the iFLEX-Adder unit. Measure the output power at the end of the output fiber (OF6). The power throughput should be better than 25% of the free space laser power.
- Turn off the first laser and perform the same steps for the other lasers.
- The alignment of the laser subsystem is now complete. The system has exceptional long-term stability and typically does not require realignment, unless the configuration has changed.

15.8.3 Step 3: alignment of DualView and QuadView simultaneous imaging systems

- Place the test glass slide with precision etched chromium mask (Photometrics) on the sample stage. Focus on the sample with a 20× air objective.
- Switch the emission light path to the side port connected to the QuadView optic.
- Adjust the exposure time of the eVolve-512 camera and observe the image in live view.
- Set the QuadView to Bypass mode and replace its filter cube by the Bypass cube.
- Adjust the four Entrance Aperture slits so that they are just at the edges of the camera image. If necessary, change the rotation of the camera relative to the QuadView to ensure that the slits are parallel to the image edges.
- Set the QuadView to QuadView mode.
- Adjust the second set of slits so that the visible image is in the center and exactly 50% of the size of the full image.
- Replace the filter cube with the fluorescence cube.
- Four subimages with different intensity appear instead of the single small (50%) image.
- Adjust for each subimage the position (up–down and left–right) so that they are located in their respective quadrants of the camera image and are nonoverlapping with each other. The grid lines of the test specimen facilitate the fine adjustment of the channels relative to each other.
- The alignment of the QuadView unit is now complete.

15.8.4 Step 4: calibration of the multiwell plate geometry

- Place the microplate on top of the stage insert.
- Select a low magnification (10×) air objective and focus on the well bottom.

- Enter the plate calibration mode of the AxioVision Mark&Find module.
- Select the multiwell plate geometry template matching the microplate.
- Calibrate the plate geometry using the provided seven-point routine.

15.8.5 Step 5: adjust TIRF excitation

- Select the TIRF focus adjustment tool in the objective turret.
- Change the filter cube turret to the TIRF adjustment cube.
- Turn on the 642 nm laser and set the power to 5 mW.
- Observe the beam with the AxioCam MRm and adjust the TIRF focus knob on the TIRF3 slider to minimize the spot size.
- Select the TIRF angle adjustment tool in the objective turret.
- Observe the beam with the AxioCam and adjust the TIRF angle micrometer to move the spot to a position approximately 20% close to the image bottom. (The exact position has to be determined from a sample with fluorescent beads and should be used to set this angle. Refer to the Zeiss TIRF3 manual.)

15.8.6 Step 6: sample for subpixel alignment of color channels

- Prepare a sample with immobilized 100 nm TetraSpec beads (Invitrogen) on poly-D-lysine (PDL)-coated glass-bottom microplate dish (SensoPlate) or culture dish (CellView) (Greiner).
- Image surface-immobilized beads with simultaneous laser (TIRF) excitation (488 and 561 nm). Use images to calculate image registration transformations described in Step 11.

15.8.7 Step 7: alignment of the CRISP focus-stabilization (autofocus) system

- The extension tube adjustment and prism adjustment typically have to be performed at the installation of the CRISP unit only.
- Start the ASI Console software, connect to the CRISP controller, and start the three-step adjustment routine.
- Manually optimize the detector lateral adjustment to maximize the dithering difference signal on the ASI controller.
- Finish the adjustment routine with the Gain Setting step.
- Generate the Focus Curve using the control button and check if the vertical axis is in the linear part of the curve and well separated from the extremes.

15.8.8 Step 8: preparation of antibleaching and antiblinking imaging buffer system

15.8.8.1 Materials

Trolox, ($\pm$)-6-hydroxy-2,5,7,8-tetramethylchromane-2-carboxylic acid (Aldrich 238813, 97%, MW = 250.29 g/mol)

Methyl viologen dichloride hydrate (Aldrich 856177, MW(anhydrous) = 257.16 g/mol, unspecified water content)
D-(+)-glucose (Sigma G5767, MW = 180.2 g/mol)
Glucose oxidase from *Aspergillus niger* type VII (Sigma G2133, 195.7 U/mg)
or
CAT: catalase, from *Aspergillus niger* (Sigma C3515, ammonium sulfate suspension 28.73 mg/mL, 9473 U/mg, 272 U/μL)
Ultrafree-MC, centrifugal filtration unit with 0.22 μm microporous PVDF membrane (Millipore)
Millex-GP syringe filter, 0.22 μm, polyethersulfone, 33 mm (Millipore)
5 mL plastic syringe (BD 309646, Fisher)

15.8.8.2 Solutions and buffers

200 mM TX:
50 mg Trolox.
950 μL DMSO.
Aliquot and store at −80 °C for up to 6 months or short-term at −20 °C. (Sensitive to oxidation!)

200 mM MV:
25.7 mg methyl viologen dichloride hydrate.
475 μL purified water.
Filter with Ultrafree-MC spin filter.
Store at −20 °C.

GOD:
27.8 mg 195.7 U/mg glucose oxidase (final concentration 10.9 U/μL).
150 μL purified water.
Mix with pipette to dissolve protein.
Add 322 μL 78% glycerol (final concentration 50%).
Filter with Ultrafree-MC spin filter.
Store at −20 °C.

10% GLUCOSE:
500 mg D-(+)-glucose.
4.5 mL purified water.
Filter with syringe filter.
Aliquot (500 μL) and store at −20 °C.

15.8.8.3 Procedure

- Complex lipid/detergent mixtures may be added to enhance the stability of some GPCRs (Knepp et al., 2011); use PBS with 0.1% DM otherwise.

- Preparation of imaging buffer. Supplement 1000 μL buffer N2 (Protocol 6) with 1.6 μL 10.9 U/μL GOD, 8.5 μL 272 U/μL CAT, 5 μL 200 mM TX, 5 μL 200 mM MV, and 40 μL 10% glucose.
- The imaging buffer has to be prepared fresh for each experiment. Record the time starting with addition of glucose. In the open-well microplate filled with 20 μL sample per well, the imaging buffer is active for at least 1 h.
- Note that the glucose oxidase reaction converts glucose in glucuronic acid, which changes the pH with time. It is important to add buffers with sufficient capacity to minimize the pH change. Recently, the enzyme pyranose oxidase has been described as an alternative to glucose oxidase that generates a neutral reaction product (Swoboda et al., 2012).

15.8.9 Step 9: multiposition, multicolor, time-lapse imaging of single molecules

- Set the laser power for the three lasers 488, 561, 642 nm to 50 mW.
- A typical camera setting is exposure time 500 ms, EM gain 200, ADC range 12e^-/unit.
- It is critical to disable the CRISP autofocus unit during well-to-well moves of the stage. Transistor–transistor logic (TTL, a digital signal standard) control of the CRISP controller is possible with the AxioVision software.
- The multichannel, Z-stack, time-lapse module enables acquisition of image sequences with alternating-laser excitation (ALEX) (Kapanidis et al., 2005). Our implementation of ALEX uses time-encoded TTL pulses to communicate with an Arduino microcontroller. The microcontroller selects the different lasers and synchronizes them with the camera exposure output signal.

15.8.10 Step 10: image processing for optimal detection of single molecules

- Particle analysis and postprocessing of single-molecule trajectories are beyond the scope of this chapter, and we refer to reviews of the algorithms used in this field (Serge, Bertaux, Rigneault, & Marguet, 2008, Rolfe et al., 2011).

Acknowledgments

We gratefully acknowledge the NovoNordisk Foundation Center for Basic Metabolic Research for financial support. The Tri-Institutional Training Program in Chemical Biology supported HT and the Danica Foundation supported TH.

References

Axelrod, D., Burghardt, T. P., & Thompson, N. L. (1984). Total internal-reflection fluorescence. *Annual Review of Biophysics and Bioengineering, 13*, 247–268.

Blanchard, S. C. (2009). Single-molecule observations of ribosome function. *Current Opinion in Structural Biology, 19*, 103–109.

Cordes, T., Maiser, A., Steinhauer, C., Schermelleh, L., & Tinnefeld, P. (2011). Mechanisms and advancement of antifading agents for fluorescence microscopy and single-molecule spectroscopy. *Physical Chemistry Chemical Physics*, *13*, 6699–6709.

Franke, R. R., Sakmar, T. P., Oprian, D. D., & Khorana, H. G. (1988). A single amino-acid substitution in rhodopsin (lysine-248→leucine) prevents activation of transducin. *Journal of Biological Chemistry*, *263*, 2119–2122.

Huber, T., & Sakmar, T. P. (2011). Escaping the flatlands: New approaches for studying the dynamic assembly and activation of GPCR signaling complexes. *Trends in Pharmacological Sciences*, *32*, 410–419.

Joo, C., Balci, H., Ishitsuka, Y., Buranachai, C., & Ha, T. (2008). Advances in single-molecule fluorescence methods for molecular biology. *Annual Review of Biochemistry*, *77*, 51–76.

Kapanidis, A. N., Laurence, T. A., Lee, N. K., Margeat, E., Kong, X. X., & Weiss, S. (2005). Alternating-laser excitation of single molecules. *Accounts of Chemical Research*, *38*, 523–533.

Knepp, A. M., Grunbeck, A., Banerjee, S., Sakmar, T. P., & Huber, T. (2011). Direct measurement of thermal stability of expressed CCR5 and stabilization by small molecule ligands. *Biochemistry*, *50*, 502–511.

Naganathan, S., Ye, S. X., Sakmar, T. P., & Huber, T. (2013). Site-specific epitope tagging of G protein-coupled receptors by bioorthogonal modification of a genetically encoded unnatural amino acid. *Biochemistry*, *52*, 1028–1036.

Oprian, D. D., Molday, R. S., Kaufman, R. J., & Khorana, H. G. (1987). Expression of a synthetic bovine rhodopsin gene in monkey kidney-cells. *Proceedings of the National Academy of Sciences of the United States of America*, *84*, 8874–8878.

Rolfe, D. J., McLachlan, C. I., Hirsch, M., Needham, S. R., Tynan, C. J., Webb, S. E. D., et al. (2011). Automated multidimensional single molecule fluorescence microscopy feature detection and tracking. *European Biophysics Journal*, *40*, 1167–1186.

Roy, R., Hohng, S., & Ha, T. (2008). A practical guide to single-molecule FRET. *Nature Methods*, *5*, 507–516.

Sage, D., Neumann, F. R., Hediger, F., Gasser, S. M., & Unser, M. (2005). Automatic tracking of individual fluorescence particles: Application to the study of chromosome dynamics. *IEEE Transactions on Image Processing*, *14*, 1372–1383.

Sakmar, T. P., Franke, R. R., & Khorana, H. G. (1989). Glutamic acid-113 serves as the retinylidene Schiff-base counterion in bovine rhodopsin. *Proceedings of the National Academy of Sciences of the United States of America*, *86*, 8309–8313.

Serge, A., Bertaux, N., Rigneault, H., & Marguet, D. (2008). Dynamic multiple-target tracing to probe spatiotemporal cartography of cell membranes. *Nature Methods*, *5*, 687–694.

Swoboda, M., Henig, J., Cheng, H. M., Brugger, D., Haltrich, D., Plumere, N., et al. (2012). Enzymatic oxygen scavenging for photostability without pH drop in single-molecule experiments. *ACS Nano*, *6*, 6364–6369.

Vogelsang, J., Kasper, R., Steinhauer, C., Person, B., Heilemann, M., Sauer, M., et al. (2008). A reducing and oxidizing system minimizes photobleaching and blinking of fluorescent dyes. *Angewandte Chemie International Edition*, *47*, 5465–5469.

Weiss, S. (1999). Fluorescence spectroscopy of single biomolecules. *Science*, *283*, 1676–1683.

Ye, S. X., Huber, T., Vogel, R., & Sakmar, T. P. (2009). FTIR analysis of GPCR activation using azido probes. *Nature Chemical Biology*, *5*, 397–399.

Ye, S. X., Kohrer, C., Huber, T., Kazmi, M., Sachdev, P., Yan, E. C. Y., et al. (2008). Site-specific incorporation of keto amino acids into functional G protein-coupled receptors using unnatural amino acid mutagenesis. *Journal of Biological Chemistry*, *283*, 1525–1533.

CHAPTER 16

Analysis of EGF Receptor Oligomerization by Homo-FRET

Cecilia de Heus[*], Nivard Kagie[†], Raimond Heukers[*], Paul M.P. van Bergen en Henegouwen[*], and Hans C. Gerritsen[†]

[*]*Cell Biology, Department of Biology, Science Faculty, Utrecht University, Utrecht, Netherlands*
[†]*Molecular Biophysics, Department of Soft Condensed Matter and Biophysics, Science Faculty, Utrecht University, Utrecht, Netherlands*

CHAPTER OUTLINE

Introduction	306
16.1 Theory Homo-FRET Quantification	309
16.1.1 Steady-State Fluorescence Anisotropy Imaging	309
16.1.2 Time-Resolved Fluorescence Anisotropy Imaging	311
16.2 Materials	313
16.2.1 Plasmid Constructs	313
16.2.2 Cell Lines	315
16.2.3 Microscope	315
16.2.3.1 Confocal Time-Resolved and Steady-State Fluorescence Anisotropy Imaging Microscopy	315
16.3 Methods	316
16.3.1 Slide Preparation	316
16.3.1.1 Reference Measurements	316
16.3.1.2 Predimerization Measurements	316
16.3.1.3 Ligand-induced Oligomerization	317
16.3.2 Analysis	317
16.4 Discussion	317
Acknowledgments	319
References	320

Abstract

Growth factor receptors are present in the plasma membrane of resting cells as monomers or (pre)dimers. Ligand binding results in higher-order oligomerization of ligand–receptor complexes. To study the regulation of receptor clustering, several experimental techniques have been developed in the last decades. However, many involve invasive approaches that are likely to disturb the integrity of the membrane, thereby affecting receptor interactions. In this chapter, we describe the use of a noninvasive approach to study receptor dimerization and oligomerization. This method is based upon the Förster energy transfer between identical adjacent fluorescent proteins (homo-FRET) and is determined by analyzing the change in fluorescence anisotropy. Homo-FRET takes place within a distance of 10 nm, making this an excellent approach for studying receptor–receptor interactions in intact cells. After excitation of monomeric GFP (mGFP) with polarized light, limiting anisotropy values (r_{inf}) of the emitted light are determined, where proteins with known cluster sizes are used as references. Dimerization and oligomerization of the epidermal growth factor receptor (EGFR) in response to ligand binding is determined by using receptors that have been fused with mGFP at their C-terminus. In this chapter, we describe the involved technology and discuss the feasibility of homo-FRET experiments for the determination of cluster sizes of growth factor receptors like EGFR.

INTRODUCTION

The epidermal growth factor receptor (EGFR), also called ErbB1 or Her1, is a member of the ErbB single-pass transmembrane tyrosine kinase receptor family (Ullrich & Schlessinger, 1990; Yarden & Sliwkowski, 2001). Activation of EGFR and its family members is involved in cell growth, cell proliferation, and migration. Many cancer types show overexpression or deregulation of EGFR, and it is therefore a well-studied receptor and an attractive anticancer drug target (Oliveira, van Bergen en Henegouwen, Storm, & Schiffelers, 2006; Sorkin & Goh, 2009). EGFR is composed of an extracellular domain, a transmembrane domain, and an intracellular C-terminal tail containing the tyrosine kinase and several sites involved in posttranslational modifications and signaling (Jorissen et al., 2003). The extracellular part of EGFR contains four domains of which domain I and III are involved in ligand binding and domain II in receptor dimerization. More than 20 different ligands are known for receptors from the ErbB family of which EGF is the most studied one. Ligand binding induces conformational changes of the ectodomain, which results in not only receptor dimerization but also even receptor oligomerization or clustering (Clayton et al., 2005). As a consequence of these changes, cross-phosphorylation of its C-terminal tail occurs (Jura et al., 2009). Interestingly, EGFR can form homo- or heterodimers with other ErbB family members and this already happens in resting cells. These inactive dimers on the plasma membrane are the so-called receptor predimers. Although the structure of receptor dimers is clear from crystal structures, the mechanism of receptor oligomerization remains obscure (Ferguson, 2008). Crystal structures have shown head to head interactions between receptors, suggesting the involvement of these sequences in oligomerization. Recently, also kinase activity was shown to be required for receptor clustering, which suggests the involvement of signaling in this process (Clayton et al., 2005; Hofman et al., 2010). For example, the production of phosphatidic acid by phospholipase D2 (PLD2) has been demonstrated to be required for EGFR clustering (Ariotti et al., 2010). Besides receptor oligomerization and downstream signaling toward cell proliferation, activation of EGF receptors results in a rapid internalization mainly via clathrin-mediated endocytosis. The intracellular transport trajectory finally ends up in lysosomes where the ligand-induced signaling is terminated by degradation of both ligand and receptors (Goh, Huang, Kim, Gygi, & Sorkin, 2010; Sorkin & Goh, 2009).

For studying the clustering of EGF receptors, different experimental techniques have been developed. Dimerization of EGFR, for example, can be determined using chemical cross-linking and detection by SDS-PAGE, by co-IP with differentially tagged EGFR, or by fluorescence complementation (Moriki, Maruyama, & Maruyama, 2001; Yu, Sharma, Takahashi, Iwamoto, & Mekada, 2002; Zhu, Iaria, Orchard, Walker, & Burgess, 2003). Studying receptor clustering is more difficult and demands other experimental strategies. Receptor clustering was initially investigated using electron microscopy (EM), which can be used at high resolution, enabling visualization of gold-labeled EGFR (van Belzen et al., 1988). Analysis of cluster formation can be done by determining particle distances, either by hand or automatically using Ripley's function (Hancock & Prior, 2005). However, limitations exist with respect to determining cluster size, which is dependent on the size of the gold particle and the antibody ($\sim$10 nm) (Hancock & Prior, 2005). Another approach for determining receptor clustering involves fluorescence light microscopy. However, conventional microscopy has a resolution limit of $\sim$200 nm, which makes this technique not very suitable for measuring the formation of small receptor clusters. To overcome this resolution limit problem, more advanced light microscope techniques were developed, which do provide information about small cluster sizes. One of the first examples was described by Gadella and coworkers and included time-resolved fluorescence microscopy based on Förster resonance energy transfer (FRET) analyzed by fluorescence lifetime imaging microscopy (FLIM) (Gadella & Jovin, 1995).

FRET is based on the energy transfer between a donor and an acceptor fluorophore. For FRET to occur, the donor emission spectrum and acceptor excitation spectrum need sufficient spectral overlap and the two fluorophores should be within a distance of $\sim$10 nm. Because energy transfer only occurs when fluorophores are within 10 nm of each other, the detection of FRET can be used to study receptor dimerization or small cluster formation. There are different methods to detect FRET, for example, by measuring changes in the donor/acceptor emission intensity ratio, changes in fluorescence lifetime (FLIM), or differences in anisotropy. The donor/acceptor ratio is determined by measuring the emission of both fluorophores. However, this donor/acceptor ratio is dependent on the concentrations of both fluorophores. With FLIM, the time that the donor fluorophore is in its excited state is measured. This lifetime can be determined by measuring the fluorescence decay of the donor probe after excitation by using a short laser pulse. The fluorescence lifetime changes upon its environment and also when energy transfer occurs. This method is less dependent on local concentrations of both probes and is therefore a more robust way to detect FRET (Chen, Mills, & Periasamy, 2003; Hofman et al., 2008).

FRET-FLIM was previously used successfully to determine EGFR cluster formation by Clayton et al. (2005). EGFR-GFP in combination with anti-phosphotyrosine-Alexa555 was used to label fixed and permeabilized cells. Using FRET-FLIM in combination with image correlation spectroscopy (ICS), Clayton et al. found an average cluster size of 2.2 in the absence of EGF and an average of 3.7 in the presence of EGF, suggesting the formation of higher-order clusters upon ligand binding. A limitation of this method is that it assumes the cell to be stationary and therefore only an average of the different cluster sizes in a cell can be determined instead of determining the subcellular distribution of the different cluster sizes. To solve this, Saffarian et al. studied EGFR clustering by a method based on fluorescence intensity distribution analysis (FIDA) in live cells for quantifying receptor clustering on cell membranes (Saffarian, Li, Elson, & Pike, 2007). Using this technique, the authors find an average cluster size of 1.3 in unstimulated cells, which is in contrast to results from Clayton et al. The difference between the two values might be explained by different measurement conditions, live cells versus fixed cells, and measuring the clustering distribution with FIDA instead of determining an average. A limitation of the FIDA method is that intensity fluctuations might occur during the measurements, which might disturb the clustering measurements. Furthermore, the FIDA method is relatively slow and may therefore not be suitable for measuring of EGFR clustering after EGF stimulation because of the fast internalization of EGFR. More recently, a number and brightness (N&B) analysis technique was used to study EGFR clustering, which demonstrated the formation of up to pentameric EGFR clusters (Sako, Minoghchi, & Yanagida, 2000). More recently, we introduced a method based on homo-FRET to analyze EGFR clustering. With this method, EGFR clustering can be accurately measured on intact cells using only a single label (EGFR-mGFP) (Bader, Hofman, Voortman, en Henegouwen, & Gerritsen, 2009; Hofman et al., 2010).

For homo-FRET, the same fluorophore for donor and acceptor is used, which has the advantage that the detection of FRET is not dependent on concentrations of donor and acceptor. Another advantage of only one fluorophore is that sample preparation is simplified. However, homo-FRET does require a different quantification method because homo-FRET does not result in a change in the emission spectrum or in fluorescence lifetime, resulting in FRET to stay unnoticed. The decreased lifetime of the donor fluorophore in homo-FRET is compensated by the excitation by the acceptor fluorophore. The here applied analysis of energy transfer is based upon the FRET-induced loss of anisotropy. When fluorophores are excited with polarized light, only the subset that is in the right orientation will be excited. In the case of large molecules that are unable to spin on the time scale of the fluorescence lifetime, the emitted light by this subset of fluorophores is also polarized, albeit with a slightly different angular distribution. In the case of two fluorophores in close proximity, there is a high probability that they are in different orientations. Upon excitation of one fluorophore, the neighboring one could be excited through FRET, resulting in the emission of light with this different orientation. Therefore, the energy transfer between two identical fluorophores with slightly different orientation result in depolarization of the emission light; the emitted light is more isotropically distributed. Therefore, quantification of the fluorescence anisotropy can be used to determine homo-FRET.

The homo-FRET method can be calibrated using reference measurements on protein clusters of known size. Subsequently, EGFR cluster sizes can be determined directly from measured fluorescence anisotropy values. As a tool for the reference measurements, we have used FKBP-mGFP, which dimerizes via its ligand AP20187. Similarly, 2xFKBP-mGFP can form oligomers by the addition of AP20187. Clusters made with the FKBP constructs can therefore be either monomeric in the absence of their ligand or dimers or oligomers in the presence of ligand. The level of anisotropy of monomers, dimers, or oligomers can be used as a reference for cluster size measurements. This homo-FRET method can be employed to study the clustering behavior of EGFR for fundamental research and anticancer drug research. For example, it has been shown that EGFR can also form inactive dimers in the absence of EGF; these predimers are formed independently of the C-terminal tail. In contrast, EGF-dependent oligomerization depends on tyrosine kinase activity and more particular on the nine tyrosine residues in the C-terminal tail of the receptor (Hofman et al., 2010).

16.1 THEORY HOMO-FRET QUANTIFICATION
16.1.1 Steady-state fluorescence anisotropy imaging

In homo-FRET experiments, a linearly polarized excitation laser beam is used to excite the fluorophores. Next, the intensities of the parallel and perpendicular polarized fluorescence are detected. Here, parallel and perpendicular are defined with respect to the polarization direction of the excitation laser. From these intensities of the fluorescence, the anisotropy can be calculated using Eq. (16.1). Upon homo-FRET between fluorophores, the anisotropy decreases because the energy transfer between two fluorophores results in emission by a molecule with a different orientation than the molecule directly excited by the laser (Fig. 16.1) (Bader et al., 2009; Lakowicz, 2006; Runnels & Scarlata, 1995):

$$r(t) = \frac{I_{\text{par}} - I_{\text{per}}}{I_{\text{par}} - 2I_{\text{per}}} \qquad (16.1)$$

Besides homo-FRET, other processes can affect the fluorescence anisotropy, such as rotational diffusion of fluorophores. If the excited fluorophore changes its orientation in the time period between excitation and emission, the anisotropy will be lowered. However, in case of a large, slowly rotating molecule like a green fluorescent protein (GFP), the depolarization due to rotations is negligible (Bader, Hofman, van Bergen en Henegouwen, & Gerritsen, 2007). The rotation correlation time of fluorescent proteins is typically in the order of 20 ns, while the fluorescence lifetime is usually <2.5 ns. The fluorophores in the sample can be separated into two groups, fluorophores that are directly excited by the polarized excitation light and fluorophores that are indirectly excited by the energy transfer involved in homo-FRET. Runnels and

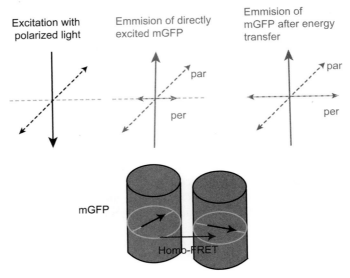

FIGURE 16.1

Schematic representation of homo-FRET in a dimer of monomeric GFP (mGFP), causing depolarization of the polarized excitation light. (For color version of this figure, the reader is referred to the online version of this chapter.)

Scarlata (1995) introduced a simple model to calculate cluster sizes from the steady-state anisotropy (r_{ss}) (Eq. 16.2):

$$r_{ss} = r_{mono} \frac{1+\omega\tau}{1+N\omega\tau} + r_{et} \frac{(N-1)\omega\tau}{1+N\omega\tau} \quad (16.2)$$

Here, r_{mono} is the anisotropy of monomers, r_{et} is the anisotropy after energy transfer, N is the number of molecules in the cluster, and ω is defined as

$$\omega = (R_0/R)^6/\tau$$

where R_0 is the Förster distance, R is the distance between two fluorophore, and τ the fluorescence lifetime. The homo-FRET rate ω can be determined from time-resolved anisotropy experiments. In steady-state anisotropy experiments, the rate is either estimated or assumed to be much faster than the rate of fluorescence. In the latter case, $\omega\tau \to \infty$ (under these conditions $E=1$) and Eq. (16.2) reduces to Eq. (16.3):

$$r_{ss} = r_{mono} \frac{1}{N} + r_{et} \frac{N-1}{N} \quad (16.3)$$

In the original work by Runnels and Scarlatta, they presented an example where anisotropy after a single energy transfer step is about zero. This further simplifies Eq. (16.2) to

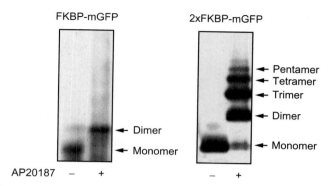

FIGURE 16.2

Native PAGE analysis of dimerization constructs. Cells expressing FKBP-mGFP or 2xFKBP-mGFP, either incubated with AP20187 or mock-treated, were lysed under nonreducing and nondenaturing conditions. After size separation on a native PAGE gel and blotting to PVDF membrane, proteins were detected with anti-GFP antibodies.

Adapted from Bader et al. (2009).

$$r_{ss} = r_{mono} \frac{1}{N} \quad (16.4)$$

However, our experimental work indicated that the assumptions mentioned earlier are not correct for our applications. Often, the energy transfer efficiency is not high enough to ignore and the anisotropy after energy transfer is not equal or close to zero. Therefore, Eq. (16.3) and (16.4) are not generally applicable. Determination of r_{et} is not trivial; therefore, we opted for using reference measurements to relate anisotropy with cluster sizes. This also takes into account effects due to an energy transfer that is not equal or close to 100%. We employ reference constructs to produce clusters of known size. To this end, FKBP-mGFP and 2xFKBP-mGFP constructs were used; they form monomers in the absence of AP20187 but dimers and oligomers, respectively, when AP20187 is added (Fig. 16.2).

Relative steady-state depolarization values due to clustering, r/r_{mono}, of the reference constructs were determined (Fig. 16.3) (Bader et al., 2009). The value of r/r_{mono} for dimers was 0.86 and for oligomers 0.78. Using these reference values, steady-state anisotropy can be converted into an average cluster size on a scale of $N=1$, $N=2$, and $N \geq 3$.

16.1.2 Time-resolved fluorescence anisotropy imaging

In time-resolved fluorescence anisotropy (imaging), the anisotropy decay is measured. This method provides more detailed information about the homo-FRET process than steady-state anisotropy measurements. For large molecules or proteins, the influence of rotation can be neglected and the decay due to homo-FRET can be described by Eq. (16.5) (Gautier et al., 2001; Tanaka & Mataga, 1979):

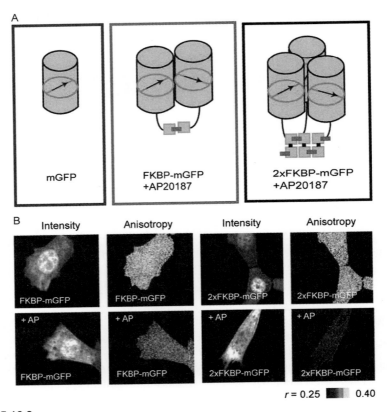

FIGURE 16.3

(A) Model for the clustering of FKBP-mGFP constructs when AP2018 is added. (B) Steady-state fluorescence images and anisotropy images of FKBP-mGFP constructs. (See color plate.)

Results adapted from Bader et al. (2009) and Hofman et al. (2010).

$$r_{\text{homo-FRET}}(t) = (r_{\text{mono}} - r_{\text{inf}})e^{-2\omega t} + r_{\text{inf}} \quad (16.5)$$

The limiting anisotropy r_{inf} is usually reached within nanoseconds (Fig. 16.4). At r_{inf}, homo-FRET has occurred multiple times and the probability of emitting a photon is equal for all fluorophores.

In this case, the limiting anisotropy is a direct measure of the cluster size that is not affected by the energy transfer rate ω, and it can be determined using Eq. (16.6) (Bader et al., 2007):

$$r_{\text{inf}} = \frac{r_{\text{mono}}}{N} + r_{\text{et}}\frac{N-1}{N} \quad (16.6)$$

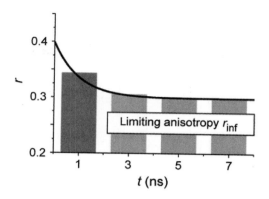

FIGURE 16.4

A schematic presentation of typical anisotropy decay due to homo-FRET. The limiting anisotropy (rinf) is reached within a few nanoseconds and is dependent on the efficiency of energy transfer.

Adapted from Bader et al. (2007).

Importantly, the value of r_{inf} only depends on the cluster size of EGFR, while r_{ss} also depends on the homo-FRET efficiency. Therefore, the clustering can be more accurately determined, without the generally unknown contribution to the transfer efficiency, by time-resolved anisotropy imaging. We determined r_{inf} values from time-resolved measurements on cells expressing EGFR-FKBP-mGFP and EGFR-2xFKBP-mGFP in the presence and absence of AP20187 (Fig. 16.5). This resulted in relative changes in limiting anisotropy for dimers $r/r_{inf} = 0.82$ and for oligomers $r/r_{inf} = 0.72$. These relative changes are almost ~30% larger than in the steady-state measurement. The difference is the result of the contribution from the initial part of the anisotropy decay. This again indicates the advantage of time-resolved measurements over steady-state measurements. Finally, we note that the measurement of the anisotropy decays allows direct determination of the energy transfer rate ω. In the earlier examples, an energy transfer rate ω of 70% can be estimated in the case of dimers. Using the Förster distance R_0 for homo energy transfer between two GFPs of about 4.65 (Sharma et al., 2004), a distance of ~4 nm is found between two GFPs.

16.2 MATERIALS

16.2.1 Plasmid constructs

For homo-FRET measurements, a monomeric GFP (mGFP) was used as fluorophore because normal GFP tends to cluster when expressed in high concentrations with a K_D of 110 µM (Zacharias, Violin, Newton, & Tsien, 2002). mGFP has a point mutation at position 206, resulting in a decrease in self-association and therefore more reliable clustering measurements (Zacharias et al., 2002). This mGFP construct can

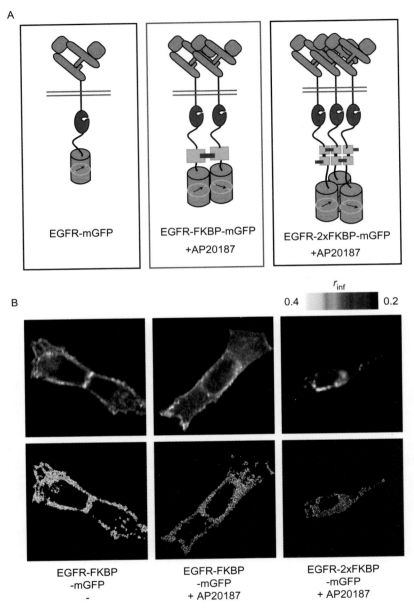

FIGURE 16.5

(A) Model for EGFR-FKBP-mGFP clustering in the presence of AP20187. (B) Fluorescence and time-resolved anisotropy images of EGFR-FKBP-mGFP constructs. (See color plate.)

Results adapted from Hofman et al. (2010).

be used for control measurements and also for fusion to other proteins to measure clustering. For the reference measurements, several constructs can be used; for cytoplasmic proteins, mGFP is fused to FKBP: 1xFKBP-mGFP or 2xFKBP-mGFP. FKBP encoding cDNA can be obtained from the vector pC$_4$-Fv1E (pC$_4$-1xFKBP, Ariad, now the Clontech iDimerize™ Inducible Homodimer System). FKBP can be dimerized by binding of the ligand AP20187. As a result, 1xFKBP-mGFP will dimerize, while a 2xFKBP-mGFP construct will form trimers, tetramers, etc. (Fig. 16.1). For calibration of EGFR clustering, three different constructs were made: EGFR-mGFP, EFGR-FKBP-mGFP, and EGFR-2xFKBP-mGFP. All these constructs were used for transient transfections. For the production of stable cell lines, the constructs were subcloned into a pcDNA3.1 vector, which contains a Zeocin resistance gene. All constructs were amplified in *E. coli*.

16.2.2 Cell lines

All EGFR-expressing cells can be used for measuring EGFR clustering using homo-FRET. We normally use HeLa and A431 cells and NIH3T3 clone 2.2 cells that do not express EGFR, which were transfected with human EGFR (Her14 cells). All cells were cultured at 37 °C and 5% CO_2 in Dulbecco's modified eagle's medium (PAA, Cambridge) containing 7% fetal bovine serum albumin, 2 mM L-glutamine, 100 U/ml penicillin, and 100 μg/ml streptomycin. For the establishment of stable cell lines expressing FKBP-mGFP, 2xFKBP-mGFP, EGFR-mGFP, EGFR-FKBP-mGFP, and EGFR-2xFKBP-mGFP, the cells were transfected with 2 μg plasmid together with 6 μl FuGENE-6 according to the manufacture's protocol and then grown under selective conditions (500 μM Zeocin) for at least 8 weeks.

16.2.3 Microscope

16.2.3.1 Confocal time-resolved and steady-state fluorescence anisotropy imaging microscopy

A confocal microscope (Nikon C1, Japan) was equipped with a 473 nm solid-state diode laser (Becker & Hickl, BDL-473-SMC) with a pulse width of 60 ps and a repetition rate of 80 MHz. The laser light was polarized by a linear polarizer (Meadowlark, Frederick, CO, USA) and focused by a 60× water immersion objective (N.A. 1.2, Nikon). The high NA objectives, like the one used here, result in reduced values for the anisotropy. This does not, however, affect the cluster sizes when reference constructs are used to calibrate cluster sizes. Fluorescence emission was selected by a 515/30 nm band-pass filter. The emitted light was then split in a parallel and a perpendicular channel with respect to the excitation light by a broadband polarizing beam splitter cube (PBS, OptoSigma, Santa Ana, CA, USA). Signals were detected with two high quantum efficiency photomultiplier tubes (Hamamatsu H7422P-40) connected to two time-gated fluorescence lifetime imaging modules (LIMO (De Grauw & Gerritsen, 2001), Nikon Instruments BV, Badhoevedorp, the Netherlands). For each pixel in the image, the LIMOs collect photons in four 2 ns-wide consecutive time gates. Two synchronized LIMOs are used to capture the time-resolved anisotropy decays of each polarization direction. An acquisition time of 3 ms per pixel was chosen and a threshold of 300 counts was applied. At this dwell time, the maximum number of counts per pixel amounted to ~4500. All images covered an area of 50×50 µm at 160×160 pixels.

Determination of the time-resolved anisotropy using the two separate detection channels requires careful synchronization in time and correction for their difference in sensitivity. The former was achieved by using reference dye with fast decay time (aqueous solution of Rose Bengal, $\tau = 70$ ps). The correction factor for the difference in transmission (C_{tr}) between both channels was determined by recording an anisotropy image of an aqueous solution of fluorescein. In this case, fast rotation of the fluorophore will result in emission that is completely depolarized before the second gate opens. Consequently, the anisotropy in the last three gates will be zero, and $I_{par} = C_{tr} \cdot I_{per}$. Finally, a correction was applied to correct for small difference in gate width between the parallel and perpendicular channel. These differences were determined using a sample containing 10 µM GFP monomers in 50/50 glycerol/buffer. In this solution, rotation of the fluorophore is much slower than the fluorescence lifetime. Consequently, the anisotropy will remain constant at the r_0 level. For every gate, a correction factor was determined that ensures that the anisotropy in this gate is identical to the steady-state anisotropy. The absence of concentration-dependent homo-FRET in the GFP solution was checked by lowering the concentration; no increase in anisotropy was observed. Steady-state anisotropy images were obtained by summation of the intensities in all four gates.

16.3 METHODS

16.3.1 Slide preparation

16.3.1.1 Reference measurements

NIH3T3 2.2 cells stably transfected with cDNA encoding the indicated reference protein were seeded on glass coverslips and the day after treated with 1 µM AP20187 for 1–2 h at 37 °C to allow controlled dimerization and oligomerization. After treatment, cells were fixed with 4% paraformaldehyde (PFA) and autofluorescence was quenched by 100 mM glycine/PBS. Finally, the coverslips were embedded in 10% polyvinyl-alcohol (Mowiol), and the anisotropy values were determined and used as reference for the determination of cluster size. Samples can be stored at −4 °C until further use.

16.3.1.2 Predimerization measurements

NIH3T3 2.2 cells were transfected with cDNA encoding either mGFP or EGFR-mGFP. The cells were seeded on 18 mm glass coverslips and allowed to adhere overnight. Cells were fixed with 4% PFA followed by the quenching of autofluorescence with 100 mM glycine/PBS for 10 min. The coverslips were embedded in Mowiol and stored at 4 or −20 °C until further use. However, determination of anisotropy immediately after embedment is preferred. These treatments were previously shown not to influence the clustering behavior (Bancroft & Stevens, 1982). To exclude any effect of the fused mGFP, a nice control is to express EGFR-mGFP in cells that contain endogenous EGFR like A431 and Her14. In these cells, hetero-predimers between EGFR-mGFP and endogenous EGFR will cause the anisotropy to be similar to the mGFP control.

16.3.1.3 Ligand-induced oligomerization

Cells were seeded on an 18 mm glass coverslip. The day after, cells were serum-starved overnight in medium containing 0.5% serum. After serum starvation, cells were treated with 8 nM EGF for 10 min at 37 °C. Then, cells were fixed with 4% PFA for 20 min at room temperature and autofluorescence was quenched with 100 mM glycine/PBS for 10 min. Finally, the coverslips were embedded in Mowiol and used for measurement immediately or stored at 4 or −20 °C until further use.

16.3.2 Analysis

In the steady-state anisotropy experiments, cluster sizes can be determined by direct comparison of the measured anisotropy values with the reference anisotropies found from the reference measurements on the FKBP constructs. Here, the steady-state anisotropies are calculated by integrating the intensities of all four time gates for both polarization directions. In addition, the difference in transmission between the two polarization channels can be corrected for by multiplying I_{per} by C_{tr}. In the time-resolved anisotropy experiments, r_{inf} values are extracted from the time-gated fluorescence anisotropy images. Cluster sizes can again be obtained by comparison with the reference measurements on FKBP constructs. The anisotropy values in the last gates are, in good approximation, equal to r_{inf}, provided that E is high enough. This condition is met for the three last gates provided that $E > 0.5$. Four-gate anisotropy decays are created by binning the intensities I_{par} and I_{per} per gate in regions of interest. In the anisotropy imaging experiments, a threshold of $I_{par,inf} + 2I_{per,inf} > 300$ counts was applied to all images. In theory, this number of counts corresponds to a standard deviation in the anisotropy of 0.05 (Bader et al., 2009; Hofman et al., 2010). An example of cluster size images derived from homo-FRET images is shown in Fig. 16.6A, where a resting cell before EGF stimulation is shown. Here, cluster sizes are small, but a clear population of predimers is visible; ~40% of the EGFR is present as dimer. After stimulating the cell with EGF for 10 min average, cluster sizes have increased up to >2; and a significant fraction of the clusters are of size >3 (Fig. 16.6B).

16.4 DISCUSSION

Studying EGFR dimerization and oligomerization is important to completely understand the signaling and trafficking mechanism of EGFR. EGFR dimerization has been studied by several biochemical studies like chemical cross-linking, co-IP with

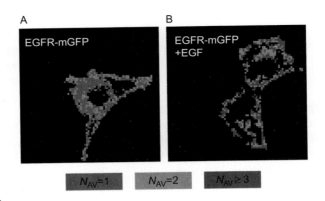

FIGURE 16.6

(A) Homo-FRET-based cluster size image before stimulation with EGF and (B) after stimulation with 8 nM EGF. (See color plate.)

Results adapted from Hofman et al. (2010).

differentially labeled EGFR, and also crystallography to determine the monomer and dimer structure. Although there is already a lot known about the dimerization of EGFR, the mechanism of oligomerization remains obscure. Advanced microscopy techniques were developed to study EGFR oligomerization; initially EM with gold-labeled EGFR was used. The cluster size that can be determined by EM depends on the labeling of EGFR, in other words the size of the gold particles and antibody used (10 nm). Advanced fluorescence microscopy techniques were developed like FRET-FLIM, FIDA, and N&B analysis. These techniques are able to measure small cluster sizes and are complementary techniques. FRET-FLIM is rapid and accurate, while FIDA is able to measure cluster size in life cells. FIDA measures the clustering on the plasma membrane of cells and is a slow method; therefore, it is not suitable for measuring EGF-induced EGFR clustering in time. EGFR will internalize before the FIDA method is able to determine the cluster sizes on the plasma membrane. N&B uses the same principle as fluorescence correlation spectroscopy (FCS) measurements and measures the distribution and association states of molecules in live cells (Nagy, Claus, Jovin, & Arndt-Jovin, 2010). This technique demonstrated the concentration-dependent predimer formation. In contrast, FIDA is able to give information about the distribution of cluster sizes within a single pixel. FIDA and FCS, however, need high illumination intensities for accurate measurements, which can cause photodamage to the proteins in the cell. To study in more detail the EGF-induced EGFR clustering, FRET was used and more specifically the homo-FRET method. With this method, small EGFR clusters can be accurately determined and the cluster size distribution in one cell can be determined within single pixels. The advantage of homo-FRET instead of hetero-FRET is that the detection of FRET is not dependent on the concentrations of donor and acceptor, which simplifies the sample preparation. Homo-FRET is detected by a change in anisotropy, which is a different method than used for measuring hetero-FRET. Anisotropy is a measure for the depolarization of the linear polarized excitation beam. To be able to convert the obtained anisotropy values into cluster sizes, we use reference proteins like EGFR-FKBP-mGFP constructs, which form clusters of known sizes when AP20187 is added. With this method, cluster sizes of EGFR in the order of $N=1$, $N=2$, or $N \geq 3$ can be determined. A conventional fluorescence microscope can be converted in a homo-FRET-detecting system when a set of polarizers is integrated and sufficient signal is generated. The limitation of this method is that it is up till now only able to distinguish monomers, dimers, and oligomers. This method is not able to measure the precise amount of molecules in higher-order clusters. Homo-FRET was used to study the predimer formation of EGFR and showed to be receptor concentration-independent by homo-FRET measurements. This is in contrast to what was found by the N&B analyses (Hofman et al., 2010; Nagy et al., 2010). This difference might be explained by the different sample preparations as homo-FRET was measured in fixed cells and N&B analysis in live cells. We note that the N&B method does not include the immobile fraction of the receptors, which is known to exist. Another explanation for the difference could be that the N&B analysis cannot distinguish the cluster size distribution within a single pixel and homo-FRET can. It might be that the amount of predimers changes upon receptor concentration but that there are still predimers formed when there are low EGFR concentrations present. This is just one example of contradicting results of EGFR clustering measurements. Combining these different methods is required to overcome these contradictions. Most probably, the most ideal way to measure clustering is by combining FRET studies with high-resolution microscopy methods like FIDA, N&B, EM, or others because in that situation, you measure both the larger cluster organizations and the smaller subclusters using homo-FRET. The super resolution microscopy technique STED would be a good possibility to combine with FRET because the resolution of STED can be up to 20 nm, which is slightly higher than homo-FRET, which makes it possible when the two methods are combined to measure small clusters simultaneously with larger clusters. STED can also be applied in life cells and gives clustering information about a whole cell, which is not the case with, for example, single-molecule tracking where only a set of clusters can be analyzed (Pellett et al., 2011).

In this chapter, we have described the homo-FRET method for measuring EGFR clustering and compared them with other experimental techniques. This method can equally be applied for other proteins, provided their functioning is not affected by the fusion to mGFP. Essential are the reference proteins, which are mGFP fused to FKBP for cytoplasmic proteins and FKBP fused to a receptor can be used as reference for the analysis of receptor clustering. Homo-FRET has the advantage of simple sample preparation and accurate cluster size determination. However, homo-FRET is not able to detect larger cluster organization.

Acknowledgments

The authors wish to thank Dr. Erik G. Hofman and Dr. Arjan N. Bader for their fruitful discussions.

References

Ariotti, N., Liang, H., Xu, Y., Zhang, Y., Yonekubo, Y., Inder, K., et al. (2010). Epidermal growth factor receptor activation remodels the plasma membrane lipid environment to induce nanocluster formation. *Molecular and Cellular Biology*, *30*(15), 3795–3804.

Bader, A. N., Hofman, E. G., van Bergen en Henegouwen, P. M., & Gerritsen, H. C. (2007). Imaging of protein cluster sizes by means of confocal time-gated fluorescence anisotropy microscopy. *Optics Express*, *15*(11), 6934–6945.

Bader, A. N., Hofman, E. G., Voortman, J., en Henegouwen, P. M., & Gerritsen, H. C. (2009). Homo-FRET imaging enables quantification of protein cluster sizes with subcellular resolution. *Biophysical Journal*, *97*(9), 2613–2622.

Bancroft, J. D., & Stevens, A. (1982). *Theory and practice of histological techniques*. London, UK: Churchill Livingstone.

Chen, Y., Mills, J. D., & Periasamy, A. (2003). Protein localization in living cells and tissues using FRET and FLIM. *Differentiation*, *71*(9–10), 528–541.

Clayton, A. H., Walker, F., Orchard, S. G., Henderson, C., Fuchs, D., Rothacker, J., et al. (2005). Ligand-induced dimer-tetramer transition during the activation of the cell surface epidermal growth factor receptor—A multidimensional microscopy analysis. *Journal of Biological Chemistry*, *280*(34), 30392–30399.

De Grauw, C. J., & Gerritsen, H. C. (2001). Multiple time-gate module for fluorescence lifetime imaging. *Applied Spectroscopy*, *55*(6), 670.

Ferguson, K. M. (2008). Structure-based view of epidermal growth factor receptor regulation. *Annual Review of Biophysics*, *37*, 353–373.

Gadella, T. W., Jr., & Jovin, T. M. (1995). Oligomerization of epidermal growth factor receptors on A431 cells studied by time-resolved fluorescence imaging microscopy. A stereochemical model for tyrosine kinase receptor activation. *The Journal of Cell Biology*, *129*(6), 1543–1558.

Gautier, I., Tramier, M., Durieux, C., Coppey, J., Pansu, R. B., Nicolas, J. C., et al. (2001). Homo-FRET microscopy in living cells to measure monomer-dimer transition of GFP-tagged proteins. *Biophysical Journal*, *80*(6), 3000–3008.

Goh, L. K., Huang, F., Kim, W., Gygi, S., & Sorkin, A. (2010). Multiple mechanisms collectively regulate clathrin-mediated endocytosis of the epidermal growth factor receptor. *The Journal of Cell Biology*, *189*(5), 871–883.

Hancock, J. F., & Prior, I. A. (2005). Electron microscopic imaging of ras signaling domains. *Methods*, *37*(2), 165–172.

Hofman, E. G., Bader, A. N., Voortman, J., van den Heuvel, D. J., Sigismund, S., Verkleij, A. J., et al. (2010). Ligand-induced EGF receptor oligomerization is kinase-dependent and enhances internalization. *Journal of Biological Chemistry*, *285*(50), 39481–39489.

Hofman, E. G., Ruonala, M. O., Bader, A. N., van den Heuvel, D., Voortman, J., Roovers, R. C., et al. (2008). EGF induces coalescence of different lipid rafts. *Journal of Cell Science*, *121*(Pt 15), 2519–2528.

Jorissen, R. N., Walker, F., Pouliot, N., Garrett, T. P., Ward, C. W., & Burgess, A. W. (2003). Epidermal growth factor receptor: Mechanisms of activation and signalling. *Experimental Cell Research*, *284*(1), 31–53.

Jura, N., Endres, N. F., Engel, K., Deindl, S., Das, R., Lamers, M. H., et al. (2009). Mechanism for activation of the EGF receptor catalytic domain by the juxtamembrane segment. *Cell*, *137*(7), 1293–1307.

Lakowicz, J. R. (2006). In J. R. Lakowicz (Ed.), *Principles of fluorescence spectroscopy*. (3th ed.). New York: Springer.

Moriki, T., Maruyama, H., & Maruyama, I. N. (2001). Activation of preformed EGF receptor dimers by ligand-induced rotation of the transmembrane domain. *Journal of Molecular Biology, 311*(5), 1011–1026.

Nagy, P., Claus, J., Jovin, T. M., & Arndt-Jovin, D. J. (2010). Distribution of resting and ligand-bound ErbB1 and ErbB2 receptor tyrosine kinases in living cells using number and brightness analysis. *Proceedings of the National Academy of Sciences of the United States of America, 107*(38), 16524–16529.

Oliveira, S., van Bergen en Henegouwen, P. M., Storm, G., & Schiffelers, R. M. (2006). Molecular biology of epidermal growth factor receptor inhibition for cancer therapy. *Expert Opinion on Biological Therapy, 6*(6), 605–617.

Pellett, P. A., Sun, X., Gould, T. J., Rothman, J. E., Xu, M. Q., Correa, I. R., Jr., et al. (2011). Two-color STED microscopy in living cells. *Biomedical Optics Express, 2*(8), 2364–2371.

Runnels, L. W., & Scarlata, S. F. (1995). Theory and application of fluorescence homotransfer to melittin oligomerization. *Biophysical Journal, 69*(4), 1569–1583.

Saffarian, S., Li, Y., Elson, E. L., & Pike, L. J. (2007). Oligomerization of the EGF receptor investigated by live cell fluorescence intensity distribution analysis. *Biophysical Journal, 93*(3), 1021–1031.

Sako, Y., Minoghchi, S., & Yanagida, T. (2000). Single-molecule imaging of EGFR signalling on the surface of living cells. *Nature Cell Biology, 2*(3), 168–172.

Sharma, P., Varma, R., Sarasij, R. C., Ira, Gousset, K., Krishnamoorthy, G., et al. (2004). Nanoscale organization of multiple GPI-anchored proteins in living cell membranes. *Cell, 116*(4), 577–589.

Sorkin, A., & Goh, L. K. (2009). Endocytosis and intracellular trafficking of ErbBs. *Experimental Cell Research, 315*(4), 683–696.

Tanaka, F., & Mataga, N. (1979). Theory of time-dependent photo-selection in interacting fixed systems. *Photochemistry and Photobiology, 29*(6), 1091–1097.

Ullrich, A., & Schlessinger, J. (1990). Signal transduction by receptors with tyrosine kinase activity. *Cell, 61*(2), 203–212.

van Belzen, N., Rijken, P. J., Hage, W. J., de Laat, S. W., Verkleij, A. J., & Boonstra, J. (1988). Direct visualization and quantitative analysis of epidermal growth factor-induced receptor clustering. *Journal of Cellular Physiology, 134*(3), 413–420.

Yarden, Y., & Sliwkowski, M. X. (2001). Untangling the ErbB signalling network. *Nature Reviews. Molecular Cell Biology, 2*(2), 127–137.

Yu, X., Sharma, K. D., Takahashi, T., Iwamoto, R., & Mekada, E. (2002). Ligand-independent dimer formation of epidermal growth factor receptor (EGFR) is a step separable from ligand-induced EGFR signaling. *Molecular Biology of the Cell, 13*(7), 2547–2557.

Zacharias, D. A., Violin, J. D., Newton, A. C., & Tsien, R. Y. (2002). Partitioning of lipid-modified monomeric GFPs into membrane microdomains of live cells. *Science, 296*(5569), 913–916.

Zhu, H. J., Iaria, J., Orchard, S., Walker, F., & Burgess, A. W. (2003). Epidermal growth factor receptor: Association of extracellular domain negatively regulates intracellular kinase activation in the absence of ligand. *Growth Factors, 21*(1), 15–30.

CHAPTER 17

Detection of G Protein-Coupled Receptor (GPCR) Dimerization by Coimmunoprecipitation

Kamila Skieterska[1], Jolien Duchou[1], Béatrice Lintermans, and Kathleen Van Craenenbroeck

Laboratory of Eukaryotic Gene Expression and Signal Transduction (LEGEST), Ghent University-UGent, Ghent, Belgium

CHAPTER OUTLINE

Introduction	324
Specific Interactions with ER Chaperones	324
Dimerization	325
Interaction with "Gatekeepers"	325
17.1 Materials	**326**
17.2 Methods	**327**
17.2.1 Overexpression of GPCRs in Mammalian Cell Lines	327
17.2.2 Cell Lysis	327
17.2.2.1 *Mechanical Disruption*	327
17.2.2.2 *Nondenaturing Detergent-Based Lysis Methods*	329
17.2.2.3 *Denaturing-Based Lysis Methods: Urea Lysis and SDS-Lysis*	332
17.2.3 Coimmunoprecipitation	332
17.2.3.1 *Standard Coimmunoprecipitation: in Total Cell Lysate*	333
17.2.3.2 *Membrane Coimmunoprecipitation: on Living Cells*	335
17.2.4 Elution	335
17.2.5 Detection of Interacting Partners	337
17.2.6 Controls	337
17.3 Discussion	**338**
Acknowledgments	339
References	339

---

[1]Both authors contributed equally to this work.

Abstract

With 356 members in the human genome, G protein-coupled receptors (GPCRs) constitute the largest family of proteins involved in signal transduction across biological membranes. GPCRs are integral membrane proteins featuring a conserved structural topology with seven transmembrane domains. By recognizing a large diversity of hormones and neurotransmitters, GPCRs mediate signal transduction pathways through their interactions with both extracellular small-molecule ligands and intracellular G proteins to initiate appropriate cellular signaling cascades. As there is a clear link between GPCRs and several disorders, GPCRs currently constitute the largest family of proteins targeted by marketed pharmaceuticals. Therefore, a detailed understanding of the biogenesis of these receptors and of GPCR–protein complex assembly can help to answer some important questions. In this chapter, we will discuss several methods to isolate GPCRs and to study, via coimmunoprecipitation, protein–protein interactions. Special attention will be given to GPCR dimerization, which often starts already in the endoplasmic reticulum and influences the maturation of the receptor. Next, we will also explain an elegant tool to study GPCR biogenesis based on the glycosylation pattern of the receptor of interest.

INTRODUCTION

G protein-coupled receptors (GPCRs) are at the interface between the extra- and intracellular environment and are thus important for cellular communication. Protein–protein interactions are integral to the organizational structure and function of cell signaling networks, and there is increasing evidence that GPCR–protein complexes are often stably assembled before expression at the plasma membrane, sometimes as soon as the GPCR is folded in the endoplasmic reticulum (ER). Furthermore, correct folding is enhanced not only by interaction with specific proteins, such as ER-resident chaperones, but also by receptor dimerization, both homo- and heterodimerization.

Specific interactions with ER chaperones

While GPCR biosynthesis and transport towards the cell surface remains poorly characterized, their exit from the ER has been defined as a crucial step in controlling their expression (Petaja-Repo, Hogue, Laperriere, Walker, & Bouvier, 2000; Van Craenenbroeck et al., 2005). There is a high concentration of molecular chaperones and enzymes involved in protein folding in the ER (Achour, Labbe-Jullie, Scott, & Marullo, 2008; Ellgaard & Helenius, 2003; Hebert, Garman, & Molinari, 2005). The conventional chaperone system is thought to also aid proper GPCR folding; it recognizes and interacts with exposed hydrophobic amino acid residues in partially folded proteins or unassembled protein complexes. Important chaperon proteins that interact with proteins and help to obtain the correct structural fold are, for example,

BiP, calnexin, calreticulin, and ERp57. Besides these classical chaperones, other proteins have been characterized typically interacting with specific GPCRs and playing a role in GPCR biogenesis; for example, receptor activity-modifying protein 1 (RAMP1) facilitates maturation of the calcitonin receptor-like receptor by interaction with the immature receptor in the ER (Luttrell, 2008).

Dimerization

A general role for GPCR dimerization in quality control and ER export has not yet been established although a plethora of studies have showed that GPCR dimerization starts in the ER. Dimerization of the metabotropic γ-aminobutyric acid (GABA$_B$) receptor and β2-adrenergic receptors was shown to be a prerequisite for its expression at the plasma membrane (Margeta-Mitrovic, Jan, & Jan, 2000; Salahpour et al., 2004). Mutants of GPCRs have been constructed by inserting an ER retention motif, causing them to be trapped in the ER. Most of these mutants will prevent ER exit of the interaction partner, suggesting that these GPCRs are interacting early in receptor biogenesis, as reviewed by Dupre, Hammad, Holland, and Wertman (2012).

Interaction with "gatekeepers"

A new concept was raised that next to the quality-control checkpoints in the ER, there is also another mechanism involved in the progression of GPCRs along the biosynthetic pathway executed by so-called gatekeepers. The role of gatekeepers seems to be the control of GPCR trafficking from the ER to the plasma membrane upon receiving a specific external stimulus (Shirvani, Gata, & Marullo, 2012). This new concept might represent an adaptive mechanism to specific physiological constraints, such as the sustained activation of the receptor.

For a detailed description of these interaction studies we refer to the cited publications, the main idea is that all these experiments make use of coimmunoprecipitation approaches as this method is ideally suited to identify interactions between proteins and is often the experiment of choice to start testing initial hypotheses. Tissues or cells endogenously expressing proteins of interest or cells artificially overexpressing the (epitope-tagged) proteins are used. We want to stress that the functional outcome of receptor activation often depends on the cellular context, and, therefore, it is important to study the GPCR complexes in endogenous tissue. Due to a shortage of high-affinity selective anti-GPCR antibodies, most studies predominantly rely on coimmunoprecipitation experiments of epitope-tagged GPCRs, transiently expressed in heterologous cell lines. The cell line used is often HEK293T and as the transfection of this cell line is an important preliminary step, we will spend some time describing the best methods to transiently transfect HEK293 cell lines resulting in high transfection efficiency.

Coimmunoprecipitation is widely used to detect GPCR–GPCR interactions biochemically. For this coimmunoprecipitation technique, an antibody, targeting a known protein, is used to isolate (immunoprecipitate) a protein complex, containing the protein of interest from a lysate extracted from cell lines, primary cells, or native tissue. The most critical step in this technique is cell lysis and solubilization of the GPCRs; detergents are needed for extraction of GPCRs from the cell membrane, but this process can lead to aggregation of the receptors, due to the hydrophobic nature of the α-helices of the GPCR or to disruption of protein–protein contacts.

After polyacrylamide gel electrophoresis and Western blotting, the members of the complex can be identified by specific antibodies.

In this chapter, we discuss general methods for carrying out and analyzing immunoprecipitation experiments and discuss the controls that should be included in all valid immunoprecipitation experiments.

A number of GPCRs need posttranslational modifications to be properly expressed and exert their normal function at the plasma membrane. N-glycosylation is a common posttranslational modification of GPCRs. During biogenesis, glycosylated GPCRs undergo a cycle of sugar chain additions and deletions by oligosaccharyltransferases and glucosidases. This is with the binding of calnexin and calreticulin until the receptor obtains the correct folding and can leave the ER for further transport to the Golgi where the glycosylation tree is further processed. The immature ER-retained receptor can often be visualized on Western blot as a band with a lower molecular weight. Therefore, detection of immature glycosylated GPCRs upon coimmunoprecipitation can indicate that the protein–protein interaction starts already in the ER (Van Craenenbroeck, 2012; Van Craenenbroeck et al., 2011).

17.1 MATERIALS

a. *Cell culture (sterile)*: Dulbecco's modified Eagle's medium (DMEM), penicillin–streptomycin, fetal calf serum (FCS), L-glutamine, EDTA–trypsin (3 mM EDTA; 0.25% trypsin in PBS pH 7.4), phosphate-buffered saline (PBS), cell culture plates, laminar flow cabinet, CO_2 incubator, and sterile pipettes
b. *Transfection (sterile)*: transfection reagents (for details, see further) and plasmid constructs encoding proteins of interest
c. *Cell lysis*: lysis buffer (for details, see further), protease, and phosphatase inhibitors (aprotinin, β-glycerophosphate, pefabloc, leupeptin), fridge or cold room, microcentrifuge, and sonicator
d. *Coimmunoprecipitation*: antibody (2 μg/sample), Protein A or G beads, SDS sample buffer, and heat block
e. *SDS-PAGE and Western blot*: acrylamide/bisacrylamide, sodium dodecyl sulfate (SDS), ammonium persulfate (APS), tetramethylethylenediamine (TEMED), 1.5 M Tris/HCl pH 8.8, 0.5 M Tris/HCl pH 6.8, electrophoresis buffer (49 mM Tris; 0.38 M glycine; 0.1% SDS), transfer buffer (47 mM Tris; 37 mM glycine; 0.03% SDS; 20% MeOH), nitrocellulose membrane, Whatman paper, polyacrylamide gel electrophoresis, and Western blot equipment (Bio-Rad)
f. *Immunodetection*: nonfat dry milk or bovine serum albumin (BSA) or commercial blocking buffer (e.g., Odyssey blocking buffer from LI-COR Biosciences, cat. nr.

927-40000), TBS buffer (100 mM Tris/HCl pH 7.5; 1.5 mM NaCl), Tween 20, primary antibody, and secondary antibody; if secondary antibody coupled to horseradish peroxidase (HRP) is used—substrate for HRP, proper detection equipment—for example, Kodak Image Station 440CF from Kodak for HRP-coupled antibodies and Odyssey from LI-COR Biosciences for infrared dye-coupled antibodies

17.2 METHODS

17.2.1 Overexpression of GPCRs in mammalian cell lines

Taken together the low expression level of GPCRs in native tissue and the toughness to raise antibodies to a lot of GPCRs, studying protein–protein interactions is mainly performed upon overexpression of the GPCRs of interest in mammalian cell lines. We perform coimmunoprecipitation (co-IP) and membrane coimmunoprecipitation (mco-IP) assays in mammalian cell lines transiently or stably overexpressing a tagged form (HA, FLAG, c-myc, etc.) of the GPCR. Different cell lines can be used: HEK293, CHO, HelaH21, COS, etc. The choice will mainly depend on the ease to obtain a high transfection efficiency. HEK293 cells are very easy to work with and a transfection efficiency of more than 95% can be achieved.

Different transfection methods including polyethylenimine (PEI), $Ca_3(PO_4)_2$, and lipofection can be used for this cell line (protocols in Box 17.1), all with their advantages and disadvantages (Table 17.1). Forty-eight hours posttransfection, the cells are collected in PBS and centrifuged for 10 min (250 g, 4 °C). The cell pellets can be stored at −70 °C for several weeks.

17.2.2 Cell lysis

A lot of different cell lysis methods were developed, including mechanical disruption and detergent-based methods, which we can subdivide in denaturing and nondenaturing methods. Next to this, combinations of different methods can be performed to achieve an optimal result. Each experiment has different requirements, causing the need to optimize the lysis method for each particular experiment.

Procedures for GPCR immunoprecipitation experiments were optimized not only to efficiently separate the hydrophobic membrane proteins from the hydrophilic proteins but also to preserve the possible protein–protein interactions.

17.2.2.1 Mechanical disruption

Collected cells are resuspended in 1 ml 20 mM Tris/HCl pH 8.0 and 1 mM EDTA and complemented with inhibitors (Table 17.3). This solution is sonicated at 20% amplitude for 10 cycles; each cycle consists of 1 s on and 5 s off. This sonication step disrupts the cell membrane and causes the release of the cellular content. When these samples are centrifuged (20 min at $42,000 \times g$), a separation between

> **Box 17.1**
>
> *Ca$_3$(PO$_4$)$_2$ transfection*
> - 2.5×10^5 cells are seeded in a 10 cm dish in 10 ml DMEM containing 10% FCS and supplemented with L-glutamine (3 mM) and penicillin (100 units/ml)/streptomycin (100 µg/ml).
> - The following day, medium on the cells is refreshed 1 h before transfection with 9 ml DMEM supplemented with 10% FCS.
> - 10 µg of DNA is used for transfection of cells in one 10 cm dish. If more than one plasmid is used for transfection of the same plate, for example, when studying protein–protein interactions, different plasmids with cDNA encoding the different proteins are need. Take the same amount of encoding plasmids for the different setups, and equalize DNA amounts among different plates by adding parental plasmid (="mock" plasmid), so that the total amount of plasmids used for each plate is the same (=10 µg).
> - Prepare per dish the following mix (in the mentioned order):
>   100 µl 2 M CaCl$_2$
>   10 µl DNA (1 µg/µl)
>   390 µl 0.1 × TE (1 mM Tris/HCl pH 7.6; 0.1 mM EDTA pH 8)
> - Add the mix dropwise to 500 µl 2 × BS/HEPES (25 mM HEPES; 274 mM NaCl; 10 mM KCl; 1.5 mM Na$_2$HPO$_4$; 12 mM dextrose pH 7.05) and aerate the complete mixture. Adding the calcium/DNA mixture to the phosphate buffer will result in small precipitates.
> - Add the mixture to the cells and mix gently.
> - Refresh the medium after at least 4 (up to 24) hours with DMEM 10% FCS.
>
> *PEI transfection*
> - 2.5×10^5 cells are seeded in a 10 cm dish in 10 ml DMEM containing 10% FCS and supplemented with L-glutamine and penicillin/streptomycin.
> - The following day, cells are refreshed with 9 ml DMEM supplemented with 2% FCS.
> - Prepare PEI stock solution:
>   Dissolve PEI (high molecular weight, water-free (25 kDa branched), Sigma Aldrich: 40,872-7): 100 mg/ml in Milli-Q water and homogenize.
>   Dilute further to 1 mg/ml and adjust pH to 7 with HCl.
>   Filter sterilize and aliquot (can be stored at −20 °C for > year).
>   Alternatively several companies sell this product.
> - Prepare per dish the following mixes:
>   Mix 1: 10 µl DNA (1 µg/µl) + 490 µl serum-free medium (without antibiotics) (=SFM).
>   Mix 2: 25 µl PEI + 475 µl SFM.
> - Add mix 2 to mix 1, vortex, and incubate for 10 min at room temperature.
> - Add the mixture to the cells and mix gently.
> - Refresh the medium after 6 h with DMEM 10% FCS.
>
> *Lipofection*
> - 2.5×10^5 cells are seeded on a 10 cm dish in 10 ml DMEM containing 10% FCS and supplemented with L-glutamine and penicillin/streptomycin.
> - The following day, cells are refreshed with 9 ml SFM.
> - Prepare per dish the following mixes:
>   Mix 1: 10 µl DNA (1 µg/µl) + 490 µl SFM
>   Mix 2: 20 µl Lipofectamine 2000 (Invitrogen) + 480 µl SFM
> - Add mix 2 to mix 1, vortex, and incubate for 45 min at room temperature.
> - Add the mixture to the cells and mix gently.
> - Refresh the medium after 6 h with DMEM 10% FCS.

Table 17.1 Advantages and Disadvantages of Different Transfection Methods for HEK293T Cells

Transfection Method	Advantages	Disadvantages
$Ca_3(PO_4)_2$	Inexpensive Easy to use	Low reproducibility[a]
PEI	High expression level of protein Inexpensive Easy to use	Low reproducibility[a] Cytotoxicity at high PEI conc. (>10 μg/ml) (Dong et al., 2007)
Lipofection	Very high efficiency Easy to use Reproducible	Expensive Not compatible with serum

[a]Commercially available analogs often give a better reproducibility.

membrane proteins and cytoplasmic proteins will take place. The pellet contains the membrane proteins, while the supernatant consists of the cytoplasmic proteins.

An alternative mechanical disruption method when no sonicator is available uses a polytron. Cells are lysed by homogenization in 7 ml 1× buffer (500 mM Tris, 50 mM KCl, 10 mM EDTA, 15 mM $CaCl_2 \cdot 2H_2O$, 40 mM $MgCl_2 \cdot 6H_2O$, pH 7.4) with a polytron (two times 10 s with incubation on ice in between) followed by centrifugation at $42,000 \times g$ for 20 min. The pellet contains the membrane fractions and the membrane proteins can be further extracted and solubilized, using a detergent-based method (see further). Membrane proteins can also be extracted directly from the cell pellet, without this homogenization step (see next paragraph). However, in our hands, this results in a less clear band resolution upon Western blot immunodetection.

17.2.2.2 Nondenaturing detergent-based lysis methods

There are a lot of possibilities concerning detergent choice for the solubilization of GPCRs (Allen, Ribeiro, Horuk, & Handel, 2009; Berger, Garcia, Lenhoff, Kaler, & Robinson, 2005; Corin et al., 2011; Grisshammer, 2009; Nekrasova, Sosinskaya, Natochin, Lancet, & Gat, 1996; Ren et al., 2009; Seddon, Curnow, & Booth, 2004), used in the different lysis buffers (Table 17.2), for example, CHAPS buffer (30 mM Tris/HCl pH 7.5; 150 mM NaCl; 1% CHAPS), TNT buffer (20 mM Tris/HCl pH 7.5; 200 mM NaCl; 0.1% Tween-20), NP-40 buffer (0.5% NP-40; 140 mM NaCl; 1.5 mM $MgCl_2$; 10 mM Tris/HCl pH 7.5), and Triton X-100 buffer (10% glycerol; 1% Triton X-100; 160 mM NaCl; 50 mM Tris/HCl pH 7.5; 1 mM EDTA; 1 mM EGTA). To prevent degradation or modification of the extracted proteins by endogenous proteases and phosphatases, specific inhibitors (10 μg/ml leupeptin, 1 mM PEFA-block, 2.5 μg/ml aprotinin, and 10 mM β-glycerol phosphate) are added to the buffer. Alternatively an inhibitor cocktail can be added (Table 17.3). Figure 17.1 gives an overview of the extraction of the dopamine D_4

Table 17.2 Types of Detergents and their Characteristics

Type of Detergent	Description of Use	Example
Ionic detergent	Very effective in the solubilization of membrane proteins but denaturing the proteins to some extent	SDS
Bile acid salts	Mild detergent and less inactivating	doc
Nonionic detergent	Mild surfactants that do not break protein–protein interactions and do not denature proteins	Triton X-100 Tween-20 DDM NP-40
Zwitterionic detergent	More harsh surfactants than the nonionic but conformation and interactions are preserved	CHAPS

Table 17.3 Function of the Specific Inhibitors

Inhibitor	Function
Leupeptin	Protease inhibitor; can inhibit cysteine, serine, and threonine peptidases
PEFA-block	Serine protease inhibitor
Aprotinin	Trypsin inhibitor
β-Glycerol phosphate	Serine and threonine phosphatase inhibitor
Protease inhibitor cocktail	Inhibits serine, cysteine, aspartic, and metalloproteases

receptor from transiently transfected HEK293 cells with different lysis methods. The conclusion for the dopamine D_4 receptor is that a RIPA buffer seemed to be the optimal choice, because it gave the best signal on SDS-PAGE (Fig. 17.1). For this reason, a detailed protocol for the RIPA lysis is provided in the following section.

17.2.2.2.1 Radioimmunoprecipitation (RIPA) lysis

The membrane fractions are resuspended in 300–500 μl of RIPA buffer (150 mM NaCl; 50 mM Tris/HCl pH 7.5; 1% NP-40; 0.1% SDS; 0.5% deoxycholic acid) with freshly added inhibitors (Table 17.3) for 1–2 h at 4 °C. This lysis buffer is based on moderate concentrations of the nonionic detergent NP-40 that will disrupt the integrity of the cell membrane, causing the extraction of membrane proteins in their native form. The buffer also contains a low concentration of the harsh denaturing anionic detergent SDS and ionic detergent sodium deoxycholate (doc). SDS and doc will help to disrupt the membrane, resulting in a higher yield of extracted membrane proteins. The concentration of these two is however quite low and will not be able to disrupt the protein–protein interactions. In addition, the detergent digitonin (0.1%)

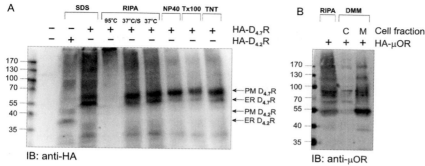

FIGURE 17.1

Comparison of different lysis methods. (A) HEK293T cells were transiently transfected with pHA-$D_{4.7}$R (additional control transfected with pHA-$D_{4.2}$R was introduced to distinguish bands of the right molecular level) to compare the efficiency of different lysis methods (SDS, RIPA, NP-40, Triton X-100 (Tx100), and TNT lysis buffer). The lysates were denatured at 37 °C in SDS sample buffer with addition of 50 mM DTT. For RIPA lysis also the influence of sonication (S) and denaturation at two different temperatures (95 and 37 °C) was tested. The receptor was visualized by immunoblotting with mouse anti-HA (16B12) and antimouse IRDye 800. PM-mature, fully glycosylated plasma membrane receptor; ER–ER-retained receptor. (B) HEK293T cells were transiently transfected with HA-μ opioid receptor (μOR), using $Ca_3(PO_4)_2$ method. Forty-eight hours after transfection cells were lysed with DMM or RIPA lysis buffers to compare efficiency of these buffers. Samples were denatured with SDS sample buffer and loaded on an SDS-PAGE gel. The protein was detected by immunoblotting with rabbit anti-μOR antibody and antirabbit IRDye 800. C-cytoplasmic proteins and M-membrane proteins.

can also be added to optimize the solubilization of the membrane proteins (Holden & Horton, 2009).

Following incubation, the lysis solution is centrifuged at $8000 \times g$ for 10 min, resulting in a pellet and the supernatant, which contains the extracted membrane proteins.

17.2.2.2.2 n-Dodecyl-β-D-maltoside (DDM) lysis

The pellet is resuspended in 0.5 ml buffer containing 20 mM Tris/HCl pH 8.0, 1 mM EDTA, 0.35 mM NaCl, 0.5% n-dodecyl-β-D-maltoside (DMM), and inhibitors (Table 17.3). Maltosides are a class of detergents often used for the solubilization of membrane proteins (Allen et al., 2009; Berger et al., 2005; Grisshammer, 2009). After centrifugation at $20,000 \times g$ for 10 min, the membrane proteins will mainly be in the supernatant fraction.

This lysis method shows a better band separation and less smear formation compared to the RIPA lysis, what can be observed in Fig. 17.1B. On top of this, there is no addition of SDS or doc in this buffer, so there is no risk for denaturation resulting in the loss of the interactions.

17.2.2.3 Denaturing-based lysis methods: urea lysis and SDS-lysis

Cells are resuspended in 6 M urea lysis buffer (20 mM Tris/HCl pH 7.5; 135 mM NaCl; 1.5 mM $MgCl_2$; 1 mM EGTA; 1% Triton X-100; 10% glycerol and 6 M urea). The presence of Triton X-100 in the buffer will help with the solubilization of the membrane proteins. Upon incubation for 30 min at room temperature, the solution is centrifuged ($8000 \times g$, 5 min), and the supernatant fraction will contain the solubilized membrane proteins. However, these proteins will be completely denatured as urea is a powerful protein denaturant that disrupts noncovalent bonds.

Another lysis method resulting in solubilization and complete denaturation of proteins is the SDS-lysis method. With this method, you resuspend the cells in SDS-lysis buffer: 62.5 mM Tris/HCl pH 6.8, 2% SDS, 10% glycerol, 0.1% bromophenol blue, and 50 mM dithiothreitol (DTT). The presence of SDS together with the DTT will result in the denaturation of all the proteins in the sample. Glycerol is added to facilitate sample loading on SDS-PAGE gel and bromophenol blue is added to visualize sample separation.

For studying protein–protein interactions, it is extremely important to keep proteins in their native form after lysis. Therefore, both methods, in which proteins are denatured during lysis procedure, are not useful for co-IP or mco-IP.

17.2.3 Coimmunoprecipitation

Coimmunoprecipitation is a common method used to study protein–protein interactions. In this technique, one of the potential interacting partners (*protein X*) is precipitated from the solution using an antibody that specifically binds to this protein X. The antibody can be directed against endogenous epitopes or against an exogenous tag (e.g., HA, FLAG, and His), added to the receptor with the help of genetic engineering methods. A comparison of advantages and disadvantages of both approaches can be found in Table 17.4. The antibody used for immunoprecipitation needs to be coupled at some moment to a solid resin. This resin is created by agarose/polyacrylamide/Sepharose or magnetic beads to which recombinant Protein A or G has been

Table 17.4 Advantages and Disadvantages of Antibodies Against Endogenous and Exogenous Epitopes Used for Immunoprecipitation Technique

Antibody Recognizing	Advantages	Disadvantages
Endogenous epitope	Isolation of protein from native tissue Does not require overexpression	Lack of good antibodies for a lot of proteins esp. GPCRs Often much lower efficiency
Exogenous epitope tag	Very good efficiency No need of having a separate antibody for each protein Availability of beads with immobilized antibody	Overexpression of protein is required No possibility to isolate proteins from native tissue

immobilized. Thanks to the high affinity of Proteins A and G for binding antibodies, a protein complex containing the antibody, the protein of interest and its interacting proteins can be isolated. The origin of the antibody that is used for IP (species of animal in which IgG was produced) defines the choice of the beads. In general, Protein A beads are used to bind and purify IgG from human and rabbit, whereas Protein G beads are recommended for isolation of IgG from human, mouse, and rat. There are also immobilized Protein A/G beads available, which combine the IgG-binding domain of both Protein A and Protein G, enabling purification of numerous IgG subclasses and species. When one of the potential interacting partners has a common tag (e.g., FLAG and HA), then it is possible to use beads immediately coupled to the specific antibody directed against the tag (FLAG-beads and HA-beads). This will save time—generally 4 h incubation of lysate with such beads is sufficient to obtain high immunoprecipitation efficiency. However, the immobilization can reduce the antibody's affinity to the antigen and prevent IP.

Typically, co-IP is performed with total cell lysates, but when studying membrane proteins, modifications of this technique can be recommended. We will first describe a standard co-IP protocol and then a protocol to isolate specifically complexes with plasma membrane expressed receptors. A schematic overview is given in Fig. 17.2.

17.2.3.1 Standard coimmunoprecipitation: in total cell lysate (Fig. 17.2A, left)

RIPA lysis seems to be one of the best choices to obtain the starting material for isolating GPCRs together with its interacting proteins, as it is an efficient cell lysis method resulting in clear bands on SDS-PAGE (Fig. 17.1). The cleared lysates are transferred to clean microcentrifuge tubes: (1) a fraction (10%) of each sample will be analyzed on SDS-PAGE to check protein expression (both expression level and molecular weight will be evaluated) and (2) to the rest of the lysate 2 µg of the antibody is added and incubated for 2–4 h at 4 °C upon rotation. Next, 20 µl of pure Protein A beads (delivered as a 50% slurry) is added per sample. These beads are first washed three times (centrifugation after each washing step: $110 \times g$, 4 °C, 1 min) with RIPA buffer supplemented with protease and phosphatase inhibitors (see paragraph about cell lysis). The washed beads are added to the lysates and incubated overnight with rotation at 4 °C.

Note: Preclearing of the lysates can be performed to remove proteins from the sample that can bind nonspecifically to the beads. To perform this step, lysates need to be incubated with the beads (without addition of antibody) for 30–60 min at 4 °C with rotation. Then, samples are centrifuged for 1 min ($3000 \times g$, 4 °C) and clear lysates are transferred to clean microcentrifuge tubes. Such precleared samples can be used for immunoprecipitation. This step is only necessary if "contaminating" proteins interfere with visualization of the proteins of interest.

The next day, beads are again washed three times in RIPA buffer (supplemented with protease and phosphatase inhibitors) to remove all proteins present in the lysate, which are not bound. When nonspecific signals are observed after Western blot

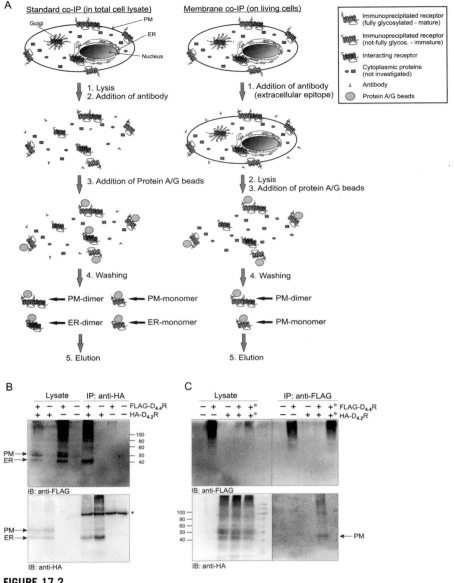

FIGURE 17.2

Comparison of standard and membrane co-IP. (A) Schematic summary of standard and membrane co-IP performed in total cell lysate or on living cells, respectively. (B) Standard co-IP performed in HEK293T cells transiently transfected with pHA-$D_{4.2}$R and/or pFLAG-$D_{4.4}$R. Immunoprecipitation (IP) was performed with mouse anti-HA (16B12). After, Western blot analysis, proteins were visualized with HRP-coupled anti-FLAG M2 or mouse anti-HA (16B12) and HRP-coupled antimouse. PM-mature, fully glycosylated plasma membrane receptor; ER–ER-retained receptor.*Signal denoting association of two heavy chains (2 × 50 kDa) of anti-HA sera. (C) Membrane co-IP was performed in HEK293T cells transiently expressing HA-$D_{4.2}$R and/or FLAG-$D_{4.4}$R, by adding anti-FLAG antibody to the living cells. Subsequently, cell lysates were made and membrane-labeled receptors immunoprecipitated. For immunoblotting, the same antibodies as in B were used. °Samples in which cells were independently transfected and mixed posttransfection. (See color plate.)

Adapted from Van Craenenbroeck et al. (2011).

immunodetection with specific antibodies, extra washing steps are recommended. Also, increasing the salt concentration in the RIPA washing buffer will make the conditions more stringent and can help to remove proteins nonspecifically bound to the beads. After this procedure, only proteins recognized by the used antibody and the interacting partners should stay attached to the beads.

17.2.3.2 Membrane coimmunoprecipitation: on living cells (Fig. 17.2A, right)

Membrane co-IP is a modification of the standard co-IP we designed to isolate proteins from the plasma membrane together with the interacting partners.

Cells (transfected or from native tissue) are collected in PBS (250 g, 4 °C, 10 min) and resuspended in 1 ml of serum-free medium. Primary antibody recognizing the extracellular epitope of the receptor is added to the cells in the dilution previously optimized for each specific antibody (examples of optimized dilutions of antibodies frequently used in our laboratory can be found in Table 17.5). We prefer to work at 4 °C to prevent receptor internalization but incubation at 37 °C is also possible. After 2 h, cells are spun down (250×g, 4 °C, 10 min) and washed once with PBS, and standard RIPA lysis protocol starts (see Section 2.2.2.1). Upon clearing the lysates, Protein A beads are immediately added and overnight incubation at 4 °C with rotation is performed. By using this modification of IP, it is possible to isolate only mature receptors present at the plasma membrane, and one can be sure that the observed interaction occurs with fully processed receptors and not in the ER or Golgi apparatus.

17.2.4 Elution

After removing all proteins that are unbound or nonspecifically bound to the beads, the elution procedure can start. Different methods allowing elution of proteins from the beads exist, but most often, SDS buffer is used as it provides the highest elution

Table 17.5 Examples of Optimized Dilutions of Primary Antibodies Used for mco-IP and Immunoblotting

Antibody	Company	Catalog Number	Dilution, mco-IP	Dilution, IB
Anti-HA.11 (16B12)	Covance	MMS-101P-1000	1:800	1:2000
Anti-HA rabbit	GeneTex	GTX29110	Not tested	1:2000
Anti-FLAG (M2)	Sigma	F3165	1:800	1:4000
Anti-c-myc (4A6)	Millipore	05-724	Not tested	1:2000
Anti-µOR (384-398)	Calbiochem	PC165L	Recognizes intracell. loop	1:2000
Anti-D_4R (D16)	Santa Cruz Biotechnology	Sc-31481	Not tested	1:100

For standard co-IP we always add 2 µg antibody per sample.

efficiency. In this paragraph, we will focus on different conditions of SDS buffer elution.

The effective elution procedure also requires optimization for the specific pair of proteins. Two factors have to be taken into consideration when the elution conditions are optimized: (1) temperature and (2) presence of reducing agent. In our experiments, most of the time, proteins are eluted at 37 °C for 10 min in SDS sample buffer (4% SDS; 50% glycerol; 0.2% bromophenol blue; 65 mM Tris/HCl pH 6.8) with addition of 50 mM DTT.

In the experiment presented in Fig. 17.3, different elution temperatures and the importance of DTT addition were tested. Protein denaturation of the lysate sample at 95 °C resulted in detection of a smear at the top of the immunoblot, which represents aggregated proteins. Denaturation at 37 °C with addition of DTT allowed for

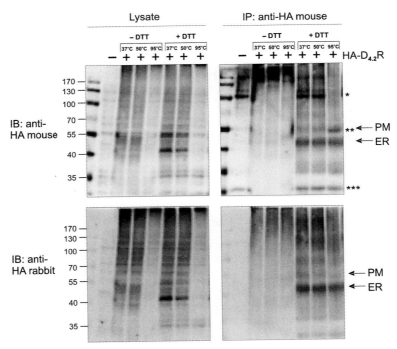

FIGURE 17.3

Comparison of different elution conditions. HEK293T cells were transiently transfected with pHA-$D_{4.2}$R. Immunoprecipitation with mouse anti-HA (16B12) was performed from total cell lysate. Proteins were eluted from the beads at three different temperatures: 37, 50, and 95 °C with or without addition of 50 mM DTT. After Western blot analysis, proteins were visualized with mouse anti-HA (16B12) and antimouse IRDye 700 or rabbit anti-HA and antirabbit IRDye 800. PM-mature, fully glycosylated plasma membrane receptor; ER–ER-retained receptor. *Association of two heavy chains (2 × 50 kDa), **one heavy chain (50 kDa), ***one light chain (25 kDa) of anti-HA sera.

separation of two forms of the dopamine D_4 receptor. The lower band of around 46 kDa represents nonmature, ER-associated receptor, and a second band of 52 kDa represents mature receptor with the fully processed N-linked glycosylation tree. For elution of proteins from the beads, presence of reducing agent seems to be extremely important, as without addition of DTT, only a smear at the top of the blot can be detected (aggregated proteins). After elution at 95 °C, less mature receptor can be isolated then after elution at nonboiling temperatures (37 or 50 °C). The strong band around 50 kDa on the upper, right panel represents immunoglobulin, which was used for IP.

17.2.5 Detection of interacting partners

Finally, when the proteins are eluted from the beads, standard SDS-PAGE can be performed to separate proteins according to their molecular weight. Next, proteins are transferred from the polyacrylamide gel to a nitrocellulose or polyvinylidine fluoride (PVDF) membrane during Western blot procedure and further immunodetection can be done with specific antibodies.

In coimmunoprecipitation studies often immunoreactive bands can be detected at the molecular level corresponding to approximately twice the molecular weight of monomeric receptors. This observation can already suggest that the receptor exists in dimerized form. Such dimer is resistant to SDS denaturation and this might indicate the involvement of hydrophobic interactions in dimerization (Salahpour, Angers, & Bouvier, 2000).

If the elution of proteins was performed at 37 °C, the IgG used for IP will appear at the height of 100 kDa (association of two antibody heavy chains), and when 95 °C is used, the antibody band will be observed at 50 kDa. This has to be taken into account when the protein under investigation has similar molecular weight, for example, a lot of class A GPCRs have a molecular weight around 50 kDa, and, therefore, we recommend for immunodetection a secondary antibody that recognizes (1) only the light chain of the primary antibody or (2) that is conformation-specific and can recognize only nondenatured antibody (antibody used for IP is denatured and will not be recognized). If primary antibody directly coupled to the enzyme (e.g., HRP) or infrared dye is available, this can also be a good alternative. Next, primary antibodies produced in another animal species than the antibody used for IP can facilitate the interpretation of the Western blot immunodetection (see Fig. 17.3—IB with anti-HA rabbit), but for close related species, the cross recognition may occur. All these procedures will help to clearly visualize the protein of interest, which otherwise would be masked by immunoglobulin band.

17.2.6 Controls

To exclude aspecific binding of proteins to the beads or antibody, it is recommended to include in the experimental setup a control sample to which no antibody or irrelevant antibody will be added during immunoprecipitation procedure. Moreover, to confirm specific coimmunoprecipitation of interacting proteins in transfected cells, the immunoprecipitation should be performed in singly (transfected with only one protein of interest, e.g., Fig. 17.2B, lanes 6 and 7) and doubly (transfected with two potential interacting partners) transfected cells. Specific interaction will be detected only in cotransfected cells (expressing both proteins).

When dimerization between receptors is studied, an additional control of specific interaction between two proteins in living cells can be added by mixing posttransfection cells that were transfected only with one protein of interest (e.g., Fig. 17.2C, lane 10 indicated by °). In this case, no interaction should be detected between the two proteins as they were not present in the same cell.

17.3 DISCUSSION

A better understanding of GPCR biogenesis can start with the unraveling of all the proteins that interact with the GPCR of interest. We have focused on the coimmunoprecipitation technique to get an initial idea about protein–protein interactions, and all methods described can also be applied to study GPCR dimerization.

The main issues are problems with GPCR solubilization, which can be hard; therefore, we have discussed different methods to tackle this problem, and aggregation of GPCRs has to be occluded as it can lead to false-positive results. Extra controls were suggested in this methodological overview.

The use of this coimmunoprecipitation technique to study GPCR complexes would benefit enormously from the availability of good and specific antibodies in order to isolate the endogenous receptor from primary cells and native tissues. Conventional antibodies are large molecules, which can generate steric hindrance; thus, antibodies are not necessarily the best tools for demonstrating and quantifying direct receptor interactions. Nanobodies can be a good alternative since they are much smaller and they can recognize epitopes usually not detected by conventional antibodies (Jahnichen et al., 2010; Rasmussen et al., 2011). Therefore, they open new perspectives in term of molecular recognition and specificity.

Next to this biochemical approach, the majority of oligomerization studies will also include biophysical approaches such as fluorescence or bioluminescence resonance energy transfer techniques (FRET or BRET) as reviewed before (Milligan, 2013; Van Craenenbroeck, 2012). Some criticisms suggest that GPCR dimerization might be promoted at relatively high receptor expression levels and potentially be at least partially an artifact of overexpression. However, studies of $\beta2$-adrenergic receptor dimerization have indicated that dimerization is unaltered over a wide range of expression levels (Mercier, Salahpour, Angers, Breit, & Bouvier, 2002).

A major advantage of FRET in view of studying biogenesis of GPCRs is that it can be combined with microscopic imaging and thus allow the study of dimerization in different subcellular compartments (Herrick-Davis, Weaver, Grinde, & Mazurkiewicz, 2006). A disadvantage of FRET is that the proteins of interest need to be tagged with a large donor or acceptor protein.

In summary, we can say that coimmunoprecipitation is a straightforward way to study protein–protein interactions and, when a good antibody is available, *in vivo* applications are possible.

Acknowledgments

KS has a predoctoral fellowship and KVC a postdoctoral fellowship from FWO (Fonds voor Wetenschappelijk onderzoek) Vlaanderen. JD has a BOF grant from Ghent University.

References

Achour, L., Labbe-Jullie, C., Scott, M. G., & Marullo, S. (2008). An escort for GPCRs: Implications for regulation of receptor density at the cell surface. *Trends in Pharmacological Sciences*, *29*(10), 528–535.

Allen, S. J., Ribeiro, S., Horuk, R., & Handel, T. M. (2009). Expression, purification and in vitro functional reconstitution of the chemokine receptor CCR1. *Protein Expression and Purification*, *66*(1), 73–81.

Berger, B. W., Garcia, R. Y., Lenhoff, A. M., Kaler, E. W., & Robinson, C. R. (2005). Relating surfactant properties to activity and solubilization of the human adenosine a3 receptor. *Biophysical Journal*, *89*(1), 452–464.

Corin, K., Baaske, P., Geissler, S., Wienken, C. J., Duhr, S., Braun, D., et al. (2011). Structure and function analyses of the purified GPCR human vomeronasal type 1 receptor 1. *Scientific Reports*, *1*, 172.

Dong, W., Li, S., Jin, G., Sun, Q., Ma, D., & Hua, Z. (2007). Efficient gene transfection into mammalian cells mediated by cross-linked polyethylenimine. *International Journal of Molecular Sciences*, *8*(2), 81–102.

Dupre, D. J., Hammad, M. M., Holland, P., & Wertman, J. (2012). Role of chaperones in g protein coupled receptor signaling complex assembly. *Subcellular Biochemistry*, *63*, 23–42.

Ellgaard, L., & Helenius, A. (2003). Quality control in the endoplasmic reticulum. *Nature Reviews Molecular Cell Biology*, *4*(3), 181–191.

Grisshammer, R. (2009). Purification of recombinant G-protein-coupled receptors. *Methods in Enzymology*, *463*, 631–645.

Hebert, D. N., Garman, S. C., & Molinari, M. (2005). The glycan code of the endoplasmic reticulum: Asparagine-linked carbohydrates as protein maturation and quality-control tags. *Trends in Cell Biology*, *15*(7), 364–370.

Herrick-Davis, K., Weaver, B. A., Grinde, E., & Mazurkiewicz, J. E. (2006). Serotonin 5-HT2C receptor homodimer biogenesis in the endoplasmic reticulum: Real-time visualization with confocal fluorescence resonance energy transfer. *Journal of Biological Chemistry*, *281*(37), 27109–27116.

Holden, P., & Horton, W. A. (2009). Crude subcellular fractionation of cultured mammalian cell lines. *BMC Research Notes*, *2*, 243.

Jahnichen, S., Blanchetot, C., Maussang, D., Gonzalez-Pajuelo, M., Chow, K. Y., Bosch, L., et al. (2010). CXCR4 nanobodies (VHH-based single variable domains) potently inhibit chemotaxis and HIV-1 replication and mobilize stem cells. *Proceedings of the National Academy of Sciences of the United States of America*, *107*(47), 20565–20570.

Luttrell, L. M. (2008). Reviews in molecular biology and biotechnology: Transmembrane signaling by G protein-coupled receptors. *Molecular Biotechnology, 39*(3), 239–264.

Margeta-Mitrovic, M., Jan, Y. N., & Jan, L. Y. (2000). A trafficking checkpoint controls $GABA_B$ receptor heterodimerization. *Neuron, 27*(1), 97–106.

Mercier, J. F., Salahpour, A., Angers, S., Breit, A., & Bouvier, M. (2002). Quantitative assessment of β_1- and β_2-adrenergic receptor homo- and heterodimerization by bioluminescence resonance energy transfer. *Journal of Biological Chemistry, 277*(47), 44925–44931.

Milligan, G. (2013). The prevalence, maintenance and relevance of GPCR oligomerization. *Molecular Pharmacology, 84*, 158–169.

Nekrasova, E., Sosinskaya, A., Natochin, M., Lancet, D., & Gat, U. (1996). Overexpression, solubilization and purification of rat and human olfactory receptors. *European Journal of Biochemistry, 238*(1), 28–37.

Petaja-Repo, U. E., Hogue, M., Laperriere, A., Walker, P., & Bouvier, M. (2000). Export from the endoplasmic reticulum represents the limiting step in the maturation and cell surface expression of the human delta opioid receptor. *Journal of Biological Chemistry, 275*(18), 13727–13736.

Rasmussen, S. G., DeVree, B. T., Zou, Y., Kruse, A. C., Chung, K. Y., Kobilka, T. S., et al. (2011). Crystal structure of the beta2 adrenergic receptor–Gs protein complex. *Nature, 477*(7366), 549–555.

Ren, H., Yu, D., Ge, B., Cook, B., Xu, Z., & Zhang, S. (2009). High-level production, solubilization and purification of synthetic human GPCR chemokine receptors CCR5, CCR3, CXCR4 and CX3CR1. *PLoS One, 4*(2), e4509.

Salahpour, A., Angers, S., & Bouvier, M. (2000). Functional significance of oligomerization of G-protein-coupled receptors. *Trends in Endocrinology and Metabolism, 11*(5), 163–168.

Salahpour, A., Angers, S., Mercier, J. F., Lagace, M., Marullo, S., & Bouvier, M. (2004). Homodimerization of the beta2-adrenergic receptor as a prerequisite for cell surface targeting. *Journal of Biological Chemistry, 279*(32), 33390–33397.

Seddon, A. M., Curnow, P., & Booth, P. J. (2004). Membrane proteins, lipids and detergents: Not just a soap opera. *Biochimica et Biophysica Acta, 1666*(1–2), 105–117.

Shirvani, H., Gata, G., & Marullo, S. (2012). Regulated GPCR trafficking to the plasma membrane: General issues and the CCR5 chemokine receptor example. *Subcellular Biochemistry, 63*, 97–111.

Van Craenenbroeck, K. (2012). GPCR oligomerization: Contribution to receptor biogenesis. *Subcellular Biochemistry, 63*, 43–65.

Van Craenenbroeck, K., Borroto-Escuela, D. O., Romero-Fernandez, W., Skieterska, K., Rondou, P., Lintermans, B., et al. (2011). Dopamine D4 receptor oligomerization—Contribution to receptor biogenesis. *FEBS Journal, 278*(8), 1333–1344.

Van Craenenbroeck, K., Clark, S. D., Cox, M. J., Oak, J. N., Liu, F., & Van Tol, H. H. (2005). Folding efficiency is rate-limiting in dopamine D_4 receptor biogenesis. *Journal of Biological Chemistry, 280*(19), 19350–19357.

CHAPTER

Lipid-Dependent GPCR Dimerization

18

Alan D. Goddard*, Patricia M. Dijkman[†], Roslin J. Adamson[†], and Anthony Watts[†]

*School of Life Sciences, University of Lincoln, Lincoln, United Kingdom
[†]Biomembrane Structure Unit, Department of Biochemistry, University of Oxford, Oxford, United Kingdom

CHAPTER OUTLINE

Introduction	342
18.1 Materials and Method	343
18.1.1 Purification of GPCRs	343
18.1.2 Labeling Strategies for FRET	344
18.1.2.1 Protocol for Site-Directed Fluorophore Labeling of Cysteines	345
18.1.3 Production of Liposomes	347
18.1.3.1 Preparing the Lipid	347
18.1.3.2 Sizing Liposomes: Sonication Versus Extrusion	347
18.1.4 Reconstitution	348
18.1.4.1 Protocol for Reconstitution Using Bio-beads	349
18.2 FRET Measurements and Analysis	351
18.3 Interpreting Results	354
Summary	355
References	355

Abstract

It has been widely demonstrated that G protein-coupled receptors (GPCRs) can form dimers both *in vivo* and *in vitro*, a process that has functional consequences. These receptor–receptor interactions take place within a phospholipid bilayer, yet, generally, little is known of the requirements for specific lipids that mediate the dimerization process. Studying this phenomenon *in vivo* is challenging due to difficulties in modulating the lipid content of cell membranes. Therefore, in this chapter, we describe techniques for reconstitution of GPCRs into model lipid bilayers of defined composition. The concentrations of specific lipids and sterols can be precisely controlled in these liposomes, as well as maintaining an appropriate lipid–protein ratio to avoid artifactual interactions. Receptor dimerization in this system is monitored via Förster resonance energy transfer (FRET), which requires the use of fluorescently labeled receptors. We therefore also include protocols for labeling with appropriate fluorophores and determining the apparent FRET efficiency, a measurement of the extent of receptor dimerization. Understanding the lipid dependence of GPCR dimerization will be key in understanding how this process is regulated in the dynamic heterogeneous environment of the cell membrane.

INTRODUCTION

All organisms rely on membrane proteins to detect and respond to changes in their external environment. G protein-coupled receptors (GPCRs) represent the largest family of mammalian cell-surface receptors, with $\sim$800 identified in humans, constituting $\sim$1% of the genome (Vassilatis et al., 2003). GPCRs generally bind an extracellular ligand and transduce, via a conformational change, this signal to an intracellular heterotrimeric G protein consisting of Gα, Gβ, and Gγ subunits. Exchange of GDP for GTP on the Gα subunit results in activation of various signaling cascades bringing about cellular responses (Oldham & Hamm, 2008). GPCRs are involved in a plethora of physiological processes and hence are important drug targets—it is estimated that $\sim$30% of pharmaceutical drugs target these receptors (Jacoby, Bouhelal, Gerspacher, & Seuwen, 2006).

An emerging theme over the last decade has been the ability of GPCRs to form dimers or higher-order oligomers with either the same GPCR or other members of this family. Such dimerization has been shown to affect both ligand binding and signaling (Ciruela et al., 2010); indeed, constitutive dimerization of class C GPCRs is essential for their activity. While GPCR dimerization has been demonstrated *in vitro* and *in vivo*, little is known of the lipid-dependence of these interactions. Membrane proteins do not exist in isolation but are influenced, both chemically and physically, by lipid, protein, and other components of the membrane including annular lipids, which interact directly with the proteins, and factors arising from the bulk of the bilayer such as lateral pressure. Lipid bilayers have a number of characteristics that are important for protein activity and are strongly influenced by their lipid composition. Membrane thickness is dependent upon the nature of the hydrocarbon chain of the lipid. The charge on the membrane is influenced by the head groups of the lipids within the bilayer. Additionally, the membrane may be asymmetric, that is, it may not have the same lipid composition within each leaflet (van Meer, 2011). Finally, lipid membranes adopt different phases that are influenced by the temperature of the environment and the lipid composition of the membrane. Bilayers are not necessarily homogenous across their surface and may form microdomains such as lipid rafts or caveolae, and it has been demonstrated that some GPCRs and G proteins localize to such regions (Patel, Murray, & Insel,

2008). Sterols such as cholesterol have also been noted to influence some GPCRs (Oates & Watts, 2011), and their presence can be modulated via, for example, the use of cyclodextrins (Oates et al., 2012).

Due to difficulties in modulating bilayer lipid composition in living cells, it is easiest to perform these studies *in vitro*. Such studies necessitate purification of active GPCRs, a process that has proved challenging. GPCRs are often unstable in detergent solution, which has been one of the main barriers to structural determination (Bill et al., 2011). However, advances made with modern detergents, thermostabilized mutants, and rapid purification strategies enable the production of a number of purified, active receptors (Zhao & Wu, 2012).

Once the receptor has been purified, it must be reconstituted into a defined lipid environment. A number of different methods exist including the use of liposomes, nanodiscs (Bayburt & Sligar, 2010), and Lipodisqs (Orwick et al., 2012). In this chapter, we will focus on the use of liposomes as it is possible that the use of small lipid discs will result in a high local GPCR concentration and hence will produce artifactual interactions. Once reconstituted, the ability of GPCRs to dimerize must be quantified. Although a number of techniques can be applied, our laboratory, and this chapter, focuses on the use of FRET. This technique can be used to indicate protein–protein interactions as the resonance energy transfer is generally limited to a distance of a few nanometers. However, FRET should be used judiciously and a number of controls must be performed to ensure that results are meaningful; such controls will be discussed in the succeeding text.

Here, we describe our approaches to GPCR reconstitution into liposomes containing defined lipid mixtures and the subsequent measurement and analysis of FRET to quantify the interaction between receptors under different lipid conditions. These techniques should be transferrable to any GPCR and also a variety of other integral or membrane-associated proteins.

18.1 MATERIALS AND METHOD
18.1.1 Purification of GPCRs

GPCRs can be purified using a number of expression systems including *E. coli*, baculovirus/insect cells, mammalian cells, yeast, and *in vitro* translation systems. The advantages and disadvantages of such systems are described elsewhere (Tapaneeyakorn, Goddard, Oates, Willis, & Watts, 2011). Our group and others have optimized expression and purification of neurotensin receptor 1 (NTS1), which binds neurotensin (NT) (Attrill, Harding, Smith, Ross, & Watts, 2009; Grisshammer, 2009; Harding et al., 2007). We use a fusion protein (FP) in which NTS1 has an N-terminal maltose-binding protein domain and a C-terminal thioredoxin domain with decahistidine tag added. Tobacco etch virus (TeV) protease sites are present at the junctions to remove the tags postpurification. In the case of fluorescently tagged variants, CFP or YFP is appended to the C-terminus of NTS1 (Harding et al., 2009) before the TeV site. Receptors are initially purified from *E. coli* using immobilized metal affinity chromatography, cleaved with TeV, and then active receptors are purified using an NT-affinity column. This process is described in detail elsewhere (Attrill et al., 2009).

18.1.2 Labeling strategies for FRET

FRET studies require the labeling of each of the dimerization partners with donor or acceptor fluorophores that have spectral overlap between their emission and absorption spectrum, respectively. When choosing donor and acceptor pairs, one must try to optimize this spectral overlap while minimizing spectral overlap of the donor and acceptor absorption spectra, and of their emission spectra, as that would lead to artifacts such as crosstalk and bleedthrough, respectively (this will be discussed in a later section).

The most common labeling strategy is the use of autofluorescent FPs, where a variant of the green fluorescent protein (GFP) is fused to the GPCR of interest through standard genetic engineering (for more detailed experimental procedure, see Harding et al. (2009)). Frequently used FRET-compatible pairs are CFP and YFP (cyan and yellow fluorescent protein), but vectors of many GFP variants with different absorption properties to suit experimental requirements are commercially available (e.g., Living Colors™, BD Biosciences, or Vivid Colors™, Life Technologies). These FPs not only are typically fused to the N- or C-terminus of the target protein to minimize problems with folding and retain GPCR functionality but also can be inserted into loop regions. Often, a linker region is introduced between the FP and the target protein to allow some flexibility between the two domains to minimize steric hindrance. Different linkers have been reported, varying in both length and content. Typically, these linkers have a high glycine and/or serine content, as these flexible and hydrophilic residues are thought to form a random coil structure and are thus unlikely to interact with the protein domains or to interfere with their folding (Evers, van Dongen, Faesen, Meijer, & Merkx, 2006).

The use of FPs has the benefit that all receptors are directly labeled during expression, and no cofactors or additional labeling and purification steps are required. FPs are relatively large (~27 kDa), however, and can impair protein function when inserted into functionally important domains (Hoffmann et al., 2005). The tendency of GFP to self-dimerize, leading to false-positive results, has also been raised as a concern (Tsien, 1998), but new "enhanced" variants of GFP, such as eCFP, do not exhibit this propensity to the same extent (Espagne et al., 2011).

Cell-free expression of GPCRs also allows fluorescently tagging receptors prior to purification, for example, by incorporation of nonnatural amino acids or by conjugation of labels to the C-terminus (Jackson, Boutell, Cooley, & He, 2004). The applicability hereof will however be dependent on the possibility for *in vitro* expression of the GPCR of interest.

Another popular technique is posttranslational site-directed labeling using small organic dyes, such as Alexa Fluor dyes (e.g., Alexa488/Alexa647) (Panchuk-

Voloshina et al., 1999) or Cy dyes (e.g., Cy3/Cy5) (Mujumdar, Ernst, Mujumdar, Lewis, & Waggoner, 1993), which are commercially available in a wide range of wavelengths and have been optimized in terms of brightness and photostability. These dyes are linked to small reactive groups, such as maleimide or succinimidyl esters, that enable coupling to sulfhydryl or amino groups, resulting in an overall small modification of the target receptor (typically <1–2 kDa). However, care must be taken that any free dye is removed as this can interfere with data interpretation. The labeling efficiency should also be determined, as this will influence the maximum FRET efficiency possible and should be corrected for. So, although these small dyes have a clear size advantage over the use of fluorescent FPs, they are experimentally more challenging, and labeling conditions need to be optimized for each individual case.

18.1.2.1 Protocol for site-directed fluorophore labeling of cysteines

Reagents

1. Purified protein stocks in detergent (0.1% dodecyl maltoside (DDM)/0.01% cholesteryl hemisuccinate (CHS) (w/v))
2. Protein buffer: 50 mM Tris–HCl pH 7.4, 100 mM NaCl, 1 mM EDTA, 0.1% DDM, 0.01% CHS (w/v), 10% glycerol (v/v)
3. Reducing agent (e.g., dithiothreitol (DTT) or Tris (2-carboxyethyl) phosphine hydrochloride (TCEP))
4. Cysteine-reactive fluorophore (e.g., maleimide-Alexa Fluor or Cy dye; typically as a dimethyl sulphoxide (DMSO) stock for long-term storage or in aqueous buffer at pH 7 for short-term storage)
5. PEGm (methoxypolyethylene glycol 5000 maleimide)

18.1.2.1.1 Preparing the receptor for labeling

One should aim to have only one label per monomer to have a homogenous population of labeled protein. Thus, a background mutant needs to be constructed in which native, accessible cysteines are removed by site-directed mutagenesis to ensure specific labeling either at one remaining native cysteines or at a newly introduced cysteine. One should rule out the possibility of nonspecific labeling of this background mutant under the same conditions used for the labeling of the final mutant. Additionally, one should check that the mutant used is still functional, through a relevant assay such as ligand binding or G protein coupling, or both.

18.1.2.1.2 Reduction of reactive site

The purified receptor is reduced prior to labeling to ensure reactivity of the cysteine residue, typically with DTT or TCEP, at a concentration in the range of 1–10 mM. It has been shown that low concentrations (0.1 mM) of the trialkylphosphine reductant TCEP interfere less with maleimide labeling than the thiol-based DTT under the same conditions (Burmeister Getz, Xiao, Chakrabarty, Cooke, & Selvin, 1999). Thus, TCEP can be favorable when labeling with maleimides if reductant removal is not optimal. TCEP has also been shown to be a stronger and faster reductant at pH < 8 (Han & Han, 1994). TCEP is more stable than DTT in the presence of metal ions such as Fe^{3+} and Ni^{2+}, which can be an important consideration when labeling protein after IMAC chromatography, while chelating agents such as EGTA adversely affect TCEP stability but increase DTT stability (Burmeister Getz et al., 1999).

The receptor is incubated with the reductant for 1–3 h at room temperature or at 4 °C depending on the stability of the protein. Then, reductant must be removed, which can be efficiently done by size-exclusion chromatography (e.g., using a 5 ml Sephadex G-25 prepacked HiTrap Desalting column, GE Healthcare, or two or more of these in tandem to increase capacity and resolution) using degassed protein buffer. Peak fractions are collected and protein concentration is determined by A_{280}.

18.1.2.1.3 Labeling and optimization thereof

Subsequently, the protein is incubated with up to 10 times molar excess of dye over protein at 4 °C for 1–3 h or overnight, depending on the observed labeling efficiency. Maleimide groups favorably react with cysteines at a pH below 7.5. At higher pH, labeling of lysines or the N-terminal amide group may occur (Brewer & Riehm, 1967; Smyth, Blumenfeld, & Konigsberg, 1964). After labeling, free dye is removed by a second round of size-exclusion chromatography (for larger dyes, the use of longer columns is advisable for optimal separation).

The exact labeling procedure (dye excess, labeling time, temperature, etc.) needs to be optimized for each case. Labeling tests using PEGm are a convenient method for optimizing maleimide labeling as they require small amounts of protein (<1 μg) and allow straightforward visualization of labeling by mobility shift assay, without the need to remove unreacted reagent (Lu & Deutsch, 2001). Reduced protein is incubated with PEGm; analogously to labeling with maleimide fluorophores, reaction with one or more cysteines will lead to an increase in the molecular mass that can be observed as a band shift via SDS-PAGE or Western blot analysis.

18.1.2.1.4 Determining labeling efficiency

The easiest method to establish the labeling efficiency is spectroscopic analysis of the ratio the A_{280} to the absorption at the fluorophore's maximum (A_{Fluor}):

$$E_{labelling} = \frac{[label]}{[receptor]} = \frac{A_{Fluor}/\varepsilon_{Fluor}}{(A_{280} - CF \times A_{Fluor})/\varepsilon_{Receptor}}$$

where ε_{Fluor} is the extinction coefficient of the fluorophore at its maximum, $\varepsilon_{Receptor}$ is the extinction coefficient of the GPCR at 280 nm, and CF is a correction factor to account for the contribution of the fluorophore to the absorption spectrum at 280 nm. For example, the contribution of AlexaFluor488 at 280 nm is 11% of its absorption at its maximum at 493 nm; hence, $CF_{Alexa488} = 0.11$. These correction factors are available from literature or can be determined experimentally from the excitation spectrum of free dye.

18.1.3 Production of liposomes

Reagents for liposome creation:

1. Stocks of appropriate lipids in chloroform (25 mg/ml)
2. Liposome buffer: 50 mM Tris pH 7.4, 100 mM NaCl, 1 mM EDTA

Once receptors have been purified, it is necessary to incorporate them into liposomes. Liposomes are spherical structures of lipids that can range in diameter from ~30 nm to ~50 μm and are classified by the number of concentric bilayers within each structure. Those that have a single layer of lipid membrane are referred to as unilamellar, whereas those with more than one layer are termed multilamellar. We tend to work with large unilamellar vesicles (LUVs) that have a size range of 100–400 nm.

18.1.3.1 Preparing the lipid

Lipids may be purchased as lyophilized powders or as liquid stocks, generally in chloroform. We find chloroform stocks easier to handle although care must be taken as the chloroform can evaporate over time, altering the concentration of lipid stock. The first step is to create a lipid mixture containing the desired constituents of the liposomes to be created. Commercially available stocks are generally available at 25 mg/ml, and we make up powdered stocks at the same concentration. The correct volumes of lipids can then be combined to create the correct molar ratio. At this point, sterols such a cholesterol can be added.

Once the lipid stock has been created, it is dried under a stream of nitrogen gas (it is preferable to avoid air to limit oxidation of the lipids) or using a rotary evaporator. In either case, a round-bottomed flask should be used and the lipid dried homogenously on the surface to create a lipid film. The film should subsequently be lyophilized overnight to ensure that any residual chloroform has been removed. Dried films can be stored under nitrogen gas at $-20\ °C$ for months.

Next, the film must be rehydrated using liposome buffer. Our standard buffer composition is illustrated although this should be determined empirically for each receptor; it is essentially the same buffer in which purification is performed but lacking detergents. Rehydration is accomplished by addition of aqueous buffer to the dried film (to a final lipid concentration of 5 mg/ml) followed by vortexing and sonication in a water bath (~1–5 min; above the phase transition temperature) to break down large lipid aggregates. The lipid film should then be freeze-thawed using liquid nitrogen 6–8 times, which helps to disrupt any multilamellar vesicles present. At this point, the lipid suspension will generally appear cloudy.

18.1.3.2 Sizing liposomes: sonication versus extrusion

Although the lipid can be used for reconstitution at this stage, it is often beneficial to size the liposomes before reconstitution. In our hands, smaller liposomes tend to be more efficient at incorporating protein. Small unilamellar vesicles (SUVs) with a diameter of 30–50 nm can be created by sonication. A water bath (above the phase transition temperature) can be used and the lipid suspension sonicated for a prolonged period of time (typically on the order of 30–60 min depending on the lipid). Alternatively, a probe sonicator can be used but care must be taken not to cause the lipids to foam; this method is more rapid than using a water bath (typically 30 s to 1 min). Probe sonication should be performed in pulses to avoid excess heating of the lipid sample, which can then lead to lipid hydrolysis. Clarification of the lipid mixture is indicative of SUV formation although the size and distribution will depend on a number of factors including lipid type, sonication power, and time. Alternatively, SUVs can be created by extrusion using a filter with a pore size of $\sim$30 nm although this can be challenging to perform for concentrated lipid stocks. However SUVs are created, they should be used immediately as their high surface tension makes them unstable leading to liposome fusion.

Our preferred method is to generate LUVs with a diameter of $\sim$100 nm via extrusion using a filter with a corresponding pore size. Lipid can be directly extruded to 100 nm or, more easily, using filters with a larger pore size, for example, 400 nm first. Clarification of the lipid solution should occur as particles are created smaller than the wavelength of visible light. Extruded liposomes can be stored at $-80\,°C$ for weeks with no apparent fusion. An excellent guide to extrusion is provided by Avanti Polar Lipids Inc. (http://www.avantilipids.com). If significant losses of lipid are observed at this stage, it is advisable to perform a phosphate assay to determine the concentration of the lipid stock used for reconstitution. To ensure that the lipid ratio is correct within the final mixture, thin layer chromatography can be used. Additionally, if required, liposomes can be sized using dynamic light scattering or a similar technique.

18.1.4 Reconstitution

Reconstitution essentially reinserts the purified GPCR of interest into a lipid bilayer of defined composition at defined lipid–protein ratios. Ideally, receptor activity is retained during the process. There are various methods of reconstitution in use; the most common of which are the dilution and dialysis method or hydrophobic adsorption using Bio-beads SM-2 (Bio-Rad) (Fig. 18.1). There are advantages to both methods, and which one to be used will depend on the nature of the sample and of the experiment for which the proteoliposomes are required.

Reconstitution by rapid dilution and dialysis can be an effective and reproducible reconstitution strategy for a relatively robust membrane protein, but the main disadvantage of this method is the time it takes for the process to be complete, as GPCRs are typically in detergent and will lose activity over time. Nevertheless, dilution and dialysis can be a relatively gentle means of reconstitution compared to Bio-beads to which both lipids and proteins can become adsorbed, changing the desired lipid–protein ratio.

Reconstitution using Bio-beads has been more reproducible in our hands (Harding et al., 2009). This is partly because of the ability to control the amount of detergent removed with each step, due to relatively well-characterized Bio-bead-to-detergent

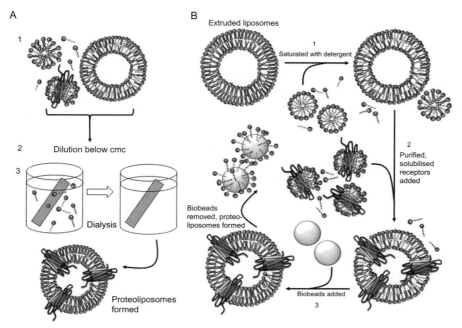

FIGURE 18.1

Reconstitution by dilution and dialysis (A) and by using Bio-beads for detergent removal (B). (A): (1) Lipids, detergent, and protein are mixed at the desired ratios and incubated for 1 h. (3) The reconstitution mixture is rapidly diluted below the cmc of the detergent. (4) The detergent monomers are removed by extensive dialysis. (B): (1) Resuspended liposomes are destabilized with detergent. (2) Detergent-solubilized protein is added to the required concentration and the mixture incubated for 1 h. (3) Detergent is removed with Bio-beads at a 10:1 Bio-bead-to-detergent ratio (w/w). (For color version of this figure, the reader is referred to the online version of this chapter.)

ratios (Rigaud et al., 1997), and also because of the relative rapidity of the method compared to dialysis, resulting in a larger proportion of functionally reconstituted receptors. However, lipid adsorption to the beads can be up to 30%, so if a low protein–lipid ratio is important, this possibility needs to be accounted for or the dialysis method used. However, size homogeneity of proteoliposomes formed using Bio-beads is far greater than for the dialysis method).

18.1.4.1 Protocol for reconstitution using Bio-beads

Reagents:

1. Stocks of extruded liposomes at 5 mg/ml in liposome buffer (50 mM Tris pH 7.4, 100 mM NaCl, 1 mM EDTA)
2. Purified protein stocks in detergent (0.1% DDM, 0.01% CHS)

3. Bio-beads SM-2 (Bio-Rad) if using Bio-bead method
4. Reagents for detergent concentration assay (optional)
5. Reagents for phospholipid concentration determination (optional)

18.1.4.1.1 Liposome preparation
Saturating the liposomes with detergent prior to reconstitution improves membrane protein incorporation into the vesicles from solution (Knol, Sjollema, & Poolman, 1998; Paternostre, Roux, & Rigaud, 1988). For each detergent and lipid mixture, there is an effective detergent–lipid molar ratio (R_{eff}) at which the lipids are saturated but not completely solubilized by the detergent. A table listing these parameters for several of the most common detergents is available in Rigaud and Levy (2003). For DDM, R_{sat} is 1 (mol/mol), and R_{sol} 1.6 (mol/mol). For a detailed description of the protocol, see Lambert, Levy, Ranck, Leblanc, and Rigaud (1998). The appropriate amount of detergent is added to the liposome solution and, in the case of DDM, allowed to incubate at room temperature with stirring for 3 h. If using alternative detergents, such as sodium cholate, incorporation into the liposomes may be more rapid.

18.1.4.1.2 Preparing Bio-beads
Bio-beads are thoroughly washed several times with methanol and then water and can be stored at 4 °C in water. Prior to use, they are washed in detergent-free buffer.

18.1.4.1.3 Incubation of lipid with receptor
After addition of the protein solution, the reaction mixture is incubated with gentle mixing for ~1 h before the addition of Bio-beads. This incubation period may be shortened if the detergent used to solubilize the lipids is deleterious to the activity of the protein.

18.1.4.1.4 Addition of Bio-beads
A wet Bio-bead-to-detergent ratio of 10 (w/w) is used, and the beads are added directly to the protein–lipid–detergent solution. The solution is incubated above the phase transition temperature of the lipids for 1–2 h and then aspirated into fresh Bio-beads and incubated with rotation overnight. Reconstitution can be monitored using turbidity measurements, and a detergent assay can be performed to confirm complete detergent removal. Proteoliposomes are isolated using sucrose density gradient centrifugation (0–35% in liposome buffer) in 5% steps. The sample is centrifuged (100,000 g; 16 h) and fractionated. Proteoliposomes may be dialyzed or pelleted to remove the sucrose.

18.1.4.1.5 Determining the lipid-to-protein ratio
The lipid-to-protein ratio used for FRET is very high (12,000:1) to avoid any artifacts in the signal due to crowding effects. Both protein and lipid are lost during the reconstitution process. Knowing the final lipid–protein ratio is useful for estimating receptor density in the proteoliposomes and thus likely resonance transfer. We use a number of ways to estimate the lipid–protein ratio in reconstituted samples. The first is a simple comparison of the A_{280} (corrected for liposome

scattering) of the protein to calculate protein concentration, compared with the lipid concentration determined from a phosphate assay (Chen, Toribara, & Warner, 1956). It is also possible to use protein assays such as the bicinchoninic acid (BCA) assay or other comparisons with bovine serum albumin, but these methods are less accurate.

18.2 FRET MEASUREMENTS AND ANALYSIS

FRET is an excellent technique for determining protein–protein interaction but care must be taken over the interpretation of results. FRET is the nonradiative transfer of energy between a donor fluorophore and an acceptor fluorophore. For a review of different fluorescence techniques for studying GPCR interactions, see Goddard and Watts (2012). FRET is distance-dependent (R^6) and as such has a defined window of operation that is suitable for biological measurements (Stryer, 1978). The efficiency of FRET between fluorophores is characterized by the Förster distance (R_0), which is the distance at which energy transfer is 50% efficient. Typically, it is only possible to measure FRET if the separation of the fluorophores is within 25% of the Förster distance. For example, the R_0 of the CFP–YFP pair is 49 Å. When CFP or YFP are linked to a receptor via a flexible linker, this will allow FRET in almost any receptor conformation (assuming a receptor diameter of ~40 Å). However, care must be taken when using small molecule fluorophores—if these have short R_0 values and are placed on distant parts of the dimer, dimerization may occur without the production of FRET.

Taking NTS1-CFP and NTS1-YFP as an example, several controls need to be generated and assayed. Empty liposomes must be produced to account for scattering and any endogenous fluorescence, which normally arises due to contamination of the solvents used in preparation of the lipids. Liposomes containing NTS1-CFP or NTS1-YFP alone should also be generated and, finally, a sample containing equimolar amounts of NTS1-CFP and NTS1-YFP. The measurements illustrated in Table 18.1 should then be made using an appropriate

Table 18.1 FRET Measurements

	Sample	Excitation (nm)	Emission Spectrum (nm)
1.	Liposomes	440	450–600
2.	Liposomes	510	520–600
3.	NTS1-CFP	440	450–600
4.	NTS1-YFP	440	450–600
5.	NTS1-YFP	510	520–600
6.	NTS1-CFP+NTS1-YFP	440	450–600
7.	NTS1-CFP+NTS1-YFP	510	520–600

Fluorescence measurements should be taken for the indicated samples and excitation and emission wavelengths. Adjust wavelength appropriately if using alternative fluorophores.

fluorimeter, for example, Varian Cary Eclipse. Note that it is possible to study more dilute samples by increasing the slit widths of the fluorimeter with the associated decrease in specificity. We tend to avoid samples where it is necessary to increase the slit widths further than 5 nm for excitation and emission. If this is the case, then more concentrated samples should be produced, for example, by ultracentrifugation followed by resuspension in a smaller volume. Low-volume cuvettes, for example, 50 μl, are ideal for these assays. Experiments should be performed in triplicate.

Once data have been collected, a number of corrections are performed (see Fig. 18.2). Firstly, the appropriate liposome background spectrum (1 or 2) is subtracted from each NTS1 containing sample of the corresponding excitation wavelength to generate spectra 3–7. This background may require scaling if the liposome fluorescence spectra have significantly different intensities to the spectra to be corrected although this should not normally be the case. Any difference likely arises due to the detergent that is present in the protein-containing samples, but not the controls, during reconstitution resulting in differently sized liposomes. It is important to scale using a region unaffected by the donor or acceptor emissions. This subtraction eliminates any effects of scattering or endogenous lipid fluorescence to create a "zero" background. Next, bleedthrough and crosstalk must be corrected for. In the following example, NTS1-CFP is the donor and NTS1-YFP the acceptor.

Bleedthrough is the emission of the donor (CFP) at the wavelength at which the acceptor (YFP) emits. This is subtracted by the following:

1. Scale the NTS1-CFP spectrum (3) so that the fluorescence at the maximum (typically ~475 nm) is equal to the fluorescence at the same wavelength in the NTS1-CFP+NTS1-YFP spectrum (6) to generate spectrum 8.
2. Subtract spectrum 8 from spectrum 6. This removes bleedthrough and the remaining signal is a combination of FRET and crosstalk (9).

Crosstalk is the direct excitation of the acceptor (YFP) with the wavelength used to excite the donor (CFP; 440 nm). This is subtracted by the following:

1. Scale the NTS1-YFP spectrum (5) so that the fluorescence at the maximum (~525 nm) is the same as in the NTS1-CFP+NTS1-YFP spectrum (7). This compares the amount of NTS1-YFP in the two samples. Determine the scaling factor required to do this (spectrum 5–spectrum 7).
2. Multiply spectrum 4 (NTS1-YFP) by this scaling factor to determine the extent of crosstalk in the NTS1-CFP+NTS1-YFP sample; this generates spectrum 10.
3. Subtract spectrum 10 (crosstalk) from spectrum 9 (FRET+crosstalk) to determine the specific FRET from this experiment (11).

Once these corrections have been made, the remaining signal is FRET. It is advisable to conduct all experiments in triplicate and to use the average of the apparent FRET efficiency values (see succeeding text).

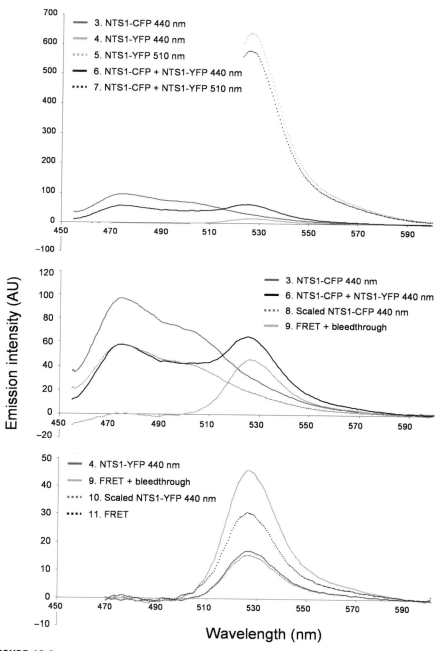

FIGURE 18.2

Processing of FRET data. After correction for liposome scattering (top panel), FRET data from reconstituted samples needs to be corrected for bleedthrough (middle panel) and crosstalk (bottom panel) as discussed in the text in more detail. (See color plate.)

18.3 INTERPRETING RESULTS

It should be noted at this point that several factors affect the amount of FRET produced. These include the distance between the fluorophores, relative fluorophore orientation, and the proportion of receptors that are dimeric (Stryer and Haugland, 1967). Therefore, if the FRET produced in different lipid environments changes, it is important to distinguish between changes in receptor architecture that affect the orientation of, or distance between, the probes and changes in the equilibrium of dimerization. If large probes with flexible linkers, such as CFP or YFP, are used, perturbations in FRET are likely due to changes in the dimerization equilibrium. However, smaller probes attached to the protein backbone are more likely to be affected by subtle changes in receptor architecture.

It is possible to quantify the FRET signal to compare different lipid environments, and this is done by calculation of apparent FRET efficiency (Harding et al., 2009; Overton and Blumer, 2002). In this case, this is simply the height of the peak in the FRET spectrum (10) compared to the total amount of YFP in the same sample (spectrum 7). Apparent FRET efficiency is also affected by the donor–acceptor ratio—we typically use a 1:1 ratio for initial assays, but higher ratios will give more efficient FRET. In a 1:1 ratio, only half of the acceptor molecules are in FRET-competent dimers due to the presence of acceptor–acceptor dimers. As the proportion of donor increases, the likelihood of an acceptor being in an acceptor–acceptor dimer decreases and FRET efficiency increases.

In Fig. 18.3, we illustrate typical results seen for the lipid-dependence of NTS1 dimerization. Brain polar lipid (BPL) is a commercially available extract (Avanti

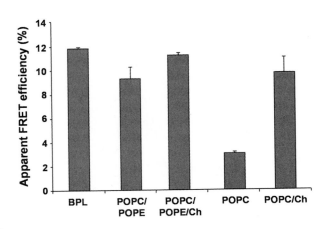

FIGURE 18.3

Lipid dependence of dimerization. Apparent FRET efficiency for NTS1 tagged with eCFP or eYFP was determined as a measure for dimerization in different lipid mixes (BPL, brain polar lipid; POPC, 1-palmitoyl-2-oleoyl-sn-glycero-3-phosphocholine; POPE, 1-palmitoyl-2-oleoyl-sn-glycero-3-phosphoethanolamine; Ch, cholesterol).

Adapted from Oates et al. (2012).

Table 18.2 Composition of Brain Polar Lipid

Component	w/w%
PC	12.6
PE	33.1
PI	4.1
PS	18.5
PA	0.8
Unknown	30.9

Details from Avanti Polar lipids. PC, phosphatidylcholine; PE, phosphatidylethanolamine; PI, phosphatidylinositol; PS, phosphatidylserine; PA, phosphatidic acid. At least some of the "unknown" is cholesterol (our observations).

Polar Lipids) that has the composition shown in Table 18.2. BPL is the best mimic of the endogenous environment of NTS1—it is unsurprising that NTS1 dimerizes efficiently in this lipid mixture. To determine the contribution of the various lipids to dimerization, we have performed FRET experiments in liposomes of defined composition. This demonstrates that presence of POPE and cholesterol strongly influences dimerization.

SUMMARY

Here, we have given details of the methods used to reconstitute a GPCR with a view to studying dimerization. The approaches could well be generic, but firstly, only few GPCRs are expressed either folded or in inclusion bodies and require refolding. Secondly, there is still too little information available at the moment to make generalizations about detergent suitability for GPCRs—we are likely to require more detergents or methods of GPCR solubilization, and some generalizations may become evident as more biophysical studies are performed that require reconstituted receptor.

References

Attrill, H., Harding, P., Smith, E., Ross, S., & Watts, A. (2009). Improved yield of a ligand-binding GPCR expressed in *E. coli* for structural studies. *Protein Expression and Purification, 64*(1), 32–38.

Bayburt, T. H., & Sligar, S. G. (2010). Membrane protein assembly into nanodiscs. *FEBS Letters, 584*(9), 1721–1727.

Bill, R. M., Henderson, P. J. F., Iwata, S., Kunji, E. R. S., Michel, H., Neutze, R., et al. (2011). Overcoming barriers to membrane protein structure determination. *Nature Biotechnology, 29*(4), 335–340.

Brewer, C. F., & Riehm, J. P. (1967). Evidence for possible nonspecific reactions between n-ethylmaleimide and proteins. *Analytical Biochemistry, 18*(2), 248–255.

Burmeister Getz, E., Xiao, M., Chakrabarty, T., Cooke, R., & Selvin, P. (1999). A comparison between the sulfhydryl reductants tris (2-carboxyethyl) phosphine and dithiothreitol for use in protein biochemistry. *Analytical Biochemistry, 273*(1), 73–80.

Chen, P. S., Jr., Toribara, T. Y., & Warner, H. (1956). Microdetermination of phosphorus. *Analytical Chemistry, 28*(11), 1756–1758.

Ciruela, F., Vallano, A., Arnau, J. M., Sanchez, S., Borroto-Escuela, D. O., Agnati, L. F., et al. (2010). G protein-coupled receptor oligomerization for what? *Journal of Receptors and Signal Transduction, 30*(5), 322–330.

Espagne, A., Erard, M., Madiona, K., Derrien, V., Jonasson, G., Lévy, B., et al. (2011). Cyan fluorescent protein carries a constitutive mutation that prevents its dimerization. *Biochemistry, 50*(4), 437–439.

Evers, T. H., van Dongen, E M W M, Faesen, A. C., Meijer, E. W., & Merkx, M. (2006). Quantitative understanding of the energy transfer between fluorescent proteins connected via flexible peptide linkers. *Biochemistry, 45*(44), 13183–13192.

Goddard, A. D., & Watts, A. (2012). Contributions of fluorescence techniques to understanding G protein-coupled receptor dimerisation. *Biophysical Reviews, 4*(4), 291–298.

Grisshammer, R. (2009). Purification of recombinant G-protein-coupled receptors. *Methods in Enzymology, 463*, 631–645.

Han, J. C., & Han, G. Y. (1994). A procedure for quantitative determination of tris (2-carboxyethyl) phosphine, an odorless reducing agent more stable and effective than dithiothreitol. *Analytical Biochemistry, 220*(1), 5–10.

Harding, P., Attrill, H., Boehringer, J., Ross, S., Wadhams, G., Smith, E., et al. (2009). Constitutive dimerization of the G-protein coupled receptor, neurotensin receptor 1, reconstituted into phospholipid bilayers. *Biophysical Journal, 96*(3), 964–973.

Harding, P., Attrill, H., Ross, S., Koeppe, J., Kapanidis, A., & Watts, A. (2007). Neurotensin receptor type 1: Escherichia coli expression, purification, characterization and biophysical study. *Biochemical Society Transactions, 35*, 760–763.

Hoffmann, C., Gaietta, G., Bünemann, M., Adams, S. R., Oberdorff-Maass, S., Behr, B., et al. (2005). A flash-based fret approach to determine G protein-coupled receptor activation in living cells. *Nature Methods, 2*(3), 171–176.

Jackson, A. M., Boutell, J., Cooley, N., & He, M. (2004). Cell-free protein synthesis for proteomics. *Briefings in Functional Genomics & Proteomics, 2*(4), 308–319.

Jacoby, E., Bouhelal, R., Gerspacher, M., & Seuwen, K. (2006). The 7 TM G-protein-coupled receptor target family. *ChemMedChem, 1*(8), 760–782.

Knol, J., Sjollema, K., & Poolman, B. (1998). Detergent-mediated reconstitution of membrane proteins. *Biochemistry, 37*(46), 16410–16415.

Lambert, O., Levy, D., Ranck, J.-L., Leblanc, G., & Rigaud, J.-L. (1998). A new "Gel-like" Phase in dodecyl maltoside-lipid mixtures: Implications in solubilization and reconstitution studies. *Biophysical Journal, 74*(2), 918–930.

Lu, J., & Deutsch, C. (2001). Pegylation: A method for assessing topological accessibilities in kv1.3. *Biochemistry, 40*(44), 13288–13301.

Mujumdar, R. B., Ernst, L. A., Mujumdar, S. R., Lewis, C. J., & Waggoner, A. S. (1993). Cyanine dye labeling reagents: Sulfoindocyanine succinimidyl esters. *Bioconjugate Chemistry, 4*(2), 105–111.

Oates, J., Faust, B., Attrill, H., Harding, P., Orwick, M., & Watts, A. (2012). The role of cholesterol on the activity and stability of neurotensin receptor 1. *Biochimica et Biophysica Acta (BBA) – Biomembranes, 1818*(9), 2228–2233.

Oates, J., & Watts, A. (2011). Uncovering the intimate relationship between lipids, cholesterol and GPCR activation. *Current Opinion in Structural Biology*, *21*(6), 802–807.

Oldham, W., & Hamm, H. (2008). Heterotrimeric G protein activation by G-protein-coupled receptors. *Nature Reviews. Molecular Cell Biology*, *9*(1), 60–71.

Orwick, M. C., Judge, P. J., Procek, J., Lindholm, L., Graziadei, A., Engel, A., et al. (2012). Detergent-free formation and physicochemical characterization of nanosized lipid–polymer complexes: Lipodisq. *Angewandte Chemie*, *124*(19), 4731–4735.

Overton, M. C., & Blumer, K. J. (2002). Use of fluorescence resonance energy transfer to analyze oligomerization of G-protein-coupled receptors expressed in yeast. *Methods*, *27*, 324–332.

Panchuk-Voloshina, N., Haugland, R. P., Bishop-Stewart, J., Bhalgat, M. K., Millard, P. J., Mao, F., et al. (1999). Alexa dyes, a series of new fluorescent dyes that yield exceptionally bright, photostable conjugates. *Journal of Histochemistry and Cytochemistry*, *47*(9), 1179–1188.

Patel, H. H., Murray, F., & Insel, P. A. (2008). G-protein-coupled receptor-signaling components in membrane raft and caveolae microdomains. *Protein–protein interactions as new drug targets*. (pp. 167–184). Berlin Heidelberg: Springer.

Paternostre, M. T., Roux, M., & Rigaud, J. L. (1988). Mechanisms of membrane protein insertion into liposomes during reconstitution procedures involving the use of detergents. 1. Solubilization of large unilamellar liposomes (prepared by reverse-phase evaporation) by triton x-100, octyl glucoside, and sodium cholate. *Biochemistry*, *27*(8), 2668–2677.

Rigaud, J.-L., & Levy, D. (2003). Reconstitution of membrane proteins into liposomes. *Methods in Enzymology*, *372*, 65–86.

Rigaud, J.-L., Mosser, G., Lacapere, J.-J., Olofsson, A., Levy, D., & Ranck, J.-L. (1997). Bio-beads: An efficient strategy for two-dimensional crystallization of membrane proteins. *Journal of Structural Biology*, *118*(3), 226–235.

Smyth, D. G., Blumenfeld, O. O., & Konigsberg, W. (1964). Reactions of n-ethylmaleimide with peptides and amino acids. *Biochemical Journal*, *91*(3), 589.

Stryer, L. (1978). Fluorescence energy transfer as a spectroscopic ruler. *Annual Review of Biochemistry*, *47*(1), 819–846.

Stryer, L., & Haugland, R. P. (1967). Energy transfer: a spectroscopic ruler. *Proceedings of the National Academy of Sciences*, *58*, 719–726.

Tapaneeyakorn, S., Goddard, A. D., Oates, J., Willis, C. L., & Watts, A. (2011). Solution- and solid-state NMR studies of GPCRs and their ligands. *Biochimica et Biophysica Acta (BBA) – Biomembranes*, *1808*(6), 1462–1475.

Tsien, R. Y. (1998). The green fluorescent protein. *Annual Review of Biochemistry*, *67*(1), 509–544.

van Meer, G. (2011). Dynamic transbilayer lipid asymmetry. *Cold Spring Harbor Perspectives in Biology*, *3*(5), http://dx.doi.org/10.1101/cshperspect.a004671.

Vassilatis, D. K., Hohmann, J. G., Zeng, H., Li, F., Ranchalis, J. E., Mortrud, M. T., et al. (2003). The G protein-coupled receptor repertoires of human and mouse. *Proceedings of the National Academy of Sciences*, *100*(8), 4903–4908.

Zhao, Q., & Wu, B.-L. (2012). Ice breaking in GPCR structural biology. *Acta Pharmacologica Sinica*, *3*, 1–11.

CHAPTER

Monitoring Peripheral Protein Oligomerization on Biological Membranes

19

Robert V. Stahelin[*,†]

[*]*Department of Biochemistry and Molecular Biology, Indiana University School of Medicine-South Bend, South Bend, Indiana, USA*
[†]*Department of Chemistry and Biochemistry, University of Notre Dame, Notre Dame, Indiana, USA*

CHAPTER OUTLINE

Introduction 360
19.1 Materials and Methods 362
    19.1.1 Cell Maintenance, Transfection, and Observation 362
    19.1.2 Cellular Imaging and Analysis 363
    19.1.3 Imaging and Analysis at the Plasma Membrane 363
19.2 Considerations 365
Summary and Conclusion 366
Acknowledgments 369
References 369

Abstract

Peripheral proteins transiently interact with cellular membranes where they regulate important cellular events such as signal transduction. A number of peripheral proteins harbor lipid-binding modules that not only bind selectively with nanomolar affinity to biological membranes but also oligomerize on the membrane surface. In some cases, specific lipid binding or specific lipid compositions can induce peripheral protein oligomerization on cellular membranes. These oligomers serve different roles in biological signaling such as regulating protein–protein interactions, induction of membrane bending, or facilitating membrane scission. A number of technologies have been employed to study protein oligomerization with fluctuation analysis of fluorescently labeled molecules recently developed for use with commercial laser-scanning microscopes. In this chapter, the approach of raster image correlation spectroscopy coupled with number and brightness (N&B) analysis to investigate protein oligomerization on cellular membranes in live cells is presented.

Important considerations are discussed for designing experiments, collecting data, and performing analysis. N&B analysis provides a robust method for assessing membrane binding and assembly properties of peripheral proteins and lipid-binding modules.

INTRODUCTION

Fluctuation analysis has been used to monitor protein oligomerization (Barnwal, Devi, Agarwal, Sharma, & Chary, 2011; Wang, Graveland-Bikker, de Kruif, & Robillard, 2004) as it serves as the basis for dynamic light scattering and fluorescence correlation spectroscopy (FCS). While dynamic light scattering requires many molecules to produce a signal, FCS can detect single molecules in complex mixtures such as live cells or solutions containing lipids and proteins. FCS was developed in the 1970s (Magde, Elson, & Webb, 1972, 1974) and can detect fluctuation changes in fluorescence while characterizing molecular brightness of the particles. FCS has become more applicable to biologists and biochemists as its use was applied to live cell studies (Berland, So, & Gratton, 1995; Schwille, Bieschke, & Oehlenschläger, 1997; Schwille, Korlach, & Webb, 1999), and more recently, scanning confocal microscopes have been used (Digman, Brown, et al., 2005; Digman, Sengupta, et al., 2005). Fluorescence intensity fluctuations are often due to protein conformation changes, protein binding to immobile or less mobile fractions, or the entry and exit of molecules through the area being imaged. In FCS, a fixed observation volume is used to detect fluctuations in fluorescence intensity as single molecules pass through the detection area. However, FCS alone does not account for the location of fluorescent particles as a function of time so a spatial correlation approach can be used in order to associate changes in fluorescent amplitude at one spot with changes in nearby positions. A change in the correlation function of FCS can be performed with raster image correlation spectroscopy (RICS) (Digman & Gratton, 2009; Digman, Stakic, & Gratton, 2013; Rossow, Sasaki, Digman, & Gratton, 2010) where the position of the observation volume is changed with respect to time. This means that the correlation is dependent upon how fast the fluorescent molecules are moving as well as how fast the observation volume is changing in a time-dependent manner.

RICS can be used to detect and measure fluorescent dynamics in complex environments such as the cellular cytoplasm and membrane organelles. While determination of protein oligomerization and protein complex stoichiometry is a challenging task in live cells, it is an important one as many biological processes often depend upon protein clustering (Choi, Zareno, Digman, Gratton, & Horwitz, 2011; Digman, Brown, Horwitz, Mantulin, & Gratton, 2008; Digman, Wiseman, Choi, Horwitz, & Gratton, 2009). Recently, determination of fluorescently labeled protein clustering has become much more robust with number and brightness (N&B) analysis (Digman, Brown, et al., 2008, Digman, Dalal, Horwitz, & Gratton, 2008). N&B

analysis is based upon moment analysis and allows for measurement of the average number of molecules as well as brightness in each pixel of a fluorescent microscopy image (Digman, Brown, et al., 2008, Digman, Dalal, Horwitz, & Gratton, 2008). N&B utilizes the first and second moment of amplitude fluctuations from the histogram of amplitude oscillations, which is originally derived from the photon-counting histogram method (Chen, Muller, So, & Gratton, 1999). The average brightness of a particle is determined from the ratio of variance to intensity at each pixel. Fluctuating particles can be determined by dividing the average intensity by the brightness at each pixel. For particles fluctuating in the focal volume, the variance is proportional to the square of the particle brightness; however, the variance of the immobile particles and detector noise is proportional to the intensity of these components. Thus, only fluorescent fluctuations that are dependent upon the mobile particles have a ratio of the variance to intensity > 1. Aggregation of mobile fluorescent particles can then be determined from a brightness map with pixel resolution.

For an electron-multiplying CCD camera, the following equations are used to compute the N&B:

$$N = \frac{(\langle I \rangle - \text{offset})^2}{\sigma^2 - \sigma_0^2} \qquad (19.1)$$

$$B = \frac{\sigma^2 - \sigma_0^2}{\langle I \rangle - \text{offset}} \qquad (19.2)$$

In these equations, N and B are the apparent number and the brightness of the molecule, $\langle I \rangle$ is the average intensity of a pixel throughout a stack of frames (time average), σ^2 is the variance, while offset and σ_0^2 are properties that depend on the camera hardware. With these parameters properly calibrated, the distribution of the brightness of each individual pixel in the image of a section of a cell can be investigated. This approach is therefore useful to study interactions of fluorescently labeled proteins that occur on cellular membranes.

Cellular membranes contain hundreds if not thousands of different lipids, which are highly dynamic. Specific lipids in the membrane bilayer (Lemmon, 2008; Stahelin, 2009) or lipid anionic charge (Yeung et al., 2006) can regulate recruitment of peripheral proteins to the membrane interface. These signaling cues are vital to membrane trafficking and signal transduction as it is estimated that nearly half of all proteins in the human genome are localized in or on membranes. A large number of peripheral proteins are recruited to cellular membranes through modular lipid-binding domains that are often found in signal transduction and membrane-trafficking proteins (Cho & Stahelin, 2005). These modular lipid-binding domains can bind, often with nanomolar affinity, to membranes containing a specific lipid ligand or recognize membrane physical properties such as charge or curvature (Lemmon, 2008). Binding to the membrane interface and restricting the protein to two dimensions reduce dimensionality and provide a platform for signal transduction to occur. In some cases, peripheral proteins may interact or oligomerize on the membrane interface, an event that is important in biological processes such as generation of membrane curvature changes (Zimmerberg & Kozlov, 2006), membrane scission (Klein, Lee, Frank, Marks, & Lemmon, 1998; Ross et al., 2011), and assembly and egress of viral proteins from the host cell (Adu-Gyamfi, Digman, Gratton, & Stahelin, 2012; Hoenen, Jung, Herwig, Groseth, & Becker, 2010; Murray et al., 2005). Because high-affinity interactions between these peripheral proteins regulate lipid signaling and trafficking events from cellular membranes, accurately characterizing lipid specificity and membrane affinity can establish how the membrane-interface signals throughout the cell. Technologies aimed at measuring the assembly and stoichiometry of lipid–protein and protein–protein interactions on or near the membrane interface are important for revealing molecular details of the underlying cellular architecture.

The biophysical approaches discussed earlier have been applied to investigate the oligomerization state of the fission protein dynamin (Ross et al., 2011) as well as oligomerization of the lipid-binding ENTH (Yoon et al., 2010) and BAR (Yoon, Zhang, & Cho, 2012) domains involved in membrane curvature generation and sensing. Additionally, they have been useful in yielding mechanistic information of assembly and oligomerization of Annexin A4 (Crosby et al., 2013), oligomerization of the voltage-dependent anion channel in response to phosphatidylglycerol binding (Betaneli, Petrov, & Schwille, 2012) as well as oligomerization, assembly, and egress of the Ebola virus matrix protein (Adu-Gyamfi et al., 2012, 2013; Soni, Adu-Gyamfi, Yong, Jee, & Stahelin, 2013). In this chapter, I will describe how RICS and N&B can be used to study assembly of peripheral proteins on biological membranes.

19.1 MATERIALS AND METHODS

19.1.1 Cell maintenance, transfection, and observation

Human embryonic kidney (HEK293) and Chinese hamster ovary-K1 (CHO-K1) cells were used to study the assembly and oligomerization of the Ebola virus matrix protein VP40 (Adu-Gyamfi et al., 2012, 2013). HEK293 cells were cultured and maintained at 37 °C in a 5% CO_2 humidified incubator supplemented with Dulbecco's Modified Eagle Medium (DMEM) (low glucose) containing 10% FBS and 1% Pen/Strep. After trypsinization, cells were transferred from a T-25 tissue culture flask to an 8-well plate used for imaging. Cells were then grown to 50–80% confluency and transfected with 1 μg DNA/dish using Lipofectamine 2000 according to the manufacturer's protocol. Cells were imaged between 12 and 20 h posttransfection.

CHO-K1 cells were cultured and maintained at 37 °C in a 5% CO_2 humidified incubator supplemented with DMEM/F12 (low glucose) containing 10% FBS and 1% Pen/Strep. After trypsinization, cells were transferred from a T-25 tissue culture flask to an 8-well plate used for imaging. Cells were then grown to 50–80% confluency and transfected with 1 μg DNA/dish using Lipofectamine LTX according to the manufacturer's protocol. Both HEK293 and CHOK-1 cells were imaged using a Zeiss LSM 710 confocal microscope using a Plan-Apochromat 63x 1.4 NA oil objective. The 488 nm line of the Ar ion laser was used for excitation of enhanced green fluorescent protein (EGFP). The laser power was maintained at 1% throughout the experiment with the emission collected through a 493–556 nm filter.

19.1.2 Cellular imaging and analysis

Investigation of peripheral protein association with cellular membranes requires fluorescent labeling of the protein of interest and a technique that can scan fast enough to detect the fluorescent protein in different pixels of an image. RICS is a good method for meeting these objectives as the scanning movement creates a space–time matrix of pixels within each image. From the stack of frames collected for a live cell, a spatial correlation map can be obtained to determine the particle's diffusion coefficient. Thus, when the spatial correlation map is obtained at different subcellular locations, the diffusion coefficient can be determined at different cellular sites. This provides information on how the membranes, organelles, or other cellular components restrict the movement of fluorescently labeled proteins. On commercial confocal microscopes, the spatial and temporal sampling time of the laser beam (pixel dwell time) is known as is the scan time between scan lines and time between images. Thus, using RICS allows for generation of spatial–temporal maps of dynamics, occurring across the living cell. RICS data was acquired on a commercial laser-scanning confocal microscope (Zeiss LSM 710 inverted microscope) using a Plan-Apochromat 63x 1.4 NA oil objective. The 488 nm line of the Ar ion laser was used for excitation of EGFP. The laser power was maintained at 1% throughout the experiment with the emission collected through a 493–556 nm filter. The data were collected as images of 256×256 pixels with a pixel dwell time of 12.6 µs. RICS analysis was done with SimFCS software using 100 frames from the series of images (Adu-Gyamfi et al., 2012).

19.1.3 Imaging and analysis at the plasma membrane

In order to more carefully and selectively image VP40 assembly on the inner leaflet of the plasma membrane, total internal reflection microscopy (TIRF) imaging was employed (Adu-Gyamfi et al., 2012, 2013). TIRF imaging was performed using a homebuilt TIRF imaging system (model No. IX81 microscope (Olympus, Melville, NY) as described previously (Ross et al., 2011). Briefly, images were collected using a Cascade 512B EMCCD camera. Samples were excited with the 488 nm line from an Ar ion laser (Melles Griot, Albuquerque, NM) through a 60×1.45 NA oil objective (Olympus). To ensure cell integrity, cells were maintained at 37 °C using a thermostated stage (Tokai Hit, Shizuoka, Japan). Images were collected at 256×256 pixels with a 50 ms exposure time per frame with 4000 total frames collected. Images were saved as 16 bit unsigned and imported into the SimFCS software (Laboratory for Fluorescence Dynamics, Irvine, CA). TIRF image series were analyzed using SimFCS (Laboratory for Fluorescence Dynamics, Irvine, CA). For N&B analysis, 512 frames were analyzed per image series. HEK293 cells expressing monomeric EGFP were used as a brightness standard and were imaged under the same conditions as EGFP-VP40 and respective mutations (see Fig. 19.1).

The brightness of the EGFP was used as the brightness of the monomer as described previously (Adu-Gyamfi et al., 2012) where the average brightness of a monomer was 0.104. This allowed for selection of oligomeric size based upon multiple of the monomer (see Figs. 19.1 and 19.2). Thus, the selection window for each species is based upon the average brightness, which will yield an average population of each species in the respective area of analysis (see Figs. 19.1 and 19.2).

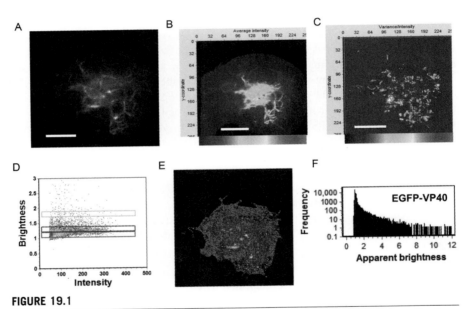

FIGURE 19.1

Brightness analysis of VP40 in HEK293 cells. (A) Membrane protrusion sites emanating from the PM were inspected with total internal reflection microscopy (TIRF) microscopy.
(B) TIRF average intensity image of a HEK293 cell transfected with plasmid expressing EGFP demonstrates sites of signal enrichment and a number of sites of membrane protrusions and viral egress. (C) Brightness image (variance/intensity) of the same cell demonstrates the enriched sites of VLP egress where significant EGFP signal (red) is detected.
(D) Brightness versus intensity plot displaying monomers (brightness of 1.104) (red box), dimers (blue box), and hexamers (green box). (E) Brightness distribution of VP40 with selected pixels from D displaying localization of monomers (red), dimers (blue), and hexamers (green). (F) Frequency versus apparent brightness plot demonstrates the extensive oligomerization of VP40 at or near the PM of HEK293 cells. The apparent brightness of a monomer is 1.104 indicating the significant frequency of a monomer but extensive enrichment of oligomers up to an apparent brightness of 12. Scale bar = 18 μm. (See color plate.)

This research was originally published in Adu-Gyamfi et al. (2013).

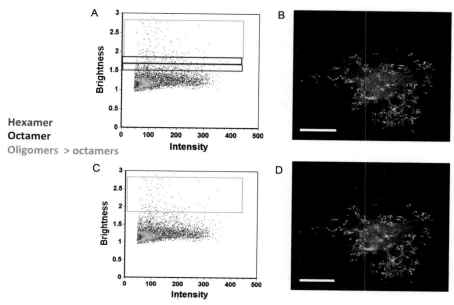

FIGURE 19.2

Brightness versus intensity analysis of EGFP-VP40 in HEK293 cells displaying oligomers. (A) Brightness versus intensity plot of the HEK293 cell shown in Fig. 19.1 highlighting hexamers (red box), octamers (blue box), and oligomers larger than octamers (green box). (B) Brightness distribution of VP40 with selected pixels from A displaying hexamer (red), octamers (blue), and oligomers larger than octamers (green). Oligomeric VP40 structures are enriched on the PM and filaments protruding from the cell PM. (C) Brightness versus intensity plot of the HEK293 cell shown in Fig. 19.1 highlighting oligomers larger than octamers (green box). (D) Brightness distribution of VP40 with selected pixels from C displaying oligomers larger than octamers (green). Scale bar = 18 μm. (For interpretation of the references to color in this figure legend, the reader is referred to the online version of this chapter.)

This research was originally published in Adu-Gyamfi et al. (2013).

19.2 CONSIDERATIONS

Cells and organelles are dynamic molecules and move during the scanning periods. To counter this issue, RICS analysis was performed with a moving average of ten frames in the SimFCS software. This method can account for motions of cells or organelles that are on the timescale of 10 s or longer. Fluorophores can bleach and although RICS analysis is independent of bleaching (because the correlation is determined in each frame), bleaching in N&B analysis can lead to problematic data even with changes as small as 10% of the average intensity. Bleaching was verified for each experiment and did not exceed 5% of the original intensity in analyzed samples. It is also important to note that N&B analysis accounts for limitations such as autofluorescence, scattering, bright immobile particles, or fast-moving particles that are dim by calculating the total variance. Variance is proportional to the particle brightness for particles fluctuating in the focal volume; however, the variance of the immobile particles, scattering, autofluorescence, and detector noise is proportional to the intensity of these components. Thus, only fluorescent fluctuations that are dependent upon the mobile particles have a ratio of the variance to intensity > 1. Brightness maps then allow for pixel resolution of the clustering of fluorescently labeled proteins (see Figs. 19.1 and 19.2).

It is also important to consider that a fluorescent molecule can move faster than the scanner during line scanning (Digman et al., 2013). This would mean that fast-moving particles would not be counted accurately and the average diffusion coefficient would be lower than the true diffusion coefficient. Digman and colleagues recommend keeping the pixel time faster and pixel size larger to detect the fast-moving particles before the next line scan (Digman et al., 2013). Typical settings of a pixel size of 0.05 μm and a slow pixel dwell time (25 μs) can detect particles diffusing in cells with diffusion coefficients of ~ 20 $\mu m^2/s$ (Digman et al., 2013). The spatial resolution of RICS is usually larger than the size of a diffraction limited area but smaller than the entire cellular cytoplasm meaning multiple boxes around a cell can be examined independently. This is advantageous for collecting large data sets from single or multiple cell measurements, which is especially important when determining N&B of particles or the molecular stoichiometry of a protein complex (Digman et al., 2009).

SUMMARY AND CONCLUSION

RICS and N&B analysis are used to monitor the diffusion of fluorescently labeled molecules in live cells or in solution. Together, they can measure the formation of protein oligomers and their stoichiometry while determining the number of molecules that are both free and in complexes. These techniques are advantageous for studying oligomerization in live cells as harsh conditions do not have to be used to lyse the cells and assess oligomerization. Imaging fluorescent fluctuations also does not require the large sample size that would be required for performing size exclusion chromatography with multiangle light scattering, which can accurately determine oligomerization state (Wyatt, 1998). Of course, creating fusion proteins with fluorescent tags can influence cellular membrane binding or oligomerization so independent methods should be employed to verify that tagged and untagged proteins behave similarly. This approach has been useful and informative for investigating the basis of assembly and oligomerization of the Ebola virus matrix protein VP40, which regulates egress of the virus from human cells (Adu-Gyamfi et al., 2012, 2013).

Oligomerization of VP40 is a critical step in the assembly and replication of the Ebola virus. Inhibition of this step of the viral life cycle has been found to

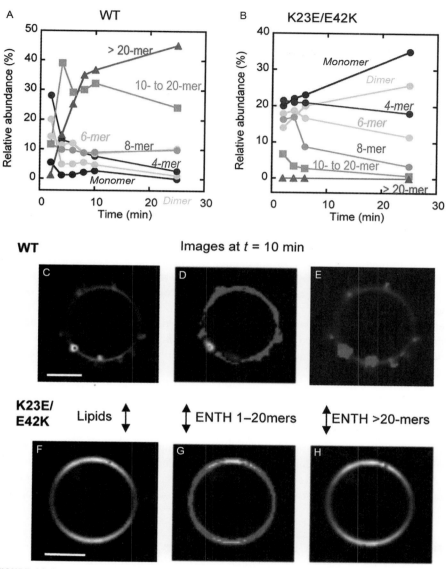

FIGURE 19.3

Number and brightness analysis of raster-scanned images of giant unilamellar vesicle (GUV) tubulation by ENTH domain. (A) The time course of the relative abundance of different aggregates for epsin 1 ENTH WT. (B) The time course of the relative abundance of different aggregates for epsin ENTH K23E/E42K mutant. (C) A representative image of POPC/POPE/POPS/PtdIns(4,5)P$_2$/Rh-PE GUV shown by Rh-PE fluorescence after a 10-min treatment with epsin 1 ENTH WT. (D) Distribution of ENTH WT aggregates (monomer to 20-mer) was superimposed onto the image of GUV. (E) Distribution of ENTH WT aggregates

Continued

downregulate the formation and release of new virus (Hoenen et al., 2010). N&B analysis demonstrated that although oligomers of VP40 were detected on the plasma membrane, the majority of plasma membrane associated VP40 was monomeric (see Figs. 19.1 and 19.2) (Adu-Gyamfi et al., 2012). This was not particularly surprising as the monomers may continuously be recruited from the cytosol to serve as building blocks for self-multimerization into larger structures required for membrane bending and plasma membrane egress. We also observed that oligomers (hexamers, octamers, and larger oligomers) were found enriched at the tips of the cells in filamentous protruding structures (see Fig. 19.2) (Adu-Gyamfi et al., 2012, 2013). Filamentous protrusion sites are thought to be the sites of egress for Ebola and other filamentous viruses such as Marburg. N&B analysis of an oligomerization-deficient mutation supported this notion as when oligomerization was reduced, detectable membrane protrusion sites were no longer observed (Adu-Gyamfi et al., 2012, 2013).

N&B analysis can also be applied to peripheral protein interactions *in vitro*. Cho and colleagues have used giant unilamellar vesicles (GUVs) to investigate the oligomerization of BAR and ENTH domains (Yoon et al., 2010, 2012) on the surface of membranes. Here, they were able to observe large oligomeric complexes that are required to induce membrane tubulation from GUVs (see Fig. 19.3).

Multiple fluorophores also can be used in live cells and provide a robust opportunity to dissect molecular complexes with the RICS and N&B methods (Digman et al., 2009, 2013), which would be a boon to monitoring protein–protein interactions or oligomerization induced by lipid membranes. This method has already been applied to resolve the interactions between endophilin and dynamin in live cells (Digman et al., 2013).

In closing, raster scanning allows for stacks of images to be generated to study assembly of fluorescently labeled proteins or observation and stoichiometry of complexes to be determined when two or more fluorescently labeled molecules are used. Because fluorescent fluctuation analysis is performed among pixels and frames in a long scan, data on particle diffusion, number of fluorescent molecules in a pixel, and their brightness can be determined. Perhaps most notably, these types of data can be collected on commercial laser-scanning microscopes sans modification of the hardware or software. The growing availability of these commercial

FIGURE 19.3—Cont'd (>20-mer) was superimposed onto the image of GUV. (F) A representative image of POPC/POPE/POPS/PtdIns(4,5)P_2/Rh-PE GUV after a 10-min treatment with epsin 1 ENTH K23E/E42K. (G) Distribution of ENTH K23E/E42K (monomer to 20-mer) was superimposed onto the image of GUV. (H) Distribution of ENTH K23E/E42K aggregates (>20-mer) was superimposed onto the image of GUV. All measurements were performed at 37 °C using POPC/POPE/POPS/PtdIns(4,5)P_2/Rh-PE (46.5:30:20:3:0.5) GUV in 20 mm Tris–HCl buffer, pH 7.4, with 0.16 m KCl solution. Protein concentration was 0.5 μm. White bars, 10 μm. (See color plate.)

This research was originally published in Yoon et al. (2010). © 2010 The American Society for Biochemistry and Molecular Biology.

instruments in core facilities may increase the use of the discussed biophysical analysis to study and understand molecular interactions and biological processes occurring on cellular membranes.

Acknowledgments

The author would like to thank Emmanuel Adu-Gyamfi, Enrico Gratton, and Michelle A. Digman for helpful discussions and their contributions to the original work. Research in the Stahelin lab is supported by the NIH (AI081077), NSF (7112361), and American Heart Association (GRNT12080254).

References

Adu-Gyamfi, E., Digman, M. A., Gratton, E., & Stahelin, R. V. (2012). Investigation of Ebola VP40 assembly and oligomerization in live cells using the number and brightness analysis. *Biophysical Journal*, *102*, 2517–2525.

Adu-Gyamfi, E., Soni, S. P., Xue, Y., Digman, M. A., Gratton, E., & Stahelin, R. V. (2013). The Ebola virus matrix protein penetrates into the plasma membrane: A key step in viral protein 40 (VP40) oligomerization and viral egress. *Journal of Biological Chemistry*, *288*, 5779–5789.

Barnwal, R. P., Devi, K. M., Agarwal, G., Sharma, Y., & Chary, K. V. (2011). Temperature-dependent oligomerization in M-crystallin: Lead or lag toward cataract, an NMR perspective. *Proteins*, *79*, 569–580.

Berland, K. M., So, P. T., & Gratton, E. (1995). Two-photon fluorescence correlation spectroscopy: Method and application to the intracellular environment. *Biophysical Journal*, *68*, 694–701.

Betaneli, V., Petrov, E. P., & Schwille, P. (2012). The role of lipids in VDAC oligomerization. *Biophysical Journal*, *102*, 523–531.

Chen, Y., Muller, J. D., So, P. T., & Gratton, E. (1999). The photon counting histogram in fluorescence fluctuation spectroscopy. *Biophysical Journal*, *77*, 553–567.

Cho, W., & Stahelin, R. V. (2005). Membrane-protein interactions in cell signaling and membrane trafficking. *Annual Review of Biophysics and Biomolecular Structure*, *34*, 119–151.

Choi, C. K., Zareno, J., Digman, M. A., Gratton, E., & Horwitz, A. R. (2011). Cross-correlated fluctuation analysis reveals phosphorylation-regulated paxillin-FAK complexes in nascent adhesions. *Biophysical Journal*, *100*, 583–592.

Crosby, K. C., Postma, M., Hink, M. A., Zeelenberg, C. H. C., Adjobo-Hermans, M. J. W., & Gadella, T. W. J. (2013). Quantitative analysis of self-association and mobility of annexin A4 at the plasma membrane. *Biophysical Journal*, *104*, 1875–1885.

Digman, M. A., Brown, C. M., Horwitz, A. R., Mantulin, W. W., & Gratton, E. (2008). Paxillin dynamics measured during adhesion assembly and disassembly by correlation spectroscopy. *Biophysical Journal*, *94*, 2819–2831.

Digman, M. A., Brown, C. M., Sengupta, P., Wiseman, P. W., Horwitz, A. R., & Gratton, E. (2005). Measuring fast dynamics in solutions and cells with a laser scanning microscope. *Biophysical Journal*, *89*, 1317–1327.

Digman, M. A., Dalal, R., Horwitz, A. F., & Gratton, E. (2008). Mapping the number of molecules and brightness in the laser scanning microscope. *Biophysical Journal*, *94*, 2320–2332.

Digman, M. A., & Gratton, E. (2009). Analysis of diffusion and binding in cells using the RICS approach. *Microscopy Research and Technique*, *72*, 323–332.

Digman, M. A., Sengupta, P., Wiseman, P. W., Brown, C. M., Horwitz, A. R., & Gratton, E. (2005). Fluctuation correlation spectroscopy with a laser-scanning microscope: Exploiting the hidden time structure. *Biophysical Journal*, *88*, L33–L36.

Digman, M. A., Stakic, M., & Gratton, E. (2013). Raster image correlation spectroscopy and number and brightness analysis. *Methods in Enzymology*, *518*, 121–144.

Digman, M. A., Wiseman, P. W., Choi, C., Horwitz, A. R., & Gratton, E. (2009). Stoichiometry of molecular complexes at adhesions in living cells. *Proceedings of the National Academy of Sciences of the United States of America*, *106*, 2170–2175.

Hoenen, T., Jung, S., Herwig, A., Groseth, A., & Becker, S. (2010). Oligomerization of Ebola virus VP40 is essential for particle morphogenesis and regulation of viral transcription. *Journal of Virology*, *84*, 7053–7063.

Klein, D. E., Lee, A., Frank, D. W., Marks, M. S., & Lemmon, M. A. (1998). The pleckstrin homology domain of dynamin isoforms require oligomerization for high affinity phosphoinositide binding. *Journal of Biological Chemistry*, *273*, 27725–27733.

Lemmon, M. A. (2008). Membrane recognition by phospholipid-binding domains. *Nature Reviews in Molecular and Cellular Biology*, *9*, 99–111.

Magde, D., Elson, E., & Webb, W. W. (1972). Thermodynamic fluctuations in a reacting system – Measurement by fluorescence correlation spectroscopy. *Physical Review Letters*, *29*, 705–708.

Magde, D., Elson, E., & Webb, W. W. (1974). Fluorescence correlation spectroscopy. II. An experimental realization. *Biopolymers*, *13*, 29–61.

Murray, P. S., Li, Z., Wang, J., Tang, C. L., Honig, B., & Murray, D. (2005). Retroviral matrix domains share electrostatic homology: Models for membrane binding function throughout the viral life cycle. *Structure*, *13*, 1521–1531.

Ross, J. A., Digman, M. A., Wang, L., Gratton, E., Albanesi, J. P., & Jameson, D. M. (2011). Oligomerization state of dynamin 2 in cell membranes using TIRF and number and brightness analysis. *Biophysical Journal*, *100*, L15–L17.

Rossow, M. J., Sasaki, J. M., Digman, M. A., & Gratton, E. (2010). Raster image correlation spectroscopy in live cells. *Nature Protocols*, *5*, 1761–1774.

Schwille, P., Bieschke, J., & Oehlenschläger, F. (1997). Kinetic investigations by fluorescence correlation spectroscopy: The analytical and diagnostic potential of diffusion studies. *Biophysical Chemistry*, *66*, 211–228.

Schwille, P., Korlach, J., & Webb, W. W. (1999). Fluorescence correlation spectroscopy with single molecule sensitivity on cell and model membranes. *Cytometry*, *36*, 176–182.

Soni, S. P., Adu-Gyamfi, E., Yong, S. S., Jee, C. S., & Stahelin, R. V. (2013). The Ebola virus matrix protein deeply penetrates the plasma membrane: An important step in viral egress. *Biophysical Journal*, *104*, 1940–1949.

Stahelin, R. V. (2009). Lipid binding domains: More than simple lipid effectors. *Journal of Lipid Research*, *50*, S299–S304.

Wang, X., Graveland-Bikker, J. F., de Kruif, C. G., & Robillard, G. T. (2004). Oligomerization of hydrophobin SC3 in solution: From soluble state to self-assembly. *Protein Science*, *13*, 810–821.

Wyatt, P. J. (1998). Submicromolar particle sizing by multiangle light scattering following fractionation. *Journal of Colloid and Interface Science, 197*, 9–20.

Yeung, T., Terebiznik, M., Yu, L., Silvius, J., Abidi, W. M., Phillips, M., et al. (2006). Receptor activation alters inner surface potential during phagocytosis. *Science, 313*, 347–351.

Yoon, Y., Tong, J., Lee, P. J., Albanese, A., Bhardwaj, N., Källberg, M., et al. (2010). Molecular basis of the potent membrane-remodeling activity of the Epsin 1 N-terminal homology domain. *Journal of Biological Chemistry, 285*, 531–540.

Yoon, Y., Zhang, X., & Cho, W. (2012). Phosphatidylinositol 4,5-bisphosphate (PtdIns(4,5)P_2) specifically induces membrane penetration and deformation by Bin/Amphiphysin/Rvs (BAR) domains. *Journal of Biological Chemistry, 287*, 34078–34090.

Zimmerberg, J., & Kozlov, M. M. (2006). How proteins produce cellular membrane curvature. *Nature Reviews in Molecular and Cellular Biology, 7*, 9–19.

CHAPTER

Single-Molecule Imaging of Receptor–Receptor Interactions

20

Kenichi G.N. Suzuki*,‡, Rinshi S. Kasai*,†, Takahiro K. Fujiwara*, and Akihiro Kusumi*,†

*Institute for Integrated Cell-Material Sciences (WPI-iCeMS), Kyoto University, Kyoto, Japan
†Institute for Frontier Medical Sciences, Kyoto University, Kyoto, Japan
‡National Centre for Biological Sciences (NCBS)/Institute for Stem Cell Biology and Regenerative Medicine (inStem), Bangalore, India

CHAPTER OUTLINE

Introduction	374
20.1 Receptor Expression and Fluorescent Labeling	376
20.1.1 Expression of Receptors in Cell Plasma Membranes at Low Density	378
20.1.2 High-efficiency Fluorescent Labeling of Receptors	378
20.2 Single-Molecule Imaging	380
20.3 Data Analysis	382
20.3.1 Determination of Receptor Monomer, Dimer, and Oligomer Fractions in Single-Color Experiments	382
20.3.2 Determination of Receptor Dimer Lifetimes in Single-Color Experiments	384
20.3.3 Colocalization Lifetimes of Receptor Dimers in Dual-Color Experiments	385
20.3.4 Outline of the Theory for Estimating Colocalization Lifetimes	386
Acknowledgments	387
References	387

Abstract

Single-molecule imaging is a powerful tool for the study of dynamic molecular interactions in living cell plasma membranes. Herein, we describe a single-molecule imaging microscopy technique that can be used to measure lifetimes and densities of receptor dimers and oligomers. This method can be performed using a total internal reflection fluorescent microscope equipped with one or two high-sensitivity cameras. For dual-color observation, two images obtained synchronously in different colors are spatially corrected and then overlaid. Receptors must be expressed at low density in cell plasma membranes because high-density expression (>2 molecules/ μm^2) creates difficulty for tracking individual fluorescent spots. In addition, the receptors should be labeled with highly photostable fluorophores at high efficiency because short photobleaching lifetimes and low labeling efficiency of receptors reduce the probability of detecting dimers and oligomers. In this chapter, we describe methods for observing and detecting colocalization of the individual fluorescent spots of receptors labeled with fluorophores via small tags and the estimation of true dimer and oligomer lifetimes after correction with photobleaching lifetimes of fluorophores.

INTRODUCTION

Receptor clustering after ligand binding is often the first key step for the induction of intracellular signaling (Kondo et al., 2012; Veatch et al., 2012). Even before ligation, many receptors, including epidermal growth factor receptors (Chung et al., 2010), several G protein-coupled receptors (GPCRs) (Hern et al., 2010; Kasai et al., 2011), and several glycosylphosphatidylinositol (GPI)-anchored receptors (Brameshuber et al., 2010; Seong, Wang, Kinoshita, & Maeda, 2013; Sharma et al., 2004; Suzuki et al., 2012), have been proposed to form dimers.

Understanding the signal transduction mechanisms of these receptors requires investigation into how these receptors are organized in resting cells and how upon ligation they are incorporated into signaling complexes.

Single-molecule imaging techniques are very important and are powerful tools to observe receptor clustering and interactions of receptors with signaling molecules in cell plasma membranes (Jaqaman et al., 2011, 2008; Nagata, Nakada, Kasai, Kusumi, & Ueda, 2013; Nguyen, Kamio, & Higuchi, 2003; Opazo et al., 2010; Sako, Minoghchi, & Yanagida, 2000; Yamagishi, Shirasaki, & Funatsu, 2013). Recent single-molecule imaging studies revealed that receptor–receptor interactions are very dynamic in steady-state cells. For example, Kasai et al. (2011) found that in live cells at 37 °C, N-formyl peptide receptor (FPR), a chemoattractant GPCR, formed transient homodimers with an equilibrium constant of 3.6 copies/μm^2 and dissociation and two-dimensional association rate constants of $11.0\ s^{-1}$ and 3.1 copies/$\mu m^2\ s^{-1}$, respectively (see Fig. 20.1A).

Furthermore, Suzuki et al. (2012) have discovered that CD59, a GPI-anchored protein (GPI-AP), forms mobile transient homodimers induced by ectodomain protein interactions and stabilized by raft–lipid interactions (Fig. 20.1B). They also found that CD59 did not form heterodimers with prolonged lifetimes, but under higher physiological expression conditions, homodimers coalesced to form hetero- and homo-GPI-anchored receptor oligomers through raft-based lipid interactions. After ligation, CD59 forms stable homo-oligomers, which induce intracellular Ca^{2+} responses dependent on GPI anchorage and cholesterol, suggesting that transient homodimers play a key role (Suzuki, 2012; Suzuki, Fujiwara, Edidin, &

Introduction

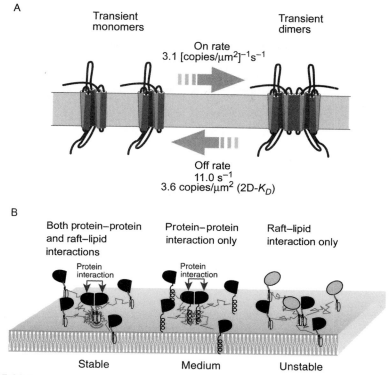

FIGURE 20.1

Dynamic equilibrium of a G protein-coupled receptor (GPCR) and a glycosylphosphatidylinositol-anchored protein (GPI-AP) among monomers, dimers, and oligomers. (A) The GPCR N-formyl peptide receptor (FPR) forms transient dimers that have three kinetic parameters: on rate, off rate, and two-dimensional association rate (2D-K_D). (B) GPI-APs form transient homodimers stabilized by both protein–protein interactions and raft–lipid interactions (left), whereas raft–lipid interactions alone without ectodomain protein interactions barely induce GPI-AP heterodimers (right). Transmembrane proteins also form transient homodimers via protein–protein interactions with shorter lifetimes (center). (For color version of this figure, the reader is referred to the online version of this chapter.)

Kusumi, 2007; Suzuki, Fujiwara, Sanematsu, et al., 2007). All of these studies have been performed by detecting the transient colocalization of two individual fluorescent spots of receptors or by measuring fluorescent intensities of individual spots in living cell membranes.

This chapter describes a protocol for single-molecule imaging and analysis of colocalizations of two individual fluorescent spots of receptors such as GPCRs and GPI-APs. Single-molecule imaging is also a basic technique for superresolution microscopies such as photoactivated localization microscopy (Owen, Williamson,

Magenau, & Gaus, 2012; Rossy, Owen, Williamson, Yang, & Gaus, 2012; Sengupta et al., 2011) and stochastic optical reconstruction microscopy (Bates, Huang, Dempsey, & Zhuang, 2007), although these modalities are outside the scope of this chapter. Along with single-fluorescent-molecule imaging, single-particle tracking of gold (Fujiwara, Ritchie, Murakoshi, Jacobson, & Kusumi, 2002; Murase et al., 2004; Suzuki, Ritchie, Kajikawa, Fujiwara, & Kusumi, 2005), latex beads (Suzuki and Sheetz, 2001; Suzuki, Sterba, & Sheetz, 2000), and quantum dots (Chen, Veracini, Benistant, & Jacobson, 2009; Lidke et al., 2004) attached to receptors have unraveled the mechanisms through which receptor diffusion is regulated by cell plasma membrane structures such as meshes of cytoskeletal actin filaments (Kalay, Fujiwara, & Kusumi, 2012; Kusumi, Fujiwara, Chadda, et al., 2012; Kusumi, Fujiwara, Morone, et al., 2012; Morone et al., 2006), lipid rafts (Kusumi, Koyama, & Suzuki, 2004; Kusumi & Suzuki, 2005; Tanaka et al., 2010), and synapses (Bannai et al., 2009). However, these bulky probes would disturb associations between receptors, and the low efficiency of receptor labeling creates challenges for the detection of receptor–receptor interactions. Therefore, this chapter focuses on single-molecule imaging of fluorophores attached to receptors via small protein tags.

20.1 RECEPTOR EXPRESSION AND FLUORESCENT LABELING

This section describes a method for expressing receptors in cell plasma membranes and labeling them with fluorophores for the observation of receptor–receptor interactions using single-molecule imaging. As mentioned in the introduction, receptor–receptor interactions can be detected by observing the colocalization of individual fluorescent spots, which represent receptor monomers. Observing the colocalization requires that conditions be optimized for receptor expression and fluorescent labeling. If the expression of receptors is high (Fig. 20.2A), the individual spots of the attached fluorophores are indistinguishable, precluding single-molecule imaging. By contrast, if the fluorescent labeling efficiency of the receptors is low and the photobleaching lifetime of the fluorophores short (Fig. 20.2B), dimerization and oligomerization of receptors are barely detectable. Therefore, observing receptor–receptor interactions using single-molecule imaging requires that the expression density of receptors be low enough to observe single molecules and fluorescent labeling efficiency of receptors be high enough to observe receptor clustering (Fig. 20.2C). If cells endogenously express receptors at a low density (<2.0 molecules/μm^2), single molecules of receptors in the cell membranes can be observed via fluorescently labeled Fab fragments or ligands attached to most of the receptors. However, this scenario is a rare one. For example, the literature reports that the physiological expression levels of GPI-APs are 3.0 and 8.2 copies/μm^2 for urokinase receptors in bovine microvascular endothelial cells and U937 cells, respectively (Mandriota et al., 1995); 10 copies/μm^2 for phospholipase A2-activating protein in HeLa TCRC-1 cells (Jemmerson, Shah, Takeya, & Fishman, 1985); 40 copies/μm^2 for decay-accelerating factor in HeLa cells

20.1 Receptor Expression and Fluorescent Labeling

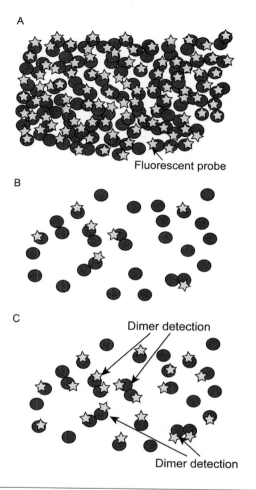

FIGURE 20.2

Receptor expression and fluorescent labeling should be optimized for detection of receptor dimers and oligomers. (A) Individual fluorescent spots are indistinguishable because receptors are expressed in cell plasma membranes at a high density (>2 molecules/µm^2). (B) Labeling efficiency of receptors with a fluorophore is too low to detect receptor dimers and oligomers. (C) Low expression density of receptors labeled with fluorophores at high efficiency is appropriate for detecting receptor dimers and oligomers. (For color version of this figure, the reader is referred to the online version of this chapter.)

(Medof, Walter, Rutgers, Knowles, & Nussenzweig, 1987); and 240 copies/µm^2 for folate receptor in MA104 cells (Rothberg, Ying, Kolhouse, Kamen, & Anderson, 1990). Therefore, we usually transfect cells with the complementary DNA (cDNA) of a receptor and express small numbers of a receptor that is not endogenously expressed in cells, as described in the succeeding text.

20.1.1 Expression of receptors in cell plasma membranes at low density

We often use Chinese hamster ovary K1 (CHO-K1) cells and T24 cells (ECV304, a subclone of T24 human bladder carcinoma) because their cell membranes are flat and we can easily observe single molecules of receptors by total internal reflection (TIRF) microscopy. If the cell membranes are exceptionally rough, the fluorescent intensities of diffusing individual receptor spots vary significantly, making quantitative analysis of the intensities difficult. These cells are cultured in HAM's F12 medium (Gibco) supplemented with 10% fetal bovine serum. Using Lipofectamine PLUS (Life Technologies), we transfect the cells with cDNAs encoding membrane receptors. cDNA sequences of GPI-APs are placed in the Epstein–Bar virus-based episomal vector pOsTet15T3—which carries tetracycline-regulated expression units, a transactivator (rtTA2-M2), and a Tet operator sequence (a Tet-on vector)—and expressed in the cells. cDNA sequences of GPCRs are placed in pcDNA3+ (Invitrogen). The cells that stably expressed the molecules of interest are selected and cloned using 800 μg/ml G418 (final concentration).

These cells are sparsely seeded (4×10^3 per coverslip) in a glass base dish (35 mmϕ with a window of 12 mmϕ, 0.15 mm thick glass; Iwaki). The expression levels of the GPI-APs are adjusted by finely controlling the concentration of doxycycline added to the cell culture medium. Specifically, the cells are grown for a day in the glass base dish and are then incubated with various concentrations of doxycycline for a day and used for single-fluorescent-molecule imaging. The doxycycline concentrations are optimized by measuring the spot number densities of fluorophores attached to the expressed receptors in cell plasma membranes. Typically, the optimized doxycycline concentrations are 1–5 ng/ml for GPI-APs and 10–50 ng/ml for transmembrane (TM) proteins. The optimal number density of receptors for single-molecule observation is 0.1–2.0 molecules/μm^2 or 500–10,000 molecules/cell.

20.1.2 High-efficiency fluorescent labeling of receptors

We usually use two methods to label membrane receptors with fluorophores. One method is to label receptors with fluorophore-conjugated Fab fragments of the antibody. These fragments are conjugated with fluorophores with an amine-reactive succinimidyl ester according to manufacturer instructions. The dye/protein (D/P) ratios of fluorophore-labeled Fab fragment are estimated with optical density measurements. When a receptor is tagged with Fab conjugated to a fluorescent probe and the D/P ratio is 1 on average, the number of fluorescent probes on a single Fab generally follows a Poisson distribution, varying mostly between 0 and 3 (no probe/Fab = ~37%, 1 probe/Fab = ~37%, 2 probes/Fab = ~19%, and 3 probes/Fab = ~7%). Therefore, whether a single fluorescent spot in a single image represents two Fab molecules or one Fab molecule conjugated with two or more fluorescent probes is indiscernible. In addition, fluorophore-conjugated Fab is applicable to the observation of receptor–receptor interactions only when Fab has high affinity to receptors, because high labeling efficiency of receptors with fluorophores is required for this experiment. Furthermore, for high-efficiency fluorescent labeling of receptors, cells must be incubated with fluorescent Fab fragments with a concentration much higher than the K_d value.

Another method is to label receptors with fluorophores via small protein tags. Of course, we can observe the colocalization of individual fluorescent spots of receptors fused with monomeric GFP (mGFP, A206K mutant). However, photobleaching must be carefully taken into account because the photobleaching lifetime of mGFP is shorter than 1 s when single molecules of mGFP are observed at a high signal-to-noise ratio. The short photobleaching lifetime makes precise estimation of colocalization lifetimes difficult. Therefore, instead of mGFP, we often used protein tags, especially Halo7-tag (Promega) (Keuning, Janssen, & Witholt, 1985) and acyl carrier protein (ACP)-tag (New England Biolabs) (Meyer et al., 2006; Rock & Cronan, 1979; Sielaff et al., 2006), which are fused with receptors and covalently conjugated with photostable organic dye molecules. These tags can be labeled with fluorophores at a precise 1:1 mol ratio. Furthermore, neither Halo7 nor ACP form dimers and therefore do not induce receptor clustering. Notably, however, when these tags are inserted at the extracellular side of receptors, the signal peptides of the receptors sometimes need to be replaced by others for expression of the receptors in cell plasma membranes. Table 20.1 shows the characteristics for these tags.

Halo7-tag (molecular weight = 33 kDa) is a modified haloalkane dehalogenase designed to bind covalently to synthetic ligands that are suicide substrates.

Table 20.1 Characteristics of Tag Proteins Suitable for Single-Molecule Imaging of Receptor–Receptor Interactions

Characteristics	Halo7	Acyl Carrier Protein (ACP)
Molecular weight	33 kDa	8 kDa
Reaction with ligand	The ligand is a suicide substrate for the tag	The ligand is covalently bound to the tag with enzyme
Fluorophores for labeling at cytoplasmic site	Rhodamine 110 Tetramethylrhodamine	None
Fluorophores for labeling at extracellular site	ATTO488, Alexa488 Cy3, Dy547 ATTO594, Alexa594, etc.	Same as Halo7
Signal peptide at N-terminus	The original signal peptide of receptors sometimes needs to be replaced by others	Same as Halo7

Fluorescently labeled receptors tagged with Halo7 can be labeled at high efficiency (>80%) (Suzuki et al., 2012). To maintain receptor functions, some linkers with lengths between 15 and 21 amino acids should be inserted between Halo7-tag and receptors. Because the fluorophore-conjugated Halo ligand is often membrane-permeable if the fluorophore molecules are membrane-permeable, Halo7-tag at the cytoplasmic site of receptors can also be labeled with membrane-permeable fluorescent ligands. The receptors are labeled with fluorescent halo ligands according to manufacturer instructions (Promega), but for single-molecule imaging, cells are incubated with 50 nM Halo ligand conjugated with fluorophores for 15 min to label the extracellular Halo7 of receptors and for 60 min to label cytoplasmic Halo7 (Suzuki et al., 2012). Under these conditions, more than 80% of the Halo7-tag can be fluorescently labeled.

ACP-tag (molecular weight = 8 kDa) is a small protein tag based on ACP from *Escherichia coli*. ACP-tag can be enzymatically conjugated with fluorophores using the substrates derived from coenzyme A (CoA). Because CoA and fluorescent CoA substrates are membrane-impermeable, receptors tagged with ACP at its extracellular surface can be fluorescently labeled. More than 95% of ACP tagged at the extracellular site of receptors can be labeled with fluorophores (Meyer et al., 2006). The receptors are labeled with fluorescent ACP ligand according to manufacturer instructions (New England Biolabs), but note that for single-molecule imaging, cells are incubated with 50 nM ACP ligand conjugated with fluorophores for 15 min (Suzuki et al., 2012).

20.2 SINGLE-MOLECULE IMAGING

Single-fluorescent-molecule tracking is performed using an objective lens-type TIRF microscope (Tokunaga, Imamoto, & Sakata-Sogawa, 2008; Tokunaga, Kitamura, Saito, Iwane, & Yanagida, 1997) constructed on an inverted microscope. Figure 20.3 shows a schematic diagram of the TIRF microscope we built to perform single- or dual-color imaging at the level of single molecules. We used an Olympus IX-70 microscope as the base with a modified mirror turret to allow side entry of the excitation laser beams to the microscope. A laser beam attenuated with neutral density filters and circularly polarized by a quarter-wave plate is expanded by two lenses (L1, $f=15$ mm; L2, $f=150$ mm for 488 nm excitation or L1, $f=20$ mm; L2, $f=80$ mm for 594 nm excitation), focused at the back focal plane of the objective lens with an L3 lens ($f=350$ mm), and steered onto the microscope at the edge of an objective lens with a high numerical aperture (Plan Apo 100×; numerical aperture = 1.49; Olympus). A single- or dual-color dichroic mirror (Chroma Technology) is also used.

The laser beam is totally internally reflected at the coverslip–medium interface, and an evanescent field (1/e penetration depth = ~100 nm) is formed on the surface of the coverslip. The basal membrane is locally illuminated with this evanescent field. The emission from the fluorophore is collected by the objective and passes the single- or dual-color excitation path dichroic. For dual-color observation, the emission is split into two imaging systems by an observation path dichroic mirror.

20.2 Single-Molecule Imaging

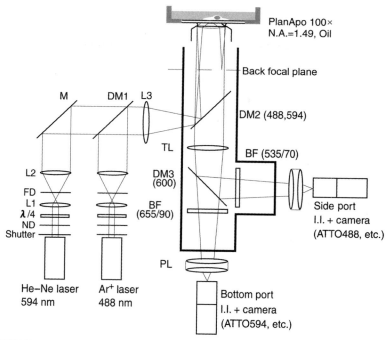

FIGURE 20.3

Microscope setup for simultaneous, dual-color imaging of two fluorescent single molecules of different species. For the observations of other dyes, the microscope is also equipped with 532 and 642 nm lasers. The excitation arm consists of the following optical components: BP, band-pass filter; DM, dichroic mirror; FD, field diaphragm; L1 and L2, working as a 10× beam expander (L1, $f=15$ mm; L2, $f=150$ mm) for 488 nm excitation or as a 4× beam expander (L1, $f=20$ mm; L2, $f=80$ mm) for 594 nm excitation; L3, focusing lens ($f=350$ mm); M, mirror; ND, neutral density filter; r/4, quarter-wave plate; and S, electronic shutter. The two-color fluorescence emission signal is split by a dichroic mirror (DM3) and detected by two cameras at the side and bottom ports. BF, barrier filter; I.I., image intensifier; PL, projection lens (2×); and TL, tube lens (1× or 2×).

The image in each arm is projected onto the photocathode of the image intensifier with a two-stage microchannel plate (C8600-03, Hamamatsu Photonics), which is lens coupled to the camera. We use Hamamatsu electron bombardment charge-coupled cameras (C7190-23). The cameras on the two detecting arms (side and bottom, two cameras of the same model number produced in the same batch placed on each arm) are synchronized frame by frame by coupling the sync out of one camera to the trigger of the second (gen-locked). Observations are performed only for individual fluorescent spots located within the central region (20 μm in diameter) of the illuminated area, in which nonuniformities appear to be small. The camera images are stored on a digital videotape (8PDV-184ME, Sony) for postexperiment spatial synchronization.

20.3 DATA ANALYSIS
20.3.1 Determination of receptor monomer, dimer, and oligomer fractions in single-color experiments

This section describes a method for measuring fractions of receptor monomers, dimers, and oligomers by determining fluorescence intensities of individual spots of receptors. A typical single-frame image of single molecules of fluorescently labeled FPR is shown in Fig. 20.4A. Fluorescent spots in the image are identified using a

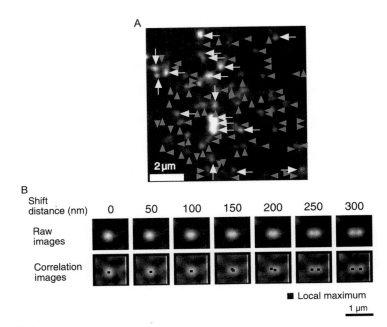

FIGURE 20.4

Identification of individual spots. (A) Typical single-frame total internal reflection image of single FPR molecules labeled with Alexa594-formyl peptide. Individual spots are identified by taking the cross correlation of the observed image with a reference image of a single-fluorescent-molecule spot. Arrowheads and arrows indicate the spots with monomeric and dimeric intensities, respectively. (B) The image cross correlation method, in addition to identifying the fluorescent spots, finds the local peaks in the correlation image, determining whether an observed spot represents one unresolvable spot or two resolvable spots (bottom, spatial resolution). A representative image of a single-molecule intensity spot is superimposed on itself but at systematically varied shift distances between the two images (from 0 to 300 nm, every 10 nm along the x-axis). Top panel shows raw images; the bottom panel displays cross correlation images, with dots indicating local peaks. For each shift distance, the result of one or two peaks is registered in the cross correlation images (bottom). (See color plate.)

custom computer program, which takes the cross correlation of the observed image using a reference two-dimensional Gaussian function with a full width of 200 nm (Fujiwara et al., 2002). In addition to identifying fluorescent spots, this method generates the local peaks in the correlation image (Fig. 20.4B) and thus determines whether an observed spot represents one unresolvable spot or two resolvable spots. In this manner, the number density of distinguishable spots can be determined.

Because the number of dyes attached to individual anti-Fab fragments exhibits Poisson distributions, the absolute number of molecules located within a single colocalized spot cannot be determined using fluorescent Fab fragments. Because ACP- or Halo7-tagged receptor molecules can be labeled with dyes at a 1:1 D/P ratio and high efficiency, the absolute number of molecules in a single colocalized spot can be determined by measuring the fluorescence intensities of individual spots of ACP- or Halo7-tagged receptors.

After each individual spot in the image is identified, the fluorescence intensities of the identified spots are determined, yielding the histograms. Figure 20.5A shows the distribution of fluorescence intensities of individual spots of ACP-CD59 labeled with Dy547 in CHO-K1 cells. The histograms are fitted by the sum of three log-normal functions (Mutch et al., 2007) and provide the spot fractions of monomers, dimers, and tetramers via comparison with the histogram for the fluorescence intensities of individual spots of monomer reference molecules. To determine the fluorescence intensity distribution of monomers, we often measure the individual spot intensities of ACP- or Halo7-tagged proteins linked to the TM domains of nonraft molecule low-density lipoprotein receptors (ACP-TM or Halo-TM), which are

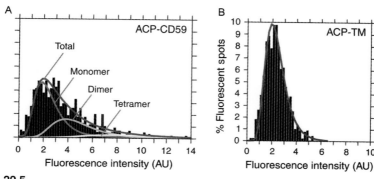

FIGURE 20.5

Fractions of monomers, homodimers, and homo-oligomers of acyl carrier protein (ACP)-CD59. (A) Distributions of the signal intensities of individual spots of Dy547-labeled ACP-CD59 expressed at densities of 0.90 ($n=788$) copies/μm^2 and fitted with three log-normal functions that represent the distributions for monomers, homodimers, and apparent homotetramers of ACP-CD59. This result was obtainable because more than 95% of ACP was fluorescently labeled. (B) Distribution of the signal intensities of individual spots of Dy547-labeled ACP transmembrane (ACP-TM; nonraft monomer reference protein; 0.16 spots/μm^2, $n=868$ spots) fitted with a single log-normal function. (See color plate.)

sparsely expressed in cell plasma membranes. Figure 20.5B shows the distribution of fluorescence intensities of individual ACP-TM spots.

20.3.2 Determination of receptor dimer lifetimes in single-color experiments

This section describes the method for determining receptor dimer lifetimes in cell plasma membranes by observing the colocalization of individual fluorescent spots. Figure 20.6A and B shows a typical video sequence and trajectories, respectively, indicating that virtually all CD59 molecules tagged with Cy3-Fab undergo diffusion and frequent colocalization and codiffusion events often lasting longer than incidental approaches. Separation into monomers follows these events. Similar to CD59,

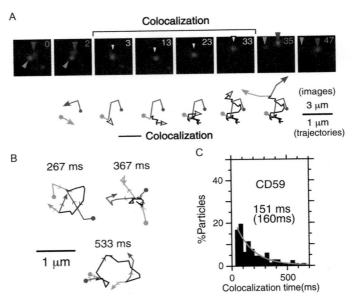

FIGURE 20.6

Transient homo-colocalization of CD59 in steady-state cells. (A) Representative single-molecule fluorescent image sequence of two diffusing CD59 molecules (Cy3-Fab-labeled CD59, arrowheads) undergoing transient colocalization and codiffusion. The number in the upper-right corner in each image indicates the number of frames (recorded at video rate). Trajectories are shown on a threefold magnified scale: black trajectory designates molecules during colocalization. (B) Trajectories of two Cy3-Fab–labeled CD59 spots that exhibited transient colocalization and codiffusion. Periods of colocalization (black) are indicated in milliseconds. (C) The histogram shows the distribution of colocalization duration of two CD59 fluorescent spots. It was generated by measuring the duration of each colocalization event and fitted with an exponential function. The decay time and the lifetime corrected with the photobleaching lifetime are shown in the box. (See color plate.)

many other receptors exhibit this frequent colocalization, codiffusion, and separation (Hern et al., 2010; Kasai et al., 2011; Suzuki et al., 2012).

For determining the colocalization duration, the histogram of the durations of individual colocalization events is fitted with a single exponential decay curve, using nonlinear least-squares fitting by the Levenberg–Marquardt algorithm, provided in the MicroCal Origin 7.5 package (Fig. 20.6C). The colocalized durations are then corrected for the photobleaching lifetimes of the probes, based on the equations described in 20.3.3, thus giving τ_{CCD}. The log-rank test is a useful statistical survival analysis for examining whether distributions of colocalization lifetimes are distinguishable.

20.3.3 Colocalization lifetimes of receptor dimers in dual-color experiments

In two-color simultaneous single-fluorescent-molecule tracking experiments that use two fluorophores—such as pairs of Alexa488 and Alexa594, rhodamine 110 and tetramethylrhodamine, or ATTO488 and ATTO594—the two full images synchronously obtained in different colors are spatially corrected and overlaid (Koyama-Honda et al., 2005). A predefined mask is simultaneously imaged through each path to determine the individual path properties. This mask has a lattice of 1 μm diameter optical holes spaced 5 μm apart. Using this grid, we can map a point imaged through one of the imaging paths to the same point imaged through the other path. In this manner, a third-order spline fit can be used to correct for translations and rotations between the two observation arms and the optical distortions and inhomogeneities inherent in both the optical paths and each camera. The details of this procedure have been published previously (Koyama-Honda et al., 2005).

In two-color simultaneous single-fluorescent-molecule tracking experiments, the distance between the two molecules can be measured directly from the coordinates (x and y positions) of each molecule (as in photoactivated localization and stochastic optical reconstruction microscopies), which can be determined independently in each image in different colors (Koyama-Honda et al., 2005; Suzuki, Fujiwara, Edidin, et al., 2007; Suzuki et al., 2012). Even when pairs of different colored molecules are known to be truly associated, the probability of scoring the two molecules as associated is limited by the localization accuracies of each molecule and the accuracies of superimposing the two images. Based on a method developed previously (Koyama-Honda et al., 2005; Suzuki, Fujiwara, Edidin, et al., 2007) and the accuracies determined herein, we found that for truly associated molecules, the probability of scoring the two molecules as associated increases to 99% using the criterion that the molecules lie within 240 nm of each other. Therefore, we used this criterion as the definition of colocalization in simultaneous two-color single-molecule observations. The distance of 240 nm coincided with the definition of colocalization in single-color experiments. Given this coincidence, we defined the colocalization of two fluorescent molecules as the event in which the two fluorescent spots representing these molecules become localized within 240 nm of each other.

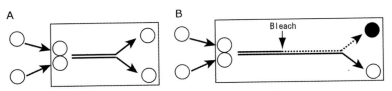

FIGURE 20.7

Two ways in which two fluorescent molecules in a single spot lose their colocalization. (A) Two spots become colocalized and separated without photobleaching. (B) One of the two spots is photobleached during colocalization.

20.3.4 Outline of the theory for estimating colocalization lifetimes

This section provides a protocol for correcting colocalization lifetimes with the photobleaching lifetimes of fluorophores. Two fluorescent spots that become colocalized lose their colocalization in one of the following ways: (1) They become separated after some time owing to dissociation and independent diffusion (Fig. 20.7A), or (2) one of the two spots is photobleached (Fig. 20.7B). We first consider the case in which two-color fluorescent probes are used and spots with different colors are colocalized. The dis-colocalization process can be expressed by the differential equation:

$$\frac{d[D]}{dt} = -k_{b1}[D] - k_{b2}[D] - k_{off}[D] \tag{20.1}$$

where $[D]$ indicates the number density of the colocalized fluorescent spots, k_{b1} and k_{b2} are the photobleaching lifetimes for the two fluorescent probes, and k_{off} represents the off rate for the two colocalized spots (which includes the time needed for the dissociation of two molecules plus that required for the two molecules to be separated by more than 240 nm owing to diffusion, but the latter is likely to be considerably shorter than the former). Eq. (20.1) can readily be solved, giving

$$[D] = D_0 \, exp[-(k_{b1} + k_{b2} + k_{off})t] \tag{20.2}$$

where D_0 is an integration constant.

Therefore, the apparent rate constant for dis-colocalization, k_{app}, which can be measured directly with colocalization experiments, can be written as follows:

$$k_{app} = k_{b1} + k_{b2} + k_{off} \tag{20.3}$$

Each rate constant can be related to its associated time constant:

$$\tau_{app} = k_{app}^{-1} \tag{20.4}$$

$$\tau_{b1} = k_{b1}^{-1} \tag{20.5}$$

$$\tau_{b2} = k_{b2}^{-1} \tag{20.6}$$

$$\tau_{off} = k_{off}^{-1} \tag{20.7}$$

Therefore, Eq. (20.3) can be rewritten as follows:

$$\tau_{app}^{-1} = \tau_{b1}^{-1} + \tau_{b2}^{-1} + \tau_{off}^{-1} \quad (20.8)$$

Because τ_{app} is the lifetime directly determined from the histogram in single-molecule colocalization experiments, and τ_{b1} and τ_{b2} can be measured using the histograms of the photobleaching times for individual non-colocalized spots (therefore, this measurement is reliable because the photobleaching lifetimes are measured simultaneously with the apparent colocalization durations), τ_{off} can be obtained as follows:

$$\tau_{off} = \left[\tau_{app}^{-1} - \tau_{b1}^{-1} - \tau_{b2}^{-1}\right]^{-1} \quad (20.9)$$

In single-color experiments,

$$\tau_{off} = \left[\tau_{app}^{-1} - 2\tau_{b1}^{-1}\right]^{-1} \quad (20.10)$$

For example, the colocalization lifetime of Cy3-Fab-labelled CD59 was estimated to be 151 ms from the single exponential decay fitting (see Fig. 20.6C). Because the photobleaching lifetime of the Cy3 probe on Fab is 5.4 s, the true colocalization lifetime was estimated using Eq. (20.10) to be 160 ms.

Acknowledgments

This work was supported in part by Grants-in-Aid for Specific Research (B) (No. 24370055) and on Innovative Areas (No. 2311002, Deciphering sugar chain-based signals regulating integrative neuronal functions) from the Ministry of Education, Culture, Sports, Science and Technology of Japan.

References

Bannai, H., Levi, S., Schweizer, C., Inoue, T., Launey, T., Racine, V., et al. (2009). Activity-dependent tuning of inhibitory neurotransmission based on GABAAR diffusion dynamics. *Neuron, 62,* 670–682.

Bates, M., Huang, B., Dempsey, G. T., & Zhuang, X. (2007). Multicolor super-resolution imaging with photo-switchable fluorescence probes. *Science, 317,* 1749–1753.

Brameshuber, M., Weghuber, J., Ruprecht, V., Gombos, I., Horvath, I., Vigh, L., et al. (2010). Imaging of mobile long-lived nanoplatforms in the live cell plasma membrane. *Journal of Biological Chemistry, 285,* 41765–41771.

Chen, Y., Veracini, L., Benistant, C., & Jacobson, K. (2009). The transmembrane protein CBP plays a role in transiently anchoring small clusters of Thy-1, a GPI-anchored protein, to the cytoskeleton. *Journal of Cell Science, 122,* 3966–3972.

Chung, I., Akita, R., Vandlen, R., Toomre, D., Schlessinger, J., & Mellman, I. (2010). Spatial control of EGF receptor activation by reversible dimerization on living cells. *Nature, 464,* 783–787.

Fujiwara, T. K., Ritchie, K., Murakoshi, H., Jacobson, K., & Kusumi, A. (2002). Phospholipids undergo hop diffusion in compartmentalized cell membrane. *The Journal of Cell Biology*, *157*, 1071–1081.

Hern, J. A., Baig, A. H., Mashanov, G. L., Birdsall, B., Corrie, J. E. T., Lazareno, S., et al. (2010). Formation and dissociation of M1 muscarinic receptor dimers seen by total internal reflection fluorescence imaging of single molecules. *Proceedings of the National Academy of Sciences of the United States of America*, *107*, 2693–2698.

Jaqaman, K., Kuwata, H., Touret, N., Collins, R., Trimble, W. S., Danuser, G., et al. (2011). Cytoskeletal control of CD36 diffusion promotes its receptor and signaling function. *Cell*, *146*, 593–606.

Jaqaman, K., Loerke, D., Mettlen, M., Kuwata, H., Grinstein, S., Schmid, S. L., et al. (2008). Robust single-particle tracking in live-cell time-lapse sequences. *Nature Methods*, *5*, 695–702.

Jemmerson, R., Shah, N., Takeya, M., & Fishman, W. H. (1985). Characterization of the placental alkaline phosphatase-like (Nagao) isozyme on the surface of A431 human epidermoid carcinoma cells. *Cancer Research*, *45*, 282–287.

Kalay, Z., Fujiwara, T. K., & Kusumi, A. (2012). Confining domains lead to reaction bursts: Reaction kinetics in the plasma membrane. *PLoS One*, *7*, e32948.

Kasai, R. S., Suzuki, K. G. N., Prossnitz, E. R., Koyama-Honda, I., Nakada, C., Fujiwara, T. K., et al. (2011). Full characterization of GPCR monomer-dimer dynamic equilibrium by single molecule imaging. *The Journal of Cell Biology*, *192*, 463–480.

Keuning, S., Janssen, D. B., & Witholt, B. (1985). Purification and characterization of hydrolytic haloalkane dehalogenase from Xanthobacter autotrophicus GJ10. *Journal of Bacteriology*, *163*, 635–639.

Kondo, Y., Ikeda, K., Tokuda, N., Nishitani, C., Ohto, U., Akashi-Takamura, S., et al. (2012). TLR4-MD-2 complex is negatively regulated by an endogenous ligand, globotetraosylceramide. *Proceedings of the National Academy of Sciences of the United States of America*, *110*, 4714–4719.

Koyama-Honda, I., Ritchie, K., Fujiwara, T., Iino, R., Murakoshi, H., Kasai, R. S., et al. (2005). Fluorescence imaging for monitoring the colocalization of two single molecules in living cells. *Biophysical Journal*, *88*, 2126–2136.

Kusumi, A., Fujiwara, T. K., Chadda, R., Xie, M., Tsunoyama, T. A., Kalay, Z., et al. (2012). Dynamic organizing principles of the plasma membrane that regulate signal transduction: Commemorating the fortieth anniversary of Singer and Nicolson's fluid-mosaic model. *Annual Review of Cell and Developmental Biology*, *28*, 215–250.

Kusumi, A., Fujiwara, T. K., Morone, N., Yoshida, K. J., Chadda, R., Xie, M., et al. (2012). Membrane mechanisms for signal transduction: The coupling of the meso-scale raft domains to membrane-skeleton-induced compartments and dynamic protein complexes. *Seminars in Cell & Developmental Biology*, *23*, 126–144.

Kusumi, A., Koyama, I., & Suzuki, K. (2004). Molecular dynamics and interactions for creation of stimulation-induced stabilized rafts from small unstable steady-state rafts. *Traffic*, *5*, 213–230.

Kusumi, A., & Suzuki, K. (2005). Toward understanding the dynamics of membrane-raft-based molecular interactions. *Biochimica et Biophysica Acta*, *1746*, 234–251.

Lidke, D. S., Nagy, P., Heintzman, R., Arndt-Jovin, D. J., Post, J. N., Grecco, H. E., et al. (2004). Quantum dot ligands provide new insights into erbB/HER receptor-mediated signal transduction. *Nature Biotechnology*, *22*, 198–203.

Mandriota, S. J., Seghezzi, G., Vassalli, J. D., Ferrara, N., Wasi, S., Mazzieri, R., et al. (1995). Vascular endothelial growth factor increases urokinase receptor expression in vascular endothelial cells. *Journal of Biological Chemistry, 270*, 9709–9716.

Medof, E., Walter, E. I., Rutgers, J. L., Knowles, D. M., & Nussenzweig, V. (1987). Decay-accelerating factor (DAF) on epithelium and glandular cells and in body fluids. *The Journal of Experimental Medicine, 165*, 848–864.

Meyer, B. H., Segura, J. M., Martinez, K. L., Hovius, R., George, N., Johnsson, K., et al. (2006). FRET imaging reveals that functional neurokinin-1 receptors are monomeric and reside in membrane microdomains of live cells. *Proceedings of the National Academy of Sciences of the United States of America, 103*, 2138–2143.

Morone, N., Fujiwara, T., Murase, K., Kasai, R. S., Ike, H., Yuasa, S., et al. (2006). Three-dimensional reconstruction of the membrane skeleton at the plasma membrane interface by electron tomography. *The Journal of Cell Biology, 174*, 851–862.

Murase, K., Fujiwara, T., Umemura, Y., Suzuki, K., Iino, R., Yamashita, H., et al. (2004). Ultrafine membrane compartments for molecular diffusion as revealed by single molecule techniques. *Biophysical Journal, 86*, 4075–4093.

Mutch, S. A., Fujimoto, B. S., Kuyper, C. L., Kuo, J. S., Bajjalieh, S. M., & Chiu, D. T. (2007). Deconvolving single-molecule intensity distributions for quantitative microscopy measurements. *Biophysical Journal, 92*, 2926–2943.

Nagata, K., Nakada, C., Kasai, R. S., Kusumi, A., & Ueda, K. (2013). ABCA1 dimer-monomer interconversion during HDL generation revealed by single-molecule imaging. *Proceedings of the National Academy of Sciences of the United States of America, 110*, 5034–5039.

Nguyen, V. T., Kamio, Y., & Higuchi, H. (2003). Single-molecule imaging of cooperative assembly of gamma-hemolysin on erythrocyte membranes. *EMBO Journal, 22*, 4968–4979.

Opazo, P., Labrecque, S., Tigaret, C. M., Frouin, A., Wiseman, P. W., De Koninck, P., et al. (2010). CaMKII triggers the diffusional trapping of surface AMPARs through phosphorylation of stargazing. *Neuron, 67*, 239–252.

Owen, D. M., Williamson, D. J., Magenau, A., & Gaus, K. (2012). Sub-resolution lipid domains exist in the plasma membrane and regulate protein diffusion and distribution. *Nature Communications, 3*, 1256.

Rock, C. O., & Cronan, J. E., Jr. (1979). Re-evaluation of the solution structure of acyl carrier protein. *Journal of Biological Chemistry, 254*, 9778–9785.

Rossy, J., Owen, D. M., Williamson, D. J., Yang, Z., & Gaus, K. (2012). Conformational states of the kinase Lck regulate clustering in early T cell signaling. *Nature Immunology, 14*, 82–89.

Rothberg, K. G., Ying, Y., Kolhouse, J. F., Kamen, B. A., & Anderson, R. G. W. (1990). The glycophospholipid-linked folate receptor internalizes folate without entering the clathrin-coated pit endocytic pathway. *The Journal of Cell Biology, 110*, 637–649.

Sako, Y., Minoghchi, S., & Yanagida, T. (2000). Single-molecule imaging of EGFR signaling on the surface of living cells. *Nature Cell Biology, 2*, 168–172.

Sengupta, P., Jovanovic-Talisman, T., Skoko, D., Renz, M., Veatch, S. L., & Lippincott-Schwartz, J. (2011). Probing protein heterogeneity in the plasma membrane using PALM and pair correlation analysis. *Nature Methods, 8*, 969–975.

Seong, J., Wang, Y., Kinoshita, T., & Maeda, Y. (2013). Implications of lipid moiety in oligomerization and immunoreactivities of GPI-anchored proteins. *Journal of Lipid Research, 54*, 1077–1091.

Sharma, P., Varma, R., Sarasij, R. C., Ira Gousset, K., Krishnamoorthy, G., Rao, M., et al. (2004). Nanoscale organization of multiple GPI-anchored proteins in living cell membranes. *Cell*, *116*, 577–589.

Sielaff, I., Arnold, A., Godin, G., Tugulu, S., Klok, H. A., & Johnsson, K. (2006). Protein function microarrays based on self-immobilizing and self-labeling fusion proteins. *ChemBioChem*, *7*, 194–202.

Suzuki, K. G. N. (2012). Lipid rafts generate digital-like signal transduction in cell plasma membranes. *Biotechnology Journal*, *7*, 753–761.

Suzuki, K. G. N., Fujiwara, T. K., Edidin, M., & Kusumi, A. (2007). Dynamic recruitment of phospholipase Cγ at transiently immobilized GPI-anchored receptor clusters induces IP$_3$-Ca^{2+} signaling: Single-molecule tracking study 2. *The Journal of Cell Biology*, *177*, 731–742.

Suzuki, K. G. N., Fujiwara, T. K., Sanematsu, F., Iino, R., Edidin, M., & Kusumi, A. (2007). GPI-anchored receptor clusters transiently recruit Lyn and Gα for temporary cluster immobilization and Lyn activation: Single-molecule tracking study 1. *The Journal of Cell Biology*, *177*, 717–730.

Suzuki, K. G., Kasai, R. S., Hirosawa, K. M., Nemoto, Y. L., Ishibashi, M., Miwa, Y., et al. (2012). Transient GPI-anchored protein homodimers are unit for raft organization and function. *Nature Chemical Biology*, *8*, 774–783.

Suzuki, K., Ritchie, K., Kajikawa, E., Fujiwara, T., & Kusumi, A. (2005). Rapid hop diffusion of a G-protein-coupled receptor in the plasma membrane as revealed by single-molecule techniques. *Biophysical Journal*, *88*, 3659–3680.

Suzuki, K., & Sheetz, M. P. (2001). Binding of cross-linked glycosylphosphatidylinositol-anchored proteins to discrete actin-associated sites and cholesterol-dependent domains. *Biophysical Journal*, *81*, 2181–2189.

Suzuki, K., Sterba, R. E., & Sheetz, M. P. (2000). Outer membrane monolayer domains from two-dimensional surface scanning resistance measurements. *Biophysical Journal*, *79*, 448–459.

Tanaka, K. A. K., Suzuki, K. G. N., Shirai, Y. M., Shibutani, S. T., Miyahara, M. S. H., Tsuboi, H., et al. (2010). Membrane molecules mobile even after chemical fixation. *Nature Methods*, *7*, 865–866.

Tokunaga, M., Imamoto, N., & Sakata-Sogawa, K. (2008). Highly inclined thin illumination enables clear single-molecule imaging in cells. *Nature Methods*, *5*, 159–161.

Tokunaga, M., Kitamura, K., Saito, K., Iwane, A. H., & Yanagida, T. (1997). Single molecule imaging of fluorophores and enzymatic reactions achieved by objective-type total internal reflection fluorescence microscopy. *Biochemical and Biophysical Research Communications*, *235*, 47–53.

Veatch, S. L., Marchta, B. B., Shelby, S. A., Chiang, E. N., Holowka, D. A., & Baird, B. A. (2012). Correlation function quantify super-resolution images and estimate apparent clustering due to over-counting. *PLoS One*, *7*, e31457.

Yamagishi, M., Shirasaki, Y., & Funatsu, T. (2013). Single-molecule tracking of mRNA in living cells. *Methods in Molecular Biology*, *950*, 153–167.

CHAPTER 21

Visualization of TCR Nanoclusters via Immunogold Labeling, Freeze-Etching, and Surface Replication

Gina J. Fiala[*,†,‡], **María Teresa Rejas**[§], **Wolfgang W. Schamel**[*,†,1], **and Hisse M. van Santen**[¶,1]

[*]*Department of Molecular Immunology, Faculty of Biology, BIOSS Centre for Biological Signaling Studies, Albert Ludwigs University Freiburg, Freiburg, Germany*
[†]*Centre for Chronic Immunodeficiency (CCI), University Clinics Freiburg and Medical Faculty, Albert Ludwigs University Freiburg, Freiburg, Germany*
[‡]*Spemann Graduate School of Biology and Medicine (SGBM), Albert Ludwigs University Freiburg, Freiburg, Germany*
[§]*Servicio de Microscopía Electrónica, Centro Biología Molecular Severo Ochoa, Consejo Superior de Investigaciones Científicas, Universidad Autónoma de Madrid, Madrid, Spain*
[¶]*Departamento de Biología Celular e Inmunología, Centro Biología Molecular Severo Ochoa, Consejo Superior de Investigaciones Científicas, Universidad Autónoma de Madrid, Madrid, Spain*

CHAPTER OUTLINE

Introduction	393
21.1 Materials	394
21.2 Equipment	395
21.3 Methods	395
21.3.1 Cell Isolation, Purification, and Fixation	396
21.3.2 Immunogold Labeling	396
21.3.3 Cell Adherence to Freshly Exfoliated Mica Sheets	399
21.3.4 Freezing of Samples	400
21.3.4.1 Preparation of the Plunge-Freezing Unit	400
21.3.4.2 Preparation of Cryovials for Storage of the Frozen Mica Sandwiches	400
21.3.4.3 Freezing of the Samples	401

[1]Contributed equally.

21.3.5 Preparation of Cell Surface Replicas 401
    21.3.5.1 Preparation of the Freeze-Fracture Unit............................ 401
    21.3.5.2 Mounting of Samples on the Specimen Table 402
    21.3.5.3 Etching and Replication 402
21.3.6 Cleaning and Mounting of Replicas 403
21.3.7 Analysis of Metal Replicas by TEM 404

21.4 Analysis 405
21.5 Considerations 405
21.5.1 Safety Precautions 405
21.5.2 Starting Number of Cells 405
21.5.3 Colloidal Gold Conjugates 406
21.5.4 Cell Surface Replication and Replication of Fracture Surfaces 406
21.5.5 Primary Antibody and Fixation Conditions 406
21.5.6 Secondary Reagents 406
21.5.7 Freezing of Samples 407
21.5.8 Mounting of Samples in the Specimen Table 407
21.5.9 Replica Floating 407
21.5.10 Replica Stability 407
21.5.11 Replica Quality 408
21.5.12 Low Labeling Efficiency and Implications for
    Interpretation of Data 408

Acknowledgments **408**
References **408**

Abstract

T cells show high sensitivity for antigen, even though their T-cell antigen receptor (TCR) has a low affinity for its ligand, a major histocompatibility complex molecule presenting a short pathogen-derived peptide. Over the past few years, it has become clear that these paradoxical properties rely at least in part on the organization of cell surface-expressed TCRs in TCR nanoclusters. We describe a protocol, comprising immunogold labeling, cell surface replica generation, and electron microscopy (EM) analysis that allows nanoscale resolution of the distribution of TCRs and other cell surface molecules of cells grown in suspension. Unlike most of the light microscopy-based single-molecule resolution techniques, this technique permits visualization of these molecules on cell surfaces that do not adhere to an experimental support. Given the potential of adhesion-induced receptor redistributions, our technique is a relevant complement to the substrate adherence-dependent techniques. Furthermore, it does not rely on introduction of fluorescently labeled recombinant molecules and therefore allows direct analysis of nonmanipulated primary cells.

INTRODUCTION

T-cell antigen receptor (TCR) clustering was long thought to be dependent on binding of its ligand, a major histocompatibility complex (MHC) molecule presenting a pathogen-derived peptide. Early confocal microscopy studies showed that upon recognition of its antigen, the TCR concentrated in the contact area between the T cell and the cognate peptide/MHC (pMHC)-expressing antigen-presenting cell (APC), forming the immunologic synapse (IS) (Grakoui et al., 1999; Monks, Freiberg, Kupfer, Sciaky, & Kupfer, 1998). With the advance in confocal imaging techniques and the development of supported lipid bilayers incorporating pMHC molecules and adhesion molecules, it was observed that IS formation is preceded by ligand-dependent formation of microclusters, 0.5–1 μm large structures containing TCRs and downstream signaling molecules, that move and coalesce in the IS (Varma, Campi, Yokosuka, Saito, & Dustin, 2006; Yokosuka et al., 2005). However, in the last decade, it has become clear that clusters of the TCR can be detected at the cell surface of T cells in absence of any ligand. These clusters, with size ranging from tens to a few hundred nanometers and therefore referred to as TCR nanoclusters, were first shown via electron microscopic analysis of cell surface replicas of intact cells (Kumar et al., 2011; Schamel et al., 2005) and of T-cell membranes (Lillemeier et al., 2010) and more recently by superresolution light microscopy techniques (Lillemeier et al., 2010; Sherman et al., 2011).

The relevance of determining the clustering state of a receptor before ligand binding lies in the fact that preclustering can give rise to synergy in ligand binding and/or receptor output, which would result in receptor properties that cannot be explained by considering the receptor as a monomer. Theoretical and experimental studies on the chemotactic receptors in *E. coli* have shown that clustering of these receptors is essential for permitting the exquisitely sensitive chemotactic response of these bacteria (Bray, Levin, & Morton-Firth, 1998). Similar mechanisms appear to play a role in TCR-dependent signaling. Even though monomeric TCRs have a low affinity for their cognate pMHC ligand, T cells only need to detect a few pMHC molecules on the APC in order to become activated (Demotz, Grey, & Sette, 1990; Harding & Unanue, 1990; Irvine, Purbhoo, Krogsgaard, & Davis, 2002). Our own studies indicate that TCR complexes present in nanoclusters become more activated than monomeric TCRs under conditions of weak stimulation (Schamel et al., 2005) and that they play a direct role in the increase in T-cell sensitivity accompanying the transition from naive to memory T cells (Kumar et al., 2011; Molnar, Deswal, & Schamel, 2010; Molnar et al., 2012; Schamel & Alarcon, 2013). Knowing whether a receptor is present at the cell surface as a cluster before ligand binding should not only help to explain the biology of that receptor but also open new possibilities for interfering with its function for therapeutic purposes.

The technique consists of fixing cells with paraformaldehyde (PFA) to avoid any protein movements at later steps, followed by labeling with antibodies against the protein of interest and colloidal gold-coupled secondary reagents. Cells are adhered to a mica sheet and frozen ultrarapidly, followed by freeze-etching and generation of cell surface replicas by oblique deposition of a heavy metal, usually platinum, and strengthening with a layer of electron-translucent carbon. After removal of the biological material, the replica is mounted for examination in the transmission electron microscope (TEM). The replica allows the analysis of the lateral distribution of receptors at the cell surface and offers planar views of large membrane areas, which area not contacting the experimental support. Among the cell surface replication methods, freeze-etching/freeze-drying is superior to critical point-drying or air-dried membrane sheet generation in terms of structural fine detail preservation.

Unlike the superresolution light microscopy techniques, this technique permits molecular resolution of cell surface molecules on the membranes that are not in contact with the support, thereby avoiding visualization of potential clustering artifacts induced by adherence (James et al., 2011). It can directly be applied to the cell surface molecule of interest, without the need to genetically modify this protein with a fluorophore, provided that good antibodies against the protein are available. The technique can also be used to study ligand-induced receptor clustering or to analyze the role of the cytoskeleton or particular lipid components of the membrane in receptor clustering by pretreatment with the appropriate inhibitors or reagents (Schamel et al., 2005). It is, however, not compatible with dynamic observations and, given the relatively low labeling efficiency, does not allow absolute quantifications by direct gold counting.

The protocol is based on our experience with primary lymphocytes and lymphocyte cell lines that are labeled in solution before adherence to the mica sheet. This freeze-etching replica technique was originally developed for adherent cell types, and we refer the reader to this body of literature for a more extensive description of this and related techniques (Boonstra, van Bergen en Henegouwen, van Belzen, Rijken, & Verkleij, 1991; Nermut, 1995; Severs, 1995; Severs, Newman, & Shotton, 1995; van Belzen et al., 1988).

21.1 MATERIALS

1. Phosphate-buffered saline (PBS) pH 7.4
2. PFA: 4% (w/v) solution in PBS (freshly prepared or from a frozen stock thawed just before use)
3. PBS containing 0.1% (w/v) bovine serum albumin (BSA), stored at 4 °C
4. 100 µg/ml poly-L-lysine in H_2O
5. Glutaraldehyde (25% solution EM grade, TAAB Laboratories, cat no. G011)
6. Parafilm
7. Bacterial culture dishes (10 cm diameter)
8. 24-well tissue culture plate
9. Mica sheets ($3'' \times 1'' \times 0.006''$; e.g., TAAB, cat no. M054)
10. Fine-tipped tweezers: Dumont tweezers (Dumoxel/Biology) curved or straight tips

11. Fine-tipped tweezers with clamping ring to plunge samples in cryogen (Dumont tweezers, Biology clamp)
12. Rounded tweezers with insulator cover
13. Primary antibody at saturating concentration in PBS containing 0.1% BSA
14. Colloidal gold-coupled protein A (CMC Utrecht) or gold-coupled secondary antibodies (Biocell or Aurion), gold diameter of 5 or 10 nm diluted in PBS containing 0.1% BSA
15. Ethane for fast freezing of the sample
16. Liquid nitrogen for sample preparation and storage
17. Cryovials with cap (2 ml volume)
18. Thin metal rod and a spirit burner to prepare cryovials for frozen mice storage
19. Domestic bleach (sodium hypochlorite) for digestion of organic material
20. Triple-distilled water
21. Pasteur pipettes with bent, melted tip to transfer replicas during cleaning/washing
22. Copper EM bare grids (400 mesh hexagonal, Gilder G400HEX-C3)
23. Acetone to clean copper EM bare grids
24. 100% ethanol
25. Filter paper

21.2 EQUIPMENT

1. Microcentrifuge
2. Spirit lamp
3. Dissecting seeker
4. Plunge-freezing unit (Fig. 21.1) (KF-80, former Reichert–Jung)
5. Liquid nitrogen bench Dewar
6. Liquid nitrogen storage tank (optional)
7. Twin cold light source
8. Freeze-etching unit (Fig. 21.2) (BAF 060; Bal-Tec)
9. Specimen tables for freeze fracturing, modified to accommodate mica rectangles (Fig. 21.3) (Bal-Tec, BU 02 098-T)
10. Bal-Tec specimen table loading device to insert specimen tables and split mica sandwiches under liquid nitrogen (Fig. 21.4)
11. Porcelain spotting dishes (Fig. 21.5)
12. TEM, preferentially with camera with tiling option (e.g., TemCam-F416, TVIPS).

21.3 METHODS

In Table 21.1, the protocol is summarized and time estimates and points of possible interruption are given.

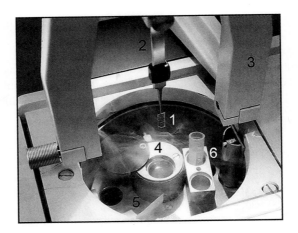

FIGURE 21.1
Freezing chamber of the KF-80 plunge-freezing unit. A mica sandwich (1) is held between tweezers with a clamping ring (2) and is inserted into the forceps injector of the unit (only the arc (3) of the injector is visible in the image). Upon shooting the sandwich in the cryogen (4), it is freed of excess of cryogen by blotting it on a piece of filter paper (5) and transferred to a perforated cryovial (6), using the tweezers (2) and taking care to maintain the sandwich all the time below the rim of the freezing chamber. (For color version of this figure, the reader is referred to the online version of this chapter.)

21.3.1 Cell isolation, purification, and fixation

Single-cell suspensions are prepared in ice-cold PBS at a concentration of 10×10^6 cells/ml. For each individual sample, not taking into account experimental replicas or the different antibodies to be used in the experiment, around 2×10^5 cells are necessary. If the cells of interest need to be purified from a population via antibody-dependent techniques (e.g., fluorescence-activated cell sorting or magnetic bead-based separation), it is necessary to achieve this by depletion of the unwanted cell population. In this way, the cells of interest will remain free of antibody at the cell surface, permitting the use of colloidal gold-conjugated protein A as the secondary reagent for the immunogold labeling. Cells are fixed by addition of an equal volume of ice-cold 4% PFA in PBS (2% final concentration) and kept on ice for 30 min, followed by two washes with ice-cold PBS to remove free PFA. Resuspend the cells after the last wash in PBS+0.1% BSA (w/v) at a concentration of 10×10^6 cells/ml and divide the cells in clean microcentrifuge tubes according to the number of different antibody stains planned.

21.3.2 Immunogold labeling

Fixed cells are labeled on ice with saturating amounts of the primary antibody in PBS+0.1% BSA for 30 min. Spin the antibody solution for 1 min at full speed in a benchtop centrifuge before adding to the cells. This removes antibody aggregates

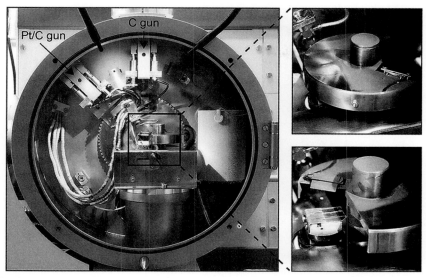

FIGURE 21.2

Frontal view of opened BAF060 freeze-etching unit. The positions of the Pt/C and C electron evaporation beam guns are indicated. The images to the right show the specimen table loaded with mica sheets during shadowing (bottom right) and the shutter in etching position (top right). (For color version of this figure, the reader is referred to the online version of this chapter.)

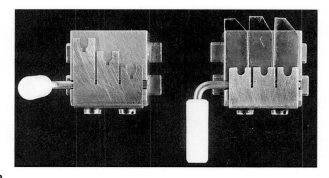

FIGURE 21.3

Modification of specimen tables. The tongues of the specimen clamp of the original specimen tables (left) were cut down (right) to allow larger areas of the sample-carrying micas to be replicated. (For color version of this figure, the reader is referred to the online version of this chapter.)

FIGURE 21.4

Specimen table loading device. The specimen table (1) with its lever in the loading position (2) is inserted into the specimen table loading device (3) and submerged in a Dewar filled with liquid nitrogen. Mica sandwiches are then deposited onto the supporting table (4), split with fine-tipped tweezers, and the sample containing mica sheet is loaded into the specimen table, using the cut corner as a reference for the correct orientation. Note that for clarity sake, the image has been taken in absence of liquid nitrogen, which under operating conditions should cover all the objects within the Dewar. (For color version of this figure, the reader is referred to the online version of this chapter.)

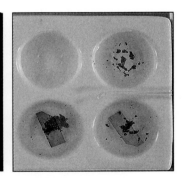

FIGURE 21.5

Floating and washing of replicas. The mica is held with a pair of fine tweezers with the replica side facing upward and submerged gently at a 30°–45° angle into domestic bleach solution. The replica should float on the surface. After overnight incubation, pieces of the replica are picked up with a molten glass pipette (middle panel) and transferred to wells filled with triple-distilled water. If upon overnight incubation the replica still consists of one big piece (right panel, lower left well), it is better to break it into smaller pieces by tapping it gently with the glass pipette. Picking up the big piece at once would result in folding over itself, making it impossible to analyze afterward. (For color version of this figure, the reader is referred to the online version of this chapter.)

Table 21.1 Summary of Procedure, Time Estimates and Points of Interruption

Step	Time	Interruption of Procedure Upon Completion of Step and Storage
Purification of cells and fixation in 2% PFA	45 min–4 h	Yes, store in PBS+0.1% PFA at 4 °C for up to 3–4 days
Immunogold labeling, preparation of mica	2 h	No
Adherence to mica and postfixation with 0.1% glutaraldehyde	2 h	Yes, store overnight in PBS at 4 °C
Ultrarapid freezing of samples	1 h	Yes, store indefinitely under liquid nitrogen
Freeze-etching and replica generation	3 h	No
Floating of replicas on domestic bleach	30 min	Yes, leave floating overnight at room temperature
Washing and mounting of replicas on grids	4 h	Yes, store indefinitely at room temperature protected from dust

that may have been present in the antibody stock. The ability of the primary antibody to recognize the antigen after 2% PFA fixation and saturating antibody concentrations should be determined beforehand (e.g., by flow cytometry). Prepare specificity controls by labeling samples with equal concentrations of isotype-matched irrelevant antibodies. Wash the cells twice with PBS+0.1% BSA, centrifuging them for 90 s in a microcentrifuge at 6000 rpm. Samples are then incubated for 1 h with 10 nm gold-conjugated protein A in PBS+0.1% BSA at previously established saturating conditions (see the succeeding text) and washed twice with PBS+0.1% BSA and once with PBS only. Resuspend cells at a concentration of 4×10^6 cells/ml in PBS.

21.3.3 Cell adherence to freshly exfoliated mica sheets

Directly after starting the first step of the immunogold labeling, mica sheets are cut in approximately 1 cm × 0.5 cm rectangular pieces and split using fine-tipped tweezers. The split mica pieces are positioned with the freshly exposed surface facing upward on a piece of parafilm, secured to the bottom of a 10 cm bacterial culture dish. Cut off the one corner of the mica with scissors (necessary for identification and orientation purposes during mounting of the mica on the specimen table) and then overlay the surface with 50 µl of a 100 µg/ml (w/v) poly-L-lysine solution, prepared in H_2O. Incubate for 1–2 h on ice. Wash the mica surface by first removing the poly-L-lysine solution and then adding and removing three times 50 µl PBS with a pipette. Be careful to avoid that the solution spills over the sides of the mica. Add directly after the last wash 50 µl of the immunogold-labeled cell suspension (corresponding to 2×10^5 cells) and incubate for 1 h on ice to allow the cells to adhere to the mica sheet. Gently remove the solution from the top of the mica by aspirating with a pipette positioned in one of the corners of the mica. Wash two times with 50 µl PBS, adding the PBS with a pipette positioned in one corner and aspirating with the pipette positioned in the diagonally opposing corner. Perform a postfixation by adding 50 µl of a freshly prepared solution of 0.1% (v/v) glutaraldehyde in PBS. Incubate for 20 min on ice and wash three times with 50 µl PBS, using the "diagonal" washing procedure described earlier. Store the samples at 4 °C in individual wells of a 24-well tissue culture plate filled with PBS until the freezing step.

21.3.4 Freezing of samples

21.3.4.1 Preparation of the plunge-freezing unit

We use a Reichert–Jung (now Leica) KF-80 plunge-freezing unit (Fig. 21.1) and liquid ethane as secondary cryogen. Alternatively, liquid propane can be used as secondary cryogen, but one has to take extra precautions as it is much more inflammable than ethane. Homemade plunge freezers or a Leica CPC unit can be used as well. For immersion fixation of the mica sandwiches, a forceps injector is used.

This ultrarapid freezing (cooling rates in excess of 10^4 °C/s) ensures good ultrastructure preservation deep as 10–20 µm, enough for the freezing of mica sandwich mounted cells, allowing us to work in the absence of cryoprotectants (Severs, 1995; Severs et al., 1995) such as glycerol, which can induce membrane artifacts and difficult etching due to its low volatility at low temperatures and under vacuum. Ultrarapid freezing can be achieved by plunge freezing or by metal block-impact freezing, while manual immersion of specimens into the cryogenic liquid does not ensure good preservation of samples that have not been cryoprotected.

Connect the freezing unit to a liquid nitrogen tank and a gas cylinder containing the secondary cryogen according to the instructions provided in the manual of the unit. Make sure that the cryogen container is inserted in its dedicated slot and place a piece of filter paper and a metal bottle cap or any other metal container that permits upright positioning of the cryovial in the freezing chamber (Fig. 21.1). Cover the chamber with its lid. Note that the chamber should be covered with the lid between each cycle of sample freezing. Bring the freezing unit to its desired operating temperature. The freezing point of ethane is -171 °C and of propane -187 °C. Select a temp just (1–2 °C) above the freezing point of the cryogen to be used. When the operating temperature has been reached, fill the cryogen container with liquid cryogen up to the top. In plunge freezers lacking temperature control, the cryogen will freeze at the temperature of liquid nitrogen. One can defrost the cryogen by inserting a metal object into the cryogen just before plunging the sample.

21.3.4.2 Preparation of cryovials for storage of the frozen mica sandwiches

During the cooling of the plunge-freezing unit, prepare the cryovials to be used for storage of the samples. Each vial can hold up to two frozen mica sandwiches, provided that they are experimental duplicates. Label the vials. Heat the tip of a dissecting seeker or similar thin metal rod in the flame of a spirit burner and punch three holes evenly spread around the side of a 2 ml cryovial, just below the thread for the screw cap. Also punch three holes through the top of the cap of the cryovial. This procedure assures that in upright position, the cryovials will keep covering the sample with liquid nitrogen during transfers and will fill up instantaneously upon submergence in liquid nitrogen.

21.3.4.3 Freezing of the samples

Once the cryogen is in the plunge freezer, fill a liquid nitrogen bench Dewar to the rim and put in sufficient cryostorage canes to hold all the samples that are going to be frozen. Fill an uncapped perforated cryovial with liquid nitrogen, and put the filled cryovial in upright position in the holding device in the chamber of the freezer unit. Take a sample containing mica from the 24-well plate using fine tweezers and carefully remove excess liquid by touching the edge of the mica against a piece of blotting paper for 2–5 s. Immediately, a freshly exfoliated piece of mica of the same size but with intact corners and lacking poly-L-lysine treatment is placed with its freshly exposed surface against the cell-containing mica surface. This mica "sandwich" is held by tweezers with a clamping ring. Load the spring of the forceps injector of the plunge-freezing unit and insert and secure the tweezers with the sample. If the plunge-freezer design allows it, turn the forceps injector away from the freezing chamber when inserting the tweezers with sample. Position the loaded injector in a vertical position above the freezing chamber, remove the cover, and immediately release the sample into the cryogen. The sample will freeze almost instantaneously. With the injector still in its down position, remove the tweezers with sample, blot away excess cryogen with the piece of filter paper located in the freezing chamber, and transfer the frozen mica sandwich into the liquid nitrogen-filled cryovial, taking care to keep the frozen mica sandwich all the time below the rim of the freezing chamber. Screw the perforated cap on the cryovial using two pairs of tweezers, transfer the cryovial to the liquid nitrogen-filled Dewar, and snap the cryovial in a storage cane. The samples can be stored in a liquid nitrogen storage tank until the moment of the cell surface replica generation.

To minimize the risk of ice recrystallization, care must be taken to avoid accidental warming of the frozen specimens. Manipulation of the frozen samples should be done using fine tweezers whose tips have been precooled in liquid nitrogen and you should work fast.

21.3.5 Preparation of cell surface replicas

21.3.5.1 Preparation of the freeze-fracture unit

Precise details of the freeze-fracture unit, evacuation and cooling down, gun cleaning and adjustments, etc. will vary between different models of freeze-fracture units and should be conducted according to manufacturer's instructions. However, the procedures and recommendations based on our experience with Balzers BAF 400T (Severs, 2007) and BAF060 machines (Fig. 21.2, left) should be of general applicability. Carefully clean the electron evaporation beam guns and insert the electron beam gun loaded with a platinum–carbon rod (95% Pt; 5% C) in the gun holder of the vacuum chamber at an angle of 45° and the electron gun loaded with the carbon rod at 90°. Switch on the unit and evacuate the vacuum chamber. Cool down the freeze-fracture unit to $-150\ °C$ and wait until a vacuum of at least 10^{-6} mbar (usually 5×10^{-7} mbar in the BF060) is reached. This process will take ~2 h. It is advisable to always carry out a trial coating prior to the actual freeze-etching and shadowing of the samples in order to check that the evaporation devices function correctly. In case of using new pieces of carbon rods, degassing is needed before making the first evaporation.

21.3.5.2 Mounting of samples on the specimen table

The samples need to be mounted on a specimen table before they can be introduced in the freeze-fracture unit. We modified the specimen clamp of a specimen table with a retaining spring for 0.3×0.8 mm gold specimen carriers so that a larger area of the mica rectangles containing the labeled cells gets exposed (Fig. 21.3). Insert the lever for the retaining clamps in the specimen table, and position it in the loading position (Fig. 21.3, right). Fill a bench Dewar with liquid nitrogen and completely submerge the specimen table loading device with the specimen table in place as well as a metal container to hold the cryotubes with mica sandwiches (Fig. 21.4) and wait until the loading device and specimen table have completely cooled down (i.e., when the liquid nitrogen stops bubbling). Make sure that the specimen table and loading device remain covered with liquid nitrogen during the complete mounting procedure. Place a cold light source with two flexible light guides around this setup. Transfer a cryovial containing the frozen mica sandwich into the metal container using insulated rounded tweezers. Open the cryovial under liquid nitrogen using two pairs of rounded tweezers and let the mica sandwich slip out of vial onto the sample supporting platform of the loading device. Submerge two pairs of fine-tipped Dumont tweezers into the liquid nitrogen to let them cool down. Warm tweezers may cause crystallization of vitreous ice in and around the cells on the mica, which in turn may damage the cells. Use one pair to hold the mica sandwich (taking care to keep it under the liquid nitrogen), and use the other pair of tweezers to separate the two mica halves. Identify the mica containing the labeled cells by means of the cut corner, and slide it with the cell containing side upward under the retaining clamp of the specimen table. Discard the other half. Depending on their size, 3–4 mica rectangles will fit on a single table, but care should be taken that the micas do not overlap. Use another pair of fine-tipped tweezers to split and position the next sample. Alternatively, thoroughly dry the same set of tweezers with a conventional hair dryer before reusing them. Do not forget to precool tweezers before handling the micas. When all micas have been positioned on the specimen table, immobilize the micas by turning the lever into the locking position (Fig. 21.3, left) and then removing it.

21.3.5.3 Etching and replication

The specimen table with the frozen samples is quickly transferred to the precooled temperature-controlled specimen stage within the high vacuum chamber of the freeze-fracture unit using the loading/retrieval manipulator provided with the

BAF060 unit. Try to keep distance and time of transfer as short as possible, in order to minimize condensation and ice crystal formation on the samples. When using freeze-fracture units without an air lock (such as the Balzers BAF 400T), vacuum of the vacuum chamber will have to be broken right before introducing sample and vacuum will then have to be reestablished before continuing the procedure. Position the cold shutter carrying the knife above the specimen table (Fig. 21.2, right) and raise the temperature of the specimen table up to $-80\ °C$, while the temperature of the shutter remains at $-150\ °C$. This permits sublimation of water molecules to form in the surface of the cells with the shutter functioning as a cold trap. Proceed with etching for 12 min to sublime surface ice.

The specimens are then shadowed with a 2 nm thick layer of atomic platinum from the Pt/C-loaded electron evaporation beam gun. This physically fragile electron-dense replica is strengthened by a uniform 20 nm thick electron-translucent carbon layer using the C-loaded gun. The thickness of the deposited layers is measured with a quartz crystal film thickness monitor included in the freeze-fracture unit. The specimen table can then be removed from the unit using the loading/retrieval device after which one should immediately proceed with cleaning and mounting of the replicas (see the succeeding text). To perform more than one round of replica generation, it is convenient to have a clean and dry replacement specimen table ready. In general, the carbon gun can be used twice before it needs to be reloaded. The Pt/C gun can be used several times before it is exhausted, but you must check the position of the Pt with respect to the tungsten filament after each session.

21.3.6 Cleaning and mounting of replicas

Remove specimen table from the freeze-fracture unit. Immediately upon retrieval, shadowed micas are removed from the specimen table, and while holding them with tweezers at the end that was positioned under the holding clamps of the cold table and with the coated surface facing upward, they are very slowly submerged at a 30°–45° angle in a well of a porcelain spotting dish filled with undiluted domestic bleach (Fig. 21.5, left). The replica should come off the mica piece at the surface of bleach solution and float. It is normal that the replica breaks in smaller pieces upon floating. However, if it disintegrates in barely visible pieces, the sample cannot be processed any further and has been lost (also see section 21.5.9, replica floating). Replica pieces are incubated at room temperature overnight floating on bleach for digestion of organic material. To remove attached organic material, floating replica pieces are washed two times with triple-distilled H_2O by carefully transferring them onto the surface of water-filled wells using a molten tip Pasteur pipette (Fig. 21.5, center). Perform a third wash for 1 h. Pick up single-layer replica pieces with bare 400 mesh hexagonal copper grids directly from the water surface. Any trace of water is carefully removed with pointed strips of filter paper applied to the edge of the grid and grids are placed on tissue paper for desiccation. Watch out to pick up unfolded replicas as doubling will impair transparency of the sample during TEM observation.

Grids should be scrupulously cleaned with acetone before picking up the replicas to ensure firm attachment of the replica to the grid.

21.3.7 Analysis of metal replicas by TEM

Replicas mounted on EM grids are examined using a TEM. Replicas are of intrinsically high contrast and observation is undertaken with a moderate beam current at an accelerating voltage of 80 kV. The resolution of the replica is limited by the dimensions of the metal grains being typically 1.5–2 nm for a platinum–carbon replica. Grids are examined first at low magnification (around 1 K) to inspect the quality of the replica. High-quality replicas have a fine granular aspect and show no signs of doubling. Areas with replicated cells can first be identified at low magnification (around 1 K) by a bulge (caused by the nuclear area of the original cell) sticking out of the plane of the replica. Augmenting magnification to around 5 K will show a surrounding area that is slightly elevated relative to the scaffold area and that is also part of the cell body. For T lymphocytes, this area is more prominent in previously activated cells as compared to naive cells. If the cells have been labeled before adherence to the mica sheet, the border of this area should coincide with a transition from a gold-labeled to an unlabeled surface. Direct counting of gold particles can be done in the EM at a magnification of 25–30 K. If the microscope is equipped with a camera that has a tiling function, it is more convenient to take overlapping images of the cell at 8–12 K magnification (depending on the size of colloidal gold conjugate used, the cell body size, the resolution of the camera, and the maximum number of overlapping images that the camera and associated software can handle) and generate a composite image of the whole cell body that can be analyzed off-line (Fig. 21.6).

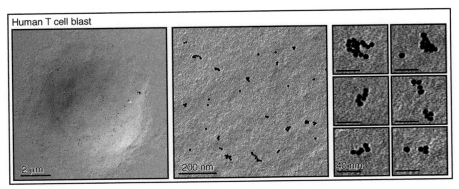

FIGURE 21.6

Analysis of cell surface replicas by transmission electron microscopy. Cell surface replicas of human T-cell blasts after anti-TCR-immunogold labeling are shown. Note the prominent bulge representing the nuclear area-covering membrane and the flatter area surrounding it. The cell overview picture (left panel) was generated from nine partially overlapping photos taken at 12 K magnification, and enlargements are shown in the middle and right panels.

21.4 ANALYSIS

For each cell replica, groups of gold particles are counted and categorized according to the number of gold particles. Gold counts and the size distribution of observed clusters are written down for each individual cell in a prepared table. When analyzing replicas of cells labeled with antibodies specific for the TCR, we consider that gold particles are part of the same cluster when they are adjacent or when the distance between two particles is less than 10 nm (i.e., the diameter of a single gold particle), taking into account that the diameter of the TCR is around 10 nm (Arechaga et al., 2010) and that the labeling efficiency is low. As a starting point, we record the gold particle distribution of 10 cell replicas of each condition (background controls and experimental groups) and determine statistical significance of differences (e.g., using two-tailed T-tests and chi-square tests). Automated data collection methods and computational methods to determine cell surface receptor distribution have been described (Zhang et al., 2006) but require careful filtering of the images to discriminate gold particles from small holes and other irregularities that cause electron-dense areas.

21.5 CONSIDERATIONS

21.5.1 Safety precautions

Use a fume hood and wear gloves and eye protection when working with PFA and glutaraldehyde. Dispose of waste according to institutional guidelines. Wear eye protection and cryogloves when handling liquid nitrogen. Extreme care must be taken to eliminate any possibility of explosion hazard when working with flammable gases. The liquefied cryogens should be removed from the unit on completion of the freezing work and disposed of after each experiment by evaporation within the fume hood (ethane) or burned in a specific burner (propane).

21.5.2 Starting number of cells

The starting numbers provided are based on our experience with naive and activated primary B and T cells and B- and T-cell lines. For other cell types grown in suspension, this may have to be adjusted empirically. Unlabeled, PFA-fixed cells can be adhered to freshly cleaved and poly-L-lysine-treated micas, and upon glutaraldehyde fixation and washing, the mica sheets can be observed under a phase-contrast light microscope. Cells should not be layered on top of each other, as this complicates the ability to discern individual cells. Too few cells on the mica, especially if they are very round cells (such as naive lymphocytes), may lead to very high surface tension around the replica of the actual cells, causing the cell replica to pop out. A cell density resulting in cells spaced at distances of approximately one cell diameter should avoid these problems. Adherent cells could in principle be grown to confluency, but if discrimination of individual cells is critical for the interpretation of the data, it would be better to grow them to a subconfluent density.

21.5.3 Colloidal gold conjugates

We have almost exclusively used colloidal gold reagents with a gold particle size of 10 nm. Gold particles of 5 nm are more difficult to detect unambiguously due to the often irregular surface and high contrast of the replicas. Use preferably freshly purchased gold conjugates as long-term storage might increase aggregation of the reagent. The aggregation state of the gold reagent can be tested by adsorbing diluted suspensions of gold conjugates onto collodion/carbon-coated EM grids and analyzing in TEM.

21.5.4 Cell surface replication and replication of fracture surfaces

Upon attachment to a support lymphocytes will, in contrast to adherent cells, not form a continuous monolayer. This discontinuity makes it more difficult to detach the metal replica from the mica sheet. Covering the cells with another mica sheet, thereby creating a mica sandwich, allows formation of a continuous thin layer of buffer that afterward facilitates the generation of a continuous replica that can be floated in the cleansing solution. When you split the frozen mica sandwich, most cells remain attached to the poly-L-lysine-coated mica and will be surface replicated, with most of the gold particles stabilized by the replica. Some cells can be fractured by the splitting of the mica sandwich. If the cell is fractured through the lipid bilayer of the nonadhered membrane, the gold label will be lost, even though the replica has a very similar appearance to nonfractured cells. On the other hand, if the fracture plane follows the plasma membrane attached to the mica surface, one will not observe the bulge caused by the nuclear area, but the gold particles will remain attached to the exoplasmic hemimembrane under the replica and can be visualized and quantified in the TEM. It should however be taken into account that the membrane represented by this replica was in contact with the mica support, which could give rise to a change in distribution of the cell surface molecules studied as compared to the nonadhered membrane.

21.5.5 Primary antibody and fixation conditions

If primary antibodies do not work under the fixation conditions used, one could try to lower the concentration of PFA to 1% or omit the prefixation step. However, cells should then be kept all the time on ice and could be labeled in presence of 0.02% of NaN_3, in order to reduce the risk of antibody-induced capping of the cell surface receptor.

21.5.6 Secondary reagents

Dilutions of the colloidal gold-coupled secondary reagent can be tested on a batch of cells stained with saturating concentrations of the primary antibody (tested by flow cytometry), followed by replica generation and analysis. Use an isotype control antibody to verify the amount of background staining. For the 10 nm colloidal gold-conjugated protein A, we have routinely used a threefold lower dilution than the one recommended by the manufacturer (CMC Utrecht).

21.5.7 Freezing of samples

The cryogen container should be filled to the rim with secondary cryogen to avoid precooling of the specimen in the gaseous layer of ethane that otherwise forms between the surface of the cryogen and the upper rim of the container. For the same reason, one should only position the forceps injector, holding the tweezers and mica sandwich, above the freezing chamber for the time needed to plunge the sample. Remove excess of the liquid ethane adhering to the mica with the filter paper before transferring to the liquid nitrogen-filled cryotube. The ethane will freeze in liquid nitrogen, making splitting of the mica sandwich just before loading on the specimen table much more difficult.

21.5.8 Mounting of samples in the specimen table

Take care that the bench Dewar, specimen table, loading device, and tweezers are clean and dry to avoid contamination of the sample and condensation of water onto the sample, which can give rise to subsequent ice crystal formation. We store specimen tables in a 56 °C stove to assure that they are completely dry before use. Do not breathe directly on the liquid nitrogen surface while mounting the micas on the table, in order to avoid that the liquid nitrogen will reach its boiling point. This would give rise to bubbles that impair visualization of the samples under the liquid nitrogen surface.

21.5.9 Replica floating

If the replica does not detach from the mica upon submergence in the bleach solution, one can scrape the edges of the mica with Dumont tweezers. Alternatively, use the tweezers to scrape a grid pattern on the surface of the replica to obtain smaller pieces of replica that may float more easily. One could in principle leave the mica sheet with the replica still attached on the bottom of the well for digestion of the organic material, but in our hands, floating of an unfolded replica piece is rare under these conditions. An oily and inconsistent appearance of the replica indicates a too thin carbon layer.

21.5.10 Replica stability

Extra support for the replicas can be provided by precoating the EM grids with a Formvar or carbon layer. This can be especially helpful when replicating small and rounded naive T lymphocytes, which have a tendency to pop out of the replica. If the replica pieces are tiny, they can be lost in the vacuum of the TEM column if repeatedly mounted in the TEM stage. It is therefore advisable to image all the relevant cells of a single replica in one session and perform quantification off-line.

21.5.11 Replica quality

Replicas should have a smooth appearance. If they look pale and lack contrast, the Pt deposition is insufficient. If the replica is too dark, there is an excess of Pt/C making it difficult to discern gold particles.

21.5.12 Low labeling efficiency and implications for interpretation of data

The total amount of gold particles we have quantified per cell indicated that we obtained labeling efficiencies that reach around 5–15%. This has therefore not allowed us to directly infer the distribution of the TCR or any other surface protein from the distribution of the corresponding gold particles. However, the technique can be used to compare receptor distribution between different populations of cells. In the case of the TCR, we have obtained conformation of the validity of its comparative power by analyzing the TCR distribution of two T-cell populations by this technique and by blue native PAGE (Kumar et al., 2011) and comparing methyl-β-cyclodextrin-treated and nontreated cells (Schamel et al., 2005).

If the staining efficiency is calculated, one can get some information by Monte Carlo simulations, in order to include or to exclude possible distributions of the surface protein (Fiala, Kaschek, Blumenthal, Reth, Timmer, & Schamel, under revision).

Acknowledgments

We thank Jose Antonio Pérez Gracia from the "servicio de diseño gráfico y fotografía" of the CBMSO for providing the photographs used in the figures. We thank Milagros Guerra and Lara Rodenstein from the "servicio de microscopía electronica" of the CBMSO for excellent technical assistance. This study was supported in part by the Excellence Initiative of the German Research Foundation (GSC-4, Spemann Graduate School) and EXC294 (BIOSS) to GJF and WWS; the EU through grant FP7/2007-2013 (SYBILLA) to GJF, and WWS; and grant BFU2009-08009 from the "Ministerio de Ciencia e Innovación" to HMvS.

References

Arechaga, I., Swamy, M., Abia, D., Schamel, W. A., Alarcon, B., & Valpuesta, J. M. (2010). Structural characterization of the TCR complex by electron microscopy. *International Immunology, 22*, 897–903.

Boonstra, J., van Bergen en Henegouwen, P. M., van Belzen, N., Rijken, P. J., & Verkleij, A. J. (1991). Immunogold labelling in combination with cryoultramicrotomy, freeze-etching, and label-fracture. *Journal of Microscopy, 161*, 135–147.

Bray, D., Levin, M. D., & Morton-Firth, C. J. (1998). Receptor clustering as a cellular mechanism to control sensitivity. *Nature, 393,* 85–88.

Demotz, S., Grey, H. M., & Sette, A. (1990). The minimal number of class II MHC-antigen complexes needed for T cell activation. *Science, 249,* 1028–1030.

Fiala, G. J., Kaschek, D., Blumenthal, B., Reth, M., Timmer, J., & Schamel, W. W. A. Pre-clustering of the B cell antigen receptor demonstrated by mathematically extended electron microscopy. *Frontiers in B cell Biology.* Under Revision.

Grakoui, A., Bromley, S. K., Sumen, C., Davis, M. M., Shaw, A. S., & Allen, P. M. (1999). The immunological synapse: A molecular machine controlling T cell activation. *Science, 285,* 221–227.

Harding, C. V., & Unanue, E. R. (1990). Quantitation of antigen-presenting cell MHC class II/peptide complexes necessary for T-cell stimulation. *Nature, 346,* 574–576.

Irvine, D. J., Purbhoo, M. A., Krogsgaard, M., & Davis, M. M. (2002). Direct observation of ligand recognition by T cells. *Nature, 419,* 845–849.

James, J. R., McColl, J., Oliveira, M. I., Dunne, P. D., Huang, E., Jansson, A., et al. (2011). The T cell receptor triggering apparatus is composed of monovalent or monomeric proteins. *Journal of Biological Chemistry, 286,* 31993–32001.

Kumar, R., Ferez, M., Swamy, M., Arechaga, I., Rejas, M. T., Valpuesta, J. M., et al. (2011). Increased sensitivity of antigen-experienced T cells through the enrichment of oligomeric T cell receptor complexes. *Immunity, 35,* 375–387.

Lillemeier, B. F., Mortelmaier, M. A., Forstner, M. B., Huppa, J. B., Groves, J. T., & Davis, M. M. (2010). TCR and Lat are expressed on separate protein islands on T cell membranes and concatenate during activation. *Nature Immunology, 11,* 90–96.

Molnar, E., Deswal, S., & Schamel, W. W. (2010). Pre-clustered TCR complexes. *FEBS Letters, 584,* 4832–4837.

Molnar, E., Swamy, M., Holzer, M., Beck-Garcia, K., Worch, R., Thiele, C., et al. (2012). Cholesterol and sphingomyelin drive ligand-independent T-cell antigen receptor nanoclustering. *Journal of Biological Chemistry, 287,* 42664–42674.

Monks, C. R., Freiberg, B. A., Kupfer, H., Sciaky, N., & Kupfer, A. (1998). Three-dimensional segregation of supramolecular activation clusters in T cells. *Nature, 395,* 82–86.

Nermut, M. V. (1995). Manipulation of cell monolayers to reveal plasma membrane surfaces for freeze-drying and surface replication. In D.M. Shotton (Series Ed.) & N. J. Severs, & D. M. Shotton (Vol. Eds.), *Rapid freezing, freeze fracture and deep etching* (techniques in modern biomedical microscopy) (pp. 151–172). Wiley-Liss, Inc, New York.

Schamel, W. W., & Alarcon, B. (2013). Organization of the resting TCR in nanoscale oligomers. *Immunological Reviews, 251,* 13–20.

Schamel, W. W., Arechaga, I., Risueno, R. M., van Santen, H. M., Cabezas, P., Risco, C., et al. (2005). Coexistence of multivalent and monovalent TCRs explains high sensitivity and wide range of response. *The Journal of Experimental Medicine, 202,* 493–503.

Severs, N. J. (1995). Freeze-fracture cytochemistry: An explanatory survey of methods. In D. M. Shotton (Series Ed.) & N. J. Severs, & D. M. Shotton (Vol. Eds.), *Rapid freezing, freeze fracture and deep etching* (techniques in modern biomedical microscopy) (pp. 173–208). Wiley-Liss, Inc, New York.

Severs, N. J. (2007). Freeze-fracture electron microscopy. *Nature Protocols, 2,* 547–576.

Severs, N. J., Newman, T. M., & Shotton, D. M. (1995). A practical introduction to rapid freezing techniques. In D. M. Shotton (Series Ed.) & N. J. Severs, & D. M. Shotton (Vol. Eds.), *Rapid freezing, freeze fracture and deep etching* (techniques in modern biomedical microscopy) (pp. 31–49). Wiley-Liss, Inc, New York.

Sherman, E., Barr, V., Manley, S., Patterson, G., Balagopalan, L., Akpan, I., et al. (2011). Functional nanoscale organization of signaling molecules downstream of the T cell antigen receptor. *Immunity, 35*, 705–720.

van Belzen, N., Rijken, P. J., Hage, W. J., de Laat, S. W., Verkleij, A. J., & Boonstra, J. (1988). Direct visualization and quantitative analysis of epidermal growth factor-induced receptor clustering. *Journal of Cell Physiology, 134*, 413–420.

Varma, R., Campi, G., Yokosuka, T., Saito, T., & Dustin, M. L. (2006). T cell receptor-proximal signals are sustained in peripheral microclusters and terminated in the central supramolecular activation cluster. *Immunity, 25*, 117–127.

Yokosuka, T., Sakata-Sogawa, K., Kobayashi, W., Hiroshima, M., Hashimoto-Tane, A., Tokunaga, M., et al. (2005). Newly generated T cell receptor microclusters initiate and sustain T cell activation by recruitment of Zap70 and SLP-76. *Nature Immunology, 6*, 1253–1262.

Zhang, J., Leiderman, K., Pfeiffer, J. R., Wilson, B. S., Oliver, J. M., & Steinberg, S. L. (2006). Characterizing the topography of membrane receptors and signaling molecules from spatial patterns obtained using nanometer-scale electron-dense probes and electron microscopy. *Micron, 37*, 14–34.

CHAPTER 22

Identification of Multimolecular Complexes and Supercomplexes in Compartment-Selective Membrane Microdomains

Panagiotis Mitsopoulos and Joaquín Madrenas
Department of Microbiology and Immunology, McGill University, Montreal, QC, Canada

CHAPTER OUTLINE

Introduction and Rationale	413
22.1 Detergent-Insoluble Glycosphingolipid (DIG) Microdomain Isolation	415
22.1.1 Materials	415
22.1.1.1 Reagents	*415*
22.1.1.2 Equipment	*415*
22.1.1.3 Buffers and Sucrose	*415*
22.1.2 Method	417
22.1.2.1 Isolation of DIG Fractions	*417*
22.1.2.2 Immunoprecipitation of DIG Fractions	*417*
22.2 Plasma Membrane Isolation	418
22.2.1 Materials	418
22.2.1.1 Reagents	*418*
22.2.1.2 Buffers and Solutions	*418*
22.2.1.3 Equipment	*419*
22.2.2 Method	420
22.2.2.1 Silica Coating	*420*
22.2.2.2 Fractionation	*420*
22.3 Cardiolipin-Enriched Mitochondrial Membrane Microdomain Isolation	421
22.3.1 Materials	421
22.3.1.1 Reagents	*421*
22.3.1.2 Buffers	*421*
22.3.2 Method	422
22.3.2.1 Mitochondria Isolation	*422*
22.3.2.2 Cardiolipin-Enriched Microdomain Isolation	*422*

CHAPTER 22 Complexes and Supercomplexes in Membrane Microdomains

22.4 Mitochondrial Supercomplex Identification by Blue-Native Gel Electrophoresis ... 423
    22.4.1 Materials ... 423
        22.4.1.1 Reagents ... 423
        22.4.1.2 Equipment ... 424
        22.4.1.3 Buffers and Solutions ... 424
    22.4.2 Method ... 425
        22.4.2.1 Sample Preparation ... 425
        22.4.2.2 Performing Blue-Native Gel Electrophoresis ... 426
        22.4.2.3 Analysis by Western Blotting ... 426
        22.4.2.4 2D Blue-Native Polyacrylamide Gel Electrophoresis (2D BN-PAGE) ... 428
        22.4.2.5 Silver Staining ... 428
22.5 Discussion ... 428
Summary ... 429
Acknowledgments ... 429
References ... 430

Abstract

Cellular membranes contain specialized microdomains that play important roles in a wide range of cellular processes. These microdomains can be found in the plasma membrane and other membranes within the cell. Initially labeled lipid rafts and defined as being resistant to extraction by nonionic detergents and enriched in cholesterol and glycosphingolipids, we now understand that these membrane microdomains are very dynamic and heterogeneous membrane structures whose composition and function can vary widely depending on their cellular location. Indeed, though they are classically associated with the plasma membrane and have been shown to facilitate a wide variety of processes, including signal transduction and membrane trafficking, specialized membrane microdomains have also been identified in other membranes including those in the mitochondria. These mitochondrial membrane microdomains are enriched in cardiolipin, the signature phospholipid of the mitochondria, and may have important implications in metabolism by facilitating optimal assembly and function of the mitochondrial respiratory chain. Furthermore, isolation of multimolecular complexes while retaining their supramolecular interactions has been critical to the study of mitochondrial respiratory supercomplexes. Here, we discuss methods to isolate various membrane microdomains, including detergent-insoluble glycosphingolipid microdomains, mitochondrial cardiolipin-enriched microdomains, and blue-native gel electrophoresis of mitochondrial membranes.

INTRODUCTION AND RATIONALE

The fluid mosaic model of biological membranes by Singer and Nicolson (1972) described the membrane as a primarily lipid matrix with proteins distributed randomly throughout (Singer & Nicolson, 1972). However, shortly thereafter, it was proposed that membranes contain "clusters of lipids," these clusters being "quasicrystalline" regions surrounded by more freely dispersed liquid crystalline lipid molecules. This concept of membrane lipid microdomains was refined to claim that these clusters would contain "lipids in a more ordered state." More importantly, the organization of lipids into membrane microdomains provided an additional conceptual framework for membrane heterogeneity across cellular subcompartments and thus the possibility of structural and functional significance for these microdomains (Karnovsky, Kleinfeld, Hoover, & Klausner, 1982).

Since its original proposal, the concept of membrane microdomains has been better characterized at the molecular level. Under the term of lipid rafts, membrane microdomains were described as being resistant to extraction by nonionic detergents and enriched in cholesterol and glycosphingolipids (Pike, 2009). The term detergent-insoluble microdomains is more encompassing of its heterogeneity in composition than lipid rafts, as some of these microdomains may be selectively enriched for other phospholipids. For simplicity, we will be using both indistinguishably in this chapter. Lipid raft structures were thought to be stably held together by lipid–lipid interactions (Pike, 2009). It has also become clear that lipid rafts are heterogeneous and dynamic sterol- and sphingolipid-enriched nanoscale microdomains that compartmentalize cellular processes (Pike, 2009), including membrane trafficking, signal transduction, and cell polarization (Lingwood, Kaiser, Levental, & Simons, 2009). These microdomains are ordered assemblies of specific proteins in which the meta-stable resting state can be activated to coalesce by specific lipid–lipid, lipid–protein, and protein–protein interactions (Lingwood & Simons, 2010; Simons & Gerl, 2010; Simons & Sampaio, 2011).

The compartmentalization of the plasma membrane into detergent-insoluble microdomains is a very dynamic process. For example, in T cells, the formation of an immunologic synapse correlates with clustering of glycolipid-enriched microdomains. These microdomains are enriched with signaling molecules critical for the initiation and sustainability of T cell antigen receptor mediated signaling such as lck, linker for activation of T cells (LAT), and protein kinase C theta, and their biological integrity is essential for proper T cell antigen receptor mediated signaling (Bi et al., 2001; Viola, Schroeder, Sakakibara, & Lanzavecchia, 1999; Xavier, Brennan, Li, McCormack, & Seed, 1998; Zhang, Trible, & Samelson, 1998). Lipid rafts move into the immunologic synapse in response to CD3 and CD28 ligation, and formation of these microdomains is important for normal signaling leading to T cell activation. Lipid rafts also contain regulators of signaling such as CTLA-4, a negative regulator of T cell activation that partitions within lipid rafts and relocates to the immunologic synapse (Darlington et al., 2002).

In addition to the plasma membrane, detergent-insoluble microdomains have been identified in other cellular membranes including mitochondrial membranes (Christie et al., 2012; Ciarlo et al., 2010; Sorice et al., 2009). In the mitochondria, cardiolipin is the major phospholipid and evidence suggests that there exist microdomains selectively enriched in cardiolipin (Osman, Voelker, & Langer, 2011). Cardiolipin is a phospholipid that has been shown to be important for optimal function of the mitochondrial respiratory chain. Thus, it is possible that the respiratory chain operates within cardiolipin-enriched microdomains in the inner mitochondrial membrane. Isolation of these microdomains will be important for the biochemical characterization of respiratory chain complexes and supercomplexes, and to determine their biological relevance.

In this chapter, we will concentrate on the characterization of multimolecular complexes and supercomplexes in detergent-insoluble microdomains from mitochondrial membranes, specifically cardiolipin-enriched microdomains. These molecular structures are a focus of intense study, specifically understanding how they are assembled and how they ensure optimal cellular respiration. There is an increasing amount of experimental evidence suggesting that mitochondrial respiratory complexes are not simply randomly dispersed within the inner mitochondrial membrane but rather undergo ordered supramolecular associations. These mitochondrial respiratory supercomplexes were first identified by Schagger and von Jagow when they performed blue-native gel electrophoresis on digitonin-solubilized yeast and mammalian mitochondria (Schagger & Pfeiffer, 2000; Schagger & von Jagow, 1991). Since then, supercomplexes have been found in many mammalian tissues, as well as other organisms including fish, fungi, and bacteria (Lenaz & Genova, 2012). These supercomplexes mainly consist of the standard respiratory complexes I, III, and IV in various stoichiometries (e.g., $I–III_2–IV_{1-4}$), though there may be some cases in which complex II is involved (Acin-Perez, Fernandez-Silva, Peleato, Perez-Martos, & Enriquez, 2008). Arrangement of respiratory complexes into supercomplexes has been suggested to aid in substrate channeling and may be important for optimal activity of the respiratory chain. Furthermore, there is evidence that cardiolipin may be necessary for supercomplex assembly (Pfeiffer et al., 2003). The reader is referred to several excellent reviews on mitochondrial respiratory supercomplexes for further details (Lenaz & Genova, 2009, 2012).

Here, we will describe methodology for isolating various membrane microdomains, including detergent-insoluble glycosphingolipid microdomains from plasma membranes for biochemical characterization, and methods for isolating cardiolipin-enriched mitochondrial membranes and native protein complexes including the mitochondrial respiratory chain supercomplexes. These techniques will provide the tools necessary to effectively study multimolecular complexes and oligomeric structures in their intrinsic membrane location and explore the biological functions associated with these specialized membrane microdomains.

22.1 DETERGENT-INSOLUBLE GLYCOSPHINGOLIPID (DIG) MICRODOMAIN ISOLATION

The following protocol is designed for the isolation of detergent-insoluble glycosphingolipid microdomains from cellular membranes, in particular T lymphocytes. This protocol is modified from methods described previously (Brown & Rose, 1992; Field, Holowka, & Baird, 1995; Sargiacomo, Sudol, Tang, & Lisanti, 1993; Xavier et al., 1998; Darlington et al., 2002).

22.1.1 Materials

22.1.1.1 Reagents
MES buffer, pH 6.5
NaCl
Triton X-100
EDTA
Na_3VO_4 (prepared fresh)
PMSF (phenylmethylsulfonyl fluoride)
Aprotinin
Sucrose
Distilled deionized water (ddH_2O)

22.1.1.2 Equipment
Multi-Purpose Rotator (Scientific Industries Inc.)
Dounce homogenizer with loose-fitting plunger (Wheaton)
Pasteur pipettes
Ultracentrifuge (e.g., Optima XL-90 with SW40 rotor, Beckman)

22.1.1.3 Buffers and sucrose
- *DIG lysis buffer* (25 mM MES pH 6.5, 150 mM NaCl, 0.5% Triton X-100, 1 mM EDTA, 1 mM fresh Na_3VO_4, 1 mM PMSF, 10 μg/mL aprotinin in ice-cold ddH_2O)

FOR 30 ML
750 μL of 1 M MES buffer, pH 6.5
1500 μL of 3 M NaCl
1500 μL of 10% Triton X-100
60 μL of 0.5 M EDTA, pH 8.0
300 μL of fresh 0.1 M Na_3VO_4
300 μL of 100 mM PMSF
30 μL of 10 mg/mL aprotinin
25.56 mL of ddH_2O

FOR 50 ML
1250 μL of 1 M MES buffer, pH 6.5
2500 μL of 3 M NaCl

2500 μL of 10% Triton X-100
100 μL of 0.5 M EDTA, pH 8.0
500 μL of fresh 0.1 M Na_3VO_4
500 μL of 100 mM PMSF
50 μL of 10 mg/mL aprotinin
42.6 mL of ddH_2O

- *MBS buffer* (25 mM MES pH 6.5, 150 mM NaCl)

FOR 300 ML
7.5 mL of 1 M MES buffer, pH 6.5
15 mL of 3 M NaCl
277.5 mL of ddH_2O

- *85% (w/v) sucrose* (64 g sucrose + 40 mL MBS buffer)
- *35% (w/v) sucrose*: Dilute 85% sucrose in MBS with 1 mM EDTA and 1 mM Na_3VO_4

4 SAMPLES (21 ML)
8.65 mL of 85% sucrose
210 μL of 0.1 M Na_3VO_4
42 μL of 0.5 M EDTA
12.098 mL of MBS

6 SAMPLES (31 ML)
12.77 mL of 85% sucrose
310 μL of 0.1 M Na_3VO_4
62 μL of 0.5 M EDTA
17.858 mL of MBS

- *5% (w/v) sucrose*: Dilute 85% sucrose in MBS with 1 mM EDTA and 1 mM Na_3VO_4

4 SAMPLES (21 ML)
1.24 mL of 85% sucrose
210 μL of 0.1 M Na_3VO_4
42 μL of 0.5 M EDTA
19.508 mL of MBS

6 SAMPLES (31 ML)
1.83 mL of 85% sucrose
310 μL of 0.1 M Na_3VO_4
62 μL of 0.5 M EDTA
28.798 mL of MBS

Note: Mix sucrose mixtures at 4 °C, rotating for 2 min.

- *2× Sample buffer*
  - Dilute 4× sample buffer (refer to Section 2.1.2) with DIG lysis buffer (1:1)
  - For example, 100 μL of 4× sample buffer + 100 μL of DIG lysis buffer

22.1 Detergent-Insoluble Glycosphingolipid (DIG) Microdomain Isolation

22.1.2 Method
22.1.2.1 Isolation of DIG fractions
- Prepare 40×10^6 to 100×10^6 cells per group. For fewer cells, use less lysis buffer to keep the cell concentration high.
- Stimulate cells as appropriate.
- The following steps must be performed on ice:
  - Prepare the buffers and sucrose mixtures (see Section 1.1.3).
  - Gently resuspend cell pellet in DIG lysis buffer. Lyse for 30 min.
  - Homogenize each lysate in a prechilled Dounce homogenizer (Wheaton; 7 mL glass tube with glass plunger) by applying 10 gentle strokes with the loose-fitting plunger.
  - Mix homogenized lysate with an equal volume of 85% (w/v) sucrose and then rotate at 4 °C for 2 min.
  - Place this mixture at the bottom of a 14×95 mm Ultra-Clear ultracentrifuge tube (Beckman). Slowly overlay this mixture with 5 mL of 35% sucrose mixture. Next, carefully overlay the 35% layer with 5 mL of 5% sucrose mixture. The overlays can be made in a variety of ways (e.g., Pasteur pipette, pipette-aid set to "slow," or a gradient-former apparatus), provided it is performed carefully and consistently.
  - Centrifuge at $200,000 \times g$ for 16–18 h at 4 °C in a prechilled SW40 rotor.
  - Place tubes on ice, discard the upper 3–4 mL, then extract the opaque band and the 35/5% interface (1 mL total), and label "DIG." Discard the next 5 mL, then extract the bottom 2 mL from the tube, and label "Soluble." Alternatively, the entire tube may be divided into 12×1 mL fractions. The extraction can be performed with a 1 mL pipetman or by making a side puncture with a butterfly needle (this requires use of polyallomer soft tubes).
- Aliquot and add $4\times$ sample buffer (at a 1:4 ratio), boil at 95 °C for 5 min, and Western blot directly or immunoprecipitate the fractions and then Western blot. Alternatively, DIGs may be stored at -70 °C. DIGs (but not the soluble fraction) can be pelleted by diluting the fraction with an equal volume of MBS and then centrifuging at $20,000 \times g$ for 10 min at 4 °C. Additionally, protein can be precipitated and then assayed.

22.1.2.2 Immunoprecipitation of DIG fractions
- Precoat protein A/G beads with 1 μg antibody in 500 μL phosphate-buffered saline (PBS) for 2 h at 4 °C.
- Split the 1 mL DIG fractions into two portions of 500 μL in microfuge tubes.
- Add 500 μL MBS to each tube and then vortex.
- Spin at $20,000 \times g$ for 10–30 min at 4 °C.
- Remove and discard 950 μL of the supernatant from each tube with a pipette without disturbing the pellet.

- Add 900 μL of DIG lysis buffer to one tube, vortex, then transfer mixture to the other tube, and vortex.
- Centrifuge the beads coated with antibody (i.e., maximum speed for 3 s). Discard the supernatant and wash the antibody-coated beads once with 500 μL of DIG lysis buffer. Centrifuge again and discard the supernatant.
- Add the DIG fraction to the antibody-coated beads and then rotate gently at 4 °C for 2 h or overnight.
- Wash three times with 1 mL of DIG lysis buffer and centrifuge at maximum speed for 3 s each time.
- Add 2× sample buffer, boil at 95 °C for 5 min, and then proceed with Western blotting.
- *Note*: Soluble fractions should be immunoprecipitated directly.

22.2 PLASMA MEMBRANE ISOLATION

This protocol is designed to obtain highly purified plasma membranes for biochemical studies and also to produce an internal membrane fraction. This protocol has been adapted from Chaney and Jacobson (1983) and Spector, Goldman, and Leinwand (1998).

22.2.1 Materials
22.2.1.1 Reagents
MES buffer
NaCl
Sorbitol
Polymethacrylic acid (PAA)
Imidazole
Aprotinin
Leupeptin
PMSF
Na_3VO_4
Ludox
Histodenz
Sodium dodecyl sulfate (SDS)
β-Mercaptoethanol
Tris
Glycerol
Bromophenol blue
ddH_2O

22.2.1.2 Buffers and solutions
- PMCB (plasma membrane coating buffer)
  - 20 mM MES

- 150 mM NaCl
- 280 mM sorbitol (182.19 g/mol)
- pH range: 5.0–5.5
- PMCB pH 6–7
  - Before adjusting the pH of the PMCB earlier, take a 50 mL aliquot (it should be pH 7).
- PAA (require 5 mL per sample)
  - 1 mg/mL PAA in PMCB (pH 6–7)
  - First, prepare 50 mg/mL stock in water (dissolves slowly) and then dilute accordingly.
- Lysis buffer (prepare 50 mL for two samples)
  - 2.5 mM imidazole, pH 7.0
  - 10 µg/mL aprotinin
  - 10 µg/mL leupeptin
  - PMSF and Na_3VO_4 can be added as well
- Ludox
  - Make 5% (v/v) mixture of ludox in PMCB (5 mL per sample).
- Histodenz (Nycodenz)
  - Measure 10 g powder and add 5.5 mL lysis buffer.
  - Vortex repeatedly, takes ~30–60 min to dissolve.
  - Take 7 mL of this and top up to 10 mL with lysis buffer.
  - This solution can be frozen as a stock, but ensure it completely dissolves upon thawing.
- 4× Sample buffer (prepare 50 mL, can be stored at room temperature)
  - 8% SDS
  - 8% β-Mercaptoethanol
  - 250 mM Tris, pH 6.8
  - 40% Glycerol (v/v)
  - 2% Bromophenol blue
  - ddH_2O
- PBS, pH 7.4
  - 8.0 g of NaCl
  - 0.2 g of KCl
  - 1.44 g of Na_2HPO_4
  - 0.24 g of KH_2PO_4
  - Up to 1 L ddH_2O

22.2.1.3 Equipment
- Optima XL-90 ultracentrifuge with SW40 rotor (Beckman) or similar
  - This protocol is designed for an 11 mL tube. Depending on the rotor used, calculate the volume of lysate and other Histodenz layers accordingly so that they fit in the tube. Smaller rotors (e.g., SW60, Beckman) may also be used.

22.2.2 Method

22.2.2.1 Silica coating
The purpose of this method is to encapsulate the cells in ludox, cross-link with PAA to form a cast around the membrane, and then lyse the cells using a Dounce homogenizer.

- Count the cells of interest (use between 5 and 50×10^6 cells).
- Wash cells with PBS.
- Resuspend cells in 1 mL of PMCB.
- Add 5 mL of 5% ludox in a 50 mL conical tube.
- Aspirate the cells into a large syringe.
- Attach a 20-gauge needle and then add cells from syringe one drop at a time to the ludox solution while manually swirling the tube.
- Add PMCB to a final volume of 20 mL.
- Centrifuge at $900 \times g$ for 3 min and then aspirate the supernatant. Appearance of a "fluffy" pellet generally means that the cells are already being lysed or dead. This is not desired.
- Wash twice with 20 mL PMCB.
- Resuspend in 1 mL of PMCB.
- Add the silica-coated cells dropwise as described previously to 5 mL of the 1 mg/mL PAA solution in a fresh 50 mL tube while swirling manually.
- Dilute by topping up to 20 mL PMCB.
- Centrifuge at $900 \times g$ for 3 min and then discard the supernatant.
- Resuspend in 2 mL of cold lysis buffer, put on ice for 30 min, and agitate occasionally.
- Lyse cells in prechilled Dounce homogenizer using the loose-fitting pestle (ten strokes) and then the tight-fitting pestle (three strokes).
- Check that cells have lysed using a light microscope.

22.2.2.2 Fractionation
Now, the plasma membrane is very dense and can be isolated from internal membranes and nuclear debris.

- Centrifuge silica membranes ($900 \times g$ for 3 min).
- Remove fluid phase and transfer it to microfuge tubes. This contains internal membranes and the cytosolic fraction.
    - Spin the fluid phase at $20,000 \times g$ for 10 min at 4 °C.
    - The pellet is internal membranes and the new fluid phase is cytosol.
    - Add a small amount of lysis buffer to the internal membranes, plus $4 \times$ sample buffer, then boil, and use for Western blotting.
    - Take the cytosol phase and add $4 \times$ sample buffer, boil, and use for Western blotting.
- Resuspend the silica membranes in 11 mL of lysis buffer.
- Layer this onto 2 mL of 70% Histodenz in the ultracentrifuge tube.

- Place in an ultracentrifuge rotor (i.e., SW40, Beckman) and spin at 28,000 × g for 30 min in an Optima XL-90 ultracentrifuge (Beckman) or similar.
- Carefully remove and discard all fluid phase.
- Resuspend the pellet in 1 mL of lysis buffer.
- Transfer into microfuge tube and keep on ice.
- Wash three times with 1 mL of lysis buffer (spin in tabletop microcentrifuge at maximum speed (e.g., 20,000 × g) for 5 s).
- The pellet contains purified plasma membrane.
- To solubilize proteins for Western blotting:
  - Add small amount of lysis buffer (i.e., 500 μL) containing 1.0% SDS.
  - Sonicate (several short bursts, keep on ice as much as possible).
  - Boil for 5 min.
  - Pellet on tabletop centrifuge at 20,000 × g for 10 min at 4 °C.
  - The supernatant now contains solubilized proteins from the plasma membranes. Transfer the supernatant to a new tube, add 4× sample buffer, then boil, and use for Western blotting.
  - Discard the silica pellet.

22.3 CARDIOLIPIN-ENRICHED MITOCHONDRIAL MEMBRANE MICRODOMAIN ISOLATION

The following protocol is designed to isolate cardiolipin-enriched microdomains from mitochondrial membranes. This protocol has been adapted from the method introduced by Ciarlo et al. (2010).

22.3.1 Materials

22.3.1.1 Reagents
HEPES
NaCl
Triton X-100
Aprotinin
Tris–HCl, pH 8.8
SDS
EDTA
Tris, pH 7.5
Sucrose
ddH$_2$O

22.3.1.2 Buffers
- Isolation buffer
  - 250 mM sucrose

- 10 mM Tris, pH 7.5
- 1 mM EDTA
• Extraction buffer
  - 25 mM HEPES, pH 7.5
  - 150 mM NaCl
  - 1% Triton X-100
  - 10 µg/mL aprotinin
• Solubilization buffer
  - 50 mM Tris–HCl, pH 8.8
  - 1% SDS
  - 5 mM EDTA

22.3.2 Method

22.3.2.1 Mitochondria isolation

- Harvest cells of interest and isolate intact mitochondria from the cells. The Qproteome Mitochondria Isolation kit (Qiagen) is recommended. Perform all steps on ice.
- Alternatively, mitochondria can be isolated from cells by differential centrifugation following homogenization using a Dounce homogenizer. Here, we outline mitochondria isolation from primary activated T cells. This protocol has been adapted from Ogilvie, Kennaway, and Shoubridge (2005) and may vary depending on cell type.
  - All steps must be performed on ice or at 4 °C.
  - Harvest cells of interest and wash twice with ice-cold PBS (refer to Section 2.1.2), centrifuging each time at $380 \times g$ for 5 min at 4 °C.
  - Resuspend cell pellet in 5 mL of ice-cold isolation buffer and transfer to a prechilled Dounce homogenizer.
  - Homogenize cells on ice by applying 20 strokes using the tight-fit pestle of the Dounce homogenizer. The number of strokes should be optimized for each cell type.
  - Transfer homogenate into four microfuge tubes and centrifuge twice at $600 \times g$ for 10 min at 4 °C.
  - Collect the supernatants and transfer to new microfuge tubes.
  - Centrifuge at $10,000 \times g$ for 20 min at 4 °C. The pellet contains the mitochondria. The supernatant (i.e., cytosol) can be collected for quality control analysis (e.g., measurement of cytochrome C protein levels).

22.3.2.2 Cardiolipin-enriched microdomain isolation

- Lyse mitochondrial pellet by resuspending in 1 mL of extraction buffer and then incubate for 20 min on ice.
- Collect lysates and centrifuge at $20,000 \times g$ in a tabletop microcentrifuge for 2 min at 4 °C.

- Collect the supernatants containing Triton X-100-soluble material. Centrifuge the pellets a second time (20,000 × g for 30 sec at 4 °C) to remove the remaining soluble material.
- Solubilize pellets in 100 μL solubilization buffer.
- Shear DNA by passage through a 22-gauge needle.
- Analyze both Triton X-100-soluble and Triton X-100-insoluble materials by Western blot. Load the fraction samples by volume.

22.4 MITOCHONDRIAL SUPERCOMPLEX IDENTIFICATION BY BLUE-NATIVE GEL ELECTROPHORESIS

Blue-native gel electrophoresis was first applied to digitonin-solubilized mitochondrial membranes by Schagger and Pfeiffer (2000) leading to the identification of mitochondrial respiratory supercomplexes. Blue-native refers to the fact that the separated protein complexes are not denatured but rather retain enzymatic activity while their electrophoretic separation relies on binding of the dye Serva Blue G to the protein. Treatment of samples with the mild detergent digitonin allows for many protein–protein interactions to remain intact, allowing for analysis of supramolecular associations of the respiratory complexes and other proteins. Blue-native samples are run on a first-dimension nondenaturing polyacrylamide gel to separate protein complexes based on size. Analysis of the components of each multiprotein complex from the first-dimension gel can be achieved by cutting the appropriate gel slice and running it on a denaturing second-dimension polyacrylamide gel followed by Western blotting. Silver staining may also be performed on first- or second-dimension gels to visualize multiprotein complexes or their components, respectively. This technique can be used to gather a considerable amount of information on multiprotein complexes and mitochondrial respiratory supercomplexes in particular. For this reason, blue-native gel electrophoresis is now one of the most commonly used techniques for studying mitochondrial respiratory supercomplexes. The following protocol for the isolation of mitochondrial respiratory supercomplexes by blue-native gel electrophoresis has been adapted from Schagger and von Jagow (1991) and Sasarman, Antonicka, and Shoubridge (2008).

22.4.1 Materials
22.4.1.1 Reagents
Aminocaproic acid
Bis-tris
Tricine
Serva Blue G
Acrylamide/bisacrylamide
EDTA

n-Dodecyl β-D-maltoside (DDM)
Ammonium persulfate
Glycerol
TEMED
ddH$_2$O

22.4.1.2 Equipment
Gradient former
Peristaltic pump
Gel electrophoresis apparatus

22.4.1.3 Buffers and solutions
- 3× Gel buffer (1.5 M aminocaproic acid, 150 mM Bis-tris, pH 7.0).
  - 19.68 g aminocaproic acid
  - 3.14 g Bis-tris
  - Up to 100 mL ddH$_2$O
- Cathode buffer (15 mM Bis-tris, 50 mM tricine, pH 7.0).
  - 3.14 g Bis-tris
  - 8.96 g tricine
  - Up to 1 L ddH$_2$O
- Blue cathode buffer.
  - 100 mL cathode buffer
  - 0.02 g Serva Blue G
- Anode buffer (50 mM Bis-tris, pH 7.0).
  - 20.93 g Bis-tris
  - Up to 2 L ddH$_2$O
- Acrylamide/bisacrylamide mix (48% acrylamide and 1.5% bisacrylamide (99.5T,C)).
  - 24.0 g acrylamide
  - 0.75 g bisacrylamide
  - Up to 50 mL ddH$_2$O
- Membrane buffer.
  - 0.5 mL 3× gel buffer
  - 0.5 mL 2 M aminocaproic acid
  - 4 μL 500 mM EDTA
- SBG sample buffer (0.75 M aminocaproic acid, 5% Serva Blue G).
  - 3.75 mL 2 M aminocaproic acid
  - 0.5 g Serva Blue G
  - Up to 10 mL ddH$_2$O
- 2 M aminocaproic acid.
  - 13.12 g aminocaproic acid
  - 50 mL ddH$_2$O
- 10% DDM (0.1 g/mL).

- 10% Ammonium persulfate (0.1 g/mL).
- Serva Blue G causes a negative charge shift of the proteins and is able to remain tightly bound to the protein (Schagger & von Jagow, 1991).
- Aminocaproic acid supports the solubilizing properties of neutral detergents and allows for omission of any salt that would impair electrophoresis (Schagger & von Jagow, 1991).
- Bis-tris is a base with a pK in the slightly acidic range (pK 6.5–6.8) and is thus able to stabilize pH 7.5 in the gel (Schagger & von Jagow, 1991).
- Tricine was found to empirically have the optimum pK (8.15) and the appropriate relative mobility at pH 7.5 to allow protein separation within polyacrylamide gradient gels (Schagger & von Jagow, 1991).

22.4.2 Method

22.4.2.1 Sample preparation

- Prepare samples for blue-native gel electrophoresis. Use isolated mitochondria from cells or tissues (refer to Section 3.2.1). Alternatively, isolation of mitochondria from cells in suspension may not be necessary due to the high concentration of mitochondrial respiratory complexes relative to other cellular proteins.
- Wash cells/mitochondria once with PBS (refer to Section 2.1.2) and then resuspend in ice-cold PBS.
- Measure the protein concentration: take an aliquot of cell suspension, sonicate or use detergent (e.g., CHAPS), and measure protein concentration (e.g., using BCA Protein Assay kit (Pierce)).
- Add more PBS and then centrifuge the cells.
- Resuspend cells in PBS to a final concentration of 5 mg/mL according to the protein concentration measured earlier. Place cells in microfuge tube.
- Add an equal volume of digitonin (4 mg/mL) and incubate on ice for 5 min (final concentration of cells is 2.5 mg/mL, final concentration of digitonin is 2 mg/mL, and digitonin/protein ratio is 0.8).
- Following 5 min incubation, add PBS to a final volume of 1.5 mL.
- Centrifuge at $10,000 \times g$ for 10 min.
- Resuspend pellet (i.e., membranes) in membrane buffer (half the volume of PBS used to bring the sample to 5 mg/mL previously).
- Add 10% DDM to a final concentration of 1% (e.g., 1/10 volume of membrane buffer added in the previous step) and then incubate on ice for 15 min.
- Centrifuge at $20,000 \times g$ for 20 min at 4 °C.
- Place supernatant in a new microfuge tube and measure the protein concentration (e.g., using BCA Protein Assay Kit (Pierce)).
- Add SBG sample buffer (e.g., half the volume of 10% DDM added previously).
- Apply the samples to a blue-native gel for electrophoresis. Alternatively, samples may be stored at −20 °C.

22.4.2.2 Performing blue-native gel electrophoresis
- Prepare 6% and 15% resolving gel mixture according to Table 22.1.
- Fill the tubing and 1 cm of cassette height with ddH$_2$O.
- Fill the front reservoir of the gradient mixer with 2.8 mL of 6% gel mixture and the rear reservoir with 2.3 mL of 15% gel mixture and then place the conical insert into the rear reservoir.
- Fill the gel cassette by means of underlaying the gel under water using a peristaltic pump at speed 9.
- After polymerization, wash gel surface with 1× gel buffer.
- Prepare 2.5 mL of 4% stacking gel mixture according to Table 22.1, then insert comb, and fill to the top with stacking gel mixture.
- After polymerization, remove comb and wash wells with 1× GB.
- Fill the wells with blue cathode buffer and underlay the samples into the wells (10–30 μg protein per well).
- Place gel cassettes into the apparatus and fill upper and lower reservoir with blue cathode buffer and anode buffer, respectively.
- Run for 15 min at 40 V and then at 80 V until the dye reaches 2/3 of the gel. Replace the blue cathode buffer with cathode buffer and continue the electrophoresis until the dye front reaches the end of the gel.
- Remove the gel from the plates and proceed with either (i) Western blotting, (ii) 2D gel electrophoresis, or (iii) silver staining (Fig. 22.1).

22.4.2.3 Analysis by Western blotting
- For Western blotting, place gels in Towbin semidry transfer buffer for 20 min on a rocker.
- Place a polyvinylidene fluoride (PVDF) membrane in methanol for 1 min, rinse three times with ddH$_2$O, and then place in Towbin semidry transfer buffer for 20 min on a shaker.

Table 22.1 Preparation of Gradient Gels for Blue-Native Gel Electrophoresis

	4% Stacking Gel		6% Resolving Gel		15% Resolving Gel	
	2.5 mL	5.0 mL	2.5 mL	5.0 mL	2.5 mL	5.0 mL
ddH$_2$O	1.45 mL	2.87 mL	2.72 mL	5.44 mL	0.84 mL	1.68 mL
3× Gel buffer	0.82 mL	1.64 mL	1.65 mL	3.3 mL	1.65 mL	3.3 mL
Acrylamide/bisacrylamide	0.2 mL	0.4 mL	0.6 mL	1.2 mL	1.5 mL	3.0 mL
Glycerol	–	–	–	–	1 mL	2 mL
Ammonium persulfate	30 μL	60 μL	30 μL	60 μL	5 μL	10 μL
TEMED	3 μL	6 μL	2 μL	4 μL	2 μL	4 μL

Volumes listed are sufficient for 1 or 2 gels (i.e., 2.5 or 5.0 mL, respectively).

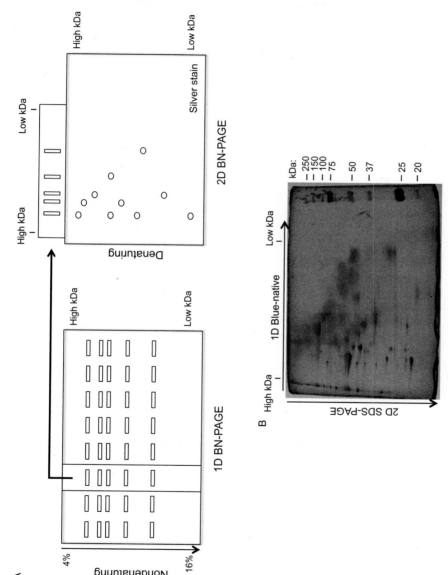

FIGURE 22.1

Procedure for two-dimensional blue-native gel electrophoresis (2D BN-PAGE). (A) Nondenatured samples are loaded into the wells of a nondenaturing 4–16% gradient polyacrylamide gel. Following electrophoresis, a lane of interest is cut from the gel and this gel slice is placed in denaturing sample buffer prior to its placement atop a denaturing polyacrylamide gel (e.g., 10%). Following second-dimension PAGE, the gel can be silver-stained to visualize the proteins that comprise the first-dimension multiprotein complex bands. Alternatively, the proteins from the gel can be transferred to a PVDF membrane for Western blot detection of specific proteins (not shown). (B) Silver stain of 2D BN-PAGE of isolated mitochondria from adult C57BL/6 mouse brain. A molecular weight ladder is included on the far right-hand portion of the gel.

- Prepare stack on a semidry transfer unit as follows: thick blot paper, thin blot paper, PVDF membrane, gel, thin blot paper, thick blot paper. Ensure there are no bubbles between any of the layers to avoid poor transfer. Transfer for 42 min, 16 V, 0.26 limit (or 0.52 limit for two gels).
- Proceed with immunoblotting using specific antibodies. *Note*: Some antibodies are able to detect proteins in their native conformation. An alternative is to denature the gel prior to membrane transfer by placing it in SDS containing 1× sample buffer (refer to Section 2.1.2) on a shaker for 30 min, boil briefly in a microwave (e.g., 10 s), and place on a shaker for an additional 15 min. This may allow binding by antibodies that require the protein to be denatured.

22.4.2.4 2D blue-native polyacrylamide gel electrophoresis (2D BN-PAGE)
- For 2D gel electrophoresis, cut the gel slice of interest from the 1D blue-native gel.
- Place the gel slice in denaturing SDS-containing 1× sample buffer on a shaker for 30 min, boil briefly in a microwave (e.g., 10 s), and place on a shaker for an additional 15 min.
- Prepare SDS-containing 10% polyacrylamide gels using a comb designed to hold a gel slice and a separate well to load the molecular weight ladder.
- Proceed with polyacrylamide gel electrophoresis as normal, then transfer to a PVDF membrane, and blot with the antibody of interest.

22.4.2.5 Silver staining
- For silver staining of 1D or 2D blue-native gels, the Silver Stain Plus Kit (Bio-Rad) is recommended.

22.5 DISCUSSION

Identification of multimolecular complexes within the biochemical microenvironment in which they function is a first step towards characterizing their functional role and targeting them for therapeutic purposes. The biophysical environment in which assembly of the complexes takes place may in itself not only define critical stages for their formation but also focus the search for molecular interactions. We have discussed here several methods to isolate specialized membrane microdomains from various membranes of the cell, including the plasma membrane and mitochondrial membranes, and the analysis of multimolecular complexes and supercomplexes in cardiolipin-enriched microdomains in the inner mitochondrial membrane.

A few aspects regarding the outlined protocols deserve further consideration. As reported by others, in our hands, mitochondrial respiratory supercomplex isolation is best achieved using the detergent digitonin since it is sufficiently mild to preserve the supramolecular interactions of multichain protein complexes (Acin-Perez et al., 2008). Some supercomplexes (and all individual complexes) can be extracted using other detergents, namely, Triton X-100, NP-40 (Igepal CA-630), Tween-20, and

DDM (Acin-Perez et al., 2008). As a control to measure the total amount of individual complexes, we recommend treating the samples of interest with DDM since all individual complexes are detectable by Western blotting following this treatment with no detectable proportion associated with supercomplexes. Meanwhile, the detergents Brij-96V, cholate, Empigen BB, perfluorooctanoic acid, and CHAPS are largely unable to extract any individual complexes or supercomplexes (Acin-Perez et al., 2008).

It is also important to keep in mind that the detection of an array of molecules within the fraction of detergent-insoluble microdomains of a given compartment does not necessarily correlate with colocalization of these molecules within the same microdomains. This is due to the heterogeneity of lipid rafts, determined by many factors including ectodomain interactions between surface receptors (Pike, 2004; Wang, Gunning, Kelley, & Ratnam, 2002).

Finally, we acknowledge that there is an ongoing debate about the biological significance of the biochemical preparation of detergent-insoluble microdomains and the multimolecular complexes that are isolated within these compartments. There is still controversy on whether these domains do actually exist as discrete entities *in vivo* or rather result from the biochemical manipulation of the cells and their membranes. This debate is apparent when discussing the existence and biological significance of mitochondrial respiratory supercomplexes. Assessment of the different sides of this argument goes beyond the scope of this methodological paper.

SUMMARY

It is becoming increasingly apparent that there are specialized membrane microdomains in many biological membranes, including the plasma and mitochondrial membranes, and that these microdomains are heterogeneous with varying compositions. The support for the existence of these microdomains in a natural setting and their formation, independently of biochemical manipulation or as an experimental artifact, is still under debate. However, through the use of different approaches, including those outlined in this chapter, it is becoming increasingly clear that these specialized membrane microdomains are biologically relevant in distinct ways depending on their location and composition. It is hoped that the techniques detailed here will be helpful to explore novel models to test and verify the functional correlates that will support their biological significance.

Acknowledgments

We thank Dr. Eric Shoubridge (McGill University, Montréal, QC) for helpful comments and discussions in the development of protocols for mitochondrial isolation and 2D BN-PAGE. Work at the Madrenas laboratory is funded by the CIHR. J.M. holds a Tier I Canada Research Chair in Human Immunology.

References

Acin-Perez, R., Fernandez-Silva, P., Peleato, M. L., Perez-Martos, A., & Enriquez, J. A. (2008). Respiratory active mitochondrial supercomplexes. *Molecular Cell, 32*(4), 529–539. http://dx.doi.org/10.1016/j.molcel.2008.10.021.

Bi, K., Tanaka, Y., Coudronniere, N., Sugie, K., Hong, S., van Stipdonk, M. J., et al. (2001). Antigen-induced translocation of PKC-theta to membrane rafts is required for T cell activation. *Nature Immunology, 2*(6), 556–563. http://dx.doi.org/10.1038/88765.

Brown, D. A., & Rose, J. K. (1992). Sorting of GPI-anchored proteins to glycolipid-enriched membrane subdomains during transport to the apical cell surface. *Cell, 68*(3), 533–544.

Chaney, L. K., & Jacobson, B. S. (1983). Coating cells with colloidal silica for high yield isolation of plasma membrane sheets and identification of transmembrane proteins. *Journal of Biological Chemistry, 258*(16), 10062–10072.

Christie, D. A., Mitsopoulos, P., Blagih, J., Dunn, S. D., St-Pierre, J., Jones, R. G., et al. (2012). Stomatin-like protein 2 deficiency in T cells is associated with altered mitochondrial respiration and defective CD4+ T cell responses. *Journal of Immunology, 189*(9), 4349–4360. http://dx.doi.org/10.4049/jimmunol.1103829.

Ciarlo, L., Manganelli, V., Garofalo, T., Matarrese, P., Tinari, A., Misasi, R., et al. (2010). Association of fission proteins with mitochondrial raft-like domains. *Cell Death and Differentiation, 17*(6), 1047–1058. http://dx.doi.org/10.1038/cdd.2009.208.

Darlington, P. J., Baroja, M. L., Chau, T. A., Siu, E., Ling, V., Carreno, B. M., et al. (2002). Surface cytotoxic T lymphocyte-associated antigen 4 partitions within lipid rafts and relocates to the immunological synapse under conditions of inhibition of T cell activation. *Journal of Experimental Medicine, 195*(10), 1337–1347.

Field, K. A., Holowka, D., & Baird, B. (1995). Fc epsilon RI-mediated recruitment of p53/56lyn to detergent-resistant membrane domains accompanies cellular signaling. *Proceedings of the National Academy of Sciences of the United States of America, 92*(20), 9201–9205.

Karnovsky, M. J., Kleinfeld, A. M., Hoover, R. L., & Klausner, R. D. (1982). The concept of lipid domains in membranes. *Journal of Cell Biology, 94*(1), 1–6.

Lenaz, G., & Genova, M. L. (2009). Structural and functional organization of the mitochondrial respiratory chain: A dynamic super-assembly. *International Journal of Biochemistry & Cell Biology, 41*(10), 1750–1772.

Lenaz, G., & Genova, M. L. (2012). Supramolecular organisation of the mitochondrial respiratory chain: A new challenge for the mechanism and control of oxidative phosphorylation. *Advances in Experimental Medicine and Biology, 748*, 107–144. http://dx.doi.org/10.1007/978-1-4614-3573-0_5.

Lingwood, D., Kaiser, H. J., Levental, I., & Simons, K. (2009). Lipid rafts as functional heterogeneity in cell membranes. *Biochemical Society Transactions, 37*(Pt 5), 955–960. http://dx.doi.org/10.1042/BST0370955.

Lingwood, D., & Simons, K. (2010). Lipid rafts as a membrane-organizing principle. *Science, 327*(5961), 46–50. http://dx.doi.org/10.1126/science.1174621.

Ogilvie, I., Kennaway, N. G., & Shoubridge, E. A. (2005). A molecular chaperone for mitochondrial complex I assembly is mutated in a progressive encephalopathy. *Journal of Clinical Investigation, 115*(10), 2784–2792. http://dx.doi.org/10.1172/JCI26020.

Osman, C., Voelker, D. R., & Langer, T. (2011). Making heads or tails of phospholipids in mitochondria. *Journal of Cell Biology, 192*(1), 7–16. http://dx.doi.org/10.1083/jcb.201006159.

Pfeiffer, K., Gohil, V., Stuart, R. A., Hunte, C., Brandt, U., Greenberg, M. L., et al. (2003). Cardiolipin stabilizes respiratory chain supercomplexes. *Journal of Biological Chemistry*, 278(52), 52873–52880. http://dx.doi.org/10.1074/jbc.M308366200.

Pike, L. J. (2004). Lipid rafts: Heterogeneity on the high seas. *Biochemical Journal*, 378(Pt 2), 281–292. http://dx.doi.org/10.1042/BJ20031672.

Pike, L. J. (2009). The challenge of lipid rafts. *Journal of Lipid Research*, 50(Suppl.), S323–S328. http://dx.doi.org/10.1194/jlr.R800040-JLR200.

Sargiacomo, M., Sudol, M., Tang, Z., & Lisanti, M. P. (1993). Signal transducing molecules and glycosyl-phosphatidylinositol-linked proteins form a caveolin-rich insoluble complex in MDCK cells. *Journal of Cell Biology*, 122(4), 789–807.

Sasarman, F., Antonicka, H., & Shoubridge, E. A. (2008). The A3243G tRNALeu(UUR) MELAS mutation causes amino acid misincorporation and a combined respiratory chain assembly defect partially suppressed by overexpression of EFTu and EFG2. *Human Molecular Genetics*, 17(23), 3697–3707. http://dx.doi.org/10.1093/hmg/ddn265.

Schagger, H., & Pfeiffer, K. (2000). Supercomplexes in the respiratory chains of yeast and mammalian mitochondria. *EMBO Journal*, 19(8), 1777–1783. http://dx.doi.org/10.1093/emboj/19.8.1777.

Schagger, H., & von Jagow, G. (1991). Blue native electrophoresis for isolation of membrane protein complexes in enzymatically active form. *Analytical Biochemistry*, 199(2), 223–231.

Simons, K., & Gerl, M. J. (2010). Revitalizing membrane rafts: New tools and insights. *Nature Reviews Molecular Cell Biology*, 11(10), 688–699. http://dx.doi.org/10.1038/nrm2977.

Simons, K., & Sampaio, J. L. (2011). Membrane organization and lipid rafts. *Cold Spring Harbor Perspectives in Biology*, 3(10), a004697. http://dx.doi.org/10.1101/cshperspect.a004697.

Singer, S. J., & Nicolson, G. L. (1972). The fluid mosaic model of the structure of cell membranes. *Science*, 175(4023), 720–731.

Sorice, M., Manganelli, V., Matarrese, P., Tinari, A., Misasi, R., Malorni, W., et al. (2009). Cardiolipin-enriched raft-like microdomains are essential activating platforms for apoptotic signals on mitochondria. *FEBS Letters*, 583(15), 2447–2450. http://dx.doi.org/10.1016/j.febslet.2009.07.018.

Spector, D. L., Goldman, R. D., & Leinwand, L. A. (1998). *Cells: A laboratory manual*. Woodbury, NY: Cold Spring Harbor Laboratory Press.

Viola, A., Schroeder, S., Sakakibara, Y., & Lanzavecchia, A. (1999). T lymphocyte costimulation mediated by reorganization of membrane microdomains. *Science*, 283(5402), 680–682.

Wang, J., Gunning, W., Kelley, K. M., & Ratnam, M. (2002). Evidence for segregation of heterologous GPI-anchored proteins into separate lipid rafts within the plasma membrane. *Journal of Membrane Biology*, 189(1), 35–43. http://dx.doi.org/10.1007/s00232-002-1002-z.

Xavier, R., Brennan, T., Li, Q., McCormack, C., & Seed, B. (1998). Membrane compartmentation is required for efficient T cell activation. *Immunity*, 8(6), 723–732.

Zhang, W., Trible, R. P., & Samelson, L. E. (1998). LAT palmitoylation: Its essential role in membrane microdomain targeting and tyrosine phosphorylation during T cell activation. *Immunity*, 9(2), 239–246.

CHAPTER

G Protein-Coupled Receptor Transactivation: From Molecules to Mice

23

Kim C. Jonas*, Adolfo Rivero-Müller[†], Ilpo T. Huhtaniemi*,[†], and Aylin C. Hanyaloglu*

*Department of Surgery and Cancer, Institute of Reproductive and Developmental Biology, Imperial College London, London, United Kingdom
[†]Department of Physiology, Institute of Biomedicine, University of Turku, Turku, Finland

CHAPTER OUTLINE

Introduction	434
23.1 Materials	435
23.2 Methods	436
23.2.1 Protein–Protein Interactions	437
23.2.2 Functional Measurement of Receptor Transactivation *in Vitro*	437
23.2.3 Studying Transactivation *in Vivo*	440
23.2.3.1 Generation of BAC Constructs	440
23.2.3.2 Embryo Generation, Detection of BAC Transgenic Offspring, and Breeding Strategy	442
23.2.3.3 Analysis of Male Reproductive Tract and Testes	442
23.2.3.4 Fertility	446
23.3 Discussion	447
Summary	448
Acknowledgments	448
References	448

Abstract

G protein-coupled receptors (GPCRs) mediate a diverse range of physiological functions via activation of complex signaling systems. Organization of GPCRs in to dimers and oligomers provides a mechanism for both signal diversity and specificity in cellular responses, yet our understanding of the physiological significance of dimerization, particularly homodimerization, has not been forthcoming. This chapter will describe how we have investigated the physiological importance of GPCR homodimerization, using the luteinizing hormone/chorionic gonadotropin receptor as a model GPCR. Using transactivation as a mode of assessing receptor dimerization, we describe our cellular system and functional assays for assessment of transactivation *in vitro* and detail our strategy for generating a mouse model to assess GPCR transactivation *in vivo*.

INTRODUCTION

G protein-coupled receptors (GPCRs) act as key core communicators of extracellular signals within a wide variety of physiological systems. The question of how diversity and specificity is achieved when many receptor signaling pathways converge on common downstream pathways is highly pertinent for this superfamily of signaling receptors. This is driven by our more current understanding of the increasing complexity of these receptor systems and the problems that can impede effectiveness of current therapeutic ligands and unwanted side effects (Smith, Bennett, & Milligan, 2011).

One mechanism that has emerged in studies over the past 15 years that influences receptor signal specificity and diversity is the organization of GPCRs in a multitude of complexes, existing not only as monomers but also as dimers and higher-order oligomers that function to diversify receptor functionality. Formation of receptor homodimers and heterodimers can impact all aspects of GPCR biology, including alterations in pharmacology, plasma membrane expression, signal transduction, and receptor trafficking (reviewed by Lohse, 2010). However, GPCR di/oligomerization has been a highly debated issue (Chabre & le Maire, 2005; Fotiadis et al., 2006; Milligan, 2013), and while there is extensive evidence for receptor di/oligomerization *in vitro*, the functional significance of this phenomenon *in vivo* is still uncertain. The clearest evidence for the physiological role of GPCR dimerization was first obtained for class C GPCRs, such as $GABA_B$, taste (T1R1-3), metabotropic glutamate, and Ca^{2+}-sensing receptors (Kaupmann et al., 1998; Pin et al., 2004). Although a key question is to understand the requirement and *in vivo* significance of GPCR dimerization, such data on the large class A/rhodopsin-like GPCRs, in particular homodimerization, have not been forthcoming. From now on, the term "dimerization" will be used, although it refers to both dimers and higher-order oligomers.

Transactivation, also termed functional complementation or intermolecular cooperativity, has been employed to study the requirement of GPCR dimerization on receptor signaling. Receptor transactivation involves coexpression of two distinct mutants that on their own are nonfunctional but when coexpressed can "rescue" the functional activity of that receptor (Fig. 23.1), with the binding of ligand to one receptor protomer within the dimer, which communicates with neighboring protomer to propagate signal. The phenomenon of transactivation has been demonstrated for several GPCRs, including $GABA_B$, D2 dopamine, opioid, thyrotropin-releasing hormone, thyroid-stimulating hormone, follicle-stimulating hormone, and luteinizing hormone/chorionic gonadotropin receptors (LHCGRs), using and employing distinct

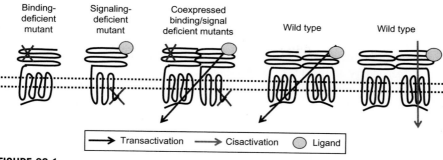

FIGURE 23.1

Schematic showing modes of GPCR activation. Ligand-binding-deficient and signaling-deficient mutant receptors are unable to function; however, when coexpressed, they can interact to transactivate and functionally rescue their respective defects. The WT receptor is able to function by both transactivation and self- or cisactivation. (For color version of this figure, the reader is referred to the online version of this chapter.)

readouts of ligand-induced receptor activation from receptor–G protein coupling, second messenger generation, to receptor phosphorylation and trafficking (Han, Moreira, Urizar, Weinstein, & Javitch, 2009; Ji et al., 2004; Lee, Ji, & Ji, 2002; Lee, Ji, Ryu, et al., 2002; Monnier et al., 2011; Pascal & Milligan, 2005; Song, Jones, & Hinkle, 2007; Urizar et al., 2005).

This chapter will describe how we have investigated GPCR dimerization via transactivation, using the LHCGR as a model receptor. This class A GPCR plays a central role in the regulation of reproductive functions: sex steroid production, ovulation, and maintenance of early pregnancy (Menon & Menon, 2012). Gonadotropins are commonly used in the treatment of a variety of conditions, such as male and female hypogonadism and infertility, and ovarian induction upon *in vitro* fertilization treatments. Conversely, blockage of gonadotropin receptor function is central in hormonal contraception, in treatment of precocious puberty, and various hormone-dependent malignancies, for example, breast and prostate cancer (reviewed by Limonta et al., 2012). For the following methods, we will discuss the use of two mutant LHCGRs that we have shown to undergo transactivation or intermolecular functional cooperativity, to rescue their respective functional defects. We will detail the exploratory work conducted from confirmation of receptor interactions to *in vitro* functional analysis and further to generation of an *in vivo* model, demonstrating that GPCR dimerization is physiologically relevant.

23.1 MATERIALS

The following are key materials required to assess LHCGR transactivation *in vitro*:

1. 10 cm cell culture-treated dishes, 12-well plates and 96-well plates (Corning, UK)

2. Lipofectamine® 2000 (Invitrogen, UK)
3. Anti-HA.11 antibody (used at 1:300 dilution, raised in mouse, Covance c/o Cambridge Bioscience, UK), anti-FLAG antibody (used at 1:500 dilution, raised in rabbit, Sigma-Aldrich, UK), and anti-rabbit Alexa 488 and anti-mouse Alexa 647 secondary antibodies (1:1000 dilution, Invitrogen, UK)
4. Dulbecco's Modified Eagle Media (D5976) supplemented with 10% fetal bovine serum with or without 100 IU penicillin/0.1 mg streptomycin for maintenance of cells and transfections, respectively (all Sigma-Aldrich, UK), and OptiMEM (Invitrogen, UK)
5. Geneticin (Invitrogen, UK) for cell line selection
6. Recombinant human chorionic gonadotropin (hCG) (e.g., National Peptides and Hormones Program, USA)
7. *CRE-luciferase (CRE-luc)* reporter gene construct in *pcDNA3.1* and *pRL-CMV* reporter gene transfection efficiency control (Promega, UK)
8. Steadylite plus reagent (Perkin-Elmer, UK) and coelenterazine (Promega, UK)

23.2 METHODS

For the following methods, we will describe the use of two specific mutant LHCGRs that we have shown to undergo transactivation to rescue their respective functional defects (Rivero-Müller et al., 2010). The ligand-binding-deficient mutant LHCGR (LHCGR^{B-}) contains a point mutation in the extracellular region, cysteine 22 to alanine (C22A), rendering it unable to bind the ligands LH or hCG (Lee, Ji, & Ji, 2002; Lee, Ji, Ryu, et al., 2002). The signaling-deficient mutant LHCGR (LHCGR^{S-}) contains the deletion valine 533-alanine 689, corresponding to transmembrane helices 6 and 7 (Fig. 23.1). The LHCGR^{S-} can bind ligand with the same efficacy as the wild type (WT) LHCGR but is unable to activate G protein signaling, as determined by measurement of generation of the second messenger cAMP.

Our rationale for choosing these specific LHCGR^{B-} and LHCGR^{S-} mutants was driven by our aim to ultimately test whether transactivation of LHCGR could result in the rescue of LHCGR function in *LhCGR* knockout (LuRKO) animals. We felt it was imperative to use mutant LHCGRs that upon expression alone were completely devoid of any ligand-binding or signaling activity. On this basis, we selected the LHCGR^{B-} C22A mutant initially as it had previously been shown by our collaborators to be ligand-binding deficient but able to traffic to the cell surface (Lee, Ji, & Ji, 2002; Lee, Ji, Ryu, et al., 2002) yet shown to undergo transactivation and functional rescue when coexpressed with other signal-deficient or impaired receptors. Studies in our own laboratories also confirmed the functional properties of this LHCGR^{B-} to be devoid of ligand binding, and so it fulfilled our stringent criteria. We conducted functional assays on several previously described LHCGR^{S-} and found the tested mutant LHCGRs retained partial ligand-induced intracellular signaling and so for the purpose of our *in vivo* study could not be used. In order to be completely certain we would obtain a LHCGR^{S-} totally devoid of ligand-induced signal generation, we took the approach of deleting TM6/7 as this has been previously described to be a "hotspot" for inactivating mutations of LHCGR (Huhtaniemi & Themmen, 2000) and is essential for G protein-dependent coupling (Sangkuhl, Schulz, Schultz, & Schöneberg, 2002). This resulted in the desired mutant capable of ligand binding but deficient of ligand-induced signaling. As this $LHCGR^{S-}$ contained a large TM deletion, we were concerned on the impact this would have on trafficking to the cell surface. Hence, to aid plasma membrane targeting, we inserted into its 5′-end the bovine prolactin leading sequence, as this has been previously shown to aid trafficking of impaired receptors to the cell surface (Osuga, Kudo, Kaipia, Kobilka, & Hsueh, 1997). As antibodies to most GPCRs have issues with affinity and specificity, we introduced differential N-terminal tags to the receptors for monitoring of receptor expression, with $LHCGR^{B-}$ HA tagged and $LHCGR^{S-}$ FLAG tagged. For all experiments described detailing LHCGR transactivation, we have conducted parallel experiments with an HA-tagged WT LHCGR as a model of cis- and transactivation (Fig. 23.1).

An important point to reemphasize is that before assessing transactivation, it is essential to ensure that the phenotypic defect of the individual mutant receptors used in transactivation assays is as expected. Our approach has been to conduct ligand-binding assays and second messenger functional assays coupled to rigorous assessment of cell surface expression of the individually expressed $LHCGR^{B-}$ and $LHCGR^{S-}$. Once established, experiments to investigate receptor transactivation can proceed.

23.2.1 Protein–protein interactions

Receptor transactivation requires the physical interaction of receptors; therefore, when beginning to investigate the possibility of receptor transactivation, it is prudent to assess whether the receptors in question can associate. As this chapter will focus more on the assessment of the functional aspects of receptor transactivation, detailed methodology for determining protein–protein interactions will not be discussed. However, we and others have shown the $LHCGR^{B-}$ and $LHCGR^{S-}$ to associate using coimmunoprecipitation (Rivero-Müller et al., 2010) and bioluminescence resonance energy transfer (BRET), respectively (Zhang, Guan, & Segaloff, 2012).

23.2.2 Functional measurement of receptor transactivation *in vitro*

To assess the functional rescue by transactivation *in vitro*, we have tested various methods for coexpressing the $LHCGR^{B-}$ and $LHCGR^{S-}$ in human embryonic kidney (HEK293) cells. In our experience, the degree of functional rescue observed correlates to the level of receptor coexpression. For that purpose, we will briefly describe how we obtained stable cell lines to carry out our studies into LHCGR transactivation.

To obtain stable cell lines to assess the functional rescue through transactivation of $LHCGR^{B-}$ and $LHCGR^{S-}$, we have used Lipofectamine® 2000 transfection reagent and a standard Geneticin selection protocol. After plating HEK293 cells onto a

10 cm dish, to obtain stable expression of either WT LHCGR or LHCGR^{S-}, we transfect cells with 12 μg of either *Lhcgr*$^{S-}$ or WT *Lhcgr* plasmid DNA and for cell lines that coexpress LHCGR^{B-} and LHCGR^{S-} with 12 μg *Lhcgr*$^{B-}$ and 12 μg *Lhcgr*$^{S-}$ as per manufacturer's instructions. Geneticin selection (with 1 mg/ml in standard culture media) is begun 48 h posttransfection and we replace media containing Geneticin every 48 h thereafter. After 5–7 days, it is usual to observe large quantities of cell death and be left with a few cells, which will form distinct colonies. Following the initial cell death, we subsequently reduce Geneticin concentration to 0.5 mg/ml. We select colonies, expand, and screen for receptor cell surface expression using flow cytometry. We utilize the N-terminal tags of the receptors for detection of cell surface receptor expression and fluorescent secondary Alexa 488 and Alexa 647 antibodies for detection of FLAG and HA.11 primary antibodies, respectively. When screening, we try to select for stable cell lines that are monoclonal with cells forming a distinct single peak of mean fluorescence. If the receptor expression level is not uniform among the cell population, further cell sorting can be conducted to obtain a pure clone.

As the LHCGR primarily couples to Gαs to mediate its physiological functions (reviewed by Menon & Menon, 2012), we routinely use assays that measure the second messenger cAMP. We utilize three methods for measuring ligand-induced cAMP accumulation: the measurement of cAMP directly using EIA kits (Assay Designs), live cAMP kinetics using the GloSensor™ technology (Promega), and a reporter gene system of the cAMP response element fused to firefly luciferase (CRE-luc). As CRE-luc activity is the assay we routinely use, we will describe our protocol for assessment of cAMP using the CRE-luc system:

1. Plate HEK293 stably expressing WT LHCGR and LHCGR^{S-} or coexpressing LHCGR^{B-}/LHCGR^{S-} cells into 96-well plates to ensure a confluency of 80–90%.
2. The following day, transfect each well with 40 ng of *CRE-luc* plasmid and 5 ng of *pRL-CMV* plasmid. We use 0.5 μl/well Lipofectamine® 2000 as a transfection reagent and 2 × 25 μl OptiMEM as a diluent for both plasmid DNA and Lipofectamine® 2000 and follow manufacturer's instructions for the transfection procedure.
Note: the cotransfection of *pRL-CMV* plasmid serves as a control for transfection efficiency, which can be used for normalization of data.
3. For experiments using single stable LHCGR^{S-}, transient transfection of *Lhcgr*$^{B-}$ is also needed. We add 100 ng/well of *Lhcgr*$^{B-}$ plasmid DNA into the transfection mix containing *CRE-luc* and *pRL-CMV* plasmids and 100 ng/well of empty *pcDNA3.1* plasmid for a transfection control into WT LHCGR functional comparisons.
4. The following day, remove media and replace with 100 μl phenol red-free DMEM containing 0 or 10 nM hCG. *Note*: it is important to use phenol red-free media as phenol red has been demonstrated to impact on the degree of signal detection with the luciferase assay.

5. Incubate at 37 °C for 4.5 h. We have tested many time points for the CRE-luc assay and have found that 4.5 h is the earliest time point that hCG-dependent LHCGR-mediated CRE-luc activity can be robustly and consistently determined. Thirty minutes prior to the end of incubation, prepare the steadylite plus solution so it has reached room temperature by the end of the incubation period.
6. Using a multistep pipette, add 100 μl of steadylite plus solution to each well, cover with foil, and place on a plate shaker for 10 min at ∼200 rpm with variable rotation.
7. Transfer the contents of the 96-well plate to a white 96-well plate and read immediately using a plate-reading luminometer.
8. To measure *pRL-CMV* reporter gene activity, add 50 μl of coelenterazine solution (10 μl 1 μg/μl coelenterazine, 5 ml 0.5 M HEPES pH 7.8, 400 μl 0.5 M EDTA) to each well, cover with foil, and place on a plate shaker for 10 min at ∼200 rpm and variable rotation.
9. Read using a plate-reading luminometer and save data.
10. To analyze data, each CRE-luc activity is normalized to its internal pRL-CMV value to control for any variations in transfection efficiency between wells. We conduct all assays in quadruplicate with a minimum of three independent experiments. Data is usually expressed as a percentage of maximum response or as a fold change over basal response.

To demonstrate that a variance in the degree of functional rescue can be observed when coexpressing receptors in either stable or transient fashion, we will describe a comparison of cAMP-generated responses using three different modes of $LHCGR^{B-}$ and $LHCGR^{S-}$ coexpression. All data shown are normalized to the WT LHCGR response generated using the same transfection conditions. When comparing signal responses of cells transiently transfected with either WT LHCGR or $LHCGR^{B-}$ with $LHCGR^{S-}$, we observe a partial functional rescue by the transactivating mutant LHCGR, approximating 57% of WT functional response (Fig. 23.2), consistent with degrees of rescue observed with other LHCGR mutants that were also transiently expressed (Ji, Lee, Song, Conn, & Ji, 2002). When employing stable cellular expression of one mutant receptor and transiently coexpressing the other, the degree of functional rescue observed increases. In stable cells expressing $LHCGR^{S-}$ transiently transfected with $LHCGR^{B-}$, a further increase in the functional rescue is observed, generating a cAMP response that is 63% of the WT response (Fig. 23.2). However, when using cell lines that stably coexpress both $LHCGR^{B-}$ and $LHCGR^{S-}$, complete restoration of cAMP-dependent signaling is observed, with the transactivating mutant receptors signaling to 100% of cells stably expressing WT LHCGR (Fig. 23.2). As stable coexpression of $LHCGR^{B-}/LHCGR^{S-}$ results in a high level of cells coexpressing receptors, it reveals that transactivation of LHCGR can facilitate functional rescue of cAMP-mediated responses by mutant LHCGRs with restoration to that of WT LHCGR. These results may explain why a recent study into LHCGR did not observe the same degree of functional rescue when using the same mutant LHCGRs (Zhang et al., 2012).

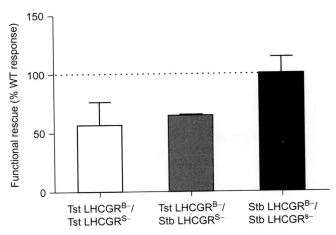

FIGURE 23.2

Functional rescue of LHCGR mutants by transactivation *in vitro*. The degree of functional rescue by transactivation of $LHCGR^{B-}$ and $LHCGR^{S-}$, as determined by CRE-luc activity, correlates with the method of receptor expression. Cells expressing receptors either stably (Stb) or transiently (Tst) as indicated, treated with hCG (10 nM) for 4.5 h. Each data point represents the mean ± SEM of three independent experiments, carried out in quadruplicate. Each data point was normalized to % of maximum WT response.

23.2.3 Studying transactivation *in vivo*

Despite the extensive data demonstrating that transactivation functionally rescues GPCR function at the cellular level, the critical question outstanding was if GPCR homodimerization, via transactivation, is a physiologically relevant mode of receptor function *in vivo*. To address this question, we employed transgenic technology to create mouse models expressing these mutants and utilizing the previously generated *LhCGR* knockout (LuRKO) mouse model (Zhang, Poutanen, Wilbertz, & Huhtaniemi, 2001), which produces a clear reproductive phenotype of delayed puberty, hypogonadism, high LH, undetectable levels of testosterone, arrested sperm development, and infertility. These phenotypic parameters provided an ideal background in which to assess whether transactivation of LHCGR by the targeted coexpression of $LHCGR^{B-}$ and $LHCGR^{S-}$ could rescue the hypogonadism and infertility of the LHCGR-null male mice, thus providing definitive evidence that receptor dimerization through transactivation is a physiologically required mechanism of LHCGR activation and signaling.

23.2.3.1 Generation of BAC constructs

The first step of investigating receptor transactivation *in vivo* was to identify which method of transgenesis to employ. We wanted to ensure that the integrity of the spatial/temporal expression pattern of $LHCGR^{B-}/LHCGR^{S-}$ was matched to that of

WT LHCGR, so we employed the use of bacterial artificial chromosomes (BAC). As BAC constructs can typically hold a DNA insert up to 350 kb (Shizuya et al., 1992), it is an ideal tool to study an entire gene of interest as it maintains the integrity of both coding and noncoding regions, ensuring that key regulatory information is not lost.

The generation of BAC constructs requires complex technical methodology, which is somewhat beyond the scope of this chapter (see a recent review by Narayanan and Chen (2011) for further information). However, we will outline in general terms our rationale and approach to generate our BAC constructs, highlighting where appropriate relevant reference materials to aid in BAC generation.

1. Source an appropriate BAC library containing your receptor gene of interest. We used the BAC high-density membrane libraries that contained the WT LHCGR from the BAC-PAC resources of Oakland Children's Hospital, clone RPCI23-18D7. Further information on BAC clones can be found from the following genomic databases:
   - National Center for Biotechnology Information
   - UCSC Human Genome browser
   - Ensembl
   - SRS Server
   - The Human BAC Clone Resource for Identification of Cancer Aberrations
   - Mouse Mapping Consortium: RPCI-23 Mouse BAC Mapping
2. Determine the location of your gene of interest from the BAC library. To identify the WT *Lhcgr*, we used DNA probes corresponding to the promoter region or exon 11 that ensured we were isolating the entire *Lhcgr* gene and not partial fragments. Using the library code and coordinates, order identified clones.
3. Clones are usually supplied in *E. coli* in the form of stab agar. Amplify the clone by culturing the *E. coli* containing the BAC in LB containing appropriate antibiotic for selection (usually chloramphenicol) and validate for your gene of interest through PCR and sequencing.
4. Recombineering (see Rivero-Müller, Lajic, and Huhtaniemi (2007) for further information). At this point, we introduced the $LHCGR^{B-}$ point mutation and $LHCGR^{S-}$ deletion into the original BAC clone. Due to the lack of specific LHCGR antibodies, we also used recombineering to insert an IRES-dsRed (for $LHCGR^{S-}$) and eCFP (for $LHCGR^{B-}$) to monitor the spatial/temporal expression of each mutant LHCGR. The internal ribosome entry site (IRES) technology avoids potential complications with directly C-terminal tagging the mutant LHCGR with fluorescent proteins that may interfere with receptor processing and trafficking, but as they are controlled by the same promoter, the encoded IRES fluorescent proteins will be transcribed and translated in the same spatial/temporal location as the mutant LHCGRs. All modifications and mutations were verified by sequencing.

23.2.3.2 Embryo generation, detection of BAC transgenic offspring, and breeding strategy

Once the BAC constructs containing mutant LHCGRs were generated, they required preparation for pronuclear injection. The BAC constructs were linearized and reconstituted in microinjection buffer (100 mM EDTA, 10 mM Tris–HCL pH 7.5, with additional 30 mM spermine and 70 mM spermidine to aid integration of the entire BAC construct into the genome) by buffer exchange. The BAC were injected into fertilized mouse oocytes (our mouse background of choice was FVB/N) and implanted into surrogate FVB/N mothers using standard methodology, as previously described (Rivero-Müller et al., 2010).

To determine the genotype of the resulting litter, we took an earlobe sample from each animal at day 15 postpartum. The genotype was determined using standard DNA extraction and PCR protocols using primers that distinguished between the WT $Lhcgr$, $Lhcgr^{B-}$, and $Lhcgr^{S-}$. Primer design is listed in Table 23.1. Mice carrying either $Lhcgr^{B-}$ or $Lhcgr^{S-}$ were retained and used for subsequent matings to obtain the homozygous LuRKO/$Lhcgr^{B-}$/$Lhcgr^{S-}$ genotype for studying LHCGR transactivation. Since the LuRKO mice are infertile, heterozygote animals were required for breeding. The breeding strategy required two distinct stages as detailed in Fig. 23.3.

23.2.3.3 Analysis of male reproductive tract and testes

Assessment of LHCGR transactivation on downstream cellular signaling events is ultimately determined by the degree of functional rescue of the infertile and hypogonadal reproductive phenotype of the LuRKO animals. To date, our studies have focused on the male LuRKO animals. In males, the primary function of the LHCGR expressed in the Leydig cells of the testes is production of testosterone (Fig. 23.4(I)). The downstream actions of LHCGR are to maintain steroidogenesis through regulating expression of key Leydig cell-specific steroidogenic genes, such as $StAR$ and $CYP17a1$, to drive testosterone production and facilitate spermatogenesis (Fig. 23.4(I), reviewed by Huhtaniemi & Alevizaki, 2007). Therefore, the best determinates of functional rescue of the LuRKO animals by LHCGR transactivation

Table 23.1 Primer Design for Genotyping of LHCGR[B−], LHCGR[S−], and LuRKO Animals

Primer ID	Sequence
IRES forward	5′-GTATTCAACAAGGGGCTGAAGG-3′
eCFP (LHCGR[B−]) reverse	5′-TTGATCCTAGCAGAAGCACAGG-3′
RFP (LHCGR[S−]) reverse	5′-CCATGGTCTTCTTCTGCATCAC-3′
WT LHCGR forward	5′-TCTGGGGATCTTGGAAATGA-3′
WT LHCGR reverse	5′-CACCTTGACACCTGGAGT-3′
LuRKO (neo) forward	5′-GGGCTCTATGGCTTCTGAGGCGGA-3′
LuRKO (neo) reverse	5′-TCTCAGGGAGGATTTGGGTATGG-3′

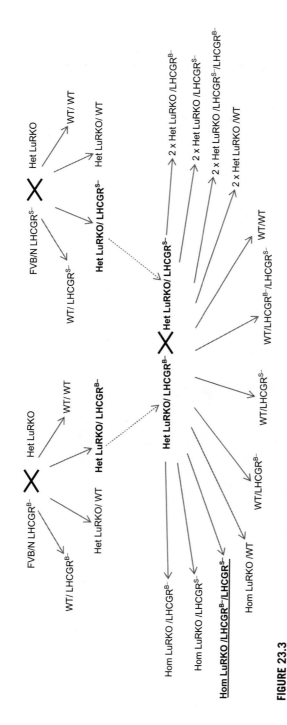

FIGURE 23.3

Breeding strategy to achieve homozygous LuRKO/LHCGR^{B-}/LHCGR^{S-} animals for *in vivo* studies of LHCGR transactivation. Hom and Het denote homozygous and heterozygous animals respectively. (For color version of this figure, the reader is referred to the online version of this chapter.)

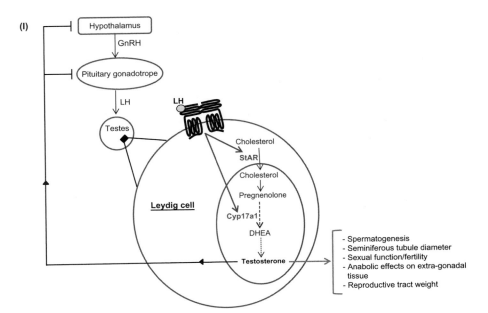

FIGURE 23.4

Mode of functional rescue by LHCGR transactivation *in vivo* (I): actions of LH/hCG via LHCGR. LH binding to LHCGR induces the expression of *StAR* and *CYP17a1*. Dashed lines represent pathways with intermediate steps not shown. (II) Morphology and histology of male reproductive tract and testis of (a) homozygous LuRKO, (b) homozygous LuRKO/LHCGR[B−], (c) homozygous LuRKO/LHCGR[S−], (d) homozygous LuRKO/LHCGR[B−]/LHCGR[S−], and (e) WT animals. Immunofluorescence shows localization of *Lhcgr*[B−] (green) and *Lhcgr*[S−] (red). (See color plate.)

Edited from Rivero-Müller et al. (2010).

are through functional, morphological, and histological analysis of the adult male gonads and reproductive tract.

The first step we undertook to assess the effects of $LHCGR^{B-}/LHCGR^{S-}$ transactivation in homozygous LuRKO animals was to examine the testes and reproductive tracts of the males. When euthanizing the animals, we used terminal anesthesia and cardiac puncture to collect cardiac blood for measurement of circulating concentrations of LH and testosterone in serum. The reproductive tract, inclusive of epididymides, seminal vesicles, and testes, were dissected and photographed, as the anabolic effects of testosterone on testis size and development of the reproductive tract are key indicators of LHCGR function. Following this, the reproductive tract and one testis were fixed using paraformaldehyde in (4% in PBS) for histological and spatiotemporal expression analysis of the $LHCGR^{B-}/LHCGR^{S-}$, and the other testis was snap frozen using liquid nitrogen for assessment of the effects of LHCGR transactivation on LH-dependent gene expression and LHCGR binding activity. As with the gross morphology of the testis and reproductive tract, the weights of these tissues are also indirect indicators of the concentration of circulating testosterone and hence LHCGR function. Examination of the dissected reproductive tract and testes revealed that expression of the single $LHCGR^{B-}$ or $LHCGR^{S-}$ in a homozygous LuRKO background had no restorative effect on the gross morphology of the reproductive tract and testis weights, despite clear detection of the expressed $LHCGR^{B-}$ or $LHCGR^{S-}$ in Leydig cells. However, coexpression of $LHCGR^{B-}$ and $LHCGR^{S-}$ in a homozygous LuRKO background showed surprising results, with the gross morphology, size, and weights of the reproductive tract and testis approximating those of WT animals (Fig. 23.4(II)). These data are indicative that transactivation of LHCGR can rescue the hypogonadal phenotype of the LuRKO animals.

For detailed histological analysis of testicular structures, paraformaldehyde-fixed testes were dehydrated, paraffin embedded, and 5 μm sections mounted onto poly-D-lysine-coated slides for use in staining protocols. We conducted immunohistochemistry and histology staining on serial sections so that the images could be directly compared. In order to interpret the functional/morphological findings of the functional rescue of LuRKO phenotype by LHCGR transactivation, it was important to undertake careful examination of the spatial/temporal expression pattern of $LHCGR^{B-}/LHCGR^{S-}$ to determine whether the expression was detected in the correct cellular compartments of the testis, that is, the Leydig cells. We determined the expression pattern of $LHCGR^{B-}$ and $LHCGR^{S-}$ using the IRES fluorescent proteins. It should be noted that although these proteins should be easily visualized without the need for immunological-based detection, the testis, as a steroidogenic tissue with a high cholesterol content, generates a high background fluorescent signal due to autofluorescence making the direct visualization of the dsRed and eCFP proteins challenging. To circumvent this, we used immunohistochemistry to detect the expression pattern of the IRES proteins, using differentially raised primary antibodies to red fluorescent protein (RFP) and green fluorescent protein (GFP) (anti-GFP antibodies recognize eCFP due to sequence similarity). We found that standard immunohistochemical methods are sufficient for visualization of RFP and eCFP. In brief, we used sodium citrate antigen retrieval, a blocking buffer of 10% normal goat serum, antibody diluent of 0.05% goat serum in PBS and primary antibody concentrations of 1:100. For a wash buffer, we used PBS and secondary antibodies of Alexa 488 to visualize eCFP and Alexa 594 to visualize RFP, both at 1:500. To visualize nuclei, we conducted the last wash step in PBS-Triton (0.1%) and counterstained the nuclei using 4′,6-diamidino-2-phenylindole dihydrochloride (DAPI). Analysis of whole testis revealed that the expression of $LHCGR^{B-}$ and $LHCGR^{S-}$ was confined to the Leydig cells, with no off-target expression, meaning that any functional rescue of the LuRKO animal phenotype could be directly correlated with $LHCGR^{B-}$/$LHCGR^{S-}$ transactivation in the correct cellular compartments (Fig. 23.4(II)). To visualize the structural and cellular components of the testis, we utilized hematoxylin and eosin staining. The most important histological analysis for determining effects of LHCGR transactivation on the functional rescue of the LuRKO animals is the size of the seminiferous tubules and Leydig cell islets and the progression/resumption of spermatogenesis to beyond arrest at the round spermatid stage as observed in the homozygous LuRKO animals. Histological analysis of the testis of the homozygous LuRKO/$LHCGR^{B-}$/$LHCGR^{S-}$ animals revealed the presence of mature sperm, indicating that spermatogenesis was restored by transactivation, and that the sizes of the seminiferous tubules and Leydig cell islets were comparable to WT animals. Taken together, these data suggested that LHCGR transactivation was sufficient to restore the gonadal function of the LuRKO to that of WT animals.

To determine the expression level of the $LHCGR^{B-}$ and $LHCGR^{S-}$, we conducted qPCR to look at mRNA expression of the mutant receptors. Although the mRNA expression levels were higher for the homozygous LuRKO/$LHCGR^{B-}$/$LHCGR^{S-}$ animals, analysis of the serum LH and testosterone showed that the circulating concentration of LH was slightly elevated in comparison to WT (but lower than in LuRKO animals). Ligand-binding assays (carried out as previously described in Rivero-Müller et al., 2010) to assess the number of $LHCGR^{S-}$ showed no significant difference to that of WT LHCGR. This indicates that the sensitivity of $LHCGR^{B-}$/$LHCGR^{S-}$ may be somewhat decreased in comparison to WT LHCGR. The restoration of testosterone production in the homozygous LuRKO/$LHCGR^{B-}$/$LHCGR^{S-}$ supports this notion, as although not statistically different from WT animals due to the large biological variation, it was slightly lower, most likely resulting from decreased negative feedback to the pituitary and hence the higher circulating concentration of LH in the transactivational model. Quantitative analysis of key LHCGR-dependent Leydig cell-specific steroidogenic factors, *StAR* and *CYP17a1*, which show diminished expression in LuRKO animals, was also restored to expression levels approximating that of WT animals in the homozygous LuRKO/$LHCGR^{B-}$/$LHCGR^{S-}$ animals. This also suggests that LHCGR transactivation is sufficient to mediate the downstream signaling requirements for the transcription/translation of LH-dependent genes.

23.2.3.4 Fertility

The ultimate proof that transactivation of $LHCGR^{B-}$/$LHCGR^{S-}$ could fulfill the functional requirement of the LHCGR *in vivo* is to test the fertility of the animals through mating experiments. For our studies, we set up breeding cages crossing homozygous LuRKO/LHCGR^{B-}/LHCGR^{S-} animals with WT females and comparing litter sizes and numbers of litters to those of WT males crossed with WT females. Our studies showed that not only were the homozygous LuRKO/LHCGR^{B-}/LHCGR^{S-} males fertile but also their fertility was comparable to WT males.

23.3 DISCUSSION

Our data outline the methods by which we have explored GPCR transactivation using the LHCGR as a model receptor. We have shown that coexpression of LHCGR^{B-} and LHCGR^{S-} *in vitro* results in receptor transactivation, and the degree of functional rescue observed is dependent on the cell surface coexpression levels of LHCGR^{B-} and LHCGR^{S-}. In a LuRKO background, using BAC transgenics to coexpress LHCGR^{B-} and LHCGR^{S-}, we have shown functional rescue of the hypogonadism and infertility of LuRKO animals. Our work presents the first model that demonstrates class A GPCR homodimerization, via transactivation, is a physiologically relevant mechanism of LHCGR functionality *in vivo*.

The structural determinates between LHCGR^{B-} and LHCGR^{S-} resulting in transactivation are yet to be conclusively demonstrated. While one study suggests that the minimal functional unit to observe transactivation is the extracellular domain of LHCGR anchored by CD8 to activate a LHCGR^{B-} (Ji et al., 2002), others have observed that the first TM domain of LHCGR is required for transactivation (Osuga, Hayashi, et al., 1997; Osuga, Kudo, et al., 1997). Whether these structural requirements reflect the dimer interface is also unclear; while there is no information on potential LHCGR dimer interfaces, BRET studies of the closely related follicle stimulating hormone receptor (FSHR) demonstrated that either the N-terminal extracellular domain or the 7TM domain was sufficient for receptor–receptor interactions (Guan et al., 2010), indicating that either transactivation or the dimer interface most likely involves multiple regions of the receptor. Transactivation appears to be sufficient for mediating LHCGR functions in male mice; however, the role of cisactivation in LHCGR signaling remains unknown.

There are certain caveats to consider when discussing receptor transactivation, and we acknowledge that LHCGR transactivation will not be observed with all LHCGR^{B-} and LHCGR^{S-} that are coexpressed. It has been previously demonstrated that the degree of functional rescue through transactivation is dependent on the position of the mutation in the LHCGR^{B-}, with mutations at the very N-terminal region of the extracellular domain favoring transactivation and mutations at the C-terminal of the extracellular domain approaching the hinge region, detrimental for transactivation (Lee, Ji, & Ji, 2002; Lee, Ji, Ryu, et al., 2002).

With the increasing sensitivity of techniques for determining G protein-dependent and G protein-independent signaling events, it is becoming increasingly evident that GPCRs can couple to a diverse array of signaling pathways. Dimerization presents a functional mechanism by which the preference for coupling to a specific pathway can be modulated. It is conceivable that regulation of the composition of a di/oligomer may change the preference of pathway activation, acting to fine tune the signal diversity of a specific receptor subtype.

SUMMARY

This chapter provides systematic methodology for assessing the functional aspects of GPCR dimerization, through receptor transactivation, using the LHCGR as a model. We have explored the *in vitro* and *in vivo* requirements for demonstrating LHCGR transactivation and discussed the impact that receptor dimerization may have on our understanding of GPCR biology.

Acknowledgments

This work is supported by BBSRC grant number BB/1008004/1, Wellcome Trust grant 082101/Z/07/Z, and Academy of Finland grant 137848.

References

Chabre, M., & le Maire, M. (2005). Monomeric G-protein-coupled receptor as a functional unit. *Biochemistry*, *44*, 9395–9403.

Fotiadis, D., Jastrzebska, B., Philippsen, A., Müller, D. J., Palczewski, K., & Engel, A. (2006). Structure of the rhodopsin dimer: A working model for G-protein-coupled receptors. *Current Opinion in Structural Biology*, *16*, 252–259.

Guan, R., Wu, X., Feng, X., Zhang, M., Hébert, T. E., & Segaloff, D. L. (2010). Structural determinants underlying constitutive dimerization of unoccupied human follitropin receptors. *Cellular Signalling*, *22*(2), 247–256.

Han, Y., Moreira, I. S., Urizar, E., Weinstein, H., & Javitch, J. A. (2009). Allosteric communication between protomers of dopamine class A GPCR dimers modulates activation. *Nature Chemical Biology*, *5*(9), 688–695.

Huhtaniemi, I., & Alevizaki, M. (2007). Mutations along the hypothalamic–pituitary–gonadal axis affecting male reproduction. *Reproductive Biomedicine Online*, *15*(6), 622–632.

Huhtaniemi, I. T., & Themmen, A. P. N. (2000). Mutations of gonadotropins and gonadotropin receptors: Elucidating the physiology and pathophysiology of pituitary–gonadal function. *Endocrine Reviews*, *21*, 551–583.

Ji, I., Lee, C., Jeoung, M., Koo, Y., Sievert, G. A., & Ji, T. H. (2004). Trans-activation of mutant follicle-stimulating hormone receptors selectively generates only one of two hormone signals. *Molecular Endocrinology*, *18*(4), 968–978.

Ji, I., Lee, C., Song, Y., Conn, P. M., & Ji, T. H. (2002). Cis- and trans-activation of hormone receptors: The LH receptor. *Molecular Endocrinology*, *16*(6), 1299–1308.

Kaupmann, K., Malitschek, B., Schuler, V., Heid, J., Froestl, W., & Beck, P. (1998). GABA (B)-receptor subtypes assemble into functional heteromeric complexes. *Nature*, *396*, 683–687.

Lee, C., Ji, I. J., & Ji, T. H. (2002). Use of defined-function mutants to access receptor–receptor interactions. *Methods*, *27*(4), 318–323.

Lee, C., Ji, I., Ryu, K., Song, Y., Conn, P. M., & Ji, T. H. (2002). Two defective heterozygous luteinizing hormone receptors can rescue hormone action. *Journal of Biological Chemistry*, *277*(18), 15795–15800.

Limonta, P., Montagnani Marelli, M., Mai, S., Motta, M., Martini, L., & Moretti, R. M. (2012). GnRH receptors in cancer: From cell biology to novel targeted therapeutic strategies. *Endocrine Reviews*, *33*(5), 784–811.

Lohse, M. J. (2010). Dimerization in GPCR mobility and signaling. *Current Opinion in Pharmacology*, *10*(1), 53–58.

Menon, K. M., & Menon, B. (2012). Structure, function and regulation of gonadotropin receptors – A perspective. *Molecular and Cellular Endocrinology*, *356*(1–2), 88–97.

Milligan, G. (2013). The prevalence, maintenance and relevance of GPCR oligomerization. *Molecular Pharmacology*, *84*(1), 158–169.

Monnier, C., Tu, H., Bourrier, E., Vol, C., Lamarque, L., Trinquet, E., et al. (2011). Transactivation between 7TM domains: Implication in heterodimeric GABAB receptor activation. *EMBO Journal*, *30*(1), 32–42.

Narayanan, K., & Chen, Q. (2011). Bacterial artificial chromosome mutagenesis using recombineering. *Journal of Biomedicine and Biotechnology*, *2011*, 971296.

Osuga, Y., Hayashi, M., Kudo, M., Conti, M., Kobilka, B., & Hsueh, A. J. (1997). Co-expression of defective luteinizing hormone receptor fragments partially reconstitutes ligand-induced signal generation. *Journal of Biological Chemistry*, *272*(40), 25006–25012.

Osuga, Y., Kudo, M., Kaipia, A., Kobilka, B., & Hsueh, A. J. (1997). Derivation of functional antagonists using N-terminal extracellular domain of gonadotropin and thyrotropin receptors. *Molecular Endocrinology*, *11*, 1659–1668.

Pascal, G., & Milligan, G. (2005). Functional complementation and the analysis of opioid receptor homodimerization. *Molecular Pharmacology*, *68*(3), 905–915.

Pin, J. P., Kniazeff, J., Goudet, C., Bessis, A. S., Liu, J., Galvez, T., et al. (2004). The activation mechanism of class-C G-protein coupled receptors. *Biology of the Cell*, *96*, 335–342.

Rivero-Müller, A., Chou, Y. Y., Ji, I., Lajic, S., Hanyaloglu, A. C., Jonas, K., et al. (2010). Rescue of defective G protein-coupled receptor function in vivo by intermolecular cooperation. *Proceedings of the National Academy of Sciences of the United States of America*, *107*(5), 2319–2324.

Rivero-Müller, A., Lajic, S., & Huhtaniemi, I. (2007). Assisted large fragment insertion by Red/ET-recombination (ALFIRE)—An alternative and enhanced method for large fragment recombineering. *Nucleic Acids Research*, *35*(10), e78.

Sangkuhl, K., Schulz, A., Schultz, G., & Schöneberg, T. (2002). Structural requirements for mutational lutropin/choriogonadotropin receptor activation. *Journal of Biological Chemistry*, *277*, 47748–47755.

Shizuya, H., Birren, B., Kim, U. J., Mancino, V., Slepak, T., Tachiiri, Y., et al. (1992). Cloning and stable maintenance of 300-kilobase-pair fragments of human DNA in Escherichia coli using an F-factor-based vector. *Proceedings of the National Academy of Sciences of the United States of America*, *89*(18), 8794–8797.

Smith, N. J., Bennett, K. A., & Milligan, G. (2011). When simple agonism is not enough: Emerging modalities of GPCR ligands. *Molecular and Cellular Endocrinology*, *331*(2), 241–247.

Song, G. J., Jones, B. W., & Hinkle, P. M. (2007). Dimerization of the thyrotropin-releasing hormone receptor potentiates hormone-dependent receptor phosphorylation. *Proceedings of the National Academy of Sciences of the United States of America*, *104*(46), 18303–18308.

Urizar, E., Montanelli, L., Loy, T., Bonomi, M., Swillens, S., Gales, C., et al. (2005). Glycoprotein hormone receptors: Link between receptor homodimerization and negative cooperativity. *EMBO Journal*, *24*(11), 1954–1964.

Zhang, M., Guan, R., & Segaloff, D. L. (2012). Revisiting and questioning functional rescue between dimerized LH receptor mutants. *Molecular Endocrinology*, *26*(4), 655–668.

Zhang, F. P., Poutanen, M., Wilbertz, J., & Huhtaniemi, I. (2001). Normal prenatal but arrested postnatal sexual development of luteinizing hormone receptor knockout (LuRKO) mice. *Molecular Endocrinology*, *15*, 172–183.

CHAPTER

Crystallization of G Protein-Coupled Receptors

24

David Salom*, Pius S. Padayatti*, and Krzysztof Palczewski*,†

*Polgenix Inc., Cleveland, Ohio, USA
†Department of Pharmacology, School of Medicine, Case Western Reserve University, Cleveland, Ohio, USA

CHAPTER OUTLINE

Introduction	452
24.1 Crystallization of Bovine Rhodopsin	453
24.1.1 Materials	453
24.1.1.1 Rod Outer Segment (ROS) Isolation with Sucrose Gradient	453
24.1.1.2 Nonyl-glucoside/Zn(OAc)$_2$ Extraction of Rhodopsin	454
24.1.1.3 Immunoaffinity Purification	454
24.1.1.4 $(NH_4)_2SO_4$-induced Phase Separation	454
24.1.1.5 Crystallization	454
24.1.2 Methods	455
24.1.2.1 ROS Isolation with Sucrose Gradient	455
24.1.2.2 Nonyl-glucoside/Zn(OAc)$_2$ Extraction of Rhodopsin	455
24.1.2.3 Immunoaffinity Purification of Rhodopsin	455
24.1.2.4 $(NH_4)_2SO_4$-induced Phase Separation	456
24.1.2.5 Crystallization	458
24.2 Crystallization of β2-AR	460
24.2.1 Materials	460
24.2.1.1 Solubilization of Membranes	460
24.2.1.2 Immunoaffinity and Gel Filtration Chromatography	461
24.2.1.3 Crystallization	461
24.2.2 Methods	461
24.2.2.1 Protein Purification	461
24.2.2.2 Mesophase Crystallization	463
24.3 Discussion	465
Acknowledgments	465
References	466

Abstract

Oligomerization is one of several mechanisms that can regulate the activity of G protein-coupled receptors (GPCRs), but little is known about the structure of GPCR oligomers. Crystallography and NMR are the only methods able to reveal the details of receptor–receptor interactions at an atomic level, and several GPCR homodimers already have been described from crystal structures. Two clusters of symmetric interfaces have been identified from these structures that concur with biochemical data, one involving helices I, II, and VIII and the other formed mainly by helices V and VI. In this chapter, we describe the protocols used in our laboratory for the crystallization of rhodopsin and the β2-adrenergic receptor (β2-AR). For bovine rhodopsin, we developed a new purification strategy including a $(NH_4)_2SO_4$-induced phase separation that proved essential to obtain crystals of photoactivated rhodopsin containing parallel dimers. Crystallization of native bovine rhodopsin was achieved by the classic vapor-diffusion technique. For β2-AR, we developed a purification strategy based on previously published protocols employing a lipidic cubic phase to obtain diffracting crystals of a β2-AR/T4-lysozyme chimera bound to the antagonist carazolol.

Abbreviations

CHS	cholesterol hemisuccinate
DDM	dodecyl-β-D-maltoside
FRAP	fluorescence recovery after photobleaching
GPCR	G protein-coupled receptor
LCP	lipidic cubic phase
MES	2-(N-morpholino) ethanesulfonic acid
NG	nonyl-β-D-glucoside
ROS	rod outer segment(s)
T4L	T4 lysozyme
β2-AR	β2-adrenergic receptor

INTRODUCTION

After a decade of heated debate, a large body of experimental data now supports the notion that G protein-coupled receptors (GPCRs) can form physiologically relevant oligomers (Milligan, 2008). Therefore, GPCR oligomerization interface(s) are potential targets for allosteric drugs to modulate GPCR function and high-resolution structures of GPCR oligomers are highly sought as templates for structure-based drug design. The first direct structural evidence of GPCR oligomerization was obtained in our laboratory by atomic-force microscopy, showing a paracrystalline arrangement of bovine rhodopsin in native membranes (Fotiadis et al., 2003). Later, parallel rhodopsin dimers were observed in crystals solved by electron crystallography at 5.5 Å (Ruprecht, Mielke, Vogel, Villa, & Schertler, 2004) and by X-ray crystallography at 3.8 Å (Lodowski et al., 2007; Salom, Le Trong, et al., 2006). Then, a parallel dimer of β2-adrenergic receptor (β2-AR)/T4L was obtained in which the interactions between the two monomers were mainly mediated by lipids (Cherezov et al., 2007). More recently, crystals of chemokine CXCR4 receptor (Wu et al., 2010), κ-opioid receptor (Wu et al., 2012), μ-opioid receptor (Manglik et al., 2012), and β1-AR (Huang, Chen, Zhang, & Huang, 2013) revealed parallel homodimer arrangements with substantial protein–protein interfaces probably reflecting functionally relevant interactions.

24.1 CRYSTALLIZATION OF BOVINE RHODOPSIN

The first crystal structure of ground-state rhodopsin, solved in our laboratory (Palczewski et al., 2000), contained antiparallel dimer interactions and the crystals were disrupted when illuminated. Next, Schertler's laboratory obtained hexagonal rhodopsin crystals (Suda, Filipek, Palczewski, Engel, & Fotiadis, 2004), (Stenkamp, 2008) also with an antiparallel monomer–monomer arrangement. Then, through modifications of rhodopsin's purification protocol, our laboratory obtained two new crystal forms (trigonal and rhombohedral) able to withstand photoactivation. This permitted the structure of an activated GPCR to be solved for the first time (Salom, Le Trong, et al., 2006; Salom, Lodowski, et al., 2006). Interestingly, one of the rhodopsin–rhodopsin interfaces in both crystal forms involving interactions between transmembrane helices I and II and helix VIII is parallel and consistent with the tridimensional model based on the paracrystalline arrangement of rhodopsin observed in native membranes (Fotiadis et al., 2004). In the first part of this chapter, we describe a protocol for the purification and crystallization of rhodopsin in trigonal form, highlighting its most innovative step, the $(NH_4)_2SO_4$-induced phase separation used to concentrate purified rhodopsin prior to crystallization.

24.1.1 Materials

All procedures are performed in a dark room under dim red light at room temperature or colder. The room is equipped with a floor centrifuge, microfuge, spectrophotometer, basic stereo microscope, and 4 °C incubator. Red filters or aluminum foil is used to cover any non-red light from the instruments. This protocol can be scaled down about 10-fold without significant loss in the final rhodopsin yield.

24.1.1.1 Rod outer segment (ROS) isolation with sucrose gradient
1. 100–150 fresh or frozen, dark-adapted, bovine retinas
2. Kuhn's buffer: 67 mM potassium phosphate, pH 7.0, 1 mM Mg(OAc)$_2$, 0.1 mM EDTA (ethylenediaminetetraacetic acid), 1 mM DTT (1,4-dithio-DL-threitol)
3. 45% sucrose in Kuhn's buffer

4. Gradient solutions, with densities to be adjusted with two hydrometers (ranges 1.060–1.130 and 1.120–1.190 g/mL)
   1.10 g/mL (∼107 mL of 45% sucrose + 93 mL Kuhn's buffer)
   1.13 g/mL (∼167 mL of 45% sucrose + 72 mL Kuhn's buffer)
   1.15 g/mL (∼205 mL of 45% sucrose + 41 mL Kuhn's buffer)
5. 40–50 mL high-speed, transparent centrifuge tubes
6. Swinging-bucket rotor able to reach $26,500 \times g$
7. 10 mL syringes with luer locks and Popper Laboratory Pipetting Needles, 14G × 6 in.
8. Funnel
9. Gauze sponges, 12 ply, 4 × 4 in.

24.1.1.2 Nonyl-glucoside/Zn(OAc)$_2$ extraction of rhodopsin
1. 0.5 M 2-(N-morpholino)ethanesulfonic acid (MES), pH 6.3–6.4
2. 1 M Zn(OAc)$_2$
3. 10% nonyl-β-D-glucoside (NG)
4. UV buffer: 1–5 mM dodecyl-β-D-maltoside (DDM), 50 mM Tris, pH 7.4, 100–150 mM NaCl, and 1 mM hydroxylamine

24.1.1.3 Immunoaffinity purification
1. 1D4 monoclonal antibody coupled to CNBr-activated Sepharose (GE Healthcare Life Sciences) or agarose (Pierce) (90–95 mL of settled gel)
2. Glass column (1–2.5 cm diameter × 20–50 cm length)
3. Peristaltic pump
4. Fraction collector
5. Washing buffer: 25–50 mM NG in 150 mM Tris, pH 7.4, 280 mM NaCl, and 6 mM KCl
6. Elution buffer: 0.5–1 mg/mL TETSQVAPA peptide in washing buffer

24.1.1.4 (NH$_4$)$_2$SO$_4$-induced phase separation
1. Solid (NH$_4$)$_2$SO$_4$
2. 40–50 mL high-speed, transparent centrifuge tubes
3. Glass rod and small magnetic rod
4. 0.5 M MES, pH 6.3–6.4
5. 2 mL, dolphin-nose bottom, microfuge tubes

24.1.1.5 Crystallization
1. 24-well, greased crystallization plates
2. Transparent microbridges
3. Basic stereo microscope, with 30× and 60× magnification
4. Thick cover slides (0.96 mm × 22 mm × 22 mm)
5. Crystallization buffer: 0–110 mM NG in 50–100 mM MES, pH 6.3–6.4, 12 mM β-mercaptoethanol, 0.1% NaN$_3$, and 2.5–5% MERPOL HCS
6. Reservoir buffer: 3–3.4 M (NH$_4$)$_2$SO$_4$ in 10–50 mM MES, pH 6.3–6.4

24.1.2 Methods

24.1.2.1 ROS isolation with sucrose gradient

The isolation of ROS from dark-adapted bovine retina essentially follows an established sucrose density gradient centrifugation procedure (Papermaster, 1982) with small modifications (Salom, Li, Zhu, Sokal, & Palczewski, 2005). Briefly, shake the retinas for 1 min in 1 volume of 45% sucrose solution and centrifuge for 5 min at $3,300 \times g$. Filter the supernatant through a gauze-lined funnel, dilute it with 1 volume of Kuhn's buffer, and centrifuge for 10 min at $13,000 \times g$. Resuspend the pellet from each tube in 1 mL of 1.10 g/mL sucrose plus 0.5 mL of Kuhn's buffer, and load the suspension into tubes containing a three-step sucrose gradient (10, 16, and 10 mL of densities 1.11, 1.13, and 1.15 g/mL, respectively). After centrifugation for 20 min at $22,000 \times g$, collect the 1.11–1.13 g/mL interface, dilute it with 1 volume of Kuhn's buffer, and recover the ROS by centrifugation at $6,500 \times g$ for 7 min. Store the ROS pellet at $-80\ °C$. Typical recovery is ~ 0.6 mg rhodopsin per retina.

24.1.2.2 Nonyl-glucoside/Zn(OAc)$_2$ extraction of rhodopsin

Rhodopsin can be selectively extracted from ROS membrane preparations by solubilizing a ROS suspension with alkyl(thio)glucosides in the presence of 2B series divalent cations, which eliminates opsin and other protein contaminants (Okada, Takeda, & Kouyama, 1998).

Resuspend the ROS pellet in ~ 1 volume of 50 mM MES, pH 6.35. Dissolve a small aliquot in UV buffer and measure its absorbance spectrum from 250 to 700 nm to estimate the rhodopsin concentration ($\varepsilon_{498nm} = 40,600\ M^{-1}\ cm^{-1}$; Spalink, Reynolds, Rentzepis, Sperling, & Applebury, 1983). Add the remaining solubilization components from their stock solutions to the ROS suspension to reach final concentrations of 5–10 mg/mL of rhodopsin, 50 mM MES, pH 6.35, 100 mM Zn(OAc)$_2$, and a NG/rhodopsin ratio of 2.2 (w/w). Mix briefly, incubate overnight at 4 °C, and remove precipitated proteins by centrifugation. The A_{280nm}/A_{500nm} ratio of the supernatant typically should be 1.8–2.0, indicating a rhodopsin purity of 80–90%. Up to 50% of rhodopsin can be lost in this step.

24.1.2.3 Immunoaffinity purification of rhodopsin

Preparation of Sepharose-immobilized 1D4 monoclonal antibody is achieved by following the CNBr-activated Sepharose manufacturer's instructions. Antirhodopsin 1D4 antibody can be produced from hybridoma supernatant and purified with a Diethylaminoethyl (DEAE)-cellulose or protein-A column. Alternatively, it can be obtained from mouse ascites fluid and purified with T-Gel (Pierce). We typically coupled the antibody at a ratio of 5 mg of 1D4 per mL of settled gel, resulting in a yield of ~ 0.5 mg purified rhodopsin per mL of gel.

This is a standard immunoaffinity chromatography step where solubilized ROS (>0.6 mg rhodopsin per mL of settled gel) are slowly loaded onto a prepacked 1D4-Sepharose column (10–20 min), the gel is washed with 5–10 column volumes of washing buffer, and rhodopsin is recovered by addition of 0.5–1 mg/mL of

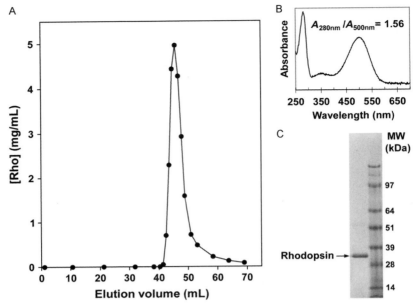

FIGURE 24.1

Immunoaffinity purification of bovine rhodopsin from ROS solubilized in NG. (A) Purified rhodopsin was eluted from a 1.5 cm × 30 cm column containing 46 mL of 1D4-Sepharose. (B) Absorption spectrum of purified rhodopsin (an aliquot was diluted in UV buffer). (C) Coomassie-stained SDS-PAGE gel of purified rhodopsin. MW markers (SeeBlue® Plus2, Invitrogen) are shown in the right lane. (For color version of this figure, the reader is referred to the online version of this chapter.)

competing peptide. The first fractions with rhodopsin start eluting at ~0.85 column volumes with the peak fraction at ~1.0 column volume (Fig. 24.1A). Elution can be achieved in 1–2 h, but to maximize the peak concentration (up to 6 mg/mL), we typically carried out the elution with 1.5 column volumes of elution buffer for 4 h in a 20–50 cm long column (Salom, Le Trong, et al., 2006). Rhodopsin's purity, assessed by the A_{280nm}/A_{500nm} ratio and electrophoresis, is >99% (Fig. 24.1B and C).

Aliquots of fractions are mixed with UV buffer to measure their rhodopsin absorbance, and those fractions with the highest absorbance are pooled to achieve a rhodopsin concentration of 1–2 mg/mL. Up to 90% of rhodopsin loaded onto the column can be recovered for the next step.

24.1.2.4 $(NH_4)_2SO_4$-induced phase separation

Addition of $(NH_4)_2SO_4$ at saturating concentrations to NG solutions induces a phase separation with a detergent-rich top phase. When rhodopsin purified in NG is treated with $(NH_4)_2SO_4$, the bottom aqueous phase appears completely colorless and rhodopsin can be effectively concentrated up to 25-fold in the top phase (Fig. 24.2A).

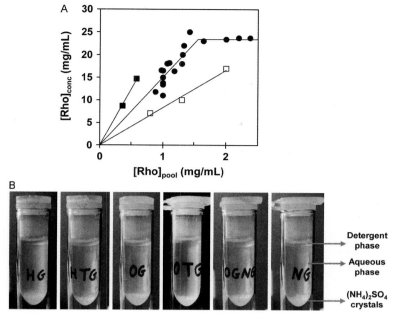

FIGURE 24.2

$(NH_4)_2SO_4$-induced phase separation of purified rhodopsin. (A) Solid $(NH_4)_2SO_4$ was added to rhodopsin purified in 25 mM NG (■), 50 mM NG (●), or 50 mM OG (□). $[Rho]_{pool}$ corresponds to the concentration of rhodopsin prior to treatment with $(NH_4)_2SO_4$, and $[Rho]_{conc}$ is the final concentration of rhodopsin after $(NH_4)_2SO_4$-induced phase separation. Modified and expanded from Salom, Le Trong, et al. (2006). (B) Phase separation in solutions containing 1% detergent in 100 mM MES, pH 6.35, after adding saturating amounts of solid $(NH_4)_2SO_4$ and overnight incubation on ice (HG, heptyl glucoside; HTG, heptyl-thio-glucoside; OG, octyl glucoside; OTG, octyl-thio-glucoside; OGNG, octyl glucoside neopentyl glycol; NG, nonyl glucoside).

- Add 0.25 volumes of 0.5 M MES, pH 6.35, to the pooled purified rhodopsin.
- In a centrifuge tube, weigh ~0.69 g of solid $(NH_4)_2SO_4$ per mL of sample. Take into account that each gram of $(NH_4)_2SO_4$ will add ~0.5 mL to the final volume; therefore, more than one tube could be needed for the entire sample.
- Add the rhodopsin sample to the centrifuge tube(s) (1.45 mL per g of $(NH_4)_2SO_4$, and stir with a small magnetic rod until the solution appears clear (15–30 min). Use a glass rod to assist mixing during the first few minutes until the magnetic rod can stir by itself. Use a tube stand or clamp to keep centrifuge tubes vertical on the magnetic stirrer.
- Incubate the tube(s) on ice for 4–7 days to allow excess $(NH_4)_2SO_4$ to crystallize out of solution. Shorter incubation periods will result in the growth of transparent $(NH_4)_2SO_4$ crystals among the rhodopsin crystals, (Fig. 24.3) thereby complicating

crystallization trials because it can be difficult to distinguish these two crystal forms under red light.
- Spin down briefly at $\sim$13,000 $\times$ g to compact the detergent/rhodopsin (det/rho) phase.
- With a wide bore 1-mL pipette tip, transfer the top, viscous det/rho phase to microfuge tube(s) (preferably, 2 mL dolphin-nose tubes). Sometimes, the whole det/rho phase holds together and can be transferred to the tube in one move. But often the pipette tip fills up with the $(NH_4)_2SO_4$ phase and just a fraction of the det/rho phase. When the microfuge tubes are full, a brief spin will condense the det/rho phase and the bottom phase can be removed with a gel-loading tip. This process must be repeated several times until all the det/rho phase is transferred to the microfuge tubes and all possible $(NH_4)_2SO_4$ has been removed from underneath. About 70% of the initial rhodopsin is recovered after this step.

Notes: This process is slightly less efficient when rhodopsin is purified in octyl glucoside (OG) (Fig. 24.2A). We tested 16 detergents commonly used for membrane protein purification and found that only alkyl(thio)glucoside detergents could be concentrated to a top, detergent-rich phase with saturated $(NH_4)_2SO_4$ (Fig. 24.2B). Therefore, in principle, any short-chain glucoside detergent could potentially be used to purify membrane proteins prior to their concentration by this $(NH_4)_2SO_4$ treatment.

24.1.2.5 Crystallization
The concentrated rhodopsin sample can be used directly for vapor-diffusion crystallization trials, without mixing with reservoir buffer. However, due to its viscosity, the sample is easier to handle if diluted with $\sim$1 volume of crystallization buffer, another buffered solution, or just water. Different components, concentrations, and volumes of the diluting aqueous solution were tested as crystallization variables and a significant percentage supported crystal growth. A sitting drop over a hanging drop format is preferred. The higher the rhodopsin concentration, the lower the reservoir $(NH_4)_2SO_4$ concentration needed to obtain crystals.

Trigonal rhodopsin crystals can be observed after 1 week of incubation at 4 °C and they grow to full size (>100 μm) in 3–4 weeks. These crystals are very stable and retain their morphology and red color for years when kept at 4 °C in the dark.

For photoactivation, a microbridge containing a drop with several crystals (or a few crystals in mother liquor) should be transferred to a new 24-well plate with reservoir buffer and the well should be sealed. Upon illumination with $\sim$500 nm or white light, crystals quickly turn from red to yellow due to isomerization of 11-*cis*-retinal (a potent antagonist) into all-*trans*-retinal (Fig. 24.3) (rhodopsin cognate agonist). The optimal time from the start of photoactivation until crystal freezing in liquid nitrogen is $\sim$2 h after which diffraction quality slowly deteriorates. If left in a drop, crystal morphology is preserved for weeks or months but eventually crystals turn colorless.

Glucose, sucrose, paraffin oil, and several other cryoprotectants have been used successfully with rhodopsin crystals. Most of the crystals were cryoprotected by adding ~10 μL of ~3 M $(NH_4)_2SO_4$ in 50 mM MES, pH 6.35, to the crystal drops before harvesting. The presence of buffer also slows $(NH_4)_2SO_4$ crystal formation due to evaporation, a benefit especially if multiple rhodopsin crystals are being harvested from the same drop.

Notes: Rhodopsin crystals can be grown without the $(NH_4)_2SO_4$-induced phase separation step, and even without ROS isolation and NG/Zn(OAc)$_2$ extraction. Thus, rhodopsin can be solubilized directly from retinas with NG or DDM and purified with immobilized 1D4. This procedure, if done carefully, produces a sample of rhodopsin at >5 mg/mL that can be used directly for crystallization without a concentration step (Salom, Le Trong, et al., 2006). However, crystals obtained this way diffracted poorly and "melted" upon white light illumination, much like the original tetragonal rhodopsin crystals (Okada et al., 2000). Inclusion of each additional purification step resulted in an increased quality of rhodopsin crystals, and the $(NH_4)_2SO_4$-induced phase separation was essential for obtaining light-stable crystals.

MERPOL DA instead of MERPOL HCS as an additive produced rhombohedral rhodopsin crystals diffracting to 3.8 Å (Salom, Le Trong, et al., 2006). However, these crystals were more difficult to reproduce and, upon illumination, lost diffraction more dramatically than the trigonal crystals. Other additives, especially

FIGURE 24.3

Photoactivated rhodopsin crystals surrounding a colorless $(NH_4)_2SO_4$ crystal. (For color version of this figure, the reader is referred to the online version of this chapter.)

amphiphiles, had negative and positive effects on crystal and diffraction quality. However, no other crystal form was identified from these crystallization trials.

24.2 CRYSTALLIZATION OF β2-AR

Here, we describe the crystallization in a monoolein/cholesterol lipidic cubic phase (LCP) of a thermostabilized β2-AR(E122W)/T4L construct described in Hanson et al. (Alexandrov, Mileni, Chien, Hanson, & Stevens, 2008; PBD ID 3D4S) as bound to carazolol instead of timolol, with some modifications to the protocol that made crystal growth more reproducible.

The LCP method, first used to crystallize bacteriorhodopsin (Landau & Rosenbusch, 1996), already has been employed to obtain one-tenth of membrane protein structures in the Protein Data Bank (Aherne, Lyons, & Caffrey, 2012) and most structures of engineered GPCRs (Caffrey, Li, & Dukkipati, 2012). Although no strategy was found to favor a particular relative orientation of GPCRs, four of these receptors crystallized in LCP appeared as dimers in a parallel arrangement (β2-AR, CXCR4, and κ- and μ-opioid receptors) (Cherezov & Caffrey, 2007; Chun et al., 2012).

Some of the advantages found with crystallization of GPCRs in mesophase are (i) rapid crystal growth, (ii) mild temperatures used for crystal growth (~20 °C), and (iii) the fact that all crystals obtained so far are type I, formed by stacked layers of two-dimensional crystals that mimic the native membrane. Another strategy to facilitate GPCR crystallization has been to modify their sequences to (i) stabilize the receptors, (ii) remove flexible regions, (iii) enhance their expression, and (iv) increase the receptors' hydrophilic areas. Such sequence modifications include truncation of the third intracellular loop or its substitution by T4L or BRIL (thermostabilized apocytochrome b_{562}), N-terminal fusion of T4L or BRIL, truncation of long N- and C-termini, and/or addition of thermo-stabilizing mutations.

24.2.1 Materials
24.2.1.1 Solubilization of membranes
1. DDM
2. Cholesterol hemisuccinate (CHS)
3. Dounce homogenizer, 100 mL
4. Insect cell (Sf9) pellets expressing the receptor
5. 1 M 4-(2-hydroxyethyl)-1-piperazineethanesulfonic acid (HEPES), pH 7.5
6. 1 M NaCl
7. 1 M $MgCl_2$
8. 1 M KCl
9. Protease inhibitor cocktail, EDTA-free (Roche)
10. Glycopeptide *N*-glycosidase (PNGase F)
11. DNase (for example, Benzonase® nuclease)
12. Iodoacetamide

24.2.1.2 Immunoaffinity and gel filtration chromatography
1. SepSphere TM alprenolol agarose (CellMosaic, LLC, Worcester, MA)
2. Nickel affinity gel (HIS-Select Nickel Affinity Gel, Sigma, St. Louis, MO)
3. Centrifugal concentrators (100 kDa molecular weight cutoff)
4. Gel filtration column—Sepharose 6 10/300 GL column (GE Life Sciences, Piscataway, NJ)
5. Carazolol

24.2.1.3 Crystallization
1. Stock options Salt kit, HR2-245 (Hampton Research, Aliso Viejo, CA)
2. Polyethylene glycol (PEG) 400
3. 1,4-Butanediol
4. HR2-428 additive screen (Hampton Research, Aliso Viejo, CA)
5. Monoolein (Nu-Chek Prep, Inc., Elysian, MN)
6. Cholesterol
7. LCP mixing devices (Emerald Biosystems, Bainbridge Island, WA)
8. LCP sandwich screening plate (Swissci, Hampton Research, Aliso Viejo, CA)
9. Cy3 mono-NHS-reactive dye (GE Life Sciences)

24.2.2 Methods
24.2.2.1 Protein purification
24.2.2.1.1 Crude membrane preparations
Prepare insect cell pellet (from 3 L of cell culture) expressing the β2-AR(E122W)/T4L construct. All following steps should be performed on ice or at 4 °C. The purification of β2-AR is designed to be completed in 2–3 days with an extra day dedicated to crystallization.

Centrifuge the Sf9 cells in phosphate saline buffer (PBS) at $50,000 \times g$ for 15 min and homogenize the resulting cell pellet in ~300 mL of minimal buffer (10 mM HEPES, pH 7.5, 1 mM $MgCl_2$, 2 mM KCl and one protease inhibitor tablet per 50 mL) with 20 up and down strokes in a Dounce homogenizer. Centrifuge the resulting homogenate at $\sim 50,000 \times g$ for 30 min and discard the supernatant. Repeat the above process at least three times.

Homogenize the combined pellet from the last step into ~400 mL of minimal buffer supplemented with 1 M NaCl and Benzonase. Repeat the membrane washing by centrifugation as in the previous step until a tight pellet is obtained.

24.2.2.1.2 Solubilization
Homogenize the washed membranes in a final volume of 500 mL with solubilization buffer (50 mM HEPES, pH 7.5, 0.15 M NaCl, 1 mM $MgCl_2$, 2 mM KCl, and protease inhibitor cocktail). Then, add PNGase F and more Benzonase. Homogenize with 20 up and down strokes in Dounce homogenizer, add iodoacetamide (2 mg/mL), and incubate at 4 °C for 30–45 min. Add DDM and CHS from a $10\times$ stock to final concentrations of 0.5% and 0.1%, respectively, and rotate the homogenate for at least 6 h.

24.2.2.1.3 Affinity chromatography

At the end of the incubation, centrifuge the homogenate at $50,000 \times g$ for 30 min to remove insoluble material. Add 5 mL of alprenolol-agarose gel to the supernatant and rotate at 4 °C for binding overnight. Next day, recover the alprenolol-agarose gel by pouring it into an empty wide gravitation column and wash the bound protein extensively with washing buffer (150–200 mL of 50 mM HEPES, pH 7.5, 0.15 M NaCl, 0.05% DDM, and 0.01% CHS). Elute the bound receptor batchwise by competition with 2 mM alprenolol in washing buffer (twice with 10 mL and once finally with 5 mL) over a total period of 4–5 h of slow rotation. Check the purity of eluted fractions by SDS-PAGE.

Incubate the eluted protein (25 mL) with ~500 μL of nickel affinity gel and rotate for at least 2 h at 4 °C. Then, wash the nickel affinity gel extensively with 150 mL of washing buffer (50 mM HEPES, pH 7.5, 150 mM NaCl, 0.025% DDM, and 0.005% CHS containing 50 μM carazolol). (This step can also be used for ligand exchange.) Perform elution of the sample at the end of incubation by competition with 1–2 mL of 200 mM imidazole in the washing buffer. Further purify the eluted samples by gel filtration in 50 mM HEPES, pH 7.5, 100 mM NaCl, 0.025% DDM, and 0.005% CHS containing 50 μM carazolol. Concentrate the peak fractions from gel filtration (typically 1–2 mL) up to 50 mg/mL for crystallization. A typical yield from 3 L of culture for the β2-AR is 3–4 mg protein (Fig. 24.4A).

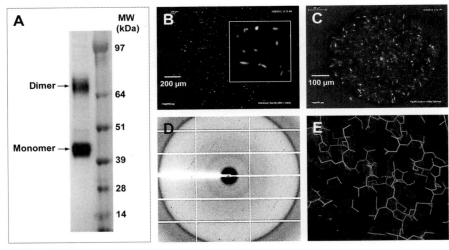

FIGURE 24.4

Crystal structure of β2-AR(E122W)/T4L. (A) SDS-PAGE analysis of a purified β2-AR(E122W)/T4L sample developed with silver staining. Both the purified monomer and dimer are present. (B, C) Representative crystals imaged under crossed polarizers. Inset in panel B shows a close-up of microcrystals. (B) Crystals grown in 0.1 M HEPES, pH 7.0, 0.15 M ammonium fluoride, 30% PEG 400, and 7% 1,4-butanediol. (C) Crystals grown in 0.1 M HEPES, pH 7.0, 0.15 M sodium sulfate, 30% PEG 400, and 7% 1,4-butanediol. (D) Representative X-ray diffraction pattern from a microcrystal with dimensions of $20 \times 10 \times 5$ μm. (E) A part of electron density of β2-AR(E122W)/T4L crystals. (See color plate.)

24.2.2.2 Mesophase crystallization
24.2.2.2.1 Preparation of monoolein/cholesterol/protein mixture

To prepare a 10:1 mol/mol monoolein/cholesterol mixture, codissolve appropriate amounts of these lipids in chloroform/methanol (2:1 v/v), evaporate the solvent with a gentle stream of nitrogen, and then keep the dried lipid under high vacuum for at least 6 h. Finally, seal the vials under an argon or nitrogen stream and store them at −20 °C for future use.

Bring the cholesterol-doped monoolein vial up to room temperature for at least an hour. Open the lipid vial and place it at 42 °C until the solid is evenly melted. Clean all syringes thoroughly with methanol first and then water. Then dry out the syringes to remove water before introducing lipids. Melt pure monoolein separately and mix it with doped monoolein if the molar monoolein/cholesterol ratio needs to be higher than 10. With a 200 μL pipette, introduce 50–75 μL of melted lipid/cholesterol mixture through the open end of the syringe with coupler secured at the other end of a 250 μL Hamilton syringe (use of an uneven coupler for LCP mixture (Caffrey & Cherezov, 2009)). Introduce the concentrated protein sample into a 100 μL syringe in a volume sufficient to fully hydrate the lipid (for monoolein at 20 °C, full hydration with water occurs at ∼40% (w/w) water). Typically, if one is using a manual setup, the volume of lipid should be adjusted to ∼22 μL and the volume of protein sample to 14 μL, which will produce enough LCP mixture to set up ∼175 nL of 96 × 2 conditions for manual screening. Both syringes are coupled through an uneven coupler and both solutions are mixed until a clear solution is obtained. Rapid disappearance of the initial turbidity upon mixing indicates a good LCP sample. Overconcentrating the detergent during centrifugal concentration of the sample often results in turbid LCP mixtures. Also, protein concentrations below 10 mg/mL often do not produce crystals. The lipid/protein mixture should appear completely clear and nonbirefringent when a drop is placed between two cover slips and observed under crossed polarizers.

24.2.2.2.2 FRAP assay

To attain buffer and precipitant conditions conducive to diffusion, nucleation, and crystal growth, it is advisable to subject a few micrograms of purified protein to a fluorescence recovery after photobleaching (FRAP) assay (Cherezov, Liu, Griffith, Hanson, & Stevens, 2008). Use a portion of purified material for labeling with Cy3 mono-NHS-reactive dye (as per manufacturer's instructions) and repurify the labeled protein on a gel filtration column to separate the free dye. Use this Cy3-labeled protein, at 1 mg/mL, to analyze the mobility of samples in a FRAP assay by employing a FRAP screen (Xu, Liu, Hanson, Stevens, & Cherezov, 2011).

Initially, a FRAP screen with 48 salts (Stock options Salt kit) combined with PEG 400 concentrations of choice (e.g., 25% in the upper 48 wells and 30% in the lower 48 wells) is set up in a single 96-well plate at pH 7.0. The same plate is then replicated but at two other pHs (e.g., pH 6.0 and 8.0). In addition, several additives can be individually set up to evaluate the effect of each on the protein diffusion rate. Alternatively, one can substitute a different PEG for PEG 400 in FRAP screens. Depending on the outcomes from such FRAP experiments, crystallization experiments are designed. Typically, only a few salts allow a significant diffusion of the receptor within the mesophase. The faster the diffusion rates for the protein in a set of conditions, the greater the chances of growing crystals under those conditions. For laboratories with no prior mesophase screening experience, we advise undertaking FRAP measurements and LCP setups using β2-AR(E122W)/T4L or human adenosine A_{2a} receptor (PDB ID 4EIY) samples, which reproducibly crystallize without the need of automation.

For a positive control in the diffusion assays, 0.2 M sodium citrate, 28% PEG 400, 0.1 M HEPES, pH 7.5, with carazolol-bound β2-AR can be used, with 0.5 M NaCl employed as a negative control (Xu et al., 2011). Conditions where we observed crystallization of β2-AR (E122W)/T4L that correlate with diffusion rates included one of the following salts: ammonium fluoride, dibasic ammonium phosphate, potassium sulfate, potassium thiocyanate, dibasic sodium phosphate dihydrate, or sodium sulfate. A screen of additives added to the salts in the preceding text (or mixture of salts) and PEG 400 after FRAP assays identified the following compounds as beneficial for the crystallization of β2-AR(E122W)/T4L: ethylene glycol, 1,6-hexanediol, 1,3-butanediol, 2-propanol, 1,4-butanediol, tert-butanol, 1,3-propanediol, and 1-propanol.

The first crystal hits obtained from initial screens are generally not suitable for X-ray diffraction. Thus, individual conditions should be expanded by extensive fine grid screens around the initial conditions. Some initial hits, despite such fine screening, did not improve, although the use of ammonium fluoride, ammonium phosphate, and sodium sulfate along with 28–30% PEG 400 and 1,4-butanediol (5–8%) produced diffraction quality crystals of a $20 \times 10 \times 5$ μm size (Fig. 24.4B and C). The most important variable in growing larger crystals is the monoolein/cholesterol ratio (cholesterol was varied from 3 to 10 mol%). Usually, adding cholesterol to monoolein slowed crystal growth from 2 days to 2 weeks.

24.2.2.2.3 Monitoring crystal growth

A detailed observation record should be kept for every crystallization tray, no matter how small the changes observed. Excellent guidelines were established by pioneers in the field (Caffrey & Cherezov, 2009). An automated plate reader is a good investment because, as in any crystallization trial, once one starts setting drops, the number of trials grows exponentially. Many setups show crystal growth over a month but plates beyond a month tend to dry out. This can be delayed by assuring that the sandwich tape has no trapped bubbles, wrapping the plates with aluminum foil, and incubating them at 22 °C.

A crossed polarizer-fitted microscope with good quality optics is all one needs to monitor the crystallization process from mesophase setups (Fig. 24.4B and C). However, other more expensive imaging options are recommended to discriminate between salt and protein crystals and avoid wasting valuable time and resources following false leads. For example, microscopes equipped with UV fluorescence (Judge, Swift, & Gonzalez, 2005), Second-order nonlinear optical imaging of chiral crystals (SONICC) (Kissick, Gualtieri, Simpson, & Cherezov, 2010), two-photon excited UV fluorescence (Madden, Dewalt, & Simpson, 2011), or two-photon excited visible fluorescence combined with second harmonic generation (Padayatti, Palczewska, Sun, Palczewski, & Salom, 2012) can help in this regard.

The first breakthrough in any mesophase crystallization trial is obtaining the initial microcrystals, no matter how small or insignificant they may appear. The next follow-up steps are to reproducibly grow microcrystals followed by optimization to obtain larger crystals. Crystal sizes as small as $10 \times 5 \times 2$ μm are adequate for X-ray diffraction data collection at modern beamlines equipped with microfocus and rastering capabilities (we used beamlines 23-ID (GMC-A) and 24-ID (NE-CAT) at APS, Argonne, IL). Moreover, the promise of free-electron laser techniques in collecting data from protein nanocrystals holds great hope for the future (Dilanian, Streltsov, Quiney, & Nugent, 2013; Kang, Lee, & Drew, 2013).

We collected a full data set from a single β2-AR(E122W)/T4L crystal. X-ray diffraction data are shown in Fig. 24.4D, with spots extending down to 3.4 Å. The resulting molecular replacement solution, along with the corresponding electron density map, is shown in Fig. 24.4E.

24.3 DISCUSSION

We have described two very different strategies for GPCR crystallization that can be adapted to the crystallization of other GPCRs.

Rhodopsin, being a stable and abundant receptor, can be obtained from native sources and most of the experimental procedures can be performed at room temperature. Trigonal crystals grow very reproducibly in a classic vapor-diffusion format. The most challenging requirement is the need to work in a dark room under dim red light.

After the crystal structure of rhodopsin was initially solved, it took 7 years to solve the structure of the first GPCR with a diffusible ligand, β2-AR/T4L (Cherezov et al., 2007). This was enabled mainly by advances in protein engineering, crystallization methods, and X-ray collection, which allowed solving the structures of at least 17 more GPCRs in the following 6 years. In addition, sequence modifications engineered into the β2-AR to enhance its crystallization can also be used as a reference, although they are often not directly transferable to other GPCRs. Therefore, before attempting the crystallization of a novel GPCR target, we recommend that multiple protein constructs be subjected to several "stress tests" (thermostability assays, expression level assays, homogeneity determination by gel filtration, purification yields, radioligand binding, FRAP assays, etc.) before the three to four best constructs are selected for crystallization trials.

Acknowledgments

The X-ray diffraction data were collected at 24-ID-C (NE-CAT) beamline at APS, Chicago. The authors thank Wenyu Sun for help with insect cell cultures and Philip Kiser for diffraction data collection.

References

Aherne, M., Lyons, J. A., & Caffrey, M. (2012). A fast, simple and robust protocol for growing crystals in the lipidic cubic phase. *Journal of Applied Crystallography*, *45*(Pt 6), 1330–1333.

Alexandrov, A. I., Mileni, M., Chien, E. Y., Hanson, M. A., & Stevens, R. C. (2008). Microscale fluorescent thermal stability assay for membrane proteins. *Structure*, *16*(3), 351–359.

Caffrey, M., & Cherezov, V. (2009). Crystallizing membrane proteins using lipidic mesophases. *Nature Protocols*, *4*(5), 706–731.

Caffrey, M., Li, D., & Dukkipati, A. (2012). Membrane protein structure determination using crystallography and lipidic mesophases: Recent advances and successes. *Biochemistry*, *51*(32), 6266–6288.

Cherezov, V., & Caffrey, M. (2007). Membrane protein crystallization in lipidic mesophases. A mechanism study using X-ray microdiffraction. *Faraday Discussions*, *136*, 195–212, discussion 213–129.

Cherezov, V., Liu, J., Griffith, M., Hanson, M. A., & Stevens, R. C. (2008). LCP-FRAP assay for pre-screening membrane proteins for in meso crystallization. *Crystal Growth & Design*, *8*(12), 4307–4315.

Cherezov, V., Rosenbaum, D. M., Hanson, M. A., Rasmussen, S. G., Thian, F. S., Kobilka, T. S., et al. (2007). High-resolution crystal structure of an engineered human beta2-adrenergic G protein-coupled receptor. *Science*, *318*(5854), 1258–1265.

Chun, E., Thompson, A. A., Liu, W., Roth, C. B., Griffith, M. T., Katritch, V., et al. (2012). Fusion partner toolchest for the stabilization and crystallization of g protein-coupled receptors. *Structure*, *20*(6), 967–976.

Dilanian, R. A., Streltsov, V. A., Quiney, H. M., & Nugent, K. A. (2013). Continuous X-ray diffractive field in protein nanocrystallography. *Acta Crystallographica. Section A*, *69*(Pt 1), 108–118.

Fotiadis, D., Liang, Y., Filipek, S., Saperstein, D. A., Engel, A., & Palczewski, K. (2003). Atomic-force microscopy: Rhodopsin dimers in native disc membranes. *Nature*, *421*(6919), 127–128.

Fotiadis, D., Liang, Y., Filipek, S., Saperstein, D. A., Engel, A., & Palczewski, K. (2004). The G protein-coupled receptor rhodopsin in the native membrane. *FEBS Letters*, *564*(3), 281–288.

Huang, J., Chen, S., Zhang, J. J., & Huang, X. Y. (2013). Crystal structure of oligomeric β1-adrenergic G protein-coupled receptors in ligand-free basal state. *Nature Structural and Molecular Biology*, *20*(4), 419–425.

Judge, R. A., Swift, K., & Gonzalez, C. (2005). An ultraviolet fluorescence-based method for identifying and distinguishing protein crystals. *Acta Crystallographica, Section D: Biological Crystallography*, *61*(Pt 1), 60–66.

Kang, H. J., Lee, C., & Drew, D. (2013). Breaking the barriers in membrane protein crystallography. *International Journal of Biochemistry & Cell Biology*, *45*(3), 636–644.

Kissick, D. J., Gualtieri, E. J., Simpson, G. J., & Cherezov, V. (2010). Nonlinear optical imaging of integral membrane protein crystals in lipidic mesophases. *Analytical Chemistry*, *82*(2), 491–497.

Landau, E. M., & Rosenbusch, J. P. (1996). Lipidic cubic phases: A novel concept for the crystallization of membrane proteins. *Proceedings of the National Academy of Sciences of the United States of America*, *93*(25), 14532–14535.

Lodowski, D. T., Salom, D., Le Trong, I., Teller, D. C., Ballesteros, J. A., Palczewski, K., et al. (2007). Crystal packing analysis of Rhodopsin crystals. *Journal of Structural Biology*, *158*(3), 455–462.

Madden, J. T., Dewalt, E. L., & Simpson, G. J. (2011). Two-photon excited UV fluorescence for protein crystal detection. *Acta Crystallographica, Section D: Biological Crystallography*, *67*(Pt 10), 839–846.

Manglik, A., Kruse, A. C., Kobilka, T. S., Thian, F. S., Mathiesen, J. M., Sunahara, R. K., et al. (2012). Crystal structure of the μ-opioid receptor bound to a morphinan antagonist. *Nature*, *485*(7398), 321–326.

Milligan, G. (2008). A day in the life of a G protein-coupled receptor: The contribution to function of G protein-coupled receptor dimerization. *British Journal of Pharmacology*, *153*(Suppl. 1), S216–S229.

Okada, T., Le Trong, I., Fox, B. A., Behnke, C. A., Stenkamp, R. E., & Palczewski, K. (2000). X-ray diffraction analysis of three-dimensional crystals of bovine rhodopsin obtained from mixed micelles. *Journal of Structural Biology*, *130*(1), 73–80.

Okada, T., Takeda, K., & Kouyama, T. (1998). Highly selective separation of rhodopsin from bovine rod outer segment membranes using combination of divalent cation and alkyl(thio)glucoside. *Photochemistry and Photobiology*, *67*(5), 495–499.

Padayatti, P., Palczewska, G., Sun, W., Palczewski, K., & Salom, D. (2012). Imaging of protein crystals with two-photon microscopy. *Biochemistry*, *51*(8), 1625–1637.

Palczewski, K., Kumasaka, T., Hori, T., Behnke, C. A., Motoshima, H., Fox, B. A., et al. (2000). Crystal structure of rhodopsin: A G protein-coupled receptor. *Science*, *289*(5480), 739–745.

Papermaster, D. S. (1982). Preparation of retinal rod outer segments. *Methods in Enzymology*, *81*, 48–52.

Ruprecht, J. J., Mielke, T., Vogel, R., Villa, C., & Schertler, G. F. (2004). Electron crystallography reveals the structure of metarhodopsin I. *EMBO Journal*, *23*(18), 3609–3620.

Salom, D., Le Trong, I., Pohl, E., Ballesteros, J. A., Stenkamp, R. E., Palczewski, K., et al. (2006). Improvements in G protein-coupled receptor purification yield light stable rhodopsin crystals. *Journal of Structural Biology*, *156*(3), 497–504.

Salom, D., Li, N., Zhu, L., Sokal, I., & Palczewski, K. (2005). Purification of the G protein-coupled receptor rhodopsin for structural studies. In S. R. George & B. F. O'Dowd (Eds.), *G protein-coupled receptor–protein interactions* (pp. 1–17). Hoboken: John Wiley & Sons, Inc.

Salom, D., Lodowski, D. T., Stenkamp, R. E., Le Trong, I., Golczak, M., Jastrzebska, B., et al. (2006). Crystal structure of a photoactivated deprotonated intermediate of rhodopsin. *Proceedings of the National Academy of Sciences of the United States of America*, *103*(44), 16123–16128.

Spalink, J. D., Reynolds, A. H., Rentzepis, P. M., Sperling, W., & Applebury, M. L. (1983). Bathorhodopsin intermediates from 11-cis-rhodopsin and 9-cis-rhodopsin. *Proceedings of the National Academy of Sciences of the United States of America*, *80*(7), 1887–1891.

Stenkamp, R. E. (2008). Alternative models for two crystal structures of bovine rhodopsin. *Acta Crystallographica, Section D: Biological Crystallography*, *D64*(Pt 8), 902–904.

Suda, K., Filipek, S., Palczewski, K., Engel, A., & Fotiadis, D. (2004). The supramolecular structure of the GPCR rhodopsin in solution and native disc membranes. *Molecular Membrane Biology*, *21*(6), 435–446.

Wu, B., Chien, E. Y., Mol, C. D., Fenalti, G., Liu, W., Katritch, V., et al. (2010). Structures of the CXCR4 chemokine GPCR with small-molecule and cyclic peptide antagonists. *Science, 330*(6007), 1066–1071.

Wu, H., Wacker, D., Mileni, M., Katritch, V., Han, G. W., Vardy, E., et al. (2012). Structure of the human kappa-opioid receptor in complex with JDTic. *Nature, 485*(7398), 327–332.

Xu, F., Liu, W., Hanson, M. A., Stevens, R. C., & Cherezov, V. (2011). Development of an automated high throughput LCP-FRAP assay to guide membrane protein crystallization in lipid mesophases. *Crystal Growth & Design, 11*(4), 1193–1201.

Index

Note: Page numbers followed by *f* indicate figures and *t* indicate tables.

A

ACP. *See* Acyl carrier protein (ACP)
Acyl carrier protein (ACP)
    Dy547-labeled ACP-CD59 expression, 383–384, 383*f*
    Halo7-tagged receptor molecules, 383
    tag proteins, 379*t*, 380
Adaptive Poisson–Boltzmann solver (APBS), 93, 101
Affinity chromatography, 462
Anisotropy fluorescence imaging
    depolarization method, 317–319
    microscope, 315–316
    steady-state, 309–311
    time-resolved, 311–313
Antibleaching and antiblinking imaging buffer system
    materials, 300–301
    procedure, 301–302
    solutions and buffers, 301
APBS. *See* Adaptive Poisson–Boltzmann solver (APBS)
Azidogroup
    opsin, 273–274
    rhodopsin, 275–276

B

BAC. *See* Bacterial artificial chromosomes (BAC)
Bacterial artificial chromosomes (BAC)
    E. coli, 441
    generation, 441–442
    genomic databases, 441
    internal ribosome entry site (IRES), 441
    Lhcgr gene, 441
    receptor transactivation *in vivo*, 440–441
    transgenic offspring and breeding strategy, 442, 442*t*, 443*f*
β2-AR
    crystallization, 461
    GPCRs, mesophase, 460
    immunoaffinity and gel filtration chromatography, 461
    LCP method, 460
    membranes solubilization, 460
    mesophase crystallization (*see* Mesophase crystallization)
    protein purification (*see* Protein purification, β2-AR)
Bio-beads
    detergent-free buffer, 350
    incubation, lipid, 350
    lipid-to-protein ratio, 350–351
    preparation, liposome, 350
    protein–lipid–detergent solution, 350
Bioluminescence resonance energy ransfer (BRET)
    acceptor/donor ratio, 144
    autofluorescence, 145
    biophysical technique, 240
    bisdeoxycoelenterazine/didehydrocoelenterazine, 144–145
    and $BRET^2$ assay (*see* $BRET^2$ assay)
    and C1 buffer, 239
    coelenterazine h, 144–145
    and CoIP, 233*f*
    competition assays, 155–156
    DeepBlueC™, 161
    diversification, 159
    donor and acceptor protein moieties, 239
    GPCR–RTK interaction (*see* GPCR–RTK interaction)
    kinetics and dose–response assays (*see* Kinetics and dose–response assays)
    light emission, 239
    luminescent and fluorescent protein, 161
    measure protein–protein interactions, 215
    nonradiative energy transfer, 143–144
    quantum dot-BRET (QD-BRET), 145–146
    receptor–receptor interactions, 144, 146
    red-shifted fluorophore, 145
    replicate measurements, 238–239
    residual light emission, 239
    saturation assay (*see* Saturation assay, BRET)
    TAS2R homo-and hetero-oligomerization, 237–238, 238*f*
    and TAS1Rs, 238
Bioorthogonal conjugation
    HEK293-F suspension cells, 273–274
    wild-type and azido-tagged rhodopsin, 275–276
Biotinylated antibodies
    and biocytin hydrazide, 282–284
    sulfo-NHS-SS-biotin, 284–285

Biotinylated-bovine serum albumin
    materials, 289
    procedure, 290–291
    solutions and buffers, 290
Bitter taste receptors. See Oligomerization
Blue-native gel electrophoresis
    2D BN-PAGE, 428
    digitonin-solubilized yeast and mammalian mitochondria, 414
    mitochondrial supercomplex (see Mitochondrial supercomplex)
Bovine rhodopsin
    crystallization, 454, 458–460
    ground-state rhodopsin, 453
    immunoaffinity purification, 454, 455–456
    nonyl-glucoside/Zn(OAc)$_2$ extraction, 454, 455
    rod outer segment (ROS) isolation, 453–454, 455
    (NH$_4$)$_2$SO$_4$-induced phase separation, 454, 456–458
    tridimensional model, 453
BRET. See Bioluminescence resonance energy transfer (BRET)
BRET^2 assay
    competitive (see Competitive BRET^2 assay)
    GPCR oligomerization, 218–219
    photobleaching and coincidence, 215
    saturation, 219

C

CAPRI. See Critical assessment of predicted interactions (CAPRI)
Cardiolipin-enriched microdomains
    buffers, 421–422
    isolation, 422–423
    mitochondria isolation, 422
    reagents, 421
cDNA. See Complementary DNA (cDNA)
Cell lysis, GPCRs
    denaturing-based lysis methods, 332
    mechanical disruption, 327–329, 330t
    nondenaturing detergent-based lysis methods, 329–331
Cell surface replicas
    etching and replication, 402–403
    freeze-fracture unit, 401–402
    specimen table, 402
CGMD. See Coarse-grained molecular dynamics (CGMD)
CG simulations. See Coarse-grained (CG) simulations
Chinese hamster ovary-K1 (CHO-K1) cells, 362–363, 378, 383–384
CHO-K1 cells. See Chinese hamster ovary-K1 (CHO-K1) cells
Cholesterol-depleting agents, lipid raft
    cholesterol oxidase, 108–109
    methyl-β-cyclodextrin, 108
CLSM. See Confocal laser scanning microscope (CLSM)
CoA. See Coenzyme A (CoA)
Coarse-grained molecular dynamics (CGMD), 98, 99
Coarse-grained (CG) simulations
    vs. all-atom (see Dimers, GPCRs;Oligomers, GPCRs)
    structures and force fields, 64–65
Coenzyme A (CoA), 380
Coimmunoprecipitation (co-IP)
    antibodies, 332–333, 332t
    anti-FLAG affinity gel, 236
    and BRET (see Bioluminescence resonance energy transfer (BRET))
    cell culture dishes, 236
    cell lysate, 333–335
    comparison, standard and membrane, 333, 334f
    and FLAG-tag, 233
    genetic engineering methods, 332–333
    GPCR oligomerization, 216, 224
    and HA-TPα, 218
    5-HT$_4$R dimers, 131
    hydrophobic GPCRs, 214–215
    and mco-IP, 327
    membrane, 335, 335t
    myc-A1R and HA-LPA1R, 218
    physical interactions, 235f
    protein A/G beads, 332–333
    protocol, 216–218
    resonance energy transfer (RET), 215
co-IP. See Coimmunoprecipitation (co-IP)
Colocalization lifetimes
    colocalized spots, 386
    dis-colocalization process, 386
    histogram, 386–387
    non-colocalized spots, 386–387
    rate constant, 386–387
    receptor dimers, 385
    single-color experiments, 386–387
Competitive BRET^2 assay
    A2AR and FGFR1 interaction, 156, 218f
    cDNA coding, 156
    GPCR–RTK interaction, 215–216
    HA-TPα and HA-A$_1$R, 220, 221f
    nonfused noninteracting protein, 219
    protocol, 219–220

receptor X-Rluc and receptor Y-GFP2, 215–216
(GFP2-GFP20)/Rluc8 BRET$_{max}$ values, 156
Complementary DNA (cDNA), 376–377, 378
Confocal fluorescence microscopy, 110
Confocal laser scanning microscope (CLSM)
  definition, 3
  SpIDA, 4, 6–8, 17
CRE-luciferase (CRE-luc) reporter gene
  cAMP accumulation, 438–439, 440f
  pcDNA3.1 and pRL-CMV, 436
Critical assessment of predicted interactions
  (CAPRI), 93–94, 300
Crude membrane preparations, 461
Crystallization
  and β2-AR (see β2-AR)
  bovine rhodopsin (see Bovine rhodopsin)
  chemokine CXCR4 receptor, 452–453
  and GPCR, 452–453, 465
  stable and abundant receptor, 465
  "stress tests", 465
Cysteines
  labeling efficiency determination, 346
  optimization and labeling, 346
  preparation, receptor, 345
  reagents, 345
  reduction, reactive site, 345–346

D

Data analysis, single-color experiments
  colocalization lifetimes (see Colocalization lifetimes)
  and dual-color experiments, 385
  receptor dimer lifetimes, 384–385
  receptor monomers, dimers and oligomers fractions, 382–384
2D Blue-native polyacrylamide gel electrophoresis (2D BN-PAGE), 428
2D BN-PAGE. See 2D Blue-native polyacrylamide gel electrophoresis (2D BN-PAGE)
3D-CTMD. See Three-dimensional continuum MD (3D-CTMD)
Detergent-insoluble glycosphingolipid (DIG)
  buffers and sucrose, 415–416
  equipment, 415
  immunoprecipitation, 417–418
  isolation, 417
  protocol, 415
  reagents, 415
Detergent-solubilized lipids
  materials, 285
  procedure, 286–287
  solutions and buffers, 285–286

Dijkstra's algorithm, 48
Dimers, GPCRs
  advantages and difficulties, approach, 174
  allosteric and biased ligands, 168–169
  allosteric modulators, 166–168, 167f
  analysis and interpretation
    allosteric communication, 172–174
    binding data, 172
    GraphPad 5 Prism, 172
    nonlinear regression analysis, 172
    orthosteric ligand binding to FP, 172–174, 173f
  cell culture and transfection, 170
  heterologous and homologous expression systems, 174
  ligand-binding complexities, 175–176
  materials, 169–170
  neurokinin NK2 receptor, 167–168
  noncompetitive antagonism, 166–167
  and oligomers (see Oligomers, GPCRs)
  orthosteric ligands, 167–168, 169
  prostaglandin F2α (PGF2α), 167–168
  radioligand binding and dissociation kinetics, 170–172
  receptor homodimer, 175
  "receptor mosaics", 175–176
  RET/PCA, 175–176
  signaling output, 175
  signaling patterns, 167–168, 169f
  "ternary complex model", 166–167
DMEM. See Dulbecco's modified Eagle's medium (DMEM)
n-Dodecyl-β-D-maltoside (DDM) lysis, 331
DualView imaging systems, 299
Dulbecco's modified Eagle's medium (DMEM), 250, 252, 253, 254

E

EGFR. See Epidermal growth factor receptor (EGFR)
Electrophoretic mobility shift assays (EMSAs)
  in vitro transcription–translation, 31–32
  materials, 31
  mutation, RXRα substantial, 31, 32f
  protein–DNA complexes, 32
Electrospray ionization mass spectrometry (ESI-MS)
  materials, 30
  protocol, 30–31
  and RARα LBD, 29, 30f
EMSAs. See Electrophoretic mobility shift assays (EMSAs)

Epidermal growth factor receptor (EGFR)
  clustering, 307
  dimerization and oligomerization, 317–319
  domain I and III, 306–307
  donor and acceptor, 308–309
  and EGFR-GFP, 308
  energy transfer, 308–309
  and FCS measurements, 317–319
  and FIDA method, 308
  fluorescence anisotropy values, 309
  FRET and FLIM, 307
  homo-FRET quantification, 309–313
  homo/heterodimers, 306–307
  posttranslational modifications and signaling, 306–307
  quantification, EGFR–eGFP, 4–6
ESI-MS. *See* Electrospray ionization mass spectrometry (ESI-MS)

F

Far-field localization techniques (STORM/PALM), 114–115, 116*f*
Far-field patterned illumination techniques (STED)
  advantage, 117
  photophysical properties, fluorophores, 115–117
  protein distribution and cluster sizes on cell membrane, 116*f*, 117
FCCS. *See* Fluorescence cross-correlation spectroscopy (FCCS)
FCS. *See* Fluorescence correlation spectroscopy (FCS)
FIDA. *See* Fluorescence intensity distribution analysis (FIDA)
FLIM. *See* Fluorescence lifetime imaging microscopy (FLIM)
Fluorescence correlation spectroscopy (FCS)
  application, 182
  autocorrelation analysis, 183
  biologists and biochemists, 360
  classical techniques, 199
  confocal microscopy, 182, 183*f*
  data analysis
    diffusion coefficient, 190–192
    molecular brightness, 192–193
  distribution and association, 317–319
  and FIDA, 317–319
  fluorescent proteins, 200
  instrument setting
    alignment and calibration, 187–188
    environment, 186
    imaging, 186
    laser, 186–187

isotonic viewing solution, 188–189
  materials, 184–185
  monitoring receptor interactions, 182
  monomeric and dimeric receptors, 194
  observation volume, 360
  and PCH, 183–184
  photobleaching, 189–190
  plasma membrane receptor, 189, 189*f*
  receptor-receptor interactions, 195
  resolving power, 198
  sample preparation
    cell type, 185
    fluorescent tag, 185
    labeling, receptor, 185–186
    plating cells, 186
    transfecting cells, 186
  SimFCS software, 363–364
  single-color, 199
  single molecules, 360
  and TIRF, 200–201
  YFP channel, 203
Fluorescence cross-correlation spectroscopy (FCCS). *See also* Receptor oligomerization, line-scanning FCCS
  cell culture and transfection, 36–37
  data acquisition, 206*f*
  data analysis, 37–39
  degree of heterodimer formation, 35–36, 36*f*
  dual-color line-scan, 199
  dyes to monitor interactions, 199
  and endoplasmic reticulum (ERs), 37
  HeLa cells, 209*t*
  materials, 36
  measurements, 37
  and PCH measurements, 202
  setting, 201*f*
Fluorescence intensity distribution analysis (FIDA)
  cluster size, 317–319
  EGFR clustering, 308
  intensity fluctuations, 308
Fluorescence lifetime imaging microscopy (FLIM)
  cluster size, 317–319
  donor/acceptor ratio, 307
  and FRET-FLIM, 308
Fluorescent probes, cyclooctynes
  azido-tagged rhodopsin and SpAAC, 277
  kinetic study and SpAAC, 278–282
  UV-Vis spectroscopy and in-gel fluorescence, 278
N-Formyl peptide receptor (FPR), 374, 375*f*, 382–383
Förster resonance energy transfer (FRET)

AlexaFluor dyes, 344–345
cell-free expression, 344
cysteines, 345–346
donor and acceptor pairs, 344
energy transfer, 307
and FLIM, 308
fluorescent proteins, 247–248
and GFP, 344
homo-FRET (see Homo-FRET microscopy)
measurements and analysis, 351–353
principles, 245–247
receptor dimerization/small cluster formation, 307
FPR. See N-Formyl peptide receptor (FPR)
Freezing
cryogen container, 407
cryovials, frozen mica sandwiches, 400–401
ice recrystallization, 401
liquid ethane, 407
liquid nitrogen-filled Dewar, 401
plunge-freezer design, 401
plunge-freezing unit, 400
tweezers, 401
FRET. See Förster resonance energy transfer (FRET)

G

Gel filtration chromatography, 461
GFP. See Green fluorescent protein (GFP)
Giant unilamellar vesicles (GUVs)
peripheral protein interactions *in vitro*, 368
raster-scanned images, 367f
Glass-bottom microtiter plate
materials, 287
procedure, 288
solutions and buffers, 287–288
Glass surface and epoxy silane
materials, 289
procedure, 289
solutions and buffers, 289
Glycosylphosphatidylinositol-anchored protein (GPI-AP)
and CD59, 374–375
cDNA sequences, 378
expression levels, 378
and GPCR, 375f
GPCR–GPCR interactions, 80–81
GPCR–lipid interactions, simulations
adaptation, membrane, 77
conventional elastic models, 76–77
3D-CTMD model, 77
hydrophobic–hydrophilic contacts, 75–76
hydrophobic mismatch, 80
MD simulations, 79–80
residual mismatch energy, 77–78
rhodopsin, 78–79
SASA calculations, 78
serotonin 5-HT2A receptor, 78–79, 79f
surface integral, 76
GPCR–RTK heteroreceptor complexes
analysis, BRET signal, 148
biochemical/pharmacological properties, 142–143
BRET signal and design, 147
cell system and coexpression, 147
design and validation, receptor, 146
donor/acceptor and substrate combination, 146
identified and study, 159
ligand-induced conformational changes translate, 156
receptor–receptor interactions
BRET fusion constructs, 149
BRET measurement, 150
calculating BRET ratio, 151
cell transfection and coexpression, BRET, 150
harvesting, transfected cells, 150
therapeutic targets, 143
GPCR–RTK interaction
calculating, 151
cell transfection and coexpression, 150
design and validation, receptor, 146
donor/acceptor and substrate combination, 146, 147t
fusion proteins, 147
generation and validation, fusion constructs, 149
harvesting transfected cells, 150
identification and study, 147
measurements, 150
ratio/signal, 148
Rluc8, 149
signal and design, 147
GPCRs. See G protein-coupled receptors (GPCRs)
GPI-AP. See Glycosylphosphatidylinositol-anchored protein (GPI-AP)
G protein-coupled receptors (GPCRs)
A1 and thromboxane A2 (TP) receptors, 216
activation, 3
$BRET^2$ assay (see $BRET^2$ assay)
cell culture and transfection, 223–224
cell lysis, 327–332
clinical drugs, 215–216
clinical significance, 263
coexpression, 222
and co-IP (see Coimmunoprecipitation)

G protein-coupled receptors (GPCRs) *(Continued)*
  controls, 337–338
  cyclic AMP assay, 220–222
  definition, 3, 64
  detection, interacting partners, 337
  dimerization, 325
  di-/oligomerization, 447
  diversification, 215
  downstream signaling modules, 244–245
  elution procedure, 335–337
  extracellular domain, 447
  extracellular signal-regulated kinase1/2 assay, 222
  $GABA_B$ receptor, 214
  gatekeepers, 325–326
  and GPI-AP, 374–375, 375f
  homo-and hetero-oligomerization, 214
  homodimers and heterodimers, 434
  homomerization and heteromerization, 244–245
  interactions, ER chaperones, 324–325
  $LHCGR^B$-and $LHCGR^S$-transactivation, 447
  and LHCGRs (*see* Luteinizing hormone/chorionic gonadotropin receptors (LHCGRs))
  lipid-dependent (*see* Lipid-dependent GPCR)
  materials, 223, 326–327, 435–436
  membranes (*see* Native-like membranes)
  molecular dynamics (MD), 64–65
  monomeric units, 244–245
  and NK1, 4–5, 5f
  oligomerization (*see* Structure-based molecular modeling approaches, GPCR oligomerization)
  oligomerization studies, 338
  overexpression, mammalian cell lines, 327, 329t
  physiological role, 434
  plasmids construction, 223
  protein–protein interactions, 338
  and PSN (*see* Protein structure network (PSN))
  reproductive functions, 435
  requirement and *in vivo* significance, 434
  and RET, 215
  RTK transactivation, 5–6
  serotonergic receptors, 136–137
  signaling pathways, 434
  signal transductions, 224
  single-molecule immunoprecipitation, 271–272
  SMD-TIRF, 270–271
  transactivation, 434–435
  visualization, 215
  Western blotting and coimmunoprecipitation, 136, 224
Graphical user interface (GUI)
  histogram parameters, 13
  MATLAB, SpIDA, 12, 12f
  SpIDA procedures, 13–15, 14f
Green fluorescent protein (GFP)
  absorption properties, 344
  and eCFP, 344
  genetic engineering, 344
GUI. *See* Graphical user interface (GUI)
GUVs. *See* Giant unilamellar vesicles (GUVs)

H

HEK293. *See* Human embryonic kidney (HEK293)
HEK293-F suspension cells
  amber codon suppression, 274
  materials, 273–274
  transfection, 274
HEK293-T
  and $pHA-D_{4.7}R$, 331f
  transfection methods, 329t
HeLa cells
  Attofluor cell chamber, 204
  growth and transfection, 202
  mTq2-and sYFP2-labeled H1 receptors, 206f, 207f, 209t
Hill–Langmuir binding isotherm model, 4–5, 5f, 16–17
Homo-FRET microscopy
  analysis, 317
  Brownian distribution, 113–114
  cell lines, 315
  fluorescence anisotropy, 110–112
  lipid rafts and receptor oligomerization, 110–112
  microscopy, 315–316
  nonradiative energy transfer, 110–112
  photobleaching, 113
  plasmid constructs, 313–315
  slide preparation, 316–317
  steady-state fluorescence anisotropy imaging, 112–113, 309–311
  time-resolved fluorescence anisotropy imaging, 111f, 311–313
Human embryonic kidney (HEK293)
  brightness analysis, VP40, 364f
  brightness *vs.* intensity analysis, EGFP-VP40, 365f
  and CHOK-1 cells, 362–363

I

Immobilization, biotinylated antibody
  materials, 292
  procedure, 293
  solutions and buffers, 292–293

Immunoaffinity
  and gel filtration chromatography, 461
  rhodopsin, 454, 455–456
Immunoblotting
  mco-IP, 335t
  mouse anti-HA, 331f
  rabbit anti-μOR antibody, 331f
In-gel fluorescence, 278

K

Kinetics and dose–response assays
  agonist-promoted FGFR1 activation, 156–159
  calculation and interpretation, 159
  cell suspension, 158
  coelenterazine, 158
  FGFR1-5-HT1A heteroreceptor complexes, 156–159, 157f
  HEK-293T cells, 158
  ligand-induced conformational changes, 156
  receptor–receptor interactions, 156

L

Laser-to-fiber and fiber-to-fiber beam system, 298–299
LBD. *See* Ligand-binding domain (LBD)
LCP method. *See* Lipidic cubic phase (LCP) method
LHCGR^{B-} and LHCGR^{S-} transactivation
  and BAC (*see* Bacterial artificial chromosomes (BAC))
  cAMP response, 439
  CRE-luc system, 438–439
  human embryonic kidney (HEK293), 437
  ligand-binding-deficient and signaling-deficient mutant receptors, 435f, 436
  ligand-induced signaling, 436–437
  LuRKO animals, 436–437, 447
  stable cell lines, 437–438
  WT functional response, 439
LhCGR knockout (LuRKO) animals
  breeding strategy, 442, 443f
  homozygous, 442–445, 444f
  Lhcgr^{B-} and *Lhcgr*$^{S-}$, 442
  LHCGR function, 436–437
  primary function, 442–445
  primer design, 442, 442t
LHCGRs. *See* Luteinizing hormone/chorionic gonadotropin receptors (LHCGRs)
Ligand-binding domain (LBD)
  AF-2, 23–24
  homo and heterodimers, 24
  positive ESI mass spectra, RARα, 29, 30f
  purification, RARα–RXRα heterodimers, 27–28, 28f
  two-hybrid assays, NR interactions, 32–35, 33f
Linear unmixing FRET (lux-FRET)
  application, 245
  buffer A, 252
  buffer B, 252
  cell culture media, 252
  cell cultures, 253
  data acquisition, 254–255
  data evaluation steps, 260, 260f
  devices, 251–252
  evaluation procedure, 261–262
  fluorescence spectrometer, 255–256
  fluorolog, 254
  and FRET (*see* Förster resonance energy transfer (FRET))
  and GPCRs, 244–245
  5-HT$_7$ and the 5-HT$_{1A}$ receptor interaction, 260, 263, 263f, 264f
  nonradiative process, 245
  oligomeric complexes, 263
  principle, 248–249
  protease inhibitors and phosphate-buffered saline stock solutions, 253
  protein-protein interaction, 263
  reagents, 249–250
  receptor-receptor interaction, 263
  reculture and seeding cells, 253
  spectral and intensity-based approaches, 245
  spinning disk microscope, 257–258
  TIRF microscope, 259–260
  transfection, 253–254
  Zeiss LSM 780, 256–257
Lipid bilayers
  CHARMM-GUI membrane builder, 74
  hydration level, 67
  structure, 70, 74
Lipid-dependent GPCR
  and brain polar lipid (BPL), 354–355, 355t
  dimerization, 342–343
  flexible linkers, 354
  and FRET (*see* Förster resonance energy transfer (FRET))
  in vitro and *in vivo* studies, 342–343
  labeling strategies, FRET, 344–346
  lipid composition, 342–343
  liposomes (*see* Liposomes)
  mammalian cell-surface receptors, 342
  purification, 343–344
  reconstitution, 348–351
  structural determination, 343

Lipidic cubic phase (LCP) method
  crystallize bacteriorhodopsin, 460
  FRAP measurements, 463–464
  monoolein/cholesterol, 460
  uneven coupler, 463
Lipid rafts
  cholesterol-depleting agents, 107–109
  colocalization studies, 109–110, 109t
  labeling methods and imaging techniques, 118–119
  membrane domains, 106–107
  micrometer-sized raft regions, 106–107
  protein diffusion, 106–107
  receptor membrane organization
    confocal fluorescence microscopy, 110
    conventional microscopy, 114
    far-field patterned illumination techniques, 115–117
    far-field localization techniques, 114–115
    homo-FRET microscopy, 110–114
    near-field techniques (NSOM), 117–118
    protein distribution and clustering, 114–115, 116f
  signaling molecules, 106–107
Liposomes
  lipid preparation, 347
  sonication vs. extrusion, 347–348
Luteinizing hormone/chorionic gonadotropin receptors (LHCGRs)
  5'-end the bovine prolactin leading sequence, 436–437
  LhCGR knockout (LuRKO), 436–437
  ligand-binding-deficient mutant, 436
  protein–protein interactions, 437
  signaling-deficient mutant, 436
  transactivation *in vitro*, 437–439
  transactivation *in vivo* (*see* Transactivation *in vivo*, LHCGR)

M

Major histocompatibility complex (MHC), 393
Male reproductive tract and testes
  histological analysis, 445–446
  immunohistochemical methods, 445–446
  LHCGR^B-and LHCGR^S-transactivation, 445–446
  LHCGR transactivation *in vivo*, 442–445, 444f
  ligand-binding assays, 446
  LuRKO animals, 442–445
  paraformaldehyde, 445
  quantitative analysis, 446
  seminiferous tubules and Leydig cell islets, 445–446
  WT animals, 446
MD simulations. *See* Molecular dynamics (MD) simulations
Membrane coimmunoprecipitation (mco-IP)
  assays, mammalian cell lines, 327
  and co-IP, 332
  and immunoblotting, 335t
"Membrane interdigitation", 68–69, 69f
Mesophase crystallization
  crystallization trial, 464–465
  FRAP assay, 463–464
  monoolein/cholesterol/protein mixture, 463
mGFP. *See* Monomeric GFP (mGFP)
MHC. *See* Major histocompatibility complex (MHC)
Microdomains
  biochemical characterization and methods, 414
  biological significance, 429
  biophysical environment, 428
  cardiolipin-enriched (*see* Cardiolipin-enriched microdomains)
  compartmentalization, 413
  detergent-insoluble, 429
  and DIG (*see* Detergent-insoluble glycosphingolipid (DIG))
  heterogeneous, 429
  individual complexes, 428–429
  lipid raft structures, 413
  microenvironment, 428
  mitochondrial supercomplex (*see* Mitochondrial supercomplex)
  plasma membrane isolation (*see* Plasma membrane)
  "quasicrystalline" regions, 413
  supercomplexes, 414
  TCR-mediated signaling, 413
Mitochondrial supercomplex
  blue-native gel electrophoresis, 426
  blue-native samples, 423
  buffers and solutions, 424–425
  2D BN-PAGE, 428
  equipment, 424
  reagents, 423–424
  sample preparation, 425
  silver staining, 423, 428
  Western blotting, 426–428
Molecular brightness, FCA
  labeling, receptor, 194
  parameter, 194–195
  and PCH, 193

photon count, 192–193
plasma membrane receptors, 192–193
residuals, curve, 191f, 193
sample histograms, 191f, 193
software packages, 192
Molecular dynamics (MD) simulations
  all-atom, 64–65, 81, 99
  definition, 96
  deformation function, 76–77
  employment, 44
  engines, 65
  GPCRs, 97–99
  ternary complex, 50–51
  workflow, PSN–MD and PSN–ENM approaches, 45–46
Monomeric GFP (mGFP)
  AP2018 addition, 312f
  cDNA encoding, 316–317
  depolarization, 310f
  EGFR-FKBP-mGFP constructs, 314f
  and FKBP-mGFP/2xFKBP, 311f
  mutation, 313–315
  photobleaching, 379
Monomers, GPCRs
  GPCR–lipid interactions, simulations, 75–80
  GPCR–membrane complexes, 74–75
  modeling, 73–74
  palmitoylation, 72–73
mTq2. See mTurquoise2 (mTq2)
mTurquoise2 (mTq2)
  cross-correlation detection, 203
  diffusion times, 204
  HeLa cells, line-FCCS analysis, 209, 209t
  mTq2-p63-sYFP2, 204–205
  and sYFP2, 202, 203, 208–209, 210

N

Native-like membranes
  biophysical parameters, 72
  fluidity, 70
  GPCR dimers and oligomers, 80–83
  GPCR monomers (see Monomers, GPCRs)
  heterogeneous mixtures, 70
  hydration level, lipid bilayers, 67
  lipid types, 67–68
  modeling physiological membranes, 65–66
  "phosphate-to-phosphate" distance, 68–69, 69f
  pure membranes, 67
  RDF, 71–72, 71f
  simulated annealing, 68
  simulations, 66–67
  structure, 69–70

Neurotensin receptor 1 (NTS1)
  excitation wavelength, 352
  and FRET efficiency, 354–355, 354f
  N-terminal maltose-binding protein, 343–344
  NTS1-CFP+NTS1-YFP spectrum, 351–352
Neutravidin
  materials, 291
  procedure, 292
  solutions and buffers, 291
NMA. See Normal mode analysis (NMA)
Nonyl-glucoside/$Zn(OAc)_2$ extraction, 454, 455
Normal mode analysis (NMA)
  applications, 99–100
  software, 100
NR–NR interactions
  EMSAs (see Electrophoretic mobility shift assays (EMSAs))
  ESI-MS (see Electrospray ionization mass spectrometry (ESI-MS))
  materials, 28
  protein crystallization, 24–27
  protocol, 29
  purification, RARα–RXRα LBD heterodimers, 27–28, 28f
NTS1. See Neurotensin receptor 1 (NTS1)
Nuclear receptors (NRs)
  definition, 22
  dimerization, 24
  and FCCS (see Fluorescence cross-correlation spectroscopy (FCCS))
  and NR-NR (see NR–NR interactions)
  structural and functional organization, 23–24, 23f
  two-hybrid assays (see Two-hybrid assays, NR LBD interactions)

O

Oligomerization
  and BRET, 237–239
  class A and B, 230
  coimmunoprecipitation, 235–237
  expression constructs generation, 233–234
  $GABA_B$ R1 and $GABA_B$ R2, 230–231
  GPCR dimerization, 231
  homo-FRET (see Homo-FRET microscopy)
  mammals, 231
  materials, 232–233
  receptor constructs, 234–235
  TAS1R and TAS2R, 230
Oligomers, GPCRs
  all-atom vs. CG simulations
    biased and nonbiased, 83
    disadvantages, 81

Oligomers, GPCRs *(Continued)*
  embedded, membranes, 82–83, 82f
  MARTINI force field, 81–82
  GPCR–GPCR interactions, 80–81
Opsin
  and 11-cis-retinal, 275
  HEK293-F suspension cells, 273–274
Orphan receptors, 22

P

Palmitoylation, 72–73
Paraformaldehyde (PFA), 393–394, 396, 405–406
PCA. *See* Protein fragment complementation assays (PCA)
PCH. *See* Photon-counting histogram (PCH)
PDB. *See* Protein Data Bank (PDB)
Peptide/MHC (pMHC), 393
Periodate-oxidized antibodies and biocytin hydrazide
  materials, 282–283
  procedure, 284
  solutions and buffers, 283
Peripheral proteins oligomerization
  advantageous, 366
  amplitude oscillations, 360–361
  assembly and stoichiometry, 361–362
  biological processes, 361–362
  cell maintenance, transfection and observation, 362–363
  cells and organelles, 365–366
  cellular imaging and analysis, 363
  cellular membranes, 361–362
  and FCS, 360
  filamentous protrusion sites, 366–368
  fluctuation analysis, 368–369
  fluorescence intensity fluctuations, 360
  high-affinity interactions, 361–362
  imaging and analysis, plasma membrane, 363–364
  laser-scanning microscopes, 368–369
  live cells, 366
  modular lipid-binding domains, 361–362
  multiangle light scattering, 366
  number and brightness (N&B) analysis, 360–361, 365–366
  pixel resolution, 366
  raster-scanned images, GUV, 367f, 368
  and RICS, 360–361
  variance, 365–366
  voltage-dependent anion channel, 362
  and VP40, 366–368
PFA. *See* Paraformaldehyde (PFA)
Phenylmethylsulfonyl fluoride (PMSF), 249, 253, 254
"Phosphate-to-phosphate" distance, 68–69
Photon-counting histogram (PCH)
  application, 199
  molecular brightness, 193
  quantitative information, fluorescent molecules, 183–184
  software packages, 192
PIE. *See* Pulsed interleaved excitation (PIE)
Plasma membrane
  imaging and analysis, 363–364
  isolation
    buffers and solutions, 418–419
    equipment, 419
    fractionation, 420–421
    protocol, 418
    reagents, 418
    silica coating, 420
  structures
    fluorescent labeling, 376–377, 377f
    HeLa TCRC-1 cells, 376–377
    high-efficiency fluorescent labeling, 378–380
    physiological expression, 376–377
    receptor expression, 376–377, 377f, 378
    receptor-receptor interactions, 376–377
pMHC. *See* Peptide/MHC (pMHC)
PMSF. *See* Phenylmethylsulfonyl fluoride (PMSF)
Protein complexation, 94
Protein crystallization, NR-NR interactions
  and LBDs, 24–26
  materials, 28
  protocol, 29
Protein Data Bank (PDB)
  CTMDapp, 77
  initial structure, 64–65
Protein fragment complementation assays (PCA), 175–176
Protein–protein docking
  available docking programs, 94–95
  CAPRI experiment, 94
  GPCR–GPCR, 95–96
  interaction, 93–94
  ligands, 95
  molecular modeling methods, 95
  reliability, 96
  small-molecular, 94
  software packages, 95
Protein purification, β2-AR
  affinity chromatography, 462
  crude membrane preparations, 461
  solubilization, 461

Protein structure graph (PSG)
  comparative hub distribution, 48, 55f
  3D and network parameter distribution, 48, 54f
  interaction, residues, 46–48
  network parameters, 47t, 48
Protein structure network (PSN)
  biomolecular, 44
  11-cis-retinal, 52, 53f
  communication pathways, dark and MII rhodopsin, 56, 58f
  global meta paths, 56, 57f
  intramolecular and intermolecular communication, 50–51
  materials, 45
  molecular dynamics (MD), 44
  PSG, 46–48
  PSN–MD method, 51–52, 53
  search, shortest communication paths, 48–49
  workflow, PSN–MD and PSN–ENM approaches, 45–46, 46f, 47t
PSG. See Protein structure graph (PSG)
PSN analysis. See Protein structure network (PSN)
Pulsed interleaved excitation (PIE), 202
Putative lipid raft markers, 109–110

Q

QB. See Quantal brightness (QB)
QuadView imaging systems, 299
Quantal brightness (QB)
  monomeric moiety
    antibody labeling, 9–10, 10f
    pharmacological agents, 11, 11f
    receptors tagged, fluorescent proteins, 9, 10f
  SpIDA, 4

R

Radial distribution function (RDF), 71–72, 71f
Raster image correlation spectroscopy (RICS)
  analysis, 363, 365–366
  application, 363
  data, 363
  fluorescent dynamics, 360–361
  and N&B, 362, 366
RDF. See Radial distribution function (RDF)
Receptor oligomerization, line-scanning FCCS
  auto-and cross-correlation analysis, 205–209, 209t
  confocal microscope and illumination, 198
  diffusion and oligomerization, 198
  dimerization, 209–210
  dual-color, 199
  endogenous protein, 210
  and FCS, 198
  fluorescence fluctuation measurements, 204–205
  fluorescent proteins, 200
  HeLa cells, 202
  membrane receptor, 198
  microscope, 200–202, 203–204
  mTq2 case (see mTurquoise2 (mTq2))
  nonfluorescent H_1R, 210
  and PCH, 198
  photons, 198
  plasma membrane, 199
  pulsed 440nm diode and continuous 514 nm Ar^+ laser, 199, 201f
  solutions, DNA constructs and cells, 199–200
Receptor tyrosine kinases (RTKs)
  and BRET (see Bioluminescence resonance energy transfer (BRET))
  and GPCRs (see GPCR–RTK heteroreceptor complexes)
  signal transduction, 142
  SpIDA (see Spatial intensity distribution analysis (SpIDA))
Reconstitution
  bio-beads, 348–349
  dilution and dialysis, 348–349, 349f
  lipid–protein ratios, 348
Replica
  cleaning and mounting, 403–404
  floating, 407
  freeze-etching and generation, 393–394
  quality, 408
Resonance energy transfer (RET), 175–176, 215
RET. See Resonance energy transfer (RET)
Retinoid X receptors (RXRs), 24–26, 27–28
Rhodopsin
  labeling scheme, 271f
  materials, 277
  procedures, 277
  wild-type and azido-tagged, 275–276
RICS. See Raster image correlation spectroscopy (RICS)
RIPA buffer
  lysis, 330–331, 331f
  protease and phosphatase inhibitors, 333
  washing buffer, 333–335
Rod outer segment (ROS) isolation
  gradients, 453–454, 455
  NG/Zn(OAc)$_2$ extraction, 459
  10% nonyl-β-D-glucoside (NG), 455–456, 456f
ROS isolation. See Rod outer segment (ROS) isolation

RTK. *See* Receptor tyrosine kinases (RTKs)
RXRs. *See* Retinoid X receptors (RXRs)

S

SASA calculations. *See* Solvent-accessible surface area (SASA)
Saturation assay, BRET
  A2AR and FGFR1 interaction, 151–153, 152f
  $BRET_{50}$, 151–153
  detection and analysis, GPCR–RTK, 153–155
Serotonin type 4 receptor dimers
  cell transfection, 127
  coimmunoprecipitation, 129–130, 131
  dimerization-screening test, 136–137
  four mouse splice variants, 127, 128f
  and GPCRs, 124
  immunofluorescence
    cell surface labeling, 132–133, 133f
    GPCRs, 131–132
    point mutation suppresses, 131–132
    prior to cell transfection, 132
  materials, 125–127
  membrane preparation and cell lysate, 127–129
  procedures, 124–125
  Qproteome Plasma Membrane Protein Kit, 136
  solubilization and deglycosylation, samples, 129
  TR-FRET technology
  Western blot and detection, 129–130, 130f
Shortest communication paths
  cluster analysis, 49
  Dijkstra's algorithm, 48
  PSN–MD and PSN–ENM, 49
Single-molecule detection (SMD)
  immunoprecipitation, GPCRs, 271–272
  site-specific labeling, 270–271, 271f
  SMD-TIRF microscopy, 272–273
Single-molecule imaging
  cameras, 381
  CD59, 374–375
  data analysis (*see* Data analysis, single-color experiments)
  GPCRs, 374, 375f
  laser beam, 380
  microchannel plate, 381
  plasma membrane structures (*see* Plasma membrane)
  receptor-receptor interactions, 374
  signaling molecules, cell plasma membranes, 374
  signal transduction mechanisms, 374
  simultaneous and dual-color imaging, 380, 381f
  stable homo-oligomers, 374–375
  superresolution microscopies, 375–376

TIRF microscope, 380
Single-molecule immunoprecipitation
  biotinylated-bovine serum albumin, 289–291
  glass-bottom microtiter plate, 287–288
  glass surface and epoxy silane, 289
  immobilization, biotinylated antibody, 292–293
  neutravidin, 291–292
Slide preparation, homo-FRET
  ligand-induced oligomerization, 317
  predimerization measurements, 316–317
  reference measurements, 316
SMD. *See* Single-molecule detection (SMD)
$(NH_4)_2SO_4$-Induced phase separation, 454, 456–458
Solubilization
  β2-AR, 461
  membranes, 460
  10% nonyl-β-D-glucoside (NG), 456–458, 456f
  ROS suspension, 455
  $(NH_4)_2SO_4$ crystallization, 457
Solvent-accessible surface area (SASA), 78
SpAAC. *See* Strain-promoted alkyne-azide cycloaddition (SpAAC)
Spatial intensity distribution analysis (SpIDA)
  advantages, 17
  analog detector calibration procedure, 16, 16f
  and CLSM, 3
  data interpretation and pharmacological analysis, 16–17
  fluorophore-conjugated antibodies, 6–8
  and GPCRs, 3
  and GUI (*see* Graphical user interface (GUI))
  image acquisitions, 11–12
  intensity histogram, imaged ROI, 4
  neurons, dopamine receptors, 6, 7f
  PCH approach, 4
  process, biological fluorescence imaging, 6, 8f
  and QB (*see* quantal brightness (QB))
  quantification, EGFR–eGFP, 4–6, 5f
  sample preparation, 8–9
SpIDA. *See* Spatial intensity distribution analysis (SpIDA)
Steady-state fluorescence anisotropy imaging
  cluster sizes, 309–310
  energy transfer, 310–311
  FKBP-mGFP constructs, AP2018, 311, 312f
  mGFP, 309, 310f
  PAGE analysis/dimerization constructs, 310–311, 311f
  rotational diffusion, fluorophores, 309–310
Strain-promoted alkyne-azide cycloaddition (SpAAC)
  azido-tagged rhodopsin, 277

and azido-tagged rhodopsin, 277
data analysis, 281–282
and in-gel fluorescence, 281
materials, silver staining, 278–280
silver staining, 281
time-series analysis, 281
Structure-based molecular modeling approaches, GPCR oligomerization
dimerization and oligomerization, 92–93
electrostatics studies
application, 100–101
software, 101
human membrane proteins, 92
MD simulation (*see* Molecular dynamics (MD) simulations)
and NMA (*see* Normal mode analysis (NMA))
protein–protein docking, 93–96
X-ray structures, 92
Sulfo-NHS-SS-biotin
materials, 284
procedure, 285
solutions and buffers, 284–285
Sweet taste receptor. *See* Oligomerization

T

TAS1R
class C GPCRs, 230
C-terminus, 237–238
in vitro analysis, 235–236
sequences, 233*f*
and TAS2R (*see* TAS2R)
taste receptors, 230
TAS2R
cloning cassette, 233*f*
homo-and hetero-oligomerization, 238*f*
in humans, 231
mammalian cell lines, 233
N-terminus, 230
protein sizes, 237
receptor-receptor complexes, 231
sst-TAS2R sequence, 233–234
T-cell antigen receptor (TCR)
adherent cells, 405–406
analysis, 405
and B-cells, 405–406
cell adherence and mica sheets, 399–400
cell isolation, purification and fixation, 396
cell surface replicas (*see* Cell surface replicas)
chemotactic response, bacteria, 393
colloidal gold conjugates, 406
confocal microscopy studies, 393
electron microscopic analysis, 393
equipment, 395, 396*f*
exoplasmic hemimembrane, 406
freezing (*see* Freezing)
frozen mica sandwich, 406
immunogold labeling, 396–399
labeling efficiency and implications, 408
ligand-induced receptor, 394
lymphocytes, 394
materials, 394–395
nanoclusters, 393
paraformaldehyde (PFA), 393–394, 405–406
primary antibody and fixation conditions, 406
replica (*see* Replica)
safety precautions, 405
secondary reagents, 406–407
specimen table, 407
and TEM, 404
TCR. *See* T-cell antigen receptor (TCR)
TEM. *See* Transmission electron microscope (TEM)
Ternary complex model, 166–167
Three-dimensional continuum MD (3D-CTMD)
analysis, 79–80
definition, 76–77
implementation, CTMDapp software, 77
Time-resolved fluorescence anisotropy imaging
cluster size, 312–313
EGFR-FKBP-mGFP clustering, 312–313, 314*f*
time-resolved measurements, 312–313
Time-resolved Förster resonance energy transfer (TR-FRET), 134–136, 135*f*
Total internal reflection fluorescence (TIRF)
antibleaching and antiblinking imaging buffer system, 300–302
brightness and photostability, 293
4-color SMD-TIRF microscope, 293, 294
CRISP focus-stabilization system, 300
DualView and QuadView, 299
excitation, 300
laser lines and optical filters, 295
laser-to-fiber and fiber-to-fiber beam system, 298–299
microscope optical setup, 295–298
motorized x-y stage, 293, 295*f*
multiwell plate geometry, 299–300
objective lens-type, 380
optimal detection, single molecules, 302
plasma membrane (PM), 363–364, 364*f*
single molecules, 302
SMD-TIRF fluorophores, 270–271
subpixel alignment, color channels, 300
Zeiss AxioVert 200M, 293

Transactivation *in vitro*, LHCGR
  cAMP kinetics, 438–439
  CRE-luc system, 438–439
  fluorescent secondary Alexa 488 and Alexa 647 antibodies, 437–438
  functional rescue, 437
  stable cell lines, 437–438
  WT response, 439, 440*f*
Transactivation *in vivo*, LHCGR
  BAC construction (*see* Bacterial artificial chromosomes (BAC))
  fertility, 446–447
  male reproductive tract and testes (*see* Male reproductive tract and testes)
  phenotypic parameters, 440
Transmission electron microscope (TEM), 404
T1R family. *See* TAS1R
T2R family. *See* TAS2R
TR-FRET. *See* Time-resolved Förster resonance energy transfer (TR-FRET)
TrkB. *See* Tropomyosin receptor kinase B (TrkB)
Tropomyosin receptor kinase B (TrkB)
  dimerization, 6
  endogenous activation, 6, 7*f*
  transactivation, endogenous neuronal, 4
Two-hybrid assays, NR LBD interactions
  cell lysis, 34
  luciferase measurement, Berthold technologies luminometer, 35
  materials, 34
  measurement, βgal activity, 35
  normalization, 35
  preparation, luciferase assay, 35
  RXRαY402A mutant, 32–33, 33*f*
  transient transfection, HeLa cells, 34

U
UV-Vis spectroscopy
  azF-containing rhodopsin, 280*f*
  characterization, purified rhodopsin, 276
  extinction coefficients, 278
  and in-gel fluorescence, 278
  labeling stoichiometry, 278

W
Western blot technique
  and ECL™, 224
  serotonin type 4 receptor dimers, 129–130, 130*f*
Wild-type and azido-tagged rhodopsin
  materials, 275
  opsin and 11-cis-retinal, 275
  purification, 275–276
  UV-Vis spectroscopy, 276

Volumes in Series

Founding Series Editor
DAVID M. PRESCOTT

Volume 1 (1964)
Methods in Cell Physiology
Edited by David M. Prescott

Volume 2 (1966)
Methods in Cell Physiology
Edited by David M. Prescott

Volume 3 (1968)
Methods in Cell Physiology
Edited by David M. Prescott

Volume 4 (1970)
Methods in Cell Physiology
Edited by David M. Prescott

Volume 5 (1972)
Methods in Cell Physiology
Edited by David M. Prescott

Volume 6 (1973)
Methods in Cell Physiology
Edited by David M. Prescott

Volume 7 (1973)
Methods in Cell Biology
Edited by David M. Prescott

Volume 8 (1974)
Methods in Cell Biology
Edited by David M. Prescott

Volume 9 (1975)
Methods in Cell Biology
Edited by David M. Prescott

Volume 10 (1975)
Methods in Cell Biology
Edited by David M. Prescott

Volume 11 (1975)
Yeast Cells
Edited by David M. Prescott

Volume 12 (1975)
Yeast Cells
Edited by David M. Prescott

Volume 13 (1976)
Methods in Cell Biology
Edited by David M. Prescott

Volume 14 (1976)
Methods in Cell Biology
Edited by David M. Prescott

Volume 15 (1977)
Methods in Cell Biology
Edited by David M. Prescott

Volume 16 (1977)
Chromatin and Chromosomal Protein Research I
Edited by Gary Stein, Janet Stein, and Lewis J. Kleinsmith

Volume 17 (1978)
Chromatin and Chromosomal Protein Research II
Edited by Gary Stein, Janet Stein, and Lewis J. Kleinsmith

Volume 18 (1978)
Chromatin and Chromosomal Protein Research III
Edited by Gary Stein, Janet Stein, and Lewis J. Kleinsmith

Volume 19 (1978)
Chromatin and Chromosomal Protein Research IV
Edited by Gary Stein, Janet Stein, and Lewis J. Kleinsmith

Volume 20 (1978)
Methods in Cell Biology
Edited by David M. Prescott

Advisory Board Chairman
KEITH R. PORTER

Volume 21A (1980)
Normal Human Tissue and Cell Culture, Part A:
   Respiratory, Cardiovascular, and Integumentary Systems
Edited by Curtis C. Harris, Benjamin F. Trump, and Gary D. Stoner

Volume 21B (1980)
Normal Human Tissue and Cell Culture, Part B: Endocrine, Urogenital, and Gastrointestinal Systems
Edited by Curtis C. Harris, Benjamin F. Trump, and Gray D. Stoner

Volume 22 (1981)
Three-Dimensional Ultrastructure in Biology
Edited by James N. Turner

Volume 23 (1981)
Basic Mechanisms of Cellular Secretion
Edited by Arthur R. Hand and Constance Oliver

Volume 24 (1982)
The Cytoskeleton, Part A: Cytoskeletal Proteins, Isolation and Characterization
Edited by Leslie Wilson

Volume 25 (1982)
The Cytoskeleton, Part B: Biological Systems and *In Vitro* Models
Edited by Leslie Wilson

Volume 26 (1982)
Prenatal Diagnosis: Cell Biological Approaches
Edited by Samuel A. Latt and Gretchen J. Darlington

Series Editor
LESLIE WILSON

Volume 27 (1986)
Echinoderm Gametes and Embryos
Edited by Thomas E. Schroeder

Volume 28 (1987)
***Dictyostelium discoideum:* Molecular Approaches to Cell Biology**
Edited by James A. Spudich

Volume 29 (1989)
Fluorescence Microscopy of Living Cells in Culture, Part A: Fluorescent Analogs, Labeling Cells, and Basic Microscopy
Edited by Yu-Li Wang and D. Lansing Taylor

Volume 30 (1989)
Fluorescence Microscopy of Living Cells in Culture, Part B: Quantitative Fluorescence Microscopy—Imaging and Spectroscopy
Edited by D. Lansing Taylor and Yu-Li Wang

Volume 31 (1989)
Vesicular Transport, Part A
Edited by Alan M. Tartakoff

Volume 32 (1989)
Vesicular Transport, Part B
Edited by Alan M. Tartakoff

Volume 33 (1990)
Flow Cytometry
Edited by Zbigniew Darzynkiewicz and Harry A. Crissman

Volume 34 (1991)
Vectorial Transport of Proteins into and across Membranes
Edited by Alan M. Tartakoff

Selected from Volumes 31, 32, and 34 (1991)
Laboratory Methods for Vesicular and Vectorial Transport
Edited by Alan M. Tartakoff

Volume 35 (1991)
Functional Organization of the Nucleus: A Laboratory Guide
Edited by Barbara A. Hamkalo and Sarah C. R. Elgin

Volume 36 (1991)
***Xenopus laevis*: Practical Uses in Cell and Molecular Biology**
Edited by Brian K. Kay and H. Benjamin Peng

Series Editors

LESLIE WILSON AND PAUL MATSUDAIRA

Volume 37 (1993)
Antibodies in Cell Biology
Edited by David J. Asai

Volume 38 (1993)
Cell Biological Applications of Confocal Microscopy
Edited by Brian Matsumoto

Volume 39 (1993)
Motility Assays for Motor Proteins
Edited by Jonathan M. Scholey

Volume 40 (1994)
A Practical Guide to the Study of Calcium in Living Cells
Edited by Richard Nuccitelli

Volume 41 (1994)
Flow Cytometry, Second Edition, Part A
Edited by Zbigniew Darzynkiewicz, J. Paul Robinson, and Harry A. Crissman

Volume 42 (1994)
Flow Cytometry, Second Edition, Part B
Edited by Zbigniew Darzynkiewicz, J. Paul Robinson, and Harry A. Crissman

Volume 43 (1994)
Protein Expression in Animal Cells
Edited by Michael G. Roth

Volume 44 (1994)
Drosophila melanogaster: **Practical Uses in Cell and Molecular Biology**
Edited by Lawrence S. B. Goldstein and Eric A. Fyrberg

Volume 45 (1994)
Microbes as Tools for Cell Biology
Edited by David G. Russell

Volume 46 (1995)
Cell Death
Edited by Lawrence M. Schwartz and Barbara A. Osborne

Volume 47 (1995)
Cilia and Flagella
Edited by William Dentler and George Witman

Volume 48 (1995)
Caenorhabditis elegans: **Modern Biological Analysis of an Organism**
Edited by Henry F. Epstein and Diane C. Shakes

Volume 49 (1995)
Methods in Plant Cell Biology, Part A
Edited by David W. Galbraith, Hans J. Bohnert, and Don P. Bourque

Volume 50 (1995)
Methods in Plant Cell Biology, Part B
Edited by David W. Galbraith, Don P. Bourque, and Hans J. Bohnert

Volume 51 (1996)
Methods in Avian Embryology
Edited by Marianne Bronner-Fraser

Volume 52 (1997)
Methods in Muscle Biology
*Edited by Charles P. Emerson, Jr.
and H. Lee Sweeney*

Volume 53 (1997)
Nuclear Structure and Function
Edited by Miguel Berrios

Volume 54 (1997)
Cumulative Index

Volume 55 (1997)
Laser Tweezers in Cell Biology
Edited by Michael P. Sheetz

Volume 56 (1998)
Video Microscopy
Edited by Greenfield Sluder and David E. Wolf

Volume 57 (1998)
Animal Cell Culture Methods
Edited by Jennie P. Mather and David Barnes

Volume 58 (1998)
Green Fluorescent Protein
Edited by Kevin F. Sullivan and Steve A. Kay

Volume 59 (1998)
The Zebrafish: Biology
*Edited by H. William Detrich III, Monte Westerfield,
and Leonard I. Zon*

Volume 60 (1998)
The Zebrafish: Genetics and Genomics
*Edited by H. William Detrich III, Monte Westerfield,
and Leonard I. Zon*

Volume 61 (1998)
Mitosis and Meiosis
Edited by Conly L. Rieder

Volume 62 (1999)
Tetrahymena thermophila
Edited by David J. Asai and James D. Forney

Volume 63 (2000)
Cytometry, Third Edition, Part A
*Edited by Zbigniew Darzynkiewicz, J. Paul Robinson,
and Harry Crissman*

Volume 64 (2000)
Cytometry, Third Edition, Part B
Edited by Zbigniew Darzynkiewicz, J. Paul Robinson, and Harry Crissman

Volume 65 (2001)
Mitochondria
Edited by Liza A. Pon and Eric A. Schon

Volume 66 (2001)
Apoptosis
Edited by Lawrence M. Schwartz and Jonathan D. Ashwell

Volume 67 (2001)
Centrosomes and Spindle Pole Bodies
Edited by Robert E. Palazzo and Trisha N. Davis

Volume 68 (2002)
Atomic Force Microscopy in Cell Biology
Edited by Bhanu P. Jena and J. K. Heinrich Hörber

Volume 69 (2002)
Methods in Cell–Matrix Adhesion
Edited by Josephine C. Adams

Volume 70 (2002)
Cell Biological Applications of Confocal Microscopy
Edited by Brian Matsumoto

Volume 71 (2003)
Neurons: Methods and Applications for Cell Biologist
Edited by Peter J. Hollenbeck and James R. Bamburg

Volume 72 (2003)
Digital Microscopy: A Second Edition of Video Microscopy
Edited by Greenfield Sluder and David E. Wolf

Volume 73 (2003)
Cumulative Index

Volume 74 (2004)
Development of Sea Urchins, Ascidians, and Other Invertebrate Deuterostomes: Experimental Approaches
Edited by Charles A. Ettensohn, Gary M. Wessel, and Gregory A. Wray

Volume 75 (2004)
Cytometry, 4th Edition: New Developments
Edited by Zbigniew Darzynkiewicz, Mario Roederer, and Hans Tanke

Volume 76 (2004)
The Zebrafish: Cellular and Developmental Biology
Edited by H. William Detrich, III, Monte Westerfield, and Leonard I. Zon

Volume 77 (2004)
The Zebrafish: Genetics, Genomics, and Informatics
Edited by William H. Detrich, III, Monte Westerfield, and Leonard I. Zon

Volume 78 (2004)
Intermediate Filament Cytoskeleton
Edited by M. Bishr Omary and Pierre A. Coulombe

Volume 79 (2007)
Cellular Electron Microscopy
Edited by J. Richard McIntosh

Volume 80 (2007)
Mitochondria, 2nd Edition
Edited by Liza A. Pon and Eric A. Schon

Volume 81 (2007)
Digital Microscopy, 3rd Edition
Edited by Greenfield Sluder and David E. Wolf

Volume 82 (2007)
Laser Manipulation of Cells and Tissues
Edited by Michael W. Berns and Karl Otto Greulich

Volume 83 (2007)
Cell Mechanics
Edited by Yu-Li Wang and Dennis E. Discher

Volume 84 (2007)
Biophysical Tools for Biologists, Volume One: In Vitro Techniques
Edited by John J. Correia and H. William Detrich, III

Volume 85 (2008)
Fluorescent Proteins
Edited by Kevin F. Sullivan

Volume 86 (2008)
Stem Cell Culture
Edited by Dr. Jennie P. Mather

Volume 87 (2008)
Avian Embryology, 2nd Edition
Edited by Dr. Marianne Bronner-Fraser

Volume 88 (2008)
Introduction to Electron Microscopy for Biologists
Edited by Prof. Terence D. Allen

Volume 89 (2008)
Biophysical Tools for Biologists, Volume Two: In Vivo Techniques
Edited by Dr. John J. Correia and Dr. H. William Detrich, III

Volume 90 (2008)
Methods in Nano Cell Biology
Edited by Bhanu P. Jena

Volume 91 (2009)
Cilia: Structure and Motility
Edited by Stephen M. King and Gregory J. Pazour

Volume 92 (2009)
Cilia: Motors and Regulation
Edited by Stephen M. King and Gregory J. Pazour

Volume 93 (2009)
Cilia: Model Organisms and Intraflagellar Transport
Edited by Stephen M. King and Gregory J. Pazour

Volume 94 (2009)
Primary Cilia
Edited by Roger D. Sloboda

Volume 95 (2010)
Microtubules, in vitro
Edited by Leslie Wilson and John J. Correia

Volume 96 (2010)
Electron Microscopy of Model Systems
Edited by Thomas Müeller-Reichert

Volume 97 (2010)
Microtubules: In Vivo
Edited by Lynne Cassimeris and Phong Tran

Volume 98 (2010)
Nuclear Mechanics & Genome Regulation
Edited by G.V. Shivashankar

Volume 99 (2010)
Calcium in Living Cells
Edited by Michael Whitaker

Volume 100 (2010)
The Zebrafish: Cellular and Developmental Biology, Part A
Edited by: H. William Detrich III, Monte Westerfield and Leonard I. Zon

Volume 101 (2011)
The Zebrafish: Cellular and Developmental Biology, Part B
Edited by: H. William Detrich III, Monte Westerfield and Leonard I. Zon

Volume 102 (2011)
Recent Advances in Cytometry, Part A: Instrumentation, Methods
Edited by Zbigniew Darzynkiewicz, Elena Holden, Alberto Orfao, William Telford and Donald Wlodkowic

Volume 103 (2011)
Recent Advances in Cytometry, Part B: Advances in Applications
Edited by Zbigniew Darzynkiewicz, Elena Holden, Alberto Orfao, Alberto Orfao and Donald Wlodkowic

Volume 104 (2011)
The Zebrafish: Genetics, Genomics and Informatics 3rd Edition
Edited by H. William Detrich III, Monte Westerfield, and Leonard I. Zon

Volume 105 (2011)
The Zebrafish: Disease Models and Chemical Screens 3rd Edition
Edited by H. William Detrich III, Monte Westerfield, and Leonard I. Zon

Volume 106 (2011)
Caenorhabditis elegans: Molecular Genetics and Development 2nd Edition
Edited by Joel H. Rothman and Andrew Singson

Volume 107 (2011)
Caenorhabditis elegans: Cell Biology and Physiology 2nd Edition
Edited by Joel H. Rothman and Andrew Singson

Volume 108 (2012)
Lipids
Edited by Gilbert Di Paolo and Markus R Wenk

Volume 109 (2012)
Tetrahymena thermophila
Edited by Kathleen Collins

Volume 110 (2012)
Methods in Cell Biology
Edited by Anand R. Asthagiri and Adam P. Arkin

Volume 111 (2012)
Methods in Cell Biology
Edited by Thomas Müler Reichart and Paul Verkade

Volume 112 (2012)
Laboratory Methods in Cell Biology
Edited by P. Michael Conn

Volume 113 (2013)
Laboratory Methods in Cell Biology
Edited by P. Michael Conn

Volume 114 (2013)
Digital Microscopy, 4th Edition
Edited by Greenfield Sluder and David E. Wolf

Volume 115 (2013)
Microtubules, *in Vitro*, 2nd Edition
Edited by John J. Correia and Leslie Wilson

Volume 116 (2013)
Lipid Droplets
Edited by H. Robert Yang and Peng Li

Color Plate

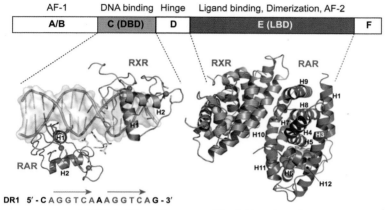

PLATE 1 (Fig. 2.1 on page 23 of this volume).

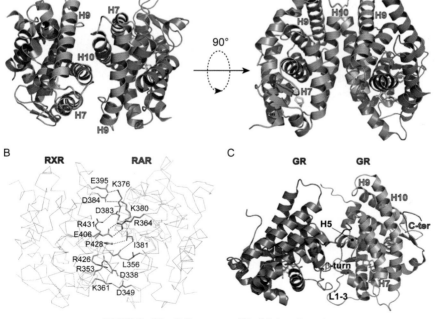

PLATE 2 (Fig. 2.2 on page 25 of this volume).

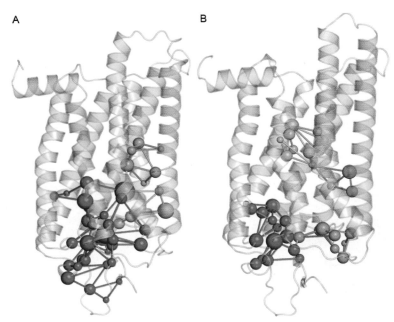

PLATE 3 (Fig. 3.2 on page 53 of this volume).

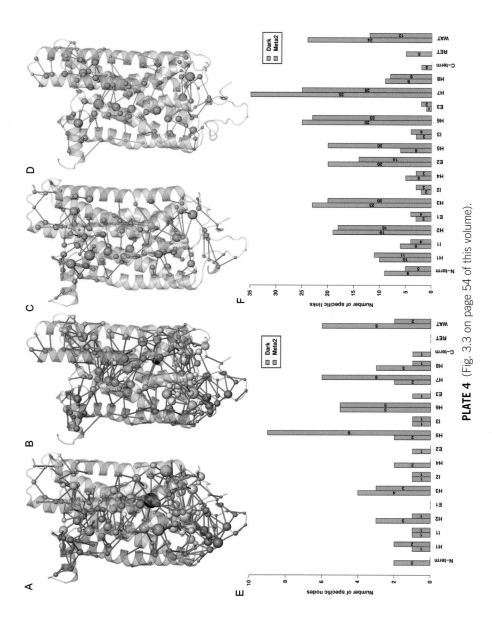

PLATE 4 (Fig. 3.3 on page 54 of this volume).

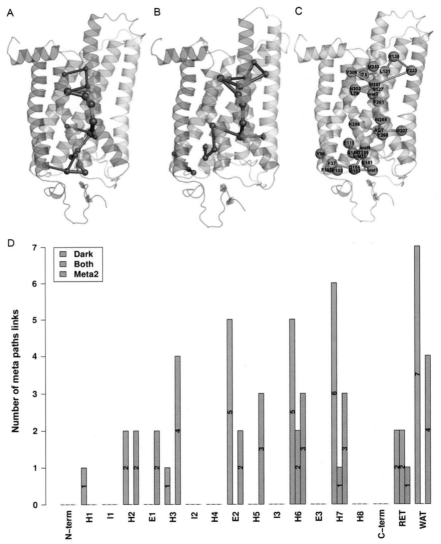

PLATE 5 (Fig. 3.5 on page 57 of this volume).

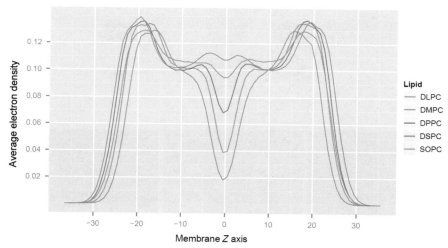

PLATE 6 (Fig. 4.1 on page 69 of this volume).

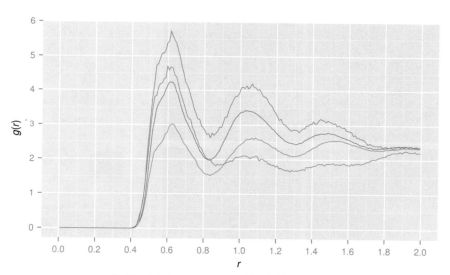

PLATE 7 (Fig. 4.2 on page 71 of this volume).

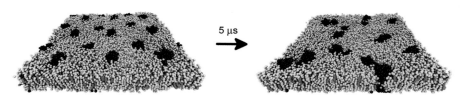

PLATE 8 (Fig. 4.4 on page 82 of this volume).

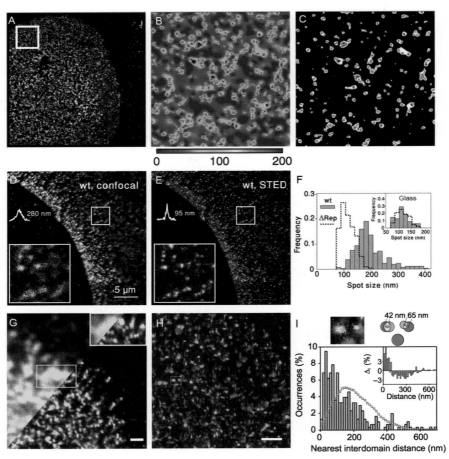

PLATE 9 (Fig. 6.2 on page 116 of this volume).

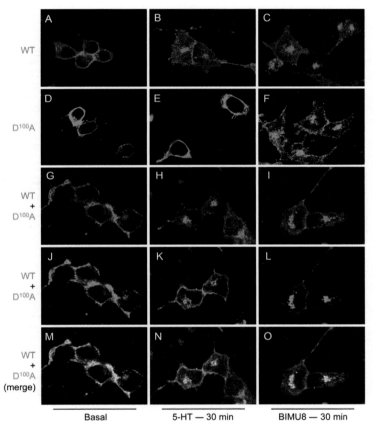

PLATE 10 (Fig. 7.3 on page 133 of this volume).

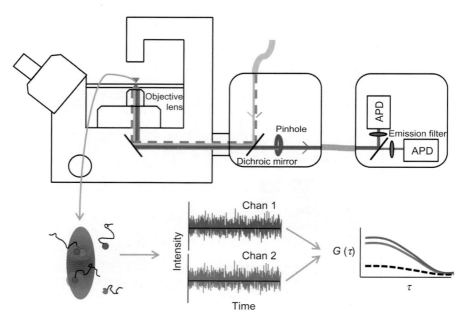

PLATE 11 (Fig. 11.1 on page 201 of this volume).

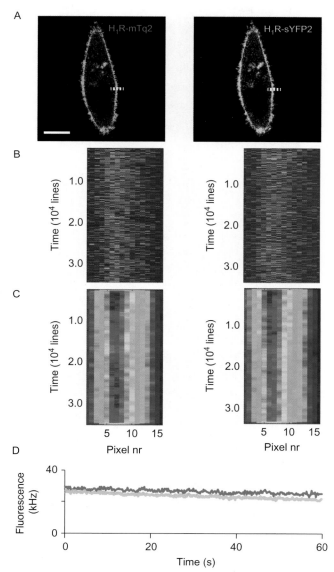

PLATE 12 (Fig. 11.2 on page 206 of this volume).

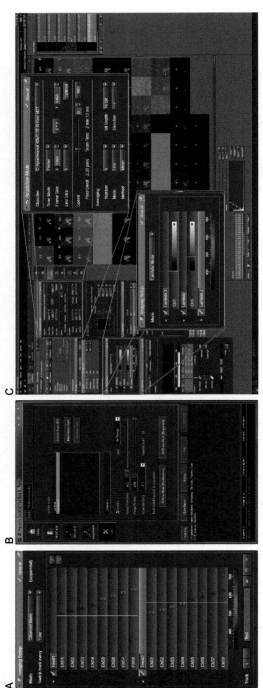

PLATE 13 (Fig. 14.4 on page 258 of this volume).

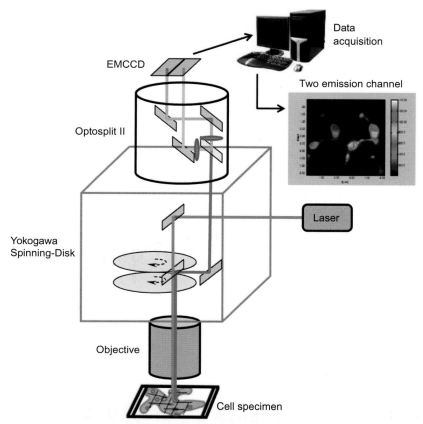

PLATE 14 (Fig. 14.5 on page 259 of this volume).

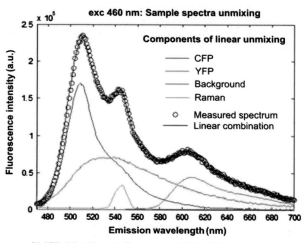

PLATE 15 (Fig. 14.7 on page 262 of this volume).

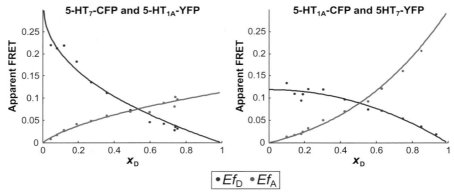

PLATE 16 (Fig. 14.8 on page 263 of this volume).

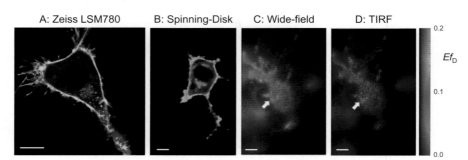

PLATE 17 (Fig. 14.9 on page 264 of this volume).

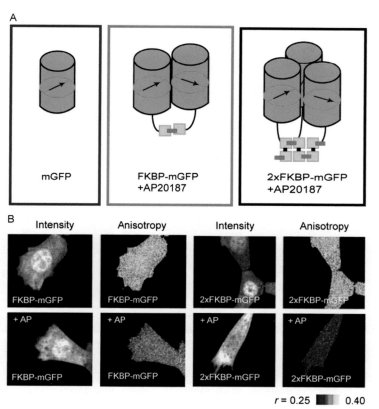

PLATE 18 (Fig. 16.3 on page 312 of this volume).

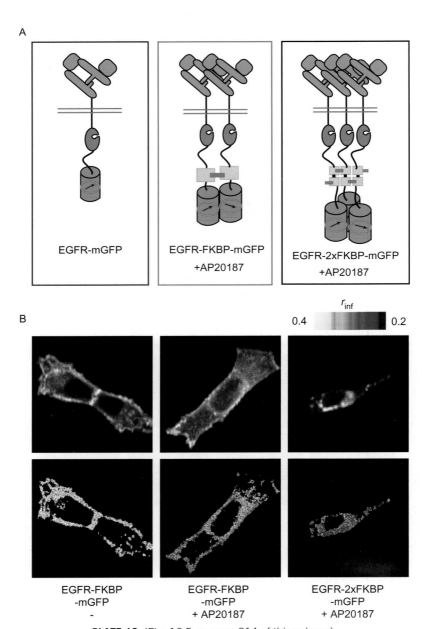

PLATE 19 (Fig. 16.5 on page 314 of this volume).

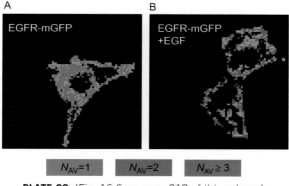

PLATE 20 (Fig. 16.6 on page 318 of this volume).

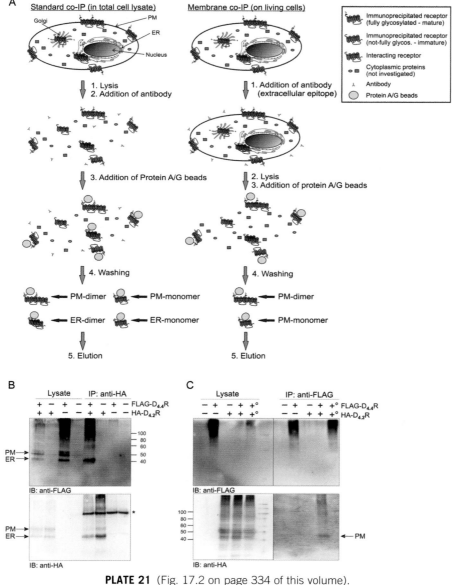

PLATE 21 (Fig. 17.2 on page 334 of this volume).

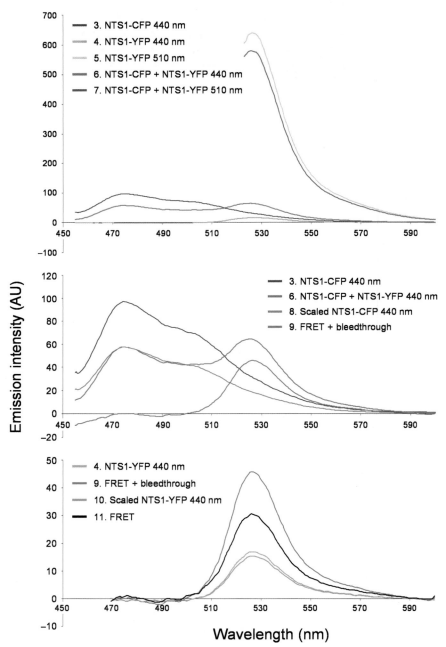

PLATE 22 (Fig. 18.2 on page 353 of this volume).

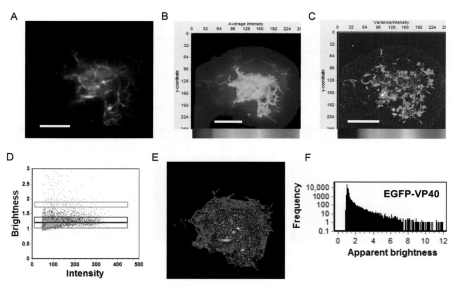

PLATE 23 (Fig. 19.1 on page 364 of this volume).

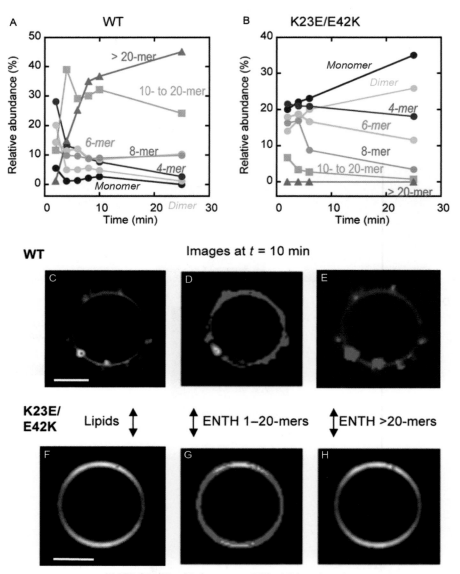

PLATE 24 (Fig. 19.3 on page 367 of this volume).

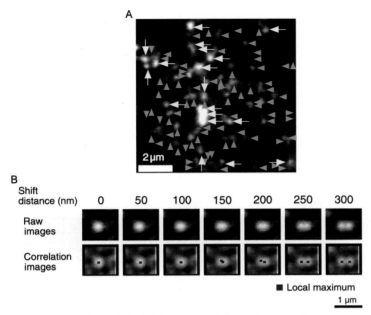

PLATE 25 (Fig. 20.4 on page 382 of this volume).

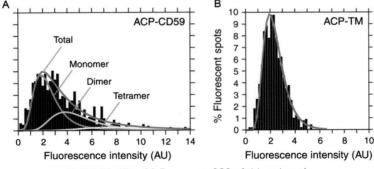

PLATE 26 (Fig. 20.5 on page 383 of this volume).

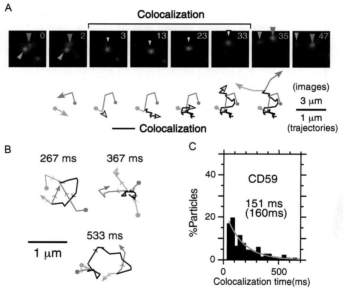

PLATE 27 (Fig. 20.6 on page 384 of this volume).

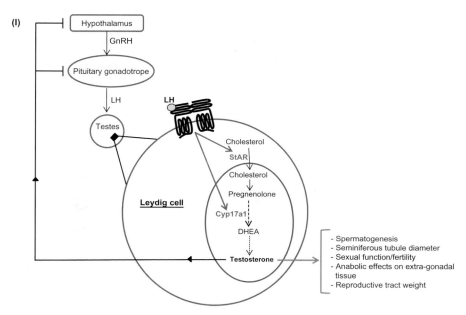
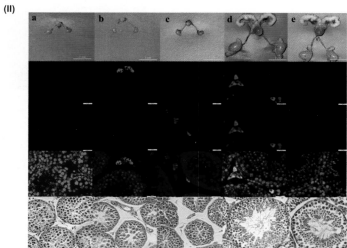

PLATE 28 (Fig. 23.4 on page 444 of this volume).

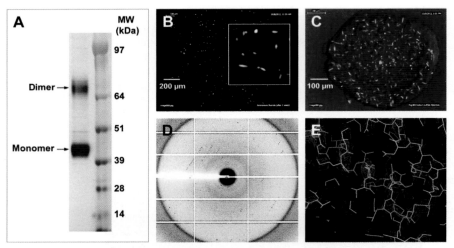

PLATE 29 (Fig. 24.4 on page 462 of this volume).